Shock Wave and High Pressure Phenomena

Series Editor-in-Chief

L. Davison, USA
Y. Horie, USA

Founding Editor

R. A. Graham, USA

Advisory Board

V. E. Fortov, Russia
Y. M. Gupta, USA
J. R. Asay, USA
G. Ben-Dor, Israel
K. Takayama, Japan
F. Lu, USA

Shock Wave and High Pressure Phenomena

L.L. Altgilbers, M.D.J. Brown, I. Grishnaev, B.M. Novac, I.R. Smith, I. Tkach, and Y. Tkach: Magnetocumulative Generators

T. Antoun, D.R. Curran, G.I. Kanel, S.V. Razorenov, and A.V. Utkin: Spall Fracture

J. Asay and M. Shahinpoor (Eds.): High-Pressure Shock Compression of Solids

S.S. Batsanov: Effects of Explosion on Materials: Modification and Synthesis Under High-Pressure Shock Compression

R. Cherét: Detonation of Condensed Explosives

L. Davison, D. Grady, and M. Shahinpoor (Eds.): High-Pressure Shock Compression of Solids II

L. Davison and M. Shahinpoor (Eds.): High-Pressure Shock Compression of Solids III

L. Davison, Y. Horie, and M. Shahinpoor (Eds.): High-Pressure Shock Compression of Solids IV

L. Davison, Y. Horie, and T. Sekine (Eds.): High-Pressure Shock Compression of Solids V

A.N. Dremin: Toward Detonation Theory

Y. Horie, L. Davison, and N.N. Thadhani (Eds.): High-Pressure Shock Compression of Solids VI

R. Graham: Solids Under High-Pressure Shock Compression

J.N. Johnson and R. Cherét (Eds.): Classic Papers in Shock Compression Science

V.F. Nesterenko: Dynamics of Heterogeneous Materials

M. Sućeska: Test Methods of Explosives

J.A. Zukas and W.P. Walters (Eds.): Explosive Effects and Applications

G.I. Kanel, S. V. Razorenov, and V.E. Fortov: Shock-Wave Phenomena and the Properties of Condensed Matter

V.E. Fortov, L.V. Altshuler, R.F. Trunin, and A.I. Funtikov: High-Pressure Shock Compression of Solids VII

L.C. Chhabildas, L. Davison, and Y. Horie (Eds.): High-Pressure Shock Compression of Solids VIII

R.P. Drake: High-Energy-Density Physics

D. Grady: Fragmentation of Rings and Shells

M.V. Zhernokletov and B.L. Glushak (Eds.): Material Properties under Intensive Dynamic Loading

G. Ben-Dor: Shock Wave Reflection Phenomena

L. Davison: Fundamentals of Shock Wave Propagation in Solids

Lee Davison

Fundamentals of Shock Wave Propagation in Solids

With 176 Figures

Springer

Lee Davison
39 Cañoncito Vista Road
Tijeras, NM 87059, USA
E-mail: leedavison@aol.com

Series Editors-in-Chief:
Lee Davison
39 Cañoncito Vista Road
Tijeras, NM 87059, USA
E-mail: leedavison@aol.com

Yasuyuki Horie
AFRL/MNME Munitions Directorate
2306 Perimeter Road
Eglin AFB, FL 32542, USA
E-mail: yasuyuki.horie@eglin.af.mil

ISBN 978-3-540-74568-6 e-ISBN 978-3-540-74569-3

DOI 10.1007/978-3-540-74569-3

Library of Congress Control Number: 2008922892

Typesetting: Camara ready by author
Production: LE-TEX Jelonek, Schmidt & Vöckler GbR, Leipzig, Germany
Cover design: WMXDesign GmbH, Heidelberg, Germany

Printed on acid-free paper

9 8 7 6 5 4 3 2 1

springer.com

Preface

My intent in writing this book is to present an introduction to the thermomechanical theory required to conduct research and pursue applications of shock physics in solid materials. Emphasis is on the range of moderate compression that can be produced by high-velocity impact or detonation of chemical explosives and in which elastoplastic responses are observed and simple equations of state are applicable. In the interest of simplicity, the presentation is restricted to plane waves producing uniaxial deformation. Although applications often involve complex multidimensional deformation fields it is necessary to begin with the simpler case. This is also the most important case because it is the usual setting of experimental research. The presentation is also restricted to theories of material response that are simple enough to permit illustrative problems to be solved with minimal recourse to numerical analysis.

The discussions are set in the context of established continuum-mechanical principles. I have endeavored to define the quantities encountered with some care and to provide equations in several convenient forms and in a way that lends itself to easy reference. Thermodynamic analysis plays an important role in continuum mechanics, and I have included a presentation of aspects of this subject that are particularly relevant to shock physics. The notation adopted is that conventional in expositions of modern continuum mechanics, insofar as possible, and variables are explained as they are encountered. Those experienced in shock physics may find some of the notation unconventional. This is unfortunate, but I have found its use necessary to ensure that statements made are precise and unambiguous. I hope that the effort required to accommodate to these changes will be rewarded by the benefits that accrue.

Shock phenomena encountered in applications are analyzed using comprehensive programs that are executed on powerful computers. In fact, the need to solve nonlinear wave-propagation problems has been a principal motivation for development of the most powerful computers of their time—the supercomputers. Solution of geometrically complex three-dimensional problems in which large deformations occur and in which the material description captures effects such as elastic–viscoplastic flow, phase transformations, fracture, chemical reaction, etc., has become routine. These powerful tools are highly effective, but it is necessary to have a sound understanding of the physical phenomena and wave interactions being simulated if the tools are to be used with confidence. An objective of this volume is to assist the reader in developing this understanding.

Following some introductory comments, we have Chap. 2 in which the kinematical and dynamical principles and equations that form the basis of our subject are discussed. The equations are presented in both Lagrangian (material) and Eulerian (spatial) form and are arranged in a way that lends itself to ready reference. There has been an effort made throughout the volume to provide equations for calculating all of the thermomechanical field quantities encountered in investigations of shock phenomena. In Chap. 3 the equations that have been developed are applied to analysis of the propagation and interaction of plane longitudinal shocks. This chapter establishes the language of the subject and provides the basis for solving elementary problems, including the design of experiments. This is the essential background for following the remainder of the volume, but a cursory reading should suffice for those who have some familiarity with the field.

Chapter 4 is the first of several addressing development of constitutive equations that describe the thermomechanical response of some important classes of materials. In this chapter, the issues are invariance to the choice of coordinates used and thermodynamic requirements imposed on theories of material response to ensure that deformation cannot cause a decrease in entropy.

Practical thermodynamic topics arising in connection with inviscid fluids are addressed in Chap. 5. We point out that solids subjected to high pressure behave much like fluids and that many problems of shock compression of solids are actually analyzed as though the material were a fluid. Topics covered include the ideal gas and Mie–Grüneisen equations of state, and the relationships among isotherms, isentropes, and Hugoniot curves. This material is extended to elastic solids in Chap. 6.

In Chap. 7 we consider the response of elastic–plastic and elastic–viscoplastic solids, beginning with a theory of small deformation and proceeding to finite elastic–plastic deformation and then to finite deformation of elastic–viscoplastic solids. In developing the finite-deformation theories we have taken the opportunity to introduce dislocation mechanics as an example of the interplay of atomistic and continuum concepts that has been important in the development of shock physics throughout much of its history. The resulting theory, although unconventional, is useful and its development instructive.

In Chaps. 8 and 9 we continue the study of wave propagation begun in Chap. 3. We begin with weak elastic waves described by linear equations and proceed to the nonlinear waves that are more typical of shock problems. In Chap. 10 we consider elastic–plastic waves, beginning with detailed consideration of those involving small deformation and proceeding on to those that are stronger and involve viscoplastic effects.

Chapter 11 contains a discussion of three theories of porous materials and the waves that propagate in them. In Chap. 12 we address fractures that can result from the tension that arises when decompression waves collide. Finally, steady detonation waves are discussed in Chap. 13.

Many of the chapters include exercises that address details of the developments presented or extend the discussion. Detailed solutions of the problems presented in the exercises are included in an appendix.

Acknowledgments. I first became interested in nonlinear wave propagation as a graduate student at the California Institute of Technology. This was followed by many years at the Sandia National Laboratories, mostly in research groups concerned with shock physics. During that time I had the pleasure and benefit of association with colleagues too numerous to name, working in most aspects of the field. Many of these people were experimentalists or specialists in the computational aspects of the subject. Some were theoreticians and some pursued applications. All were interested in the subject as a whole and interactions among them led to significant advancement of the science and technology. The community of people interested in shock compression of solids and related matters is large enough to provide a broad perspective but small enough that one can know many of its members. In recent years I have been involved with the Springer series on High-Pressure Shock Compression of Condensed Matter and have benefited greatly from discussions with authors contributing books and articles to the series. They come from laboratories situated throughout the world.

I particularly acknowledge my long association with Robert A. Graham, who was the founding editor of the series in which this book appears and with Yasuyuki Horie, a current editor and supporter of the effort that resulted in the preparation of this book. Most of all, I acknowledge the support and encouragement of my wife, Helen.

Tijeras, New Mexico
September, 2007

Lee Davison

Contents

CHAPTER 1

Introduction

Nonlinear wave propagation has been a subject of serious scientific investigation since the 1800s, but the early work was concerned mostly with gases and was almost entirely theoretical. Experimental investigations of shock compression of condensed matter have been reported since the early 1900s and military applications motivated a marked increase in interest in the subject during the 1940s and 1950s. Since this early period the field has grown in both depth and breadth and work is now being reported by hundreds of investigators working in countries scattered throughout the world.

Modern scientific work on shock compression of solids began in the United States and in the Soviet Union. This early work has been reviewed by Al'tshuler [2] and by Rice et al. [86]. Classic accounts of the subject appear in books by Courant and Friedrichs [28] and by Zel'dovich and Raizer [109]. Presentations complementary to the present book have recently been prepared by Asay and Shahinpoor (eds.) [7], Drumheller [39], Graham [50], and Kanel et al. [63].

Investigations of nonlinear wave propagation in solids have as their objective the development of methods for predicting effects of dynamic events such as high-velocity impacts and detonation of explosives. Shock-compression experiments provide access to the highest pressures attainable in a laboratory environment and lead to deformation of materials at the highest possible rate. In some cases the objective of shock-compression experiments is to establish the basis for, or to validate, theories of material response with which one can predict wave propagation and interaction phenomena. In other cases, interest lies less with wave propagation than with the effects the waves produce in the materials in which they are propagating. These effects include chemical and metallurgical changes, polymorphic phase transformations, fracture and fragmentation, etc. Experiments that can be conducted in a laboratory support development of equations of state for stresses from the acoustic range to several hundred GPa* and temperatures from a few Kelvin to tens of thousands of Kelvin. Experiments particularly concerned with effects of the rate at which the material is deformed cover the range of strain rates from about 10^3 to at least 10^9 s^{-1}. They provide information on mechanisms of inelastic deformation of both brittle and ductile

* 1 GPa = 10^9 Pa = 10^9 N/m^2

materials, rates at which chemical reactions and phase transformations proceed, etc.

A wide variety of solid materials has been investigated in shock-wave experiments; they include most elemental metals and many metallic alloys, inorganic compounds including ceramics, rocks, and minerals, soils, polymers, fiber-reinforced and laminated composites, porous solids, and explosives.

Work conducted in the lower range of stress and temperature addresses high-rate deformation processes, matters in the domain of solid-state physics, metallurgical effects, solid-state chemical reactions, electromechanical phenomena, and explosive detonation. Most material behaviors have been, or can be, subjects of shock-compression investigations. Weak shocks, those associated with stresses of a few GPa, elicit electrical responses in dielectric, piezoelectric, and ferroelectric materials. Shocks of this strength produce inelastic flow and associated changes in the microstructure of metals and interaction of these shocks can produce tensile stresses leading to nucleation and growth of microvoids and cracks followed by complete fracture of samples of the material. As one considers stronger shocks, the stress is dominated by the pressure and interest turns to determination of the equation of state of the material. Work conducted at the high extremes of pressure and temperature is directed toward determination of the equation of state of dense plasmas.

Shock physics, as is the case with most scientific endeavors, involves experimental observation and measurement, development of theoretical descriptions of these observations, and computational work to validate the theories and apply them to the solution of practical problems. A contemporary view holds that a physical phenomenon is understood if there exists a theory that allows it to be simulated numerically.

A *shock*, in the sense used here, is a propagating discontinuity of density, temperature, stress, etc. that exists in a material continuum. Not all propagating disturbances are, or evolve into, shocks. Waves that produce a decrease in mass density of the material usually spread as they propagate and are not shocks. Smooth compressive disturbances, on the other hand, usually become steeper as they propagate and evolve into shocks. The shock is a continuum concept, but discussion of shocks inevitably involves consideration of the behavior of matter at the atomic or molecular level, and sometimes at the scale of grains of a polycrystal or other mesoscopic features of the material. Both common intuition and experimental observation indicates that a propagating shock cannot be a strict mathematical discontinuity but must have some spatial extent. Examination of materials recovered after having been subjected to shock compression reveal substantial alteration of the material microstructure, indicating that the shock transition cannot be instantaneous and also that the shock-compression process is highly irregular when viewed at microscopic or mesoscopic scales, (i.e. on length scales comparable to the lattice spacing or the size of grains of polycrystalline materials). It is found, however, that the fields associated with a steadily propagating smooth wave satisfy the same conditions of mass, momentum, and energy conservation as are satisfied at a strict discontinuity. This means that it is reasonable to interpret a steady, structured wave as a shock discontinuity if its thickness is small compared to other dimensions of interest. In its idealized form the shock deforms the material instantaneously. A structured shock deforms the material at a finite rate, but one that is the maximum that can be achieved in a steady waveform. In the experimental context, it is these steady waves, whether considered to be discontinuities or smooth disturbances, that can be interpreted with the greatest confidence.

Shock-compression phenomena, particularly for weak shocks, are greatly affected by microstructural features of the material in which they propagate. Inelastic deformation of metals results from motion of dislocations and formation of twins and other defects of the crystal structure, adiabatic shear bands, etc. Often theories of material response are motivated by knowledge of the operation of these deformation mechanisms, and these theories have been quite successful in predicting continuum response.

Although our presentation represents a rather idealized, mathematical, view of shock waves, much of the content of the field of shock physics concerns experimental observations and addresses metallurgical, chemical, and other effects of impacts, explosions, and similar brief, high-pressure events in a way that focuses on the effects these stimuli produce in materials rather than on the shock itself. Nevertheless, the emphasis of this book is on shock-wave propagation and interaction and, as a consequence, is concerned more with thermomechanical phenomena than with matters in the realm of materials science.

Knowledge of the subject matter of this book is widely applied to the design of munitions, both conventional and nuclear, and to assessment of the effects of these munitions. This latter work includes design of armor to mitigate the intended effect of the munitions. Shock physics is also important for its utility in determining the high-pressure equations of state needed to study the structure of the Earth and planets and because it permits analysis of effects of events such as high-velocity collisions of asteroids with planets. Shock processes have played a significant role in the evolution of the solar system, and analysis of chemical and physical effects of shocks on natural materials has proven important in studying this evolution. Knowledge of the equation of state of planetary materials is often obtained from shock-compression experiments and is essential for understanding the structure of these bodies. Considerable effort has been devoted to development of methods by which shock-compression can be used to produce dense, strongly bonded compacts of powdered metals and inorganic compounds such as oxides and nitrides. Shock treatment to increase the defect density in powdered materials makes it easier to sinter them into strong compacts. Shock-induced chemical reactions have yielded a variety of novel compounds. Diamond abrasives can be made by shock compression of graphite. Useful bimetallic plates can be made by explosively cladding a strong and inexpensive structural metal with a layer of material resistant to corrosion such as may occur in chemical-processing vessels. Material for making laminated coins is also prepared by this method. The surface of structural parts made of some metals can be effectively hardened by detonating a sheet of explosive in contact with the part. In most cases, shock processing is expensive, which means that the product sought must be one of unusually high value if the process is to be commercially useful.

This volume, a textbook rather than a treatise, contains very few references. There are, however, many works in which extensive lists of references have been compiled. Numerous references to work conducted prior to 1979 have been cited by Davison and Graham [32] and a concise listing of important early milestones is included in the preface to Reference [7]. Graham [50] cites many review articles and papers, particularly in the areas of shock induced electrical, magnetic, and chemical phenomena. Asay and Shahinpoor have presented a substantial bibliography on shock compression of solids in Reference [7, App. A]. Russian work is reviewed in References [10,44,83], with many citations provided. The American Physical Society Topical Group on Shock Compression of Condensed Matter has held conferences on this subject in alternate years beginning in 1979 and the proceedings of these conferences provide a rich source of information on contemporary research.

It is important to mention that many topics that fall under the heading of shock compression of solids are not discussed in this book:

- Experimental methods of shock physics have been presented in considerable detail by Antoun, et al. [4], Barker et al. [9], Chhabildas and Graham [21], Graham and Asay [52], McQueen et al. [77], and by Zhernokletov and Glushak [110].
- A very important aspect of shock physics is the use of computers to simulate impacts, explosions and other stimuli that produce shock waves and the effect that they have on material bodies. A brief account of the early history of this work has been prepared by Johnson and Anderson [61]. Benson [12] and McGlaun and Yarrington [76] and have discussed the capabilities and methodology of the software used for these simulations and have provided extensive lists of references.
- An important thermomechanical response to shock compression is polymorphic phase transformation. The lattice structure of many materials changes to a denser packing when sufficient pressure is applied. Work in this area has been reviewed by Ahrens [1], Duvall and Graham [40], Funtikov and Pavlovsky [45], Kanel et al. [63], and McQueen et al. [77].
- Nunziato et al. [85] have discussed responses of viscoelastic solids.

- Electrical, magnetic, and optical responses of materials to shock compression have been reviewed by Graham [50].
- Heterogeneous materials have been discussed by Baer [8], Davison et al. (eds) [33], and Nesterenko [83].
- Information on metallurgical aspects of shock compression can be found in publications of Gray [53], Johnson [59], Meyers [80], Meyers and Aimone [81], and Zurek and Meyers [111]. The mechanical response of ceramic materials differs considerably from that of ductile materials such as metals and has been discussed by Cagnoux and Tranchet [17] and by Mashimo [75]. Shock-induced chemical reaction and material synthesis has been discussed by Batsanov [10], Graham [50], Kondo [66], Nesterenko [83], Sekine [88], and Thadhani and Aizawa [95].

CHAPTER 2

Mechanical Principles

In this chapter we discuss kinematical descriptions of motion, mathematical representation of internal force systems, and principles of balance of mass, momentum, and energy in general forms that that apply to all continuous materials. Several subsequent chapters are devoted to discussion of the constitutive relations that are used to express the distinguishing characteristics of specific materials.

A few of the basic concepts of continuum mechanics must be set forth before we turn our attention to matters specific to shock waves. Continuum mechanics is a rich and highly developed field of classical physics and mathematics. It has a long history in which concern for wave-propagation problems has played a significant role. The introduction given in this chapter is presented to establish the notation and provide a ready reference to the equations that we shall be using. Several texts and treatises on the subject are listed among the references and the reader is encouraged to refer to them for a more comprehensive presentation [72,98,102,103].

2.1 Notation

The mathematical notation adopted in this book is that in common use in contemporary expositions of continuum mechanics. Field variables are tensors, with most equations being written in indicial notation for maximum clarity. The summation convention applies to repeated indices. Commas denote differentiation with respect to the coordinate indicated, a superimposed T denotes tensor transpose, and a superimposed -1 denotes a tensor inverse. When indicial notation is not used, vectors and second-order tensors are set in boldface type. For the most part, upper-case Latin letters are used to denote quantities related to the Lagrangian (reference) configuration of the material body and lower-case Latin letters denote quantities related to the Eulerian (current or spatial) configuration of the body.

2.2 Material Bodies, Configurations, and Motions

The setting for continuum mechanics is an aggregation of matter called a *material body* (or simply a *body*) occupying a region of three dimensional Euclidean space. A *place* $\mathbf{x}$ in this space is identified by its coordinates x_i, $i = 1, 2, \text{and } 3$, in a Cartesian inertial frame, x.

The essence of a *deformable* body is that the relative positions of material particles comprising it can be altered by applied forces. Any of the possible manifestations of the body is called a *configuration*, which is simply an arrangement of its particles in the space (see Fig. 2.1). To describe these arrangements it is necessary to identify each material (continuum) particle of the body in a unique way. This is accomplished by selecting a convenient *reference configuration*. When the body is in this reference configuration, each of its particles occupies a place, $\mathbf{X}$, having coordinates X_I, $I = 1, 2, \text{and } 3$, in a Cartesian reference frame, X. The reference coordinate associated with a particle remains unchanged as the current configuration of the body is altered. It is often convenient to select the reference configuration to be the initial configuration of the body.

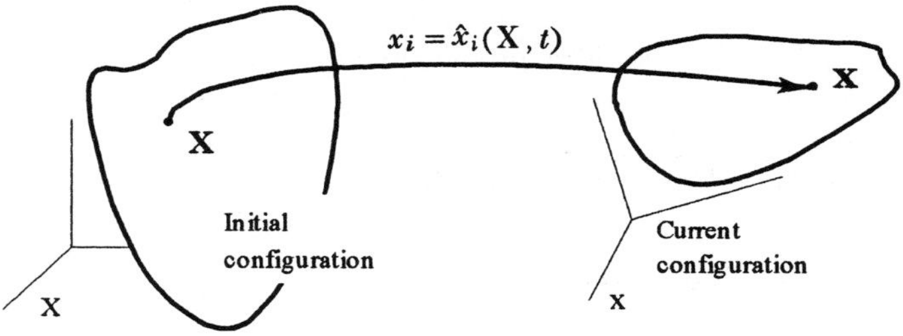

Figure 2.1. Reference and current configurations of a body

A *deformation* of the body is a continuous mapping of its reference configuration to a configuration in the spatial frame. A *motion* of the body is a continuous time-dependent sequence of configurations, and is described by functions of the form

$$x_i = \hat{x}_i(\mathbf{X}, t) \tag{2.1}$$

that give the location of each particle of the body at any time, t. It is a basic principle of continuum mechanics that a deformation cannot place two material particles in the same position at the same time, which means that Eq. 2.1 can be inverted to give equivalent expressions of the form

$$X_I = \hat{X}_I(\mathbf{x}, t)\,. \tag{2.2}$$

We shall assume, for the present, that the functions $\hat{\mathbf{x}}$ and $\hat{\mathbf{X}}$ are as smooth as necessary for the developments undertaken. The focus of this book is shock propagation, however, and we shall have to allow the smoothness assumption to be violated on isolated surfaces to accommodate motions involving shocks.

Equation 2.1 is said to form the *material description* of the motion; it (and other functions of $\mathbf{X}$) focuses attention on what happens to a given material particle. Equation 2.2, which forms the *spatial description*, focuses attention on what happens at a given place $\mathbf{x}$ in the spatial frame. We shall often follow the traditional practice of referring to the spatial description as *Eulerian* and the material description as *Lagrangian*. Although the two descriptions are equivalent, it often proves convenient to solve problems of fluid mechanics in the spatial description and problems of solid mechanics in the material description. A measurement made, for example, by attaching a thermometer or pressure transducer to the wall of a wind tunnel records Eulerian information—a history of some variable recorded as the material flows past a specific place. A measurement made by attaching an instrument to a material sample, as when a thermocouple or strain gauge is cemented to a solid body, provides Lagrangian information—a history of the variable at a specific material particle.

Several important kinematical quantities are derived from the motion 2.1. In differential form this equation becomes

$$dx_i = F_{iJ}\, dX_J\,, \tag{2.3}$$

where the two-point tensor having components

$$F_{iJ} = \frac{\partial \hat{x}_i(\mathbf{X}, t)}{\partial X_J} \tag{2.4}$$

is called the *deformation gradient*. Equation 2.3 shows how a line element $d\mathbf{X}$ in the reference configuration is rotated and/or stretched to its spatial image $d\mathbf{x}$ in the course of the motion. The inverse deformation gradient, $\mathbf{F}^{-1}$, is obtainable by inverting Eq. 2.4 or by differentiation of Eq. 2.2:

$$\overset{-1}{F}_{Ji} = \frac{\partial \hat{X}_J(\mathbf{x}, t)}{\partial x_i}\,. \tag{2.5}$$

In view of the invertiblity of the motion, we have $0 < J < \infty$, where

$$J = \det \mathbf{F} \tag{2.6}$$

is the Jacobian of the transformation 2.1. In fact, J is the ratio of the volumes that a fixed material element occupies in the spatial and reference configurations:

$$dv = J\, dV\,, \tag{2.7}$$

where dV is the volume of a material element in the reference configuration and dv is the volume of this same portion of the material in the spatial configuration.

By a direct, but tedious, calculation one can show that

$$\frac{\partial J}{\partial F_{iJ}} = J \overset{-1}{F}_{Ji} \,. \tag{2.8}$$

The polar decomposition theorem provides two ways of separating the deformation gradient into parts that can be identified with the stretch and the rotation to which a material element has been subjected:

$$F_{iJ} = R_{iI}\, U_{IJ} = V_{ij}\, R_{jJ} \,, \tag{2.9}$$

where **U** and **V** are symmetric and positive-definite and **R** is orthogonal. In these relations the *rotation tensor*, **R**, the *right stretch tensor*, **U**, and the *left stretch tensor*, **V**, are uniquely determined. The expression **RU** can be interpreted as a decomposition of **F** into a stretch followed by a rotation, whereas the expression **VR** corresponds to a process of rotation followed by stretch.

When considering material responses, we shall encounter the *right* and *left Cauchy–Green tensors*

$$C_{IJ} = F_{iI}\, F_{iJ} \tag{2.10}$$

and

$$c_{ij} = \overset{-1}{F}_{Ji}\, \overset{-1}{F}_{Jj} \,, \tag{2.11}$$

respectively.

In the absence of deformation, **F** is the identity tensor so **C** and **c** also have this property. It will prove useful to define the *Lagrangian strain tensor*

$$E_{IJ} = \tfrac{1}{2}\left(C_{IJ} - \delta_{IJ}\right) \tag{2.12}$$

and the *Eulerian strain tensor*

$$e_{ij} = \tfrac{1}{2}\left(\delta_{ij} - c_{ij}\right), \tag{2.13}$$

where $\delta_{ij} = 1$ for $i = j$ and 0 otherwise. These strain tensors vanish in the absence of deformation.

Time derivatives of the various field quantities are required for analysis of most mechanical processes. In some cases, interest lies with the change in time of a variable associated with a particular point in space (e.g. the reading on a thermometer in your home). In other cases, interest lies with change of a variable evaluated at a particular material particle (e.g. the reading on a thermometer attached to a weather balloon entrained in the wind). The partial time derivative of a function at fixed **X** is particularly important for describing the behavior of materials and is called a *material derivative*. As a general convention, a material derivative is denoted by a dot over the relevant variable:

$$\dot{\xi} \equiv \frac{\partial \xi(\mathbf{X}, t)}{\partial t}. \tag{2.14}$$

The material derivative can also be calculated for a variable expressed in the spatial description if the motion is known. Consider a function $\zeta = \hat{\zeta}(\mathbf{x}, t)$. We have

$$\dot{\zeta} = \frac{\partial \zeta(\mathbf{x}(\mathbf{X}, t), t)}{\partial t} = \frac{\partial \zeta(\mathbf{x}, t)}{\partial t} + \frac{\partial \zeta(\mathbf{x}, t)}{\partial x_i}\frac{\partial x_i(\mathbf{X}, t)}{\partial t} = \frac{\partial \zeta(\mathbf{x}, t)}{\partial t} + \frac{\partial \zeta(\mathbf{x}, t)}{\partial x_i}\dot{x}_i\,. \tag{2.15}$$

The velocity of the material particle **X** is the rate of change of its position in space, so the components of the *particle velocity vector* are

$$\dot{x}_i = \frac{\partial \hat{x}_i(\mathbf{X}, t)}{\partial t}. \tag{2.16}$$

The *particle acceleration* is simply the material derivative of the particle velocity:

$$\ddot{x}_i = \frac{\partial \dot{x}_i(\mathbf{X}, t)}{\partial t} = \frac{\partial \dot{x}_i(\mathbf{x}, t)}{\partial t} + \dot{x}_j \frac{\partial \dot{x}_i(\mathbf{x}, t)}{\partial x_j}. \tag{2.17}$$

The spatial *velocity gradient tensor*,

$$l_{ij} \equiv \dot{x}_{i,j} = \dot{F}_{iI} \overset{-1}{F}_{Ij}\ , \tag{2.18}$$

plays a role in the theories we shall be using. Most often it appears in terms of either the *stretching*, its symmetric part,

$$d_{ij} \equiv \tfrac{1}{2}(l_{ij} + l_{ji})\,, \tag{2.19}$$

or the *spin*, its antisymmetric part,

$$w_{ij} \equiv \tfrac{1}{2}(l_{ij} - l_{ji})\,, \tag{2.20}$$

so that

$$l_{ij} = d_{ij} + w_{ij}\,. \tag{2.21}$$

The material derivative of J can be shown to satisfy the equations

$$\dot{J} = (\det \mathbf{F})^{\boldsymbol{\cdot}} = (\det \mathbf{F})\,\mathrm{tr}\big(\dot{\mathbf{F}}\overset{-1}{\mathbf{F}}\big) = (\det \mathbf{F})\,\mathrm{tr}\,\mathbf{l} = J\,\dot{x}_{i,i}\,. \tag{2.22}$$

Although somewhat out of place in a section otherwise devoted to kinematical matters, it is useful to note that each element dV of matter is assigned a property called *mass*. The ratio of this mass to the volume of the element is the *mass density*, ρ_R, the mass per unit volume of material in the reference configuration. We designate by ρ the mass density of this same material element in the current configuration. The reciprocals of these quantities, the *specific volume* in the reference and current configurations, respectively, are designated v_R and v: $v_R = 1/\rho_R$ and $v = 1/\rho$. Equations 2.7 and 2.22 can be written in the forms

$$J = \frac{\rho_R}{\rho} = \frac{v}{v_R} \tag{2.23}$$

and

$$\frac{\dot{\rho}}{\rho} = -\frac{\dot{v}}{v} = -\dot{x}_{i,i} = -d_{ii}\,. \tag{2.24}$$

The *Lagrangian compression*

$$\Delta = \frac{v_R - v}{v_R} = \frac{\rho - \rho_R}{\rho} = 1 - J \tag{2.25}$$

occurs frequently in the shock-wave literature.

Uniaxial Deformation and Motion. This book is concerned with the propagation of plane longitudinal waves, for which the motion is uniaxial* and can be expressed in the especially simple form

$$x = X + U(X, t), \quad x_2 = X_2, \quad x_3 = X_3\,. \tag{2.26}$$

In describing this motion, we have chosen coincident reference and spatial coordinates oriented so that the displacement is along the 1 axes and we have written $x = x_1$ and $X = X_1$ for simplicity. We shall continue to do this throughout this book, using the subscripts only occasionally for clarity. It is apparent from the form of the equation that a material particle **X** is displaced from its reference position by an amount U in the positive x direction. There is no lateral displacement of any particle. The spatial description of the uniaxial motion equivalent to that described by Eq. 2.26 has the form

$$X = x - u(x, t), \quad X_2 = x_2, \quad X_3 = x_3\,. \tag{2.27}$$

* The term "uniaxial" is also used in discussing the extension of slender rods. That situation differs from the one considered here in that the term refers to a uniaxial stress field—it is the transverse stress that vanishes, not the transverse displacement. Transverse deformation of a slender rod occurs in the amount required to produce a state of vanishing transverse stress components. We shall conform to the usual practice of the shock-physics community in calling the motions of Eqs. 2.26 and 2.27 uniaxial motions or uniaxial strains.

When attention is restricted to motions in this class, only the coordinate in the direction normal to the plane of the wave enters the equations of the theory. The state and properties of the material points, and the imposed boundary and initial conditions, can vary with this coordinate, but are independent of the lateral position. Although the body is a three-dimensional object, all material particles in a given plane normal to the coordinate are constituted the same, experience the same conditions, and exhibit the same responses.

Uniaxial motions appear most naturally in material bodies that occupy the entire space, half of the space, or a plate of infinite lateral extent but finite thickness. The reason that this unrealizable setting is of practical importance is that the waves of interest propagate at a finite speed. This ensures that there is a known, finite time interval during which phenomena occurring near the 1 axes are completely independent of any effect of lateral boundaries located at some distance from these axes.

In the case of the uniaxial deformation of Eq. 2.26, the components of the deformation gradient are

$$F_{11}=1+\frac{\partial U(X,t)}{\partial X},\quad F_{22}=F_{33}=1,\ \text{and}\ F_{iJ}=0,\ \text{for}\ i\neq J\,, \tag{2.28}$$

the rotation tensor vanishes, and the components of the particle velocity vector are

$$\dot{x}=\frac{\partial U(X,t)}{\partial t},\quad \dot{x}_2=\dot{x}_3=0\,. \tag{2.29}$$

Evaluation of Eq. 2.18 gives the only nonvanishing component of the velocity gradient tensor as

$$l_{11}=\frac{\dot{F}_{11}}{F_{11}}, \tag{2.30}$$

so $l_{11}=d_{11}$ and $\mathbf{w}=\mathbf{0}$.

The volume change experienced during the uniaxial motion as given by Eq. 2.7, is

$$dv=F_{11}\,dV=(1+U_X)\,dV\,, \tag{2.31}$$

where we have introduced the notation $U_X\equiv\partial U(X,t)/\partial X$. For uniaxial motion Eqs. 2.23 and 2.24 take the forms

$$J=F_{11}=1+U_X=\frac{v}{v_R} \tag{2.32}$$

and

$$\frac{\dot{\rho}}{\rho}=-\frac{\dot{v}}{v}=-\frac{\dot{F}_{11}}{F_{11}}=-\frac{U_{Xt}}{1+U_X}. \tag{2.33}$$

In the special case of uniaxial deformation we have $\Delta = 1 - F_{11} = -U_X$.

From Eqs. 2.3 and 2.28 we see that a line element lying along the X axis in the reference configuration lies along the x axis in the spatial configuration, but increases in length by the factor F_{11} (i.e., $dx = F_{11}\,dX$) in the course of the motion. The length of line elements oriented perpendicular to the X axis is unaltered by the motion (i.e., $dx_2 = dX_2$, and $dx_3 = dX_3$). Line elements lying along the coordinate directions maintain their orthogonality in the course of the motion, so we call these directions *principal axes* of the deformation. A line element not perpendicular to the 1 axes is both stretched and rotated. Consider the body shown in Fig. 2.2. Using Eq. 2.3, we find that the line elements initially lying at ±45° rotate in opposite directions so that the 90° angle between them changes to $90° - \gamma$ (where $\gamma > 0$ when the thickness is increased), from which we obtain

$$\gamma = 2\arctan[(F_{11} - 1)/(F_{11} + 1)]. \tag{2.34}$$

The change in a right angle is conventionally taken as a measure of shear, so we see that the uniaxial stretching produces shear on planes making an angle with the 1 axis.

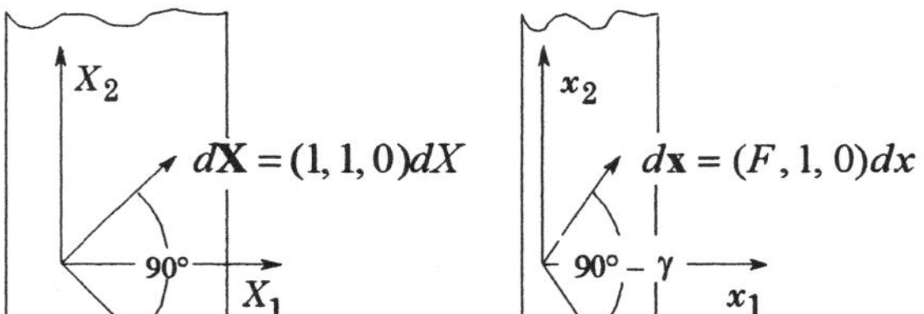

Figure 2.2. Line elements in a body subject to uniaxial deformation along the 1 axes. The angle γ is defined as positive when the plate thickness is increased, i.e., when the material is in tension.

Usually, shear is discussed using a drawing such as that shown in Fig. 2.3a, but it is easy to see that a similar figure arises when the uniaxial deformation is viewed as shown in Fig. 2.3b. Note that the uniaxial deformation involves a change of volume of the material in addition to shear, whereas the simple shear of Fig. 2.3a occurs at fixed volume. Since solids are just those materials that resist shear, the fact that shear occurs when waves of uniaxial deformation propagate through a body makes the study of solids in this context different from, and richer than, the corresponding study of inviscid fluids.

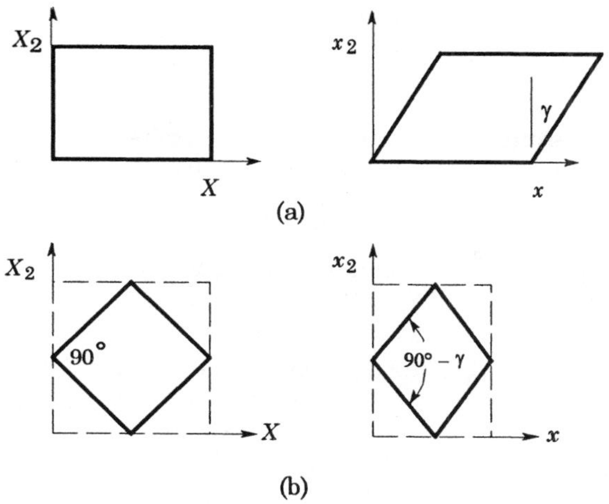

Figure 2.3. A body subject to (a) simple shear and (b) uniaxial compression. Since shear strain is measured in terms of the change in a right angle, it is easy to see that shear is produced in the uniaxial deformation.

Spatial expressions (i.e., expressions in terms of $u(x,t)$) for the particle velocity and deformation gradient that are equivalent to those already given in material form are readily obtained by differentiating Eq. 2.27. Holding X and then t fixed, we have

$$F_{11} = \frac{1}{1-u_x} \tag{2.35}$$

and

$$\dot{x} = \frac{u_t}{1-u_x} \tag{2.36}$$

for the nontrivial components of the deformation gradient and particle velocity. In writing these expressions, we have used the shorthand notation

$$u_x = \frac{\partial u(x,t)}{\partial x}, \quad u_t = \frac{\partial u(x,t)}{\partial t}. \tag{2.37}$$

Solving Eqs. 2.35 and 2.36 for the derivatives of $u(x,t)$, gives

$$u_t = \frac{\dot{x}}{F_{11}} \quad \text{and} \quad u_x = \frac{F_{11}-1}{F_{11}}. \tag{2.38}$$

Finally, equating the material and spatial expressions for $\dot{x}$ and F_{11} gives the relations

$$u_x = \frac{U_X}{1+U_X}, \quad u_t = \frac{U_t}{1+U_X} \tag{2.39}$$

and

$$U_X = \frac{u_x}{1-u_x}, \quad U_t = \frac{u_t}{1-u_x}, \tag{2.40}$$

where we have introduced the additional shorthand notation

$$U_X \equiv \frac{\partial U(X,t)}{\partial X}, \quad U_t \equiv \frac{\partial U(X,t)}{\partial t}. \tag{2.41}$$

The particle acceleration, the material derivative of the particle velocity, is given by

$$\ddot{x} = \frac{\partial \dot{x}(X,t)}{\partial t}, \tag{2.42}$$

and the spatial expression for this quantity is

$$\ddot{x} = \frac{\partial \dot{x}(x(X,t),t)}{\partial t} = \frac{\partial \dot{x}(x,t)}{\partial t} + \dot{x}(x,t)\frac{\partial \dot{x}(x,t)}{\partial x}. \tag{2.43}$$

For convenience, we record the following results of evaluating the strains, velocity gradients, etc., for the case of longitudinal motion:

$$C_{11} = (F_{11})^2 = (1+U_X)^2, \quad C_{22} = C_{33} = 1, \quad C_{IJ} = 0 \text{ for } I \neq J \tag{2.44}$$

$$b_{11} = (F_{11})^2 = (1-u_x)^{-2}, \quad b_{22} = b_{33} = 1, \quad b_{ij} = 0, \quad i \neq j \tag{2.45}$$

$$E_{11} = \tfrac{1}{2}[(F_{11})^2 - 1] = U_X + \tfrac{1}{2}(U_X)^2, \quad \text{other } E_{IJ} = 0 \tag{2.46}$$

$$e_{11} = \tfrac{1}{2}\left[1-(F_{11})^{-2}\right] = u_x - \tfrac{1}{2}(u_x)^2, \quad \text{other } e_{ij} = 0 \tag{2.47}$$

$$\dot{x} = u_t F_{11} = U_t \tag{2.48}$$

$$F_{11} = v/v_R = 1 + U_X, \quad F_{22} = F_{33} = 1, \quad \text{other } F_{iJ} = 0 \tag{2.49}$$

$$\Delta = 1 - F_{11} = -U_X = 1-(v/v_R) \tag{2.50}$$

$$l_{11} = \frac{\dot{F}_{11}}{F_{11}} = \frac{U_{Xt}}{1+U_X} = \frac{(1-u_x)u_{xt} + u_{xx}u_t}{(1-u_x)^2}, \quad \text{other } l_{ij} = 0, \tag{2.51}$$

and, because $\mathbf{l}$ is symmetrical for uniaxial motions, $\mathbf{d} = \mathbf{l}$, and $\mathbf{w} = \mathbf{0}$.

2.3 Force and Stress

The internal force system at a point in a material body is characterized by the *Cauchy stress tensor*, $\mathbf{t}(x,t)$ having components t_{ij}. Satisfaction of the principle of balance of moment of momentum requires that this stress tensor be symmetric (see, for example, [72]):

$$t_{ij} = t_{ji}\,. \tag{2.52}$$

The force per unit area, $\mathbf{t}^{(\mathbf{n})}(\mathbf{x},t)$, on a plane in the current configuration passing through the point $\mathbf{x}$ and characterized by the unit normal vector $\mathbf{n}$, has the components

$$t_i^{(\mathbf{n})}(\mathbf{x},t) = n_j\, t_{ji}(\mathbf{x},t) \tag{2.53}$$

(see Fig. 2.4). From this equation, we see that the components of the stress tensor correspond to normal and tangential forces on the faces of a unit cube aligned with the spatial coordinate frame (see Fig. 2.5). As shown in the figure, the diagonal components of the tensor correspond to normal forces on the coordinate planes and the off-diagonal components correspond to transverse (shear) forces on these planes. It is always possible to find a coordinate system (called a *principal coordinate system*) in which the off-diagonal components vanish. The components of the stress tensor expressed in this coordinate frame are called *principal stresses.*

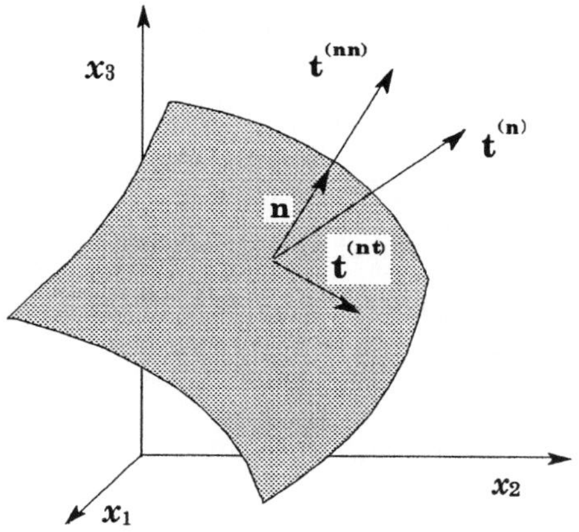

Figure 2.4. Stress vector at a point on a surface, along with its normal and tangential components

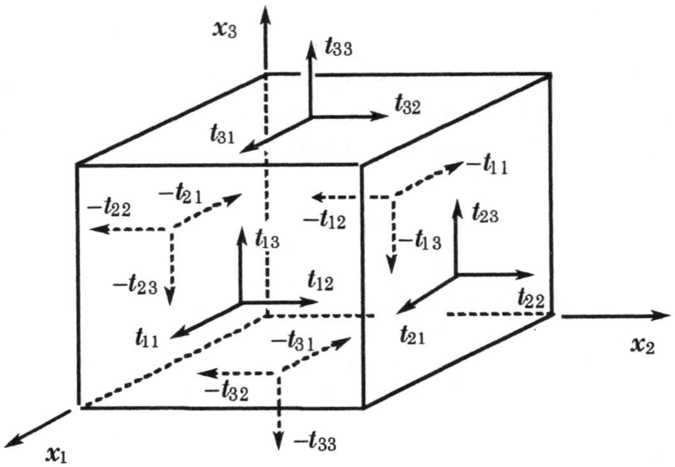

Figure 2.5. Forces associated with components of the Cauchy stress tensor

It is useful to note that Eq. 2.53, which gives the stress applied on a plane having unit normal **n**, can be decomposed into components normal and tangential to this plane. For the normal component, we have (see Fig. 2.4)

$$t_i^{(\mathbf{nn})} = t_j^{(\mathbf{n})} n_j\, n_i = t_{kj}\, n_k\, n_j\, n_i\,, \tag{2.54}$$

and, for the tangential component (the maximum shear stress on the plane),

$$t_i^{(\mathbf{nt})} = t_i^{(\mathbf{n})} - t_i^{(\mathbf{nn})} = t_{ki}\, n_k - t_{kj}\, n_k\, n_j\, n_i\ . \tag{2.55}$$

The trace of the stress tensor is invariant to changes in the coordinate frame and the scalar quantity

$$p = -\tfrac{1}{3} t_{ii}\,, \tag{2.56}$$

the (negative) average of the principal stresses, is called the *pressure*.*

The tensor $\mathbf{t}'$, having components t'_{ij} defined by the equation

$$t'_{ij} = t_{ij} - \tfrac{1}{3} t_{kk}\, \delta_{ij} = t_{ij} + p\, \delta_{ij} \tag{2.57}$$

obtained by subtracting the average stress from the Cauchy stress tensor is called the *stress-deviator* tensor, and is a measure of the shear stress present in the body. Writing the foregoing equation in the form

---

* More precisely, the *mechanical pressure*. When dissipative effects such as viscosity contribute to the stress, the mechanical pressure differs from the *thermodynamic pressure* discussed in Chap. 5.

$$t_{ij} = t'_{ij} - p\,\delta_{ij} \tag{2.58}$$

expresses the stress as the sum of a pressure and a shear term.

Let us suppose that a body in a state of uniaxial deformation is of such symmetry that the stress tensor is of the form

$$\left\| t_{ij} \right\| = \begin{Vmatrix} t_{11} & 0 & 0 \\ 0 & t_{22} & 0 \\ 0 & 0 & t_{22} \end{Vmatrix}, \tag{2.59}$$

the case in which the transverse stresses associated with the constraint that there is no lateral deformation are the same on all planes parallel to the x_1 axis. Let us consider a plane inclined so that its normal makes the angle φ with the x_1 axis. By symmetry, all such planes are equivalent, so we may take the one having the normal $\mathbf{n} = (\cos\varphi, \sin\varphi, 0)$. From Eqs. 2.53–2.56 we find that

$$\begin{aligned} \mathbf{t}^{(\mathbf{n})} &= (t_{11}\cos\varphi, \quad t_{22}\sin\varphi, \quad 0) \\ \mathbf{t}^{(\mathbf{nn})} &= (t_{11}\cos^2\varphi + t_{22}\sin^2\varphi)\,(\cos\varphi, \sin\varphi, 0) \\ \mathbf{t}^{(\mathbf{nt})} &= (t_{11} - t_{22})\cos\varphi\,\sin\varphi\,(\sin\varphi, -\cos\varphi, 0) \\ p &= -\tfrac{1}{3}(t_{11} + 2\,t_{22}) \end{aligned} \tag{2.60}$$

for this case. The coefficient $t_{11} - t_{22}$ in the third of these equations, which is a difference of principal stresses, is characteristic of shear stress. It can be shown that the magnitude of the shear stress is maximized on the planes for which $\varphi = \pm\,\pi/4$. The magnitude of the shear stress on these planes is

$$\tau_{\max} = \tfrac{1}{2}\left| t_{11} - t_{22} \right|. \tag{2.61}$$

The maximum shear stress is realized on planes lying at 45° to the x axis, so we often use the variable

$$\tau_{45^\circ} = \tfrac{1}{2}(t_{11} - t_{22}), \tag{2.62}$$

which is positive in tension and negative in compression. Combining this result with Eq. 2.60_4, we obtain the equation

$$-t_{11} = p - \tfrac{4}{3}\tau_{45^\circ}, \tag{2.63}$$

which shows that the longitudinal stress component includes both pressure and shear-stress contributions. This is a special case of the result expressed more generally by Eq. 2.58.

Another stress measure that we shall encounter is the *second Piola–Kirchhoff stress tensor*

$$T_{IJ} = \frac{v}{v_R} \overset{-1}{F}_{Ij} \overset{-1}{F}_{Ji} t_{ij} \,. \tag{2.64}$$

This stress tensor arises as the force conjugate to the strain tensor **E** in a thermodynamic theory of elastic materials (see Chap. 6). The symmetry of **t** implies the symmetry of this tensor.

2.4 Governing Equations

The mechanical behavior of the material under consideration is governed by the principles of conservation of mass, and of balance of momentum, moment of momentum, and energy. These relations are usually expressed by partial differential equations, but motions such as those in which shocks are embedded involve discontinuities that necessitate using a more general expression of these principles.

The *principle of conservation of mass* postulates that the mass of any part, $\mathcal{P}$, of a material body is unaltered by a motion taking this material to a configuration $p(t)$ at time t. Denoting the mass density (mass per unit volume) of the material when it is in its current configuration by $\rho(\mathbf{x}, t)$, the mass of $\mathcal{P}$ is given by

$$M = \int_{p(t)} \rho(\mathbf{x}, t)\, dv \,. \tag{2.65}$$

The principle of conservation of mass then takes the form

$$\dot{M} = \frac{d}{dt} \int_{p(t)} \rho(\mathbf{x}, t)\, dv = 0 \,, \tag{2.66}$$

where the equality must hold for *any* part $\mathcal{P}$ of the body.

The *principle of balance of momentum* postulates that the rate of increase of momentum of $\mathcal{P}$ is balanced by the momentum supplied by applied forces during the course of a motion taking this material to a configuration $p(t)$. The momentum of $\mathcal{P}$ can be expressed in the form

$$\int_{p(t)} \rho \dot{x}_i \, dv \,,$$

and the rate at which momentum is supplied by applied forces can be expressed in the form

$$\int_{s(t)} t_i^{(\mathbf{n})} \, ds + \int_{p(t)} \rho f_i \, dv \,,$$

where $\mathbf{f}$ is an extrinsic body force per unit mass of material, $s(t)$ is the surface of the part $\mathcal{P}$ of the body at the time t, and $\mathbf{n}(\mathbf{x},t)$ is the unit outward normal vector to the surface. The first integral accounts for the total force applied by the surface traction distribution and the second integral accounts for the total force applied to internal parts of $\mathcal{P}$ by such extrinsic means as gravity.

Using these representations, we can express the principle of balance of momentum in the form

$$\frac{d}{dt}\int_{p(t)} \rho \dot{x}_i \, dv = \int_{s(t)} t_i^{(\mathbf{n})} \, ds + \int_{p(t)} \rho f_i \, dv . \tag{2.67}$$

The *principle of balance of energy* postulates that the rate of increase of total energy of $\mathcal{P}$ is balanced by the energy supplied by forces applied to the surface, heat conducted through the surface, extrinsic forces applied within the volume of the body, and an extrinsic supply of heat deposited within the volume of the body. The energy of $\mathcal{P}$ can be expressed in the form

$$\int_{p(t)} \rho\left(\varepsilon + \tfrac{1}{2}\dot{x}_i \dot{x}_i\right) dv , \tag{2.68}$$

where $\varepsilon(\mathbf{x},t)$ is the specific internal energy (i.e., internal energy per unit mass) of the material and $\dot{x}_i \dot{x}_i / 2$ is the specific kinetic energy. The rate at which energy is transferred by applied forces, heat conduction out of the body, and an extrinsic heat supply, can be expressed in the form

$$\int_{s(t)} t_i^{(\mathbf{n})} \dot{x}_i \, ds - \int_{s(t)} q_i n_i \, ds + \int_{p(t)} \rho f_i \dot{x}_i \, dv + \int_{p(t)} \rho r \, dv . \tag{2.69}$$

In this expression, $\mathbf{q}$ is the heat flux vector such that $\mathbf{q}\cdot\mathbf{n} = q_i n_i$ is the rate at which energy is transferred out of the body through a unit area of the surface (in the current configuration) and r is the rate at which extrinsic sources supply heat to interior parts of the body (per unit mass of material). The first integral in the foregoing expression accounts for the rate at which work is done by the surface-traction distribution, the second integral accounts for the rate at which energy is conducted to the body through its surface, the third integral accounts for the rate at which work is done by extrinsic body forces and, finally, the fourth integral accounts for an extrinsic heat source.

Using these representations, we can express the principle of balance of energy in the form

$$\frac{d}{dt}\int_{p(t)} \rho\left(\varepsilon + \tfrac{1}{2}\dot{x}_i \dot{x}_i\right) dv = \int_{s(t)} \left(t_i^{(\mathbf{n})} \dot{x}_i - q_i n_i\right) ds + \int_{p(t)} \left(\rho f_i \dot{x}_i + \rho r\right) dv , \tag{2.70}$$

where the equation must hold for any part of the body.

The *principle of conservation of moment of momentum* is satisfied if, and only if, the Cauchy stress tensor is symmetric (see, for example, [72,82,100]). This requirement is imposed when the constitutive equations describing the response of a specific material are developed, so it does not lead to a field equation.

2.4.1 Differential Equations Describing Smooth Fields

Using Eq. 2.7, we can express the principle of conservation of mass, as represented by Eq. 2.66, in terms of the Lagrangian coordinates:

$$\dot{M} = \frac{d}{dt}\int_{\mathcal{P}} \rho(\mathbf{x}(\mathbf{X},t),t)\,J\,dV = 0\,. \tag{2.71}$$

Since the domain of integration is now independent of time, we can interchange the order of integration and differentiation when the flow is smooth enough for the derivative of the integrand to exist. This gives the equation

$$\int_{\mathcal{P}} (\rho J)^{\bullet}\,dV = 0\,. \tag{2.72}$$

Since this equation must hold for any part, $\mathcal{P}$, of $\mathcal{B}$, the integrand must vanish:

$$\frac{\partial}{\partial t}\big(\rho(\mathbf{X},t)J(\mathbf{X},t)\big) = \dot{\rho}J + \rho\dot{J} = 0\,, \tag{2.73}$$

or, using Eq. 2.22,

$$\dot{\rho} + \rho\,\dot{x}_{i,i} = 0\,. \tag{2.74}$$

Using Eq. 2.15, this result can be written

$$\frac{\partial\rho(\mathbf{x},t)}{\partial t} + \frac{\partial(\rho\dot{x}_i)}{\partial x_i} = 0\,, \tag{2.75}$$

or, in brief notation,

$$\rho_t + (\rho\dot{x}_i)_{,i} = 0\,. \tag{2.76}$$

It is important to note that the equation of conservation of mass in the form of Eq. 2.73 can be integrated to yield the result

$$\rho(\mathbf{X},t)\,J(\mathbf{X},t) = \vartheta(\mathbf{X})\,, \tag{2.77}$$

where $\vartheta(\mathbf{X})$ is an undetermined integration function. If the body is in its reference condition at some time t^*, then the integration function can be evaluated and we find that $\rho(\mathbf{X},t)J(\mathbf{X},t) = \rho_R(\mathbf{X})$, which we often write

$$\det \mathbf{F}(\mathbf{X},t) = \rho_R(\mathbf{X})/\rho(\mathbf{X},t)\,, \tag{2.78}$$

an integrated expression of the principle of conservation of mass. This result is the same as that of Eq. 2.23.

Under conditions of sufficient smoothness, the principle of balance of momentum, as represented by Eq. 2.67, can be transformed to the reference configuration, the material derivative taken inside the integral, and that expression manipulated to yield the result

$$0 = \int_{\mathcal{P}} \left[(\rho\, \dot{x}_i\, J)^{\cdot} - (t_{ji,j} + \rho f_i)\, J \right] dV = \int_{\mathcal{P}} \left[(\rho\, \dot{x}_i)^{\cdot} + \rho\, \dot{x}_i\, \dot{x}_{j,j} - t_{ji,j} + \rho f_i \right] J\, dV\,. \tag{2.79}$$

In obtaining this expression, we have used the divergence theorem to transform the surface integral associated with the surface traction vector to a volume integral involving the stress gradient.

As before, the requirement that this equation hold for all parts $\mathcal{P}$ of the body necessitates that the integrand vanish, a result that can be placed in the form

$$\frac{\partial}{\partial t}(\rho \dot{x}_i) + \frac{\partial}{\partial x_j}(\rho \dot{x}_j \dot{x}_i - t_{ji}) = \rho f_i\,. \tag{2.80}$$

Further manipulation of this equation, and use of Eq. 2.76 leads to the conventional expression of the principle of balance of momentum in the form of a differential equation:

$$t_{ji,j} + \rho f_i = \rho\, \ddot{x}_i\,. \tag{2.81}$$

Proceeding with the left member of Eq. 2.70 as before yields

$$\frac{\partial}{\partial t}\left[\rho\,(\varepsilon + \tfrac{1}{2}\dot{x}_i \dot{x}_i)\right] + \frac{\partial}{\partial x_j}\left[\rho\,(\varepsilon + \tfrac{1}{2}\dot{x}_i \dot{x}_i)\,\dot{x}_j - t_{ji}\dot{x}_i + q_i\right] = \rho f_i\, \dot{x}_i + \rho r\,, \tag{2.82}$$

and making use of Eqs. 2.74 and 2.81 permits this result to be placed in the more conventional form of the expression of the principle of balance of energy:

$$\rho\,\dot{\varepsilon} = t_{ij}\, \dot{F}_{iJ}\, \overset{-1}{F}_{Jj} - q_{i,i} + \rho r = t_{ij}\, \dot{x}_{i,j} - q_{i,i} + \rho r\,, \tag{2.83}$$

or, since $\mathbf{t}$ is symmetric,

$$\rho\,\dot{\varepsilon} = t_{ij}\, d_{ji} - q_{i,i} + \rho r\,. \tag{2.84}$$

The term

$$\dot{W} \equiv t_{ij}\, \dot{x}_{i,j} = t_{ij}\, d_{ij} \tag{2.85}$$

appearing in the energy equation is called the *stress power*.

Summary: Eulerian Form of the Equations of Balance. The differential equations representing the principles of conservation of mass and balance of momentum and energy are

$$\begin{gathered}\frac{\partial \rho}{\partial t}+\frac{\partial(\rho \dot{x}_i)}{\partial x_i}=0\\ \frac{\partial t_{ji}}{\partial x_j}-\rho\left(\frac{\partial \dot{x}_i}{\partial t}+\dot{x}_j\frac{\partial \dot{x}_i}{\partial x_j}\right)=-\rho f_i\\ \rho\left(\frac{\partial \varepsilon}{\partial t}+\dot{x}_i\frac{\partial \varepsilon}{\partial x_i}\right)-t_{ij}\frac{\partial \dot{x}_i}{\partial x_j}=-\frac{\partial q_i}{\partial x_i}+\rho r.\end{gathered} \tag{2.86}$$

Since the independent coordinate appearing in these equations is the Eulerian coordinate, x, they are said to be in the Eulerian form.

When Eqs. 2.86 are restricted to uniaxial motions, they take the simpler form:

$$\begin{gathered}\frac{\partial \rho}{\partial t}+\frac{\partial}{\partial x}(\rho \dot{x})=0\\ \frac{\partial t_{11}}{\partial x}-\rho\left(\frac{\partial \dot{x}}{\partial t}+\frac{\partial \dot{x}}{\partial x}\dot{x}\right)=-\rho f\\ \rho\left(\frac{\partial \varepsilon}{\partial t}+\frac{\partial \varepsilon}{\partial x}\dot{x}\right)-t_{11}\frac{\partial \dot{x}}{\partial x}=-\frac{\partial q}{\partial x}+\rho r.\end{gathered} \tag{2.87}$$

Eulerian Form of the Equations for Smooth Fields in Inviscid Fluids. Any stress present in an inviscid fluid consists only of pressure, i.e., it is of the form

$$t_{ij}=-p\,\delta_{ij}\,. \tag{2.88}$$

When the differential equations of balance are specialized to this case they become

$$\begin{gathered}\frac{\partial \rho}{\partial t}+\frac{\partial(\rho \dot{x}_i)}{\partial x_i}=0\\ \frac{\partial p}{\partial x_i}+\rho\left(\frac{\partial \dot{x}_i}{\partial t}+\dot{x}_j\frac{\partial \dot{x}_i}{\partial x_j}\right)=\rho f_i\\ \rho\left(\frac{\partial \varepsilon}{\partial t}+\dot{x}_i\frac{\partial \varepsilon}{\partial x_i}\right)+p\frac{\partial \dot{x}_i}{\partial x_i}=-\frac{\partial q_i}{\partial x_i}+\rho r,\end{gathered} \tag{2.89}$$

or, when further restricted to uniaxial motions,

$$\frac{\partial \rho}{\partial t}+\frac{\partial}{\partial x}(\rho\,\dot{x})=0$$

$$\frac{\partial p}{\partial x}+\rho\left(\frac{\partial \dot{x}}{\partial t}+\frac{\partial \dot{x}}{\partial x}\dot{x}\right)=\rho f \tag{2.90}$$

$$\rho\left(\frac{\partial \varepsilon}{\partial t}+\frac{\partial \varepsilon}{\partial x}\dot{x}\right)+p\,\frac{\partial \dot{x}}{\partial x}=-\frac{\partial q}{\partial x}+\rho r,$$

Lagrangian Form of the Conservation Equations. Although Eqs. 2.86 present the principles of conservation of mass and balance of momentum and energy in their most familiar form, it is often convenient to have equations in which the independent variables are $(\mathbf{X}, t)$ rather than $(\mathbf{x}, t)$. Transforming the independent coordinate from $\mathbf{x}$ to $\mathbf{X}$ in Eqs. 2.86 yields the result

$$\frac{\partial \rho}{\partial t}+\rho\,\overset{-1}{F_{Ii}}\frac{\partial \dot{x}_i}{\partial X_I}=0$$

$$\frac{\partial}{\partial X_J}(F_{iK}\,T_{KJ})-\rho_R\frac{\partial \dot{x}_i}{\partial t}=-\rho_R\,f_i \tag{2.91}$$

$$\rho_R\frac{\partial \varepsilon}{\partial t}-T_{IJ}\frac{\partial E_{IJ}}{\partial t}=-\frac{\partial Q_I}{\partial X_I}+\rho_R r,$$

where all of the field variables are functions of $\mathbf{X}$ and t and $Q_I = J\,X_{I,k}\,q_k$ relates the Lagrangian and Eulerian components of the heat flux vector. For the case of uniaxial strain these equations take the form

$$\frac{\partial \dot{x}}{\partial X}-\rho_R\frac{\partial v}{\partial t}=0$$

$$\frac{\partial t_{11}}{\partial X}-\rho_R\frac{\partial \dot{x}}{\partial t}=-\rho_R\,f \tag{2.92}$$

$$\rho_R\,\frac{\partial \varepsilon}{\partial t}-t_{11}\,\frac{\partial \dot{x}}{\partial X}=-\frac{\partial Q}{\partial X}+\rho_R\,r.$$

In writing these latter equations we have used the result $(\rho_R/\rho)\,T_{11}=t_{11}$. When the material is an inviscid fluid Eqs. 2.92 take the form

$$\frac{\partial \dot{x}}{\partial X}-\rho_R\frac{\partial v}{\partial t}=0$$

$$\frac{\partial p}{\partial X}+\rho_R\frac{\partial \dot{x}}{\partial t}=\rho_R\,f \tag{2.93}$$

$$\rho_R\,\frac{\partial \varepsilon}{\partial t}+p\,\frac{\partial \dot{x}}{\partial X}=-\frac{\partial Q}{\partial X}+\rho_R\,r.$$

2.4.2 Jump Equations Describing Plane Shocks

A *shock,* also called a *shock wave,* is a propagating surface at which the displacement is continuous but the mass density, particle velocity, stress, and other field variables are discontinuous. At these discontinuities, the derivatives appearing in the differential equations developed in the foregoing section do not exist, so those equations are meaningless. Nevertheless, the field quantities appearing in the integral forms of the governing equations are defined almost everywhere, so the integrals are meaningful. The discontinuities that occur at a shock are constrained by the requirement that the integral equations be satisfied in the neighborhood of the shock. In this section, jump conditions are developed for plane shocks embedded in smooth uniaxial motions. A general treatment of curved shocks is unnecessary for the analysis of the problems to be considered in the following chapters, and is beyond the scope of this book.

The integral form of the conservation laws of Eqs. 2.66, 2.67, and 2.70 for uniaxial motions is

$$\frac{d}{dt}\int_{x_a(t)}^{x_b(t)} \rho(x,t)\,dx = 0$$

$$\frac{d}{dt}\int_{x_a(t)}^{x_b(t)} \rho\dot{x}\,dx = t_{11}(x_b(t)) - t_{11}(x_a(t)) + \int_{x_a(t)}^{x_b(t)} \rho f\,dx \tag{2.94}$$

$$\frac{d}{dt}\int_{x_a(t)}^{x_b(t)} \rho(\varepsilon + \tfrac{1}{2}\dot{x}^2)\,dx = t_{11}(x_b(t))\,\dot{x}(x_b(t)) - t_{11}(x_a(t))\,\dot{x}(x_a(t)) + \int_{x_a(t)}^{x_b(t)} \rho(f\dot{x} + r)\,dx,$$

where $x_a(t) = x(X_a, t)$ and $x_b(t) = x(X_b, t)$, for $X_b > X_a$, are material surfaces. We consider application of these equations to the case of a shock embedded in a smooth flow, as shown in Fig. 2.6. Since our primary interest is in shock discontinuities we have omitted the heat flux term that would give rise to dispersion.

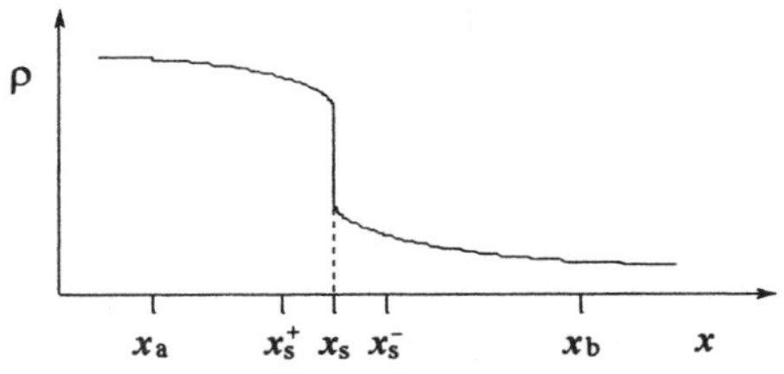

Figure 2.6. Smooth density waveform containing an embedded shock

Let us begin by considering Eq. 2.94$_1$, representing the principle of conservation of mass. If the interval of integration (x_a, x_b) is decomposed into three parts, (x_a, x_S^+), (x_S^+, x_S^-), and (x_S^-, x_b), as shown in the figure, the equation can be written

$$\frac{d}{dt}\left\{\int_{x_a(t)}^{x_S^+(t)} \rho(x,t)\,dx + \int_{x_S^+(t)}^{x_S^-(t)} \rho(x,t)\,dx + \int_{x_S^-(t)}^{x_b(t)} \rho(x,t)\,dx\right\} = 0. \tag{2.95}$$

We shall require that the surfaces x_S^+ and x_S^- move with the shock, so that

$$\frac{dx_S^+}{dt} = \frac{dx_S^-}{dt} = u_S, \tag{2.96}$$

where u_S is the Eulerian shock velocity, the velocity with which the shock moves along the x coordinate. Since the surfaces at x_a and x_b are transported with the material, we also have

$$\frac{dx_a}{dt} = \dot{x}(x_a(t), t), \quad \text{and} \quad \frac{dx_b}{dt} = \dot{x}(x_b(t), t). \tag{2.97}$$

In the limit that $x_S^- - x_S^+ \to 0$, the integral over the range (x_S^+, x_S^-) vanishes since $\rho(x,t)$ is bounded. Considering the remaining two integrals we have, by Leibnitz's theorem on differentiation of an integral with respect to a parameter,

$$\begin{aligned}&\int_{x_a(t)}^{x_S^+(t)} \frac{\partial\rho}{\partial t}dx + \rho(x_S^+, t)\frac{dx_S^+}{dt} - \rho(x_a(t), t)\frac{dx_a(t)}{dt}\\ &\quad + \int_{x_S^-(t)}^{x_b(t)} \frac{\partial\rho}{\partial t}dx + \rho(x_b(t), t)\frac{dx_b(t)}{dt} - \rho(x_S^-, t)\frac{dx_S^-(t)}{dt} = 0.\end{aligned} \tag{2.98}$$

Using Eqs. 2.96 and 2.97, we can write Eq. 2.98 in the form

$$\begin{aligned}&\int_{x_a(t)}^{x_S^+(t)} \frac{\partial\rho}{\partial t}dx + \int_{x_S^-(t)}^{x_b(t)} \frac{\partial\rho}{\partial t}dx\\ &\quad = -[\![\rho]\!]u_S + \rho(x_a(t))\,\dot{x}(x_a(t)) - \rho(x_b(t))\,\dot{x}(x_b(t)),\end{aligned} \tag{2.99}$$

where we have adopted the notation

$$[\![\xi]\!] \equiv \xi^+ - \xi^- \tag{2.100}$$

for the change from the initial value, ξ^-, to the final value, ξ^+, taken by the variable ξ as the shock passes a point. The quantity $[\![\xi]\!]$ is called the *jump* in ξ. Since $\rho(x,t)$ is smooth in each range of integration, Eq. 2.87$_1$ is satisfied in each of the ranges. Substituting this equation into Eq. 2.99 gives

$$\int_{x_a(t)}^{x_S^+(t)} \frac{\partial(\rho\dot{x})}{\partial x} dx + \int_{x_S^-(t)}^{x_b(t)} \frac{\partial(\rho\dot{x})}{\partial x} dx$$
$$= [\![\rho]\!] u_S - \rho(x_a(t))\,\dot{x}(x_a(t)) + \rho(x_b(t))\,\dot{x}(x_b(t)). \tag{2.101}$$

The integrals in this equation can be evaluated immediately and, after some cancellation, we obtain the result

$$[\![\rho]\!] u_S = [\![\rho\dot{x}]\!], \tag{2.102}$$

representing the constraint on the jumps that is imposed by the principle of conservation of mass.

Proceeding similarly with Eq. 2.94$_2$, representing the principle of conservation of momentum, we obtain

$$\frac{d}{dt}\left\{ \int_{x_a(t)}^{x_S^+(t)} \rho\dot{x}\,dx + \int_{x_S^-(t)}^{x_b(t)} \rho\dot{x}\,dx \right\}$$
$$= t_{11}(X_b, t) - t_{11}(X_a, t) + \int_{x_a(t)}^{x_S^+(t)} \rho f\,dx + \int_{x_S^-(t)}^{x_b(t)} \rho f\,dx. \tag{2.103}$$

Differentiation followed by use of Eqs. 2.96 and 2.97 allows us to write this equation in the form

$$\int_{x_a(t)}^{x_S^+(t)} \frac{\partial(\rho\dot{x})}{\partial t} dx + \int_{x_S^-(t)}^{x_b(t)} \frac{\partial(\rho\dot{x})}{\partial t} dx - \int_{x_a(t)}^{x_S^+(t)} \rho f\,dx - \int_{x_S^-(t)}^{x_b(t)} \rho f\,dx$$
$$= -[\![\rho\dot{x}]\!] u_S + t_{11}(X_b, t) - t_{11}(X_a, t) + \rho_a \dot{x}_a^2 - \rho_b \dot{x}_b^2. \tag{2.104}$$

Now, let us consider the derivative $\partial(\rho\dot{x})/\partial t$. From Eqs. 2.87$_{1,2}$ we have

$$\frac{\partial\rho}{\partial t} = -\frac{\partial(\rho\dot{x})}{\partial x}$$
$$\frac{\partial t_{11}}{\partial x} = \rho\left(\frac{\partial\dot{x}}{\partial t} + \dot{x}\frac{\partial\dot{x}}{\partial x}\right) - \rho f, \tag{2.105}$$

so we can write

$$\frac{\partial(\rho\dot{x})}{\partial t} = \frac{\partial}{\partial x}(t_{11} - \rho\dot{x}^2) + \rho f, \tag{2.106}$$

and Eq. 2.104 becomes

$$\int_{x_a(t)}^{x_S^+(t)} \frac{\partial}{\partial x}(t_{11} - \rho\dot{x}^2)dx + \int_{x_S^-(t)}^{x_b(t)} \frac{\partial}{\partial x}(t_{11} - \rho\dot{x}^2)dx$$
$$= -[\![\rho\dot{x}]\!]u_S + t_{11}(X_b, t) - t_{11}(X_a, t) + \rho_a\dot{x}_a^2 - \rho_b\dot{x}_b^2, \tag{2.107}$$

or

$$[\![\rho\dot{x}]\!]u_S = [\![-t_{11} + \rho\dot{x}^2]\!], \tag{2.108}$$

the constraint imposed on the jumps by the principle of conservation of momentum.

Similar analysis of Eq. 2.94₃ leads to the condition

$$[\![\rho(\varepsilon + \tfrac{1}{2}\dot{x}^2)]\!]u_S = [\![\rho(\varepsilon + \tfrac{1}{2}\dot{x}^2)\dot{x} - t_{11}\dot{x}]\!] \tag{2.109}$$

imposed on the jumps by the principle of balance of energy.

Eulerian Forms. In summary, the jump conditions representing the principles of conservation of mass, and balance of momentum and energy as applied to plane, longitudinal shocks are

$$\begin{aligned} [\![\rho]\!]u_S &= [\![\rho\dot{x}]\!] \\ [\![\rho\dot{x}]\!]u_S &= [\![\rho\dot{x}^2 - t_{11}]\!] \\ [\![\rho(\varepsilon + \tfrac{1}{2}\dot{x}^2)]\!]u_S &= [\![\rho(\varepsilon + \tfrac{1}{2}\dot{x}^2)\dot{x} - t_{11}\dot{x}]\!]. \end{aligned} \tag{2.110}$$

These results are easily expressed in terms of the variables u_x and u_t using Eqs. 2.16 and 2.35.

As stated, these equations relate the nine variables $\mathcal{S}^+ = \{\rho^+, t_{11}^+, \varepsilon^+, \dot{x}^+\}$, $\mathcal{S}^- = \{\rho^-, t_{11}^-, \varepsilon^-, \dot{x}^-\}$, and u_S. If we assume that the state ahead of the shock, $\mathcal{S}^-$, is known, then there remain the five variables $\mathcal{S}^+$ and u_S. Using the three jump conditions, we can express any three variables of the set $\mathcal{S}^+, u_S$ in terms of the other two.

Lagrangian Forms. The essential difference between the Eulerian and Lagrangian and expressions of the shock-jump equations is that the latter involve a Lagrangian measure of the shock velocity. The Eulerian shock velocity is the rate at which the shock advances through the spatial frame, i.e., along the x coordinate. The Lagrangian shock velocity is defined in analogous fashion as the rate at which the shock advances through the reference frame, i.e., along the X coordinate.

We begin by calculating the relationship between these two measures of shock velocity. Suppose that the shock is at the particle X at the time t and propagates a small distance to arrive at the particle $X+\Delta X$ at the time $t+\Delta t$. The Lagrangian shock velocity is then the limiting value of $U_S = \Delta X/\Delta t$. Let the initial spatial position of the particle X be x. If the material ahead of the shock has the density ρ^-, the initial position of the particle $X+\Delta X$ will be $x+(\rho_R/\rho^-)\Delta X$. If the particle velocity in the material ahead of the shock is $\dot{x}^-$, the particle $X+\Delta X$ will move a distance $\dot{x}^-\Delta t$ in the time it takes for the shock to arrive there from its initial position, X. Thus, the distance the shock travels along the x axis in the time Δt is $(\rho_R/\rho^-)\Delta X+\dot{x}^-\Delta t$. The Eulerian shock velocity is $u_S = \Delta x/\Delta t$ so we have

$$u_S = \frac{\rho_R}{\rho^-}\frac{\Delta X}{\Delta t} + \dot{x}^- = \frac{\rho_R}{\rho^-}U_S + \dot{x}^-, \tag{2.111}$$

or

$$U_S = \frac{\rho^-}{\rho_R}(u_S - \dot{x}^-). \tag{2.112}$$

When we substitute the expression $(\rho_R/\rho^-)U_S+\dot{x}^-$ for u_S in Eqs. 2.110, we obtain the shock-jump equations in the Lagrangian form

$$\begin{aligned} \rho_R U_S[\![-v]\!] &= [\![\dot{x}]\!] \\ \rho_R U_S[\![\dot{x}]\!] &= [\![-t_{11}]\!] \\ \rho_R U_S[\![\varepsilon+\tfrac{1}{2}\dot{x}^2]\!] &= [\![-t_{11}\dot{x}]\!]. \end{aligned} \tag{2.113}$$

Subsidiary Jump Equations. Elimination of the shock and particle velocities from either of Eqs. 2.110 or 2.113 yields the result

$$[\![\varepsilon]\!] = \tfrac{1}{2}(-t_{11}^+ - t_{11}^-)[\![-v]\!], \tag{2.114}$$

called the *Rankine–Hugoniot equation*, which involves only thermodynamic variables. This important equation provides the means for introducing the effect of a shock transition into a thermodynamic analysis.

With the use of Eq. 2.113_1, Eq. 2.113_3 can be placed in the form

$$[\![\varepsilon]\!] = \tfrac{1}{2}[\![\dot{x}]\!]^2 - \frac{t_{11}^-}{\rho_R U_S}[\![\dot{x}]\!]. \tag{2.115}$$

Useful expressions for the shock velocity are

$$U_S = \frac{1}{\rho_R}\left\{\frac{[\![-t_{11}]\!]}{[\![-v]\!]}\right\}^{1/2} = \frac{1}{\rho_R}\frac{[\![-t_{11}]\!]}{[\![\dot{x}]\!]}, \tag{2.116}$$

and the particle velocity jump is related to the stress and specific volume jumps by the equation

$$[\![\dot{x}]\!] = \left\{[\![-t_{11}]\!][\![-v]\!]\right\}^{1/2}. \tag{2.117}$$

For a given value of U_S Eqs. 2.116 represent lines, called *Rayleigh lines*, connecting the initial and final states in the $t_{11}-v$ and $t_{11}-\dot{x}$ planes. The equations show that steep Rayleigh lines correspond to high shock velocities.

2.4.3 Jump Equations Describing Plane Steady Waves

In the foregoing section, we studied propagating discontinuities in stress, density, etc. One might doubt that real materials could respond in this discontinuous fashion, although most would agree that a thin, but smooth, transition might be satisfactorily approximated as a discontinuity. In this section, we consider what we shall call a *steady wave* or *structured shock*. In particular, we shall consider waves in which the motion is uniaxial, as described by Eqs. 2.26 or 2.27, but with the additional constraint that the disturbance propagate unchanged in form and at a constant velocity. This means that the displacement can be described by the equation

$$U(X,t)=U(Z) \quad \text{where} \quad Z=X-Ct \tag{2.118}$$

in the Lagrangian representation, or

$$u(x,t)=U(z) \quad \text{where} \quad z=x-ct \tag{2.119}$$

in the Eulerian representation.

We intend that these waveforms be like shocks in that they represent transitions between given initial and final states, but that the transitions be smooth, as illustrated in Fig. 2.7. Accordingly, we establish the initial and final states as

$$\begin{aligned} S^-:&\ \dot{x}\to\dot{x}^- \text{ and } v\to v^- \text{ as } Z \text{ or } z\to\infty, \\ S^+:&\ \dot{x}\to\dot{x}^+ \text{ and } v\to v^+ \text{ as } Z \text{ or } z\to-\infty, \end{aligned} \tag{2.120}$$

respectively. In addition, we shall establish the limiting stress states $t_{11}\to t_{11}^-$ and $t_{11}\to t_{11}^+$ as Z or $z\to\pm\infty$, respectively.

When the displacement is in the form of a steady wave, the displacement gradient and particle velocity are simply related. We have

$$U_X=U_Z, \quad U_t=-CU_Z, \tag{2.121}$$

so

$$1-\frac{v}{v_R}=-U_Z, \quad \dot{x}=-CU_Z, \tag{2.122}$$

and

$$1-\frac{v}{v_R}=\frac{\dot{x}}{C}. \tag{2.123}$$

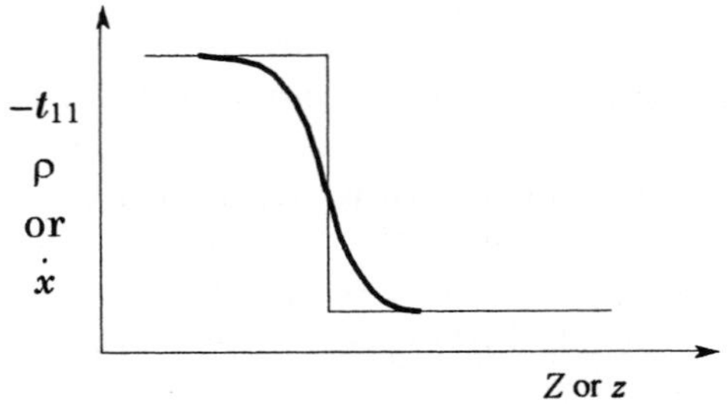

Figure 2.7. Illustration of a shock and a smooth steady wave.

The wave is analyzed using the equations of balance in Lagrangian form, Eqs. 2.92. When specialized to the case of a steady wave, the equations of balance of mass, momentum, and energy (now including heat flux) can be written as

$$\begin{aligned} &\frac{\partial}{\partial Z}(\dot{x}+\rho_R\, C v)=0 \\ &\frac{d}{dZ}(t_{11}+\rho_R\, C\dot{x})=0 \\ &\frac{d(-\rho_R\, C\varepsilon+Q)}{dZ}=t_{11}(Z)\frac{d\dot{x}}{dZ}\,. \end{aligned} \tag{2.124}$$

Since ρ_R and C are constants the first and second of these equations are immediately integrable to give

$$\begin{aligned} \dot{x}(Z)+\rho_R\, C v(Z) &= \dot{x}^- +\rho_R\, C v^- \\ t_{11}(Z)+\rho_R\, C\dot{x}(Z) &= t_{11}^- +\rho_R\, C\dot{x}^-, \end{aligned} \tag{2.125}$$

where the constants of integration have been chosen using data for the state $\mathcal{S}^-$. If Eqs. 2.125 are evaluated in the limit $Z\to-\infty$ the result can be placed in the form

$$C=-\frac{1}{\rho_R}\frac{t_{11}^+-t_{11}^-}{\dot{x}^+-\dot{x}^-}=-\frac{1}{\rho_R}\frac{\dot{x}^+-\dot{x}^-}{v^+-v^-}\,. \tag{2.126}$$

Comparison of this result with that of Eq. $2.116_{1,2}$ shows that the steady wave velocity is the same as the velocity of a shock transition between the same states so that we can write

$$C=U_S\,. \tag{2.127}$$

Noting the similarity of Eq. 2.125_2 to the shock-jump condition 2.113_2, we write the former in the form

$$\rho_R U_S [\![\dot{x}]\!] = [\![-t_{11}]\!], \tag{2.128}$$

with the understanding that the equation describes the jump from the initial state to the state at *any* point in the waveform. Applying Eq. 2.128 to a point Z_1 and then to another point Z_2 in the waveform, and subtracting the results shows that this equation actually applies to the jump between *any* two points of the steady waveform.

When we take the jump (in the foregoing sense) of Eq. 2.123 and use Eq. 2.127 we arrive at the equation

$$\rho_R U_S [\![-v]\!] = [\![\dot{x}]\!], \tag{2.129}$$

a result analogous to that of Eq. 2.113_1 except that it is applicable to the jump between any two states in the steady wave.

Equations 2.128 and 2.129 can be combined to yield

$$(\rho_R U_S)^2 [\![-v]\!] = [\![-t_{11}]\!], \tag{2.130}$$

a relation between jumps in the stress and specific volume.

Using the energy equation 2.124_2 one can show that

$$\rho_R U_S [\![\varepsilon + \tfrac{1}{2}\dot{x}^2]\!] = [\![-t_{11}\,\dot{x} + Q]\!], \tag{2.131}$$

where the equation describes the jump between any two points in the waveform.

If $Q = 0$, i.e. the material does not conduct heat, Eqs. 2.128, 2.129, and 2.131 are entirely analogous to the shock-jump equations 2.113, and the steady wavespeed, as given by Eq. 2.126 is the same as the speed of propagation of a shock transition between the same states S^- and S^+. The steady wave is the same as the shock in every way, except that it need not be a discontinuity, and it is this similarity that justifies describing it as a structured shock.

Since a shock of the sort discussed in previous chapters is a discontinuous transition between two states there can be no meaningful discussion of the path on stress–volume, stress–particle velocity, or other graphs that is followed by a material point as the wave passes. Such a path does exist for a structured shock, however, and Eqs. 2.128 and 2.129 show that the path followed by the state point is the Rayleigh line.

2.5 Concluding Remarks on Mechanical Principles

Kinematical and dynamical matters such as are discussed in this chapter are among the best-established topics covered in this book, but a few remarks on their completeness and applicability are in order.

When analyses are conducted on the basis of uniaxial motion, any solutions obtained will, of course, be of this form. Such motions cannot usually be expected to prevail at the microscale where the effects of grains in polycrystalline materials and other inhomogeneities can be expected. Phenomena such as formation of "hot spots" in shock-compressed explosives are known to have a profound affect on the initiation of chemical reactions. Solid-state chemical reactions occur in many materials that are not explosives. In both cases, these effects are attributed to mechanical and thermal irregularities in what is nominally a uniaxial motion. Evidently, uniaxial motions occur in real materials only in some average sense. These fluctuations have been little studied (but see [5,79,83]) and many of their possible effects are unknown. In addition to these microscale effects, investigations have shown that uniaxial motions can exhibit unstable responses to small perturbations. Instabilities of this sort are less important in solids than in fluids because their growth is inhibited by the ability of the solid to resist shear stress. Nevertheless, large-scale effects of these instabilities are occasionally apparent.

2.6 Exercises

2.6.1. A rod is subjected to uniaxial extension so that its length changes from L to l. For this case, strain ε is conventionally defined as the ratio of the change in its length to the length itself. For finite strains, it makes a difference whether this is interpreted to mean $\varepsilon = (l - L)/L$ or $\varepsilon = (l - L)/l$. The exercise consists in evaluating the longitudinal component of U_X, u_x, $\mathbf{F}$, $\mathbf{E}$, and $\mathbf{e}$, both exactly and in the limit of small deformation, and explaining what they measure and how they relate to one another.

2.6.2. For simple shear, as illustrated in Fig. 2.3a, relate F_{12}, E_{12}, and e_{12} to γ and explain exactly what each quantity measures. Interpret E_{22} and e_{22}. Discuss the small-strain limits of these quantities.

2.6.3. Derive linear equations for strains, etc. that are valid in the case of small deformation.

2.6.4. Propose a definition of Eulerian compression analogous to the definition of Lagrangian compression of Eq. 2.25. Derive equations relating the two measures of compression.

2.6.5 Show that symmetry of the Cauchy stress tensor is a necessary and sufficient condition for satisfaction of the requirement of balance of moment of momentum. You may find it useful to consult an elementary text on elasticity theory or continuum mechanics.

2.6.6. Derive the Lagrangian forms, Eqs. 2.91, of the field equations by transformation of the Eulerian forms of Eqs. 2.86. Hint: you may find the equation $(F_{iI}/J)_{,i} = 0$ useful [103, Eq. 18.1].

2.6.7. Derive Eqs. 2.92 from Eqs. 2.91. Hint: you may find the equation $(J F_{Ik}^{-1})_{,I} = 0$ useful [103, Eq. 18.2].

2.6.8. Equation 2.112 relates U_S and u_S using conditions in the material ahead of the shock: $U_S = \rho^-(u_S - \dot{x}^-)/\rho_R$. Show that the same relation holds with ρ^- replaced by ρ^+ and $\dot{x}^-$ replaced by $\dot{x}^+$.

2.6.9. Derive Eqs. 2.113 from Eqs. 2.110 and 2.112.

2.6.10. When a shock passes through a material, both its specific internal energy and its particle velocity change. Show that, when the material ahead of the shock is unstressed and at rest, the energy added at the shock transition is equally partitioned between internal energy and kinetic energy.

2.6.11. Derive the jump conditions of Eq. 2.110 by making an explicit calculation for a slab of material comprising two regions in which constant fields are separated by a plane longitudinal shock.

CHAPTER 3

Plane Longitudinal Shocks

Study of the propagation and interaction of plane longitudinal shocks is a principal topic of this book. A shock separates the material in which it is propagating into a part through which it has passed, called the *upstream* material or material *behind* the shock, and the material into which it is advancing, called the *downstream* material or material *ahead* of the shock. The state of the upstream material is characterized by the four variables $S^+ = \{\rho^+, t_{11}^+, \varepsilon^+, \dot{x}^+\}$ and the state of the downstream material is characterized by the variables $S^- = \{\rho^-, t_{11}^-, \varepsilon^-, \dot{x}^-\}$. Knowledge of the shock velocity, either U_S or u_S, completes the description of the wave. As we have seen, these nine variables must satisfy the three jump conditions of Eqs. 2.110 or 2.113. Usually, the values of the downstream variables, S^-, are given as initial conditions. The strength of the shock depends upon the stimulus producing it, and this stimulus is usually characterized by a boundary condition giving the value of one of the upstream variables. This leaves the shock velocity and three of the variables in the upstream set S^+ unknown. The three jump equations must be augmented by an additional equation describing the response to a shock transition of the specific material of interest. When this is done, all of the variables characterizing the transition can be determined.

3.1 Full-field Solutions

The simplest shock-propagation problem arises when a material halfspace in a uniform state is subjected to a step change in the normal stress or particle velocity applied over its surface. In this case, a shock of constant amplitude originates at the surface and propagates into the material. The downstream material remains undisturbed until the arrival of the shock, at which time it undergoes a transition to the upstream state that is determined by the imposed boundary condition and the nature of the material itself. The shock is a discontinuous transition between uniform states, as shown in Fig. 3.1. It is easy to see that, when $f = 0$, $q = 0$, and $r = 0$, any constant state $S = \{\rho, t_{11}, \varepsilon, \dot{x}\}$ satisfies the field equations 2.87 or 2.92, so satisfaction of the jump conditions ensures that the integral forms of the balance equations are satisfied. The disturbance just described can be produced and studied in the laboratory.

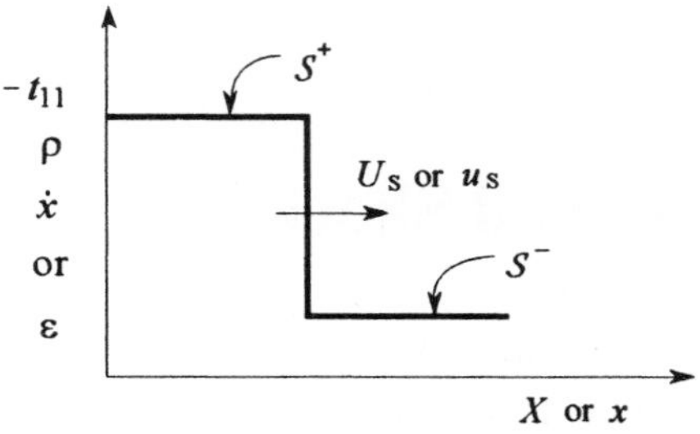

Figure 3.1. Shock transition between uniform states

3.2 Propagation of a Shock into Quiescent Material

For quiescent material we assume that $\dot{x}^- = 0$ and ρ^-, t_{11}^-, and ε^- are known. In this case, the jump conditions in Eulerian form, 2.110, become

$$\rho^+ = \frac{\rho^- u_S}{u_S - \dot{x}^+}$$

$$[\![-t_{11}]\!] = \rho^- \dot{x}^+ u_S \qquad (3.1)$$

$$[\![\varepsilon]\!] = \frac{1}{2}(\dot{x}^+)^2 + \frac{1}{\rho^- u_S}(-t_{11}^-)\, \dot{x}^+.$$

The Lagrangian form of these equations is

$$\rho^+ = \frac{\rho^- \rho_R U_S}{\rho_R U_S - \rho^- \dot{x}^+}$$

$$[\![-t_{11}]\!] = \rho_R \dot{x}^+ U_S \qquad (3.2)$$

$$[\![\varepsilon]\!] = \frac{1}{2}(\dot{x}^+)^2 + \frac{1}{\rho_R U_S}(-t_{11}^-)\, \dot{x}^+.$$

These equations show clearly that the common experimental practice of measuring $\dot{x}^+$ and u_S (or U_S) suffices to determine the remaining three variables characterizing the state of the material behind the shock.

When solids or liquids are subjected to strong shocks, it is usual to ignore the effect of pressures near one atmosphere, so we take $t_{11}^- = 0$ and $\rho^- = \rho_R$. Since the internal energy in the reference configuration can be chosen to be zero, we also have $\varepsilon^- = 0$. In this case, Eqs. 3.1 become

$$\rho^+ = \frac{\rho_R u_S}{u_S - \dot{x}^+}, \quad -t_{11}^+ = \rho_R \dot{x}^+ u_S, \quad \text{and} \quad \varepsilon^+ = \tfrac{1}{2}(\dot{x}^+)^2. \qquad (3.3)$$

Three useful relations derivable from the foregoing are

$$\dot{x}^+ = \rho_R [\![-v]\!] u_S$$

$$-t_{11}^+ = -(\rho_R\ u_S)^2 [\![-v]\!] \tag{3.4}$$

$$\varepsilon^+ = \tfrac{1}{2}(-t_{11}^+)[\![-v]\!].$$

Note that the initial condition $t_{11} = 0$ cannot be imposed on gases because they cannot exist at zero pressure.

The Lagrangian shock velocity satisfies the equation

$$U_S = (u_S - \dot{x})/(1 + U_X),$$

where the same value is obtained whether we use $\dot{x}^+$ and U_X^+ or $\dot{x}^-$ and U_X^-. Since the state of the material ahead of the shock is usually known, we adopt the latter option. If $\dot{x}^- = 0$, we have

$$U_S = \frac{u_S}{1 + U_X^-} = \frac{\rho^-}{\rho_R} u_S. \tag{3.5}$$

When the quiescent material ahead of the shock is in its reference state,

$$U_S = u_S \tag{3.6}$$

and the Lagrangian and Eulerian forms of the jump equations are identical.

It is useful to note the form taken by the jump conditions in the limit of weak shocks. When the shock propagates into a material in its reference state, $\mathcal{S}_R$, in which $t_{11}^- = 0$, $\rho^- = \rho_R$, $\varepsilon^- = 0$, and $\dot{x}^- = 0$, the lowest-order approximation to Eqs. 3.1 is

$$\rho^+ = \rho_R \left(1 + \frac{\dot{x}^+}{u_S}\right)$$

$$-t_{11}^+ = \rho_R\, u_S\, \dot{x}^+ \tag{3.7}$$

$$\varepsilon^+ = \tfrac{1}{2}(\dot{x}^+)^2,$$

where u_S is the longitudinal soundspeed in the reference state.

In the limit of small jumps in t_{11} and v, Eq. 2.116$_1$ takes the form

$$U_S = \frac{1}{\rho_R}\left(\frac{dt_{11}}{dv}\right)^{1/2}, \tag{3.8}$$

which we identify as the Lagrangian soundspeed (see Chap. 9), the speed of propagation of weak waves, often called acoustic waves, through the reference configuration.

3.3 Symmetrical Plate-impact Experiment

The simplest experiment that can be conducted to study shock propagation in a solid consists in impacting a plate of the material to be studied against a target plate of the same material, as depicted in Fig. 3.2. The plates are usually circular disks of diameter several times greater than their thickness. They are prepared to have very smooth, flat faces and are aligned so that the impact occurs simultaneously over the entire face. An experiment conducted under these conditions introduces a shock propagating from the impact face into the interior of each plate. These shocks produce the uniaxial motion of Eqs. 2.26 or 2.27 in a region near the axis of the disks during the time interval between the moment of impact and the arrival of a wave originating at the periphery of the impact face. Typical sample thicknesses are in the range 1–20 mm. The diameter is not important provided that it is sufficient to delay the arrival of the wave from the target periphery until the required measurements can be completed. The impact velocity can easily be measured to high precision and, as we shall see, yields the particle velocity of the material behind the shock with similar precision. Among the other variables entering the jump conditions, the easiest to measure is the shock velocity. Accounts of experimental methods are given in references [4,9,21,77,110].

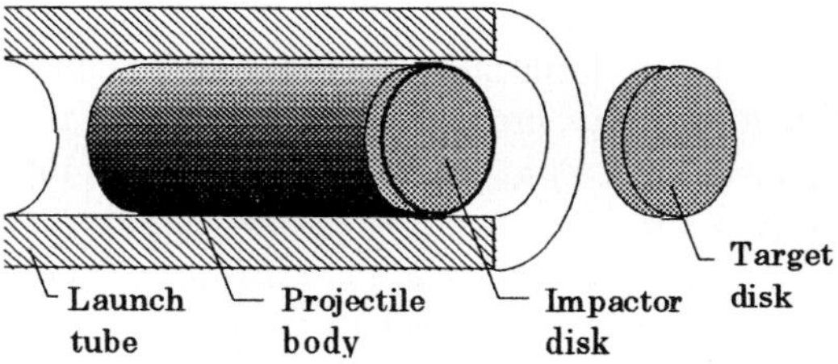

Figure. 3.2. Schematic configuration of a plate-impact experiment. The projectile guides the impactor disk as it is accelerated along the launch tube (shown in cutaway view) by compressed gas or gun propellant. It collides with the target disk placed slightly beyond the end of the tube

We shall now consider what can be learned about the material from the experiment just described. In this experiment, one of the samples, the *target*, is at rest and the other, the *impactor*, is driven into the target at a known projectile velocity, $\dot{x}_P$. The stress, mass density, and specific internal energy of both the impactor and target material have known reference-state values. Experiments of this sort are usually described by diagrams such as those shown in Fig. 3.3

The experiment that we have described is called a *symmetric-impact experiment* because the impactor and target are of the same material. It is easy to see (for example, by viewing the experiment in a coordinate frame in which the

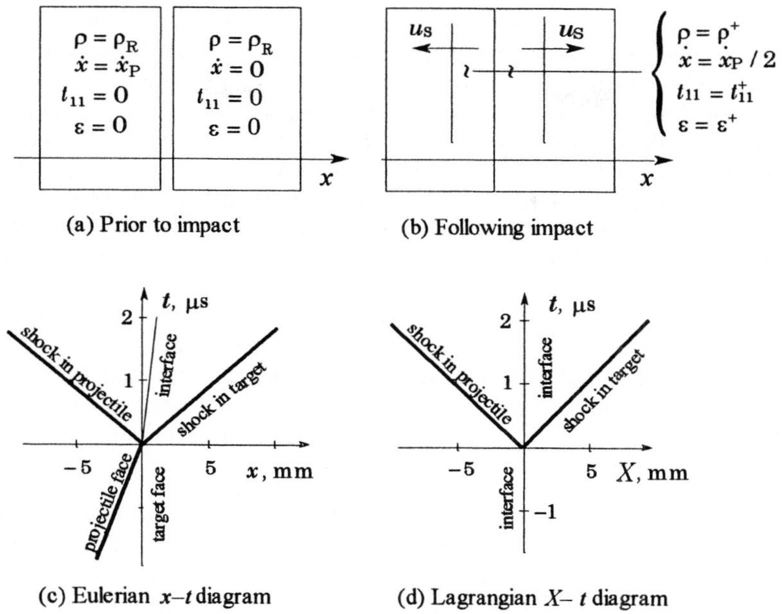

Figure 3.3. Plate-impact test configuration (side view) and space–time diagrams of the interaction. The side-view figures show the relative positions of the parts, indicate notation, etc., but the horizontal dimension is much too great in comparison to the vertical dimension for the figure to be scale representations of the actual parts

two plates approach each other at equal but oppositely directed velocities) that, because of this symmetry, the particle velocity imparted to the target plate is one-half the projectile velocity: $\dot{x}^+ = \dot{x}_P/2$. Assuming that the shock velocity has also been measured, the remaining variables comprising the state $\mathcal{S}^+$ are obtainable from Eqs. 3.1 or 3.2.

3.4 Hugoniot Relations

The three shock-jump equations contain no information characterizing the differing responses of specific materials, and this must be provided before shocks can be analyzed. If shocks of various strengths are introduced into the material and the response recorded in terms of the resulting value of one of the other variables, one obtains points on a curve describing the locus of states achievable by shock transition from the given initial state. This locus is called a *Hugoniot curve* or simply a *Hugoniot* (after H. Hugoniot [58]). The Hugoniot curve, which depends on the initial state, $\mathcal{S}^-$, is characteristic of the specific material studied. It contains the minimum amount of information about the material that suffices to solve shock-propagation problems in terms of the variables under consideration. A Hugoniot curve falls far short of a complete description of material response and additional thermodynamic information is required to determine temperature and entropy jumps at the shock or to solve other wave-propagation problems. On the other hand, a Hugoniot curve can be determined from a comprehensive material model by thermodynamic and mechanical analysis (see Chap. 5).

There are ten pairs of the variables $v, t_{11}, \dot{x}, \varepsilon$ and U_S (or u_S) that provide equivalent representations of the Hugoniot curve. Any of these representations can be transformed to any of the other possible representations using the jump conditions, so they all embody the same information. Three of these ten Hugoniots are particularly important: The $t_{11} - v$ Hugoniot or, equivalently, the $t_{11} - \rho$ or $t_{11} - \Delta$ Hugoniot, is an important thermodynamic relation, the $t_{11} - \dot{x}$ Hugoniot is used for analysis of shock interactions, and the $U_S - \dot{x}$ Hugoniot is frequently the most direct representation of shock wave measurements. Other Hugoniots, for example those involving temperature, are also of interest and are discussed in Chap. 5.

It is important to remember that a Hugoniot curve is the locus of states achievable by shock transition from a particular initial state, S^-. When the initial state is changed, the Hugoniot curve also changes. The Hugoniot is said to be *centered* on the initial state, and the one centered on the undeformed and unstressed state is called the *principal Hugoniot curve*. In the circumstances most often of interest in problems of shock propagation in condensed matter, the observed behavior is essentially unaffected by pressures of the order of one atmosphere and is not strongly sensitive to temperature variation in the range encountered in the ambient atmosphere of the laboratory. Accordingly, unless explicitly stated otherwise, a Hugoniot curve for a solid or liquid can be assumed to correspond to material in the state $t_{11}^- = 0, \varepsilon^- = 0, \rho^- = \rho_R$ and $\dot{x}^- = 0$. Because material properties are invariant to rigid translations, a Hugoniot can be recentered to any desired value of initial particle velocity by simple translation. Reflection in any line $\dot{x} = \text{const.}$ reverses the direction of shock propagation.

When additional thermodynamic information is available, a Hugoniot can be transformed to a different initial thermodynamic state. Two material response curves of particular interest in calculating shock interactions are Hugoniot curves and decompression isentropes centered at shock-compressed states. The former are called *second-shock Hugoniots.* Although calculation of these curves must await development of some thermodynamic equations in Chap. 5, we will see that these curves lie sufficiently close to the principal Hugoniot that they can often, to good approximation, be replaced by this Hugoniot. This is the course that we shall adopt in this chapter. Some comparisons that indicate the accuracy of these approximations are given in Sect. 3.7.2.

Stress–Volume Hugoniot. Figure 3.4 shows a typical $t_{11} - v$ Hugoniot, with points corresponding to a shock transition from $\mathcal{S}^-$ to $\mathcal{S}^+$ marked. The straight line connecting $\mathcal{S}^-$ and $\mathcal{S}^+$ is called the *Rayleigh line* for the transition. The shock speed is given by Eq. 2.116$_1$, which shows that it is related to the slope of the Rayleigh line: Steeper Rayleigh lines correspond to higher shock speeds. From Eq. 2.114 we see that, in the absence of heat flux and extrinsic energy deposition, the increase in specific internal energy that occurs at a shock transition is measured by the area under the Rayleigh line, as indicated by the shading on Fig. 3.4. Shock transitions are irreversible thermodynamic processes, so this jump in specific internal energy is greater than occurs in isentropic compression by the same amount. Detailed discussion of this issue must await our development of the necessary thermodynamic relations in Chap. 5. We shall find that the jump in entropy that occurs upon passage of a shock is proportional to the cube of the volume change, so the isentrope and the Hugoniot through a given point differ only slightly for small compressions.

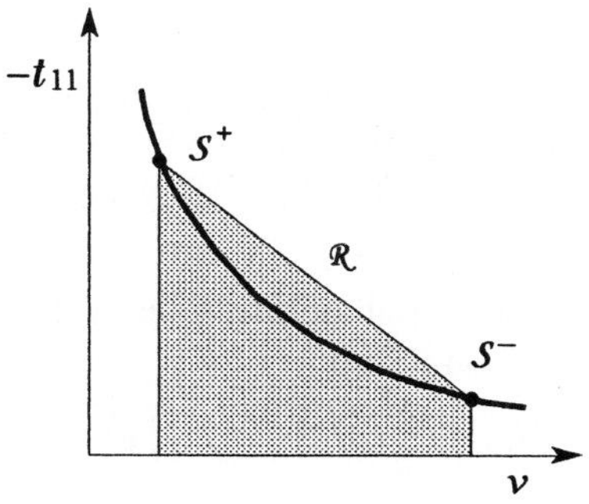

Figure 3.4. Typical stress–volume Hugoniot

Stress–Particle-Velocity Hugoniot. A typical $t_{11} - \dot{x}$ Hugoniot is illustrated in Fig. 3.5. The shock velocity is related to the slope of the Rayleigh line according to Eq. 2.116$_2$: $U_S = [-t_{11}]/(\rho_R [\dot{x}])$. The Hugoniot shown in the figure corresponds to a shock propagating in the $+X$ direction, i.e., with positive velocity. When the shock velocity is negative, the relevant Hugoniot is obtained by reflection of the one shown about the line $\dot{x} = \dot{x}^-$.

Linear $U_S - \dot{x}$ Hugoniot. When experiments such as the symmetric plate-impact experiment are conducted on a given solid over a range of projectile velocities, it is often observed that the velocity of the shock is proportional to the imparted particle velocity. The experiment is usually done on material at rest, at zero pressure, and at a specific reference temperature (usually ~300 K). Under these conditions, the linear result can be expressed in the form

$$U_S = C_B + S\dot{x}^+ , \tag{3.9}$$

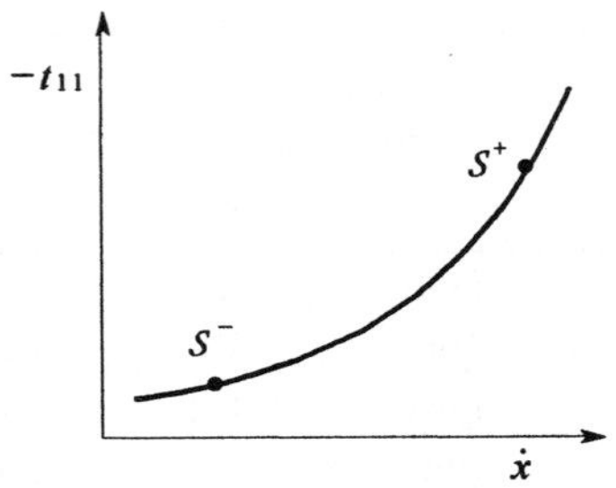

Figure 3.5. Stress–particle-velocity Hugoniot (centered on S^-) for a typical material

where C_B is a material constant that is usually found to be approximately equal to the bulk soundspeed in the material and S is a dimensionless material constant.* For most materials, which we shall call *normal materials*, the value of S is positive, typically having a value of about 1.5. A few materials become more compressible as the compression is increased over a limited range, suggesting that their behavior might be representable by Eq. 3.9 with $S < 0$. We regard this as an anomalous behavior requiring careful special attention, and assume henceforth that $S > 0$. As we shall see, only compressive shocks can propagate in material governed by Eq. 3.9 with $S > 0$. Decompression processes will be shown to produce smooth waves.

Since the properties of a material are invariant to rigid translations, Eq. 3.9 can be written in the more general form†

$$U_S = C_B(\text{sgn}[\![\dot{x}]\!]) + S\,[\![\dot{x}]\!] \tag{3.10}$$

to extend its applicability to waves propagating in either direction into material moving at a constant velocity, $\dot{x}^- \neq 0$. Although we have written Eq. 3.10 as an expression for the Lagrangian shock velocity, we recall that the Lagrangian and Eulerian shock velocities are the same for the usual conditions of the experiment in which the shock propagates into a body at rest in its reference configuration. Values of ρ_R, C_B, and S for a few materials are given in Table 3.1 and extensive compilations are available in many publications, including [10,62,74,77,105].

Equation 3.9 is one of the earliest results obtained in quantitative experimental investigations of shock compression of solids, and forms the basis of

* Measurements suggest that a quadratic term be included in the $U_S - \dot{x}$ Hugoniot for a few materials, even in the range of moderate pressures. When very high pressures are considered, a small positive quadratic term can be included in the Hugoniot to capture effects that cause it to pass through a point of vertical tangency and fold back on itself [68, p. 112].

† $\text{sgn}\,\xi \equiv 1$ for $\xi \geq 0$ and -1 for $\xi < 0$.

Table 3.1. Shock parameters for several materials, including impedance in the weak-shock limit[a]

Material	ρ_R kg/m^3	C_B m/s	S	Z_R GPa/(km/s)
Plexiglas	1186	2598	1.516	3.1
NaCl	2165	3528	1.343	7.6
Beryllium	1851	7998	1.124	14.0
Aluminum (2024)	2785	5328	1.338	14.4
Bismuth	9836	1826	1.473	17.6
Lead	11,350	2051	1.460	23.8
Copper	8930	3940	1.489	35.8
Gold	19,240	3056	1.572	58.0
Tungsten	19,224	4029	1.237	77.5
Iridium	22,484	3916	1.457	88.5
Al_2O_3	3988	11,186	1.05	44.1

[a] Data from [77].

much of the practical application of this subject. This Hugoniot is most applicable to description of the behavior of solids in the pressure range above about 5–10 GPa, where the average stress (pressure) is sufficient to render the shear stresses negligible. In this case the response of the solid to shock compression is similar to that of a fluid, a point to which we shall return later. Note that a pressure of 5 GPa is rather large compared to values usually encountered by engineers, yet it is near the lower limit of the range of interest here.

Substitution of Eq. 3.10 into Eqs. 2.113 yields the jump equations

$$\begin{aligned} [\![-v]\!] &= \frac{[\![\dot{x}]\!]}{\rho_R\left(C_B(\operatorname{sgn}[\![\dot{x}]\!]) + S[\![\dot{x}]\!]\right)} \\ p^+ &= \rho_R\left(C_B(\operatorname{sgn}[\![\dot{x}]\!]) + S[\![\dot{x}]\!]\right)[\![\dot{x}]\!] \qquad (3.11) \\ [\![\varepsilon]\!] &= \frac{1}{2}[\![\dot{x}]\!]^2 + \frac{1}{\rho_R\left(C_B(\operatorname{sgn}[\![\dot{x}]\!]) + S[\![\dot{x}]\!]\right)}, \end{aligned}$$

where we have written p in place of $-t_{11}$ in recognition of the fluid-behavior approximation usually associated with Eq. 3.10.

Equation 3.11_2 is the pressure–particle-velocity Hugoniot associated with Eq. 3.10. The pressure–volume Hugoniot associated with Eq. 3.10 can be obtained from Eqs. $3.11_{1,2}$ in the form

$$p^{+} = \frac{(\rho_R C_B)^2 (v_R - v)}{[1 - \rho_R S(v_R - v)]^2} = \frac{\rho_R C_B^2 \Delta}{(1 - S\Delta)^2}. \tag{3.12}$$

In the common special case of a compression shock propagating in the $+X$ direction into material that is at rest, undeformed, and at zero pressure Eqs. 3.11 take the simpler forms

$$\rho^{+} = \frac{\rho_R (C_B + S\dot{x}^{+})}{C_B + (S-1)\dot{x}^{+}}$$

$$p^{+} = \rho_R (C_B + S\dot{x}^{+})\dot{x}^{+}$$

$$\varepsilon^{+} = \tfrac{1}{2}(\dot{x}^{+})^2 + \frac{p^{+}\Delta^{+}}{2\rho_R} \tag{3.13}$$

and, since we also have Eq. 3.9, there are now four equations that permit any four of the unknown variables to be expressed in terms of the one remaining. This remaining variable is determined from the boundary condition and serves as a measure of the shock strength.

Plots of some Hugoniot curves for copper and aluminum alloy 2024 are given in Fig. 3.6.

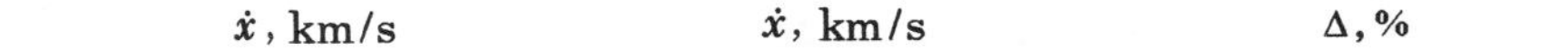

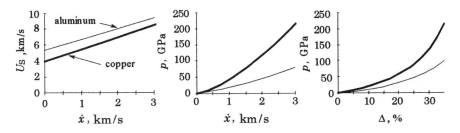

Figure 3.6. Hugoniot curves for copper and aluminum alloy 2024. The curves are plotted using parameters taken from Table 3.1

3.5 Longitudinal Stability of Plane Shocks

Thus far, we have assumed that shocks exist and we have developed jump conditions governing their behavior. Shocks do exist, of course, but smooth waves also exist. To understand whether a given stimulus will produce a shock or a smooth wave, we need to study the differential equations 2.87 or 2.92 in conjunction with the jump conditions. We defer this to Chap. 9. We can, however, conduct a simple examination of the longitudinal stability of shocks as an indication of what to expect. Suppose a shock of the form shown in

Fig. 3.7a is somehow slightly disturbed and separates into two shocks as indicated in Fig. 3.7b, i.e. a small part of the jump falls behind the main shock. If the shock of Fig. 3.7a is to be considered stable, this small shock must propagate faster than the main shock so as to overtake it and restore it to its original amplitude. To see whether or not this occurs, consider the two stress–volume Hugoniot curves in Fig. 3.8. When the Hugoniot has the concave upward shape indicated in Fig. 3.8a, the Rayleigh line connecting B to C is steeper than the one connecting A and B, so the second wave is advancing through the material faster than the leading wave and will restore the original one-step waveform. Similar consideration of the behavior of a small disturbance postulated to advance beyond the main shock, as shown in Fig. 3.7c, shows that it will be overtaken by the main shock, thus restoring the original unperturbed waveform. On this basis, we say that a shock transition from A to C is longitudinally stable. The same considerations applied to a transition from C to A indicate that this transition is unstable. The transition from A to C is a compression process whereas that from C to A is a decompression process. The stress–volume Hugoniot for most materials is of the form given in Fig. 3.8a, so compression shocks are stable and decompression shocks are unstable. We shall see in Chap. 9 that a decompression shock evolves as a smooth disturbance called a centered simple wave.

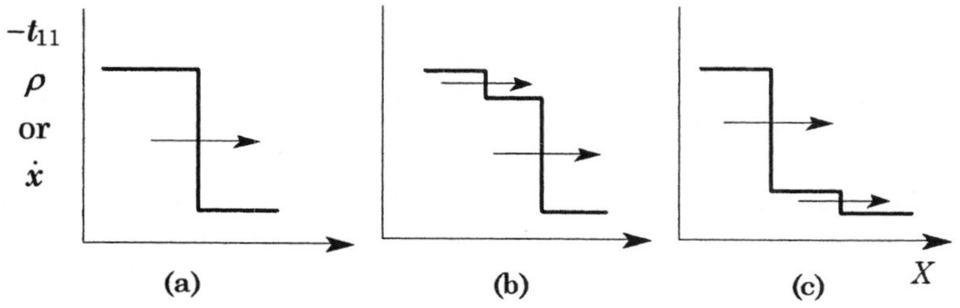

Figure 3.7. Two perturbed versions of the shock shown in part (a) of the figure are shown. In each case, the perturbation involves separation of a small part of the shock from its main portion. The drawings are made for the case in which the Hugoniot is concave upward, as illustrated in Fig. 3.8a. Part (b) represents the case in which the perturbation propagates slower than the main shock and part (c) of the figure shows the case in which the perturbation propagates faster than the main shock. the drawing is made for the case that the stress–volume Hugoniot is concave upward, as shown in Fig. 3.8a

When the compression curve has the unusual form shown in Fig. 3.8b, as it does over a limited range of state for a few materials (e.g. vitreous silica for pressures from 0 to about 3 GPa), the converse situation prevails: Compression waves spread as they propagate, whereas decompression shocks are stable and smooth decompression waves coalesce into shocks.

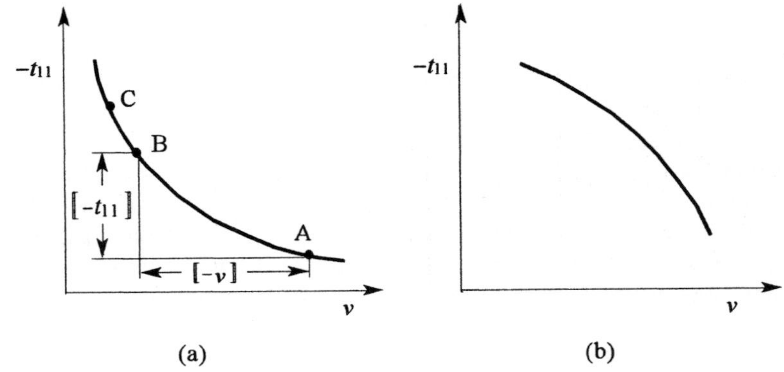

Figure 3.8. Stress–volume Hugoniot curves

Often the thermodynamic condition that the specific entropy at a material point cannot decrease as the shock passes is used to rule out decompression shocks in normal materials. As we shall see, this requirement leads to the same result that we have just obtained.

3.6 Unsymmetrical Plate-impact Experiment

In Section 3.3 we noted that planar impact of two plates produced a shock propagating away from the impact interface into the interior of each plate. A key point of that section was that, because the plates were of the same material, the target plate material was accelerated to one-half the projectile velocity and the impactor plate material was decelerated to this same velocity.

In this section, we consider the case in which the impactor (also called projectile) and target plates are of different materials, with the Hugoniot of the impactor plate being known. The situation is as depicted in Fig. 3.9. The curves shown are p–$\dot{x}$ Hugoniots of the impactor and target materials (the latter indicated schematically by a dotted line since its quantitative description is presumed unknown). The target-plate Hugoniot is oriented for a right-propagating shock and passes through $(p^-, \dot{x}^-) = (0, 0)$ since the shock will propagate from the interface into the target and produce a transition from this state to some state $(p^+, \dot{x}^+)$. The impactor-plate Hugoniot is oriented for a left-propagating (negative velocity) shock and passes through $(p, \dot{x}) = (0, \dot{x}_P)$, the initial state of the impactor material. Since the interface is in compression it remains in contact and the particle velocity behind the shock is the same for each plate. Similarly, the pressure must be the same in the material on each side of the interface since the interface itself has no mass. Therefore, the values of p and $\dot{x}$ in the region between the two shocks lie at the intersection of the two Hugoniots. Once the intersection is known, Eqs. 2.113, applied to each shock, suffice to determine the other variables and, thus, the complete solution of this problem.

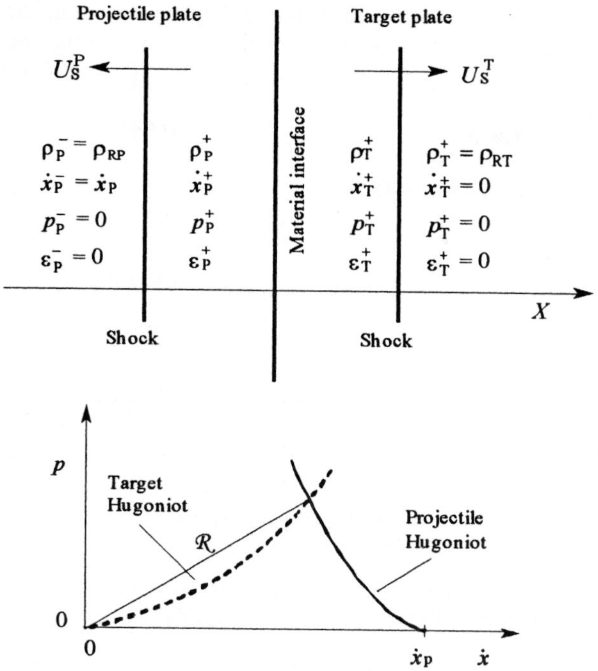

Figure 3.9. Unsymmetrical plate-impact experiment. The Hugoniot of the target plate, presumed unknown, is illustrated by a dotted line.

As an example, suppose the Hugoniot for the impactor plate is given by Eq. 3.11$_2$, with known parameter values:

$$p^+ = \rho_{RP}\left[-C_{BP} + S_P\,(\dot{x}^+ - \dot{x}_P)\right](\dot{x}^+ - \dot{x}_P)\,. \tag{3.14}$$

Then, suppose that the impact velocity, $\dot{x}_P$, and the shock velocity in the target plate, U_{ST}, are measured in the experiment. Applying Eq. 2.116$_2$ to the present situation gives

$$p^+ = \rho_{RT}\, U_{ST}\, \dot{x}^+\,. \tag{3.15}$$

Equations 3.14 and 3.15 can be solved for $\dot{x}^+$:

$$\dot{x}^+ = \dot{x}_P - \frac{\rho_{RP}\,C_{BP} + \rho_{RT}\,U_{ST}}{2S\,\rho_{RP}}\left\{\left[1 + \frac{4\,\rho_{RP}\,\rho_{RT}\,U_{ST}\,S_P\,\dot{x}_P}{(\rho_{RP}\,C_{BP} + \rho_{RT}\,U_{ST})^2}\right]^{1/2} - 1\right\}, \tag{3.16}$$

and then p^+ recovered from Eq. 3.15 to yield a point on the Hugoniot of the target plate. It is not necessary that the impactor-plate Hugoniot satisfy Eq. 3.11$_2$ or that the Hugoniot of the target plate have this or any other particular form. If the Hugoniot of the impactor plate has a form other than that of

Eq. 3.11_2, the intersection on Fig. 3.9 can be determined using the appropriate equation or by graphical or numerical methods. Conducting the experiment described for several values of $\dot{x}_P$ allows one to plot the Hugoniot of the target-plate material.

Shock Impedance. The quantity

$$Z = \rho_R \left| U_S \right| \tag{3.17}$$

is called the *shock impedance*. For acoustic waves it has the value

$$Z = \rho_R C_B , \tag{3.18}$$

and is a material constant. Since $\left| U_S \right|$ exceeds C_B by an amount that depends on the strength of the shock, Z is not a constant, but it is still a material property. According to Eq. 2.116_2,

$$[\![\dot{x}]\!] = \frac{1}{Z} [\![-t_{11}]\!] , \tag{3.19}$$

and we see that Z is a measure of the degree to which a material resists being set in motion by a shock of given strength. Similarly, we have

$$[\![-v]\!] = \frac{1}{Z^2} [\![-t_{11}]\!] , \tag{3.20}$$

which shows that Z is also a measure of the degree to which a material resists shock compression. Materials for which Z is large are difficult to set in motion or compress with a shock.

Consider the case in which an impactor plate in its reference state and moving at velocity $\dot{x}_P$ collides with a target plate at rest in its reference state. Equation 3.19 takes the forms

$$x^+ - x_P = t_{11}^+ / Z_P$$

$$x^+ = t_{11}^+ / Z_T$$

for the impactor and target plates, respectively (recall that $U_S < 0$ in the impactor plate). In writing these equations, we have used the fact that the $t_{11}^+ - \dot{x}^+$ state is the same for each plate. Solving these equations for t_{11}^+ and $\dot{x}^+$ yields

$$\begin{aligned} -t_{11}^+ &= \frac{Z_P Z_T}{Z_P + Z_T} \dot{x}_P \\ \dot{x}^+ &= \frac{Z_P}{Z_P + Z_T} \dot{x}_P . \end{aligned} \tag{3.21}$$

We see from these equations that generation of a large compressive stress in a given target plate is most easily accomplished by using a high-Z impactor. Preliminary screening for impactor candidates (or for materials required to have certain shock-generation properties for other applications) is conveniently accomplished by calculation of the impedance of various materials in the weak-shock limit using tabulated values of density and soundspeed (or the elastic constants required to determine the soundspeed). If we take the value of C_B inferred from shock experiments (see Section 3.4) as a suitable measure of soundspeed, we obtain some typical values that are shown in Table 3.1. In general, dense metals have high impedance, but we see that lead is an exception because of its low soundspeed. Aluminum oxide and beryllium are low-density materials that have high impedance by virtue of their high soundspeed. Careful exploitation of variations of this sort allows optimal selection of materials to satisfy specific application requirements.

3.7 Plane-shock Interactions

The simplest cases of shock interaction are those in which a shock propagating into material in a uniform state encounters a boundary such as an unrestrained surface or immovable wall, a material interface, or another shock. An interaction at a boundary produces a reflected wave, whereas an interaction of two shocks or an interaction at a material interface produces both reflected and transmitted waves. In this section, we shall assume that the Hugoniot curves of all materials in question are known. Some specific examples are analyzed using the linear $U_S - \dot{x}$ Hugoniot discussed in Sect. 3.4.

The basis of calculation of all shock interactions is that the longitudinal stress component and particle velocity are the same on both sides of the plane of interaction. For this reason, it is appropriate to analyze the problems in the $t_{11} - \dot{x}$ plane. Since states produced by shock compression always lie on an Hugoniot curve and, since these variables are continuous across planes of interaction, the state at this plane following the interaction can be determined by finding the intersection of the appropriate $t_{11} - \dot{x}$ Hugoniots.

3.7.1 Shock Interacting with a Material Interface

Consider the situation depicted in Fig. 3.10 in which a shock propagating into material that is unstressed and at rest encounters another material that is also unstressed and at rest. Part (a) of the figure serves simply to present the conditions prior to the shock encountering the interface. Part (b) shows the trajectory of each of the shocks and of the material interface in both the Lagrangian and Eulerian space–time planes. The broken lines are material-particle trajectories, i.e., lines having slope $\dot{x}$ in Eulerian coordinates, and the numbers designate various regions in which the state of the material is uniform.

Determination of the material response to the encounter is shown graphically on part (c) of the figure. The curves labeled A and B are Hugoniots corresponding to right-propagating shocks (shocks for which $U_S > 0$) centered on the initial state of each of the two materials. The incident shock has post-transition values $t_{11}^{(1)}$ and $\dot{x}^{(1)}$ corresponding to point 1 in the figure. Since the shock is propagating in material A, the post-transition state lies on the Hugoniot for this material. When the incident shock encounters the interface, stress is suddenly applied to material B, and we can expect a right-propagating shock to form in this material. The state behind this shock must correspond to some point 2 on the Hugoniot for material B. At this juncture two cases arise, depending on whether the shock impedance of material B is higher or lower than that of material A.

Case 1, $Z_A < Z_B$. In this case, material B reflects a shock back into material A. This left-propagating shock ($U_S < 0$) produces a transition from state 1 to state 2, which lies at the intersection of the Hugoniots A' and B. If these Hugoniots are known quantitatively, the intersection points can be calculated. The jump conditions then suffice to determine all of the other variables. Waveforms are shown at two points in time in the left part of Fig. 3.10d. A check against the stability criterion presented in Section 3.5 shows that all of the shocks discussed are stable.

Before the foregoing analysis can be carried out the Hugoniot centered at point 1 on the $t_{11} - \dot{x}$ plane shown at the left of Fig. 3.10c must be known. This second-shock Hugoniot can be calculated but, as noted previously, it lies very close to the principal Hugoniot for a large range of moderate shock strengths. Often analyses are carried out using the approximation that these Hugoniots coincide for compressive stresses exceeding the value behind the incident shock, and we shall do this here. (To calculate the interaction, this part of the principal Hugoniot must be reflected about the vertical line through state 1 in the $t_{11} - \dot{x}$ plane so that it applies to a left-propagating shock.) If Hugoniots A and B lie close together, it is clear from the figure that a gradual deviation of the second-shock Hugoniot from the principal Hugoniot leads to only a very small error when the latter is used as an approximation to the former.

Case 2, $Z_A > Z_B$. Analysis of this case proceeds in the same way as the analysis of the first case, with results as indicated on the appropriate parts of Fig. 3.10. The difference is that, because the relative position of the Hugoniots A and B is reversed, the wave reflected from material B back into material A results in decompression of material A. As we can see from the criterion of Sect. 3.5, this shock is unstable and will spread into a smooth wave. The smooth decompression process is isentropic, so the curve B' is the isentrope through state 1. Further analysis of this wave must be on the basis of the differential equations, Eqs. 2.87 or 2.92, appropriate to smooth waves. Our discussion of the smooth waveform must await Chap. 9. Nevertheless, if the

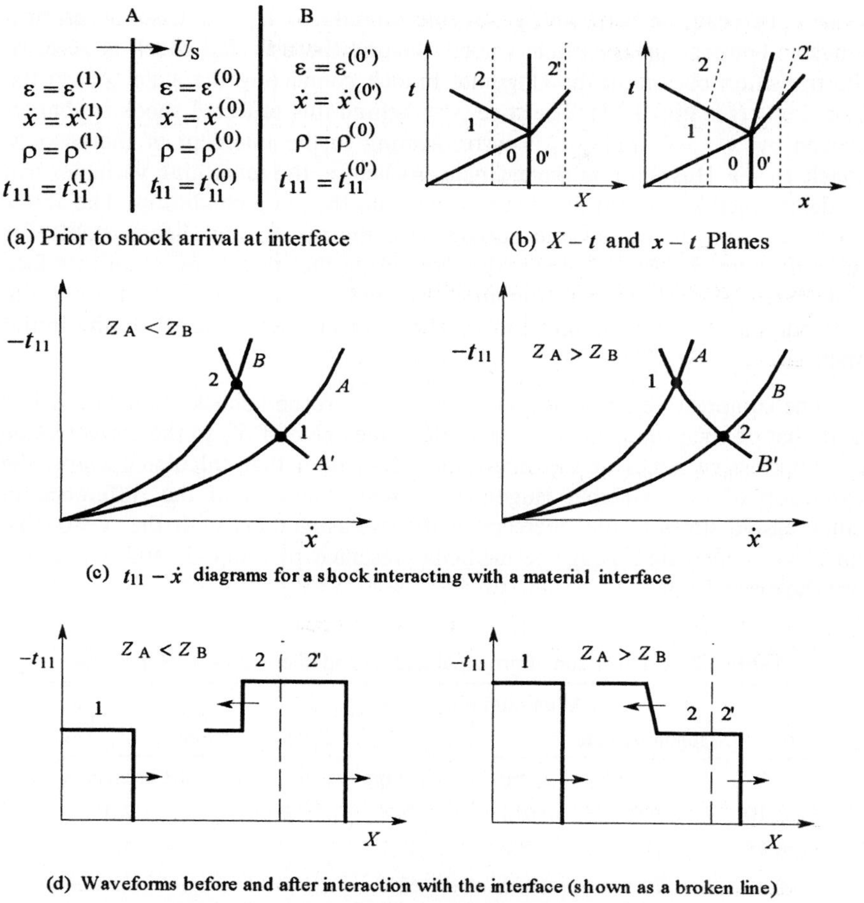

Figure 3.10. Shock encountering a material interface

spreading of the decompression wave is of less interest than the state produced as a result of the interaction, the analysis can be continued with reasonable accuracy by proceeding as though the shock were stable. This is a very common and useful procedure. As in the previous case, it is most accurate when Hugoniots A and B lie close together.

3.7.2 Shock Interaction with a Boundary

Interaction of a shock with an unrestrained (i.e. stress-free) or an immovable (also called fixed or rigidly restrained) boundary can be analyzed as a limiting case of the interaction with a material interface. In this context, the shock impedance of the "material" on the downstream side of the unrestrained boundary is zero and that on the downstream side of the immovable boundary is infinite. We consider the situation of Fig. 3.12a.

Case 1: Interaction With an Immovable Boundary. In the case of an immovable boundary, the particle velocity behind the reflected shock is zero, so the transition occurs on the Hugoniot branch connecting the state $\mathcal{S}^+$ to the line $\dot{x}=0$ (see Fig. 3.11). The state $\mathcal{S}^{++}$ behind this reflected shock is characterized by $\dot{x}^{++}=0$ and $-t_{11}^{++}$ corresponding to the intercept of the second-shock $t_{11}-\dot{x}$ Hugoniot on the ordinate. As before, the remaining variables can be determined by substituting these values into the jump conditions. The shock reflected from an immovable surface is compressive, and hence stable for normal materials. As was done when analyzing the interaction at an interface considered previously, the Hugoniot centered on the state $\mathcal{S}^+$ behind the incident shock will be assumed to overlie the Hugoniot centered on the initial state, $\mathcal{S}^-$.

The compressive stress at points on the true second-shock Hugoniot is less than that on the principal Hugoniot at the same value of $\dot{x}$, so the correct value of compressive stress at the intercept is less than that calculated using the extension of the principal Hugoniot. A few examples of the difference in compressive stress at the intercept of the two Hugoniots with the $\dot{x}=0$ line have been calculated using the methods presented in Chap. 5, and the results are shown in Table 3.2.

Table 3.2. Comparison of principal and second-shock $t_{11}-\dot{x}$ Hugoniots

Aluminum alloy 2024					
Hugoniot state			intercept stress		
Δ	$-t_{11}$, GPa	$\dot{x}$, m/s	$-t_{11}^{*}$, GPa[a]	$-t_{11}^{**}$, GPa[b]	difference, %
0.10	10.5	615	23.3	23.9	2.4
0.15	18.6	1000	43.2	44.6	3.1
0.20	29.5	1455	68.3	74.8	8.5
0.25	44.6	2000	104.1	119.0	12.5
0.30	66.2	2670	154.2	185.5	16.9
Copper					
Hugoniot state			intercept stress		
Δ	$-t_{11}$, GPa	$\dot{x}$, m/s	$-t_{11}^{*}$, GPa	$-t_{11}^{**}$, GPa	difference, %
0.10	19.2	465	43.1	44.2	2.6
0.15	34.4	760	79.6	84.2	5.5
0.20	56.1	1120	132.4	145.5	9.0
0.25	88.0	1570	209.8	241.6	13.2
0.30	135.7	2135	323.6	392.7	17.6

[a] $-t_{11}^{*}$ is the intercept on the $\dot{x}=0$ line calculated using the second-shock Hugoniot.
[b] $-t_{11}^{**}$ is the intercept on the $\dot{x}=0$ line calculated by approximating the second shock Hugoniot with the principal Hugoniot.

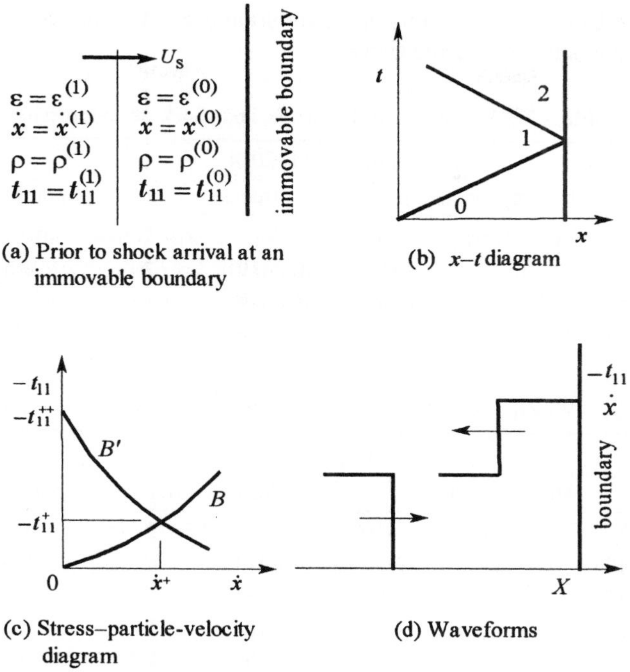

Figure 3.11. Shock interaction with an immovable boundary

Case 2: Interaction With an Unrestrained Boundary. When the incident shock encounters an unrestrained boundary, a decompression wave is reflected back into the material, as shown Fig. 3.12b and Fig. 3.12d. The process proceeds along the isentrope connecting the state S^+ to the line $-t_{11} = 0$. The decompression isentrope can be approximated by the Hugoniot and the interaction analyzed as though the decompression shock were stable.

When the decompression process is analyzed as though it were a decompression shock, the particle velocity of the material adjacent to the surface, $\dot{x}_{fs}$, is found to be twice as great as that behind the incident shock: $\dot{x}^+ = \dot{x}_{fs}/2$. Use of this approximation (called the *free-surface approximation*) facilitates determination of Hugoniot curves in experiments that do not provide information on the particle velocity as easily as plate-impact experiments. When the shock velocity is measured in addition to the velocity of the unrestrained surface, one obtains a point on the $U_S - \dot{x}$ Hugoniot. An example of an experiment usually interpreted in this way is the common one in which the shock being investigated is generated by an explosive detonated in contact with the material.

The accuracy of the free-surface approximation is indicated by some examples listed in Table 3.3. The value of one-half of the true free-surface velocity given in the table exceeds the particle velocity behind the incident shock so the free-surface velocity approximation overestimates the true value of the particle velocity behind the shock.

Table 3.3. Comparison of isentropic and shock decompression

Aluminum alloy 2024					
Hugoniot state			true $t_{11}=0$ intercept point		
Δ	$-t_{11}$, GPa	$\dot{x}$, m/s	Δ	$\dot{x}_{fs}/2$, m/s	error, %
0.10	10.5	615	–0.0019	620	0.8
0.15	18.6	1000	–0.0038	1010	1.0
0.20	29.5	1455	–0.0105	1483	1.9
0.25	44.6	2000	–0.0226	2058	2.9
0.30	66.2	2670	–0.0479	2790	4.5
Copper					
Hugoniot state			true $t_{11}=0$ intercept point		
Δ	$-t_{11}$, GPa	$\dot{x}$, m/s	Δ	$\dot{x}_{fs}/2$, m/s	error, %
0.10	19.2	465	–0.0013	468	0.6
0.15	34.4	760	–0.0051	770	1.3
0.20	56.1	1120	–0.0129	1145	2.2
0.25	88.0	1570	–0.0305	1628	3.7

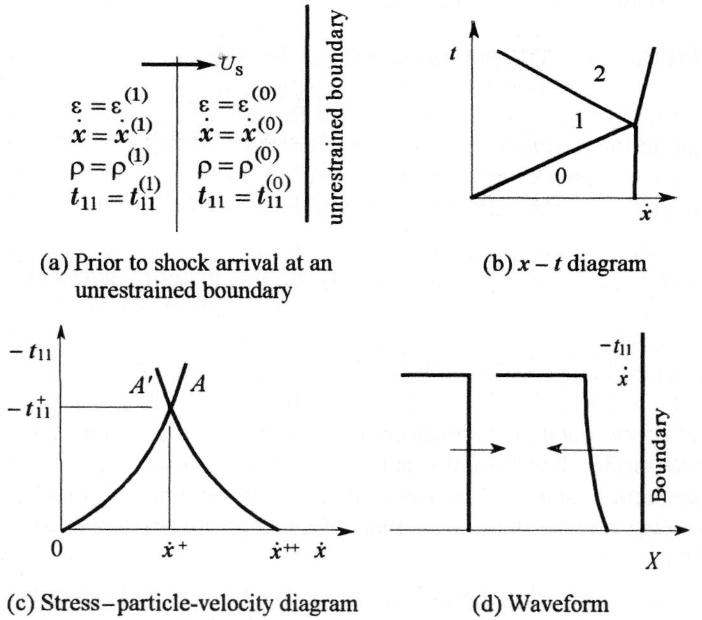

Figure 3.12. Shock interaction with unrestrained boundary.

3.7.3 Shock Interacting with Another Shock

Colliding Shocks. Consider the collision of two shocks illustrated in Fig. 3.13. In the context of a linear theory, these shocks will pass through one another without interaction. Nonlinear responses produce an interaction that modifies each shock, but we shall still describe them as transmitted shocks, as illustrated in Fig. 3.13c. We analyze the interaction in the $t_{11} - \dot{x}$ plane, as indicated in Fig. 3.13d. The state $\mathcal{S}^{++}$ behind the left-propagating shock lies on the negative-U_S branch of the Hugoniot centered on the state in which the material is unstressed and at rest. The state $\mathcal{S}^{+}$ behind the right-propagating shock lies on the positive-U_S branch of this Hugoniot. The interaction produces shocks propagating into the states $\mathcal{S}^{+}$ and $\mathcal{S}^{++}$, as indicated in Figs. 3.13b and 3.13c. The state behind each of these shocks lies on the appropriate Hugoniot. Since the stresses and particle velocities match on the two sides of the collision plane, their values correspond to the intersection point of the $t_{11} - \dot{x}$ Hugoniots, as shown on Fig. 3.13d. Stress profiles at points in time before and after the collision are shown in parts (e) and (f) of the figure.

Note that the Hugoniots centered on $\mathcal{S}^{+}$ and $\mathcal{S}^{++}$ are second-shock Hugoniots. As discussed previously, they can be approximated as reflections of the principal Hugoniot or can be calculated more accurately using the equations given in Chap. 5.

Colliding Decompression Waves. The interesting feature of the interaction of decompression waves is that it results in a reduction in compressive stress, often placing the material in a state of tension that is strong enough to produce fracture.

To understand a situation in which decompression waves arise and collide, consider the impact of two plates of finite thickness, as shown in Fig. 3.14. The figure shows an impactor plate having thickness L_P and moving in the $+X$ direction at a velocity $\dot{x}_P$ impacting a stationary target plate of thickness $L_T > L_P$. The two plates, shown at the instant of collision, are of the same material, are unstressed, and at the reference density of the material. An X–t diagram of the shock trajectories is given in part (b) of the figure and the various shock transitions are plotted in the stress–particle-velocity plane comprising part (c) of the figure.

The impact produces a right-propagating shock in the target plate and a left-propagating shock in the impactor plate. As discussed previously, this state, which we designate "state 1" is determined by the intersection of the Hugoniots through states 0 and 0' and correspond to right- and left-propagating waves, respectively.

Now, let us follow the left-propagating shock in the impactor. When it encounters the unrestrained back surface of the impactor it is reflected as a right-

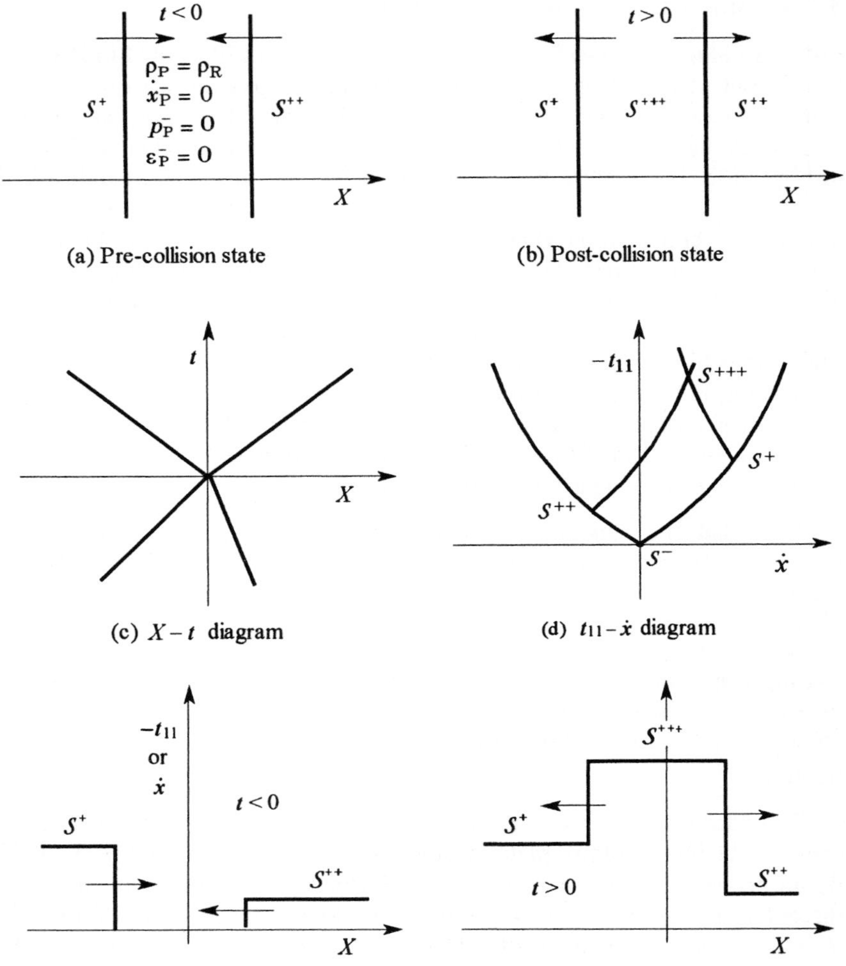

Figure 3.13. Colliding shocks

propagating decompression shock. The endstate of this transition, labeled 2', lies on the Hugoniot for right-propagating waves centered on the state 1. Since the material behind this shock is unstressed and at rest, the shock can pass over the impact interface without any interaction, thus extending state 2' into the target plate.

Let us now turn our attention to the right-propagating shock introduced into the target plate by the impact. When this wave encounters the unsupported rear boundary of the target, a reflection occurs that is analogous to that just discussed.

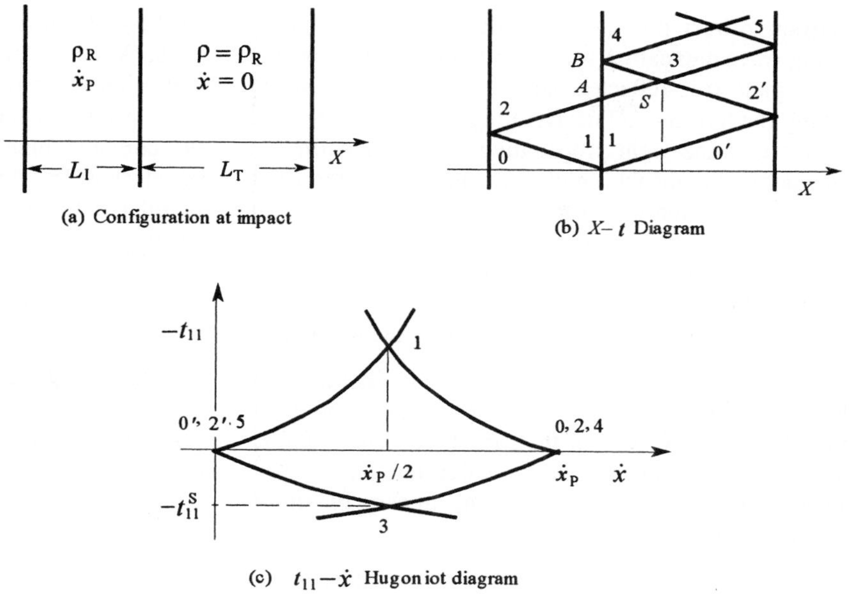

Figure 3.14. Collision of decompression waves. The figure is drawn on the assumption that the tension to which the material is subjected in state 3 is insufficient to cause fracture. If the tension is sufficiently great, a fracture called a spall is produced at the point designated S on the X–t diagram. Spallation results in detachment of a layer of material adjacent to the back face of the target plate. When it does occur the reverberating waves interact with the surface produced.

It leaves the material near the back face of the target plate in the state 2 (shown on the $t_{11}-\dot{x}$ plane) in which it is unstressed and moving in the +X direction at the velocity $\dot{x}_P$: after this reflection, the target material is accelerated to the initial velocity of the projectile. The next event is the collision of the two decompression waves. Since we have chosen the target plate to be thicker than the impactor plate, this collision will occur in the interior of the target at a distance from its back face equal to the thickness of the impactor. Since the material to the left of the collision plane is at rest and the material to the right of this plane is moving to the right at the velocity $\dot{x}_P$, the material will tend to separate. Figure 3.14b shows transmitted waves emanating from the collision plane producing a transition from the state 2' to the same $t_{11}-\dot{x}$ state 3 discussed previously. The Hugoniot for this transition is centered on state 2' and has a negative slope corresponding to the negative velocity of the shock. State 3 is defined by the intersection point of these Hugoniot curves, which we see must lie in the tensile half of the stress–particle-velocity plane at a particle velocity of $\dot{x}_P/2$. If the tension produced by the interaction is sufficient to cause the material to fracture, the target plate will separate at the collision plane.

Let us suppose that the material does not fracture. Then we can carry the analysis to later times, as indicated in the X–t plane. When the right-propagating wave emanating from the collision plane encounters the back face of the target plate it is reflected back into the material as a recompression wave relieving the tension and decelerating the material near the surface to rest. At the instant that the right-propagating wave emanating from the collision plane encounters the impact interface the material to the right of the interface has the velocity $\dot{x}_P/2$ and the material to the left is at rest. Since the interface cannot resist tension, the materials separate and a recompression wave is reflected back into the target plate. The material behind this wave is unstressed and moving at the velocity $\dot{x}_P$. These waves collide at the same plane as experienced in the previous collision, returning the material to a state of compression.

The time at which the various events discussed above occur can easily be determined from the shock velocities (as given by the jump conditions) and the plate thickness. One sees that, in the absence of fracture, the tensile state 3 persists at the collision plane for the amount of time required for two shock transits of the impactor plate. The history of stress at the collision plane $X = L_P$ is shown in Fig. 3.14d.

The problem of tensile fracture of materials is of great technological importance, and is often studied by experiments conducted as suggested by the foregoing discussion. Fractures produced in this manner are called spall fractures, and are discussed in greater detail in Chap. 12.

Contact Surfaces. In analyzing the foregoing shock interactions, we have focused on the continuity of the particle velocity and stress fields at interaction planes. It is important to note that other fields, such as specific volume and specific internal energy, need not be, and generally are not, continuous at the interaction plane. These discontinuities, which do not propagate through the material, are called *contact surfaces*. When a shock encounters a contact surface, an interaction occurs that is like the one that occurs at a material interface. We shall address this issue further after theories of material behavior have been discussed.

3.8 Exercises

3.8.1. Consider a half-space of material that is unstressed and at rest. Suppose that the stress $-t_{11}$ is applied to the boundary at $t = 0$ and maintained thereafter. Show that the energy in the material at any time $t > 0$ is equal to the work done by the applied stress in the interval from $t = 0$ to the time t.

3.8.2. Can a material described by Eq. 3.9 be compressed to arbitrarily large density by a shock?

3.8.3. What pressure, particle velocity, and compression result when a copper plate moving at 1 km/s impacts an aluminum plate at rest? Use the Hugoniot parameters of Table 3.1.

3.8.4. The highest velocity to which plates can be accelerated by the two-stage light-gas guns used in shock-physics laboratories is about 10 km/s. What pressures and compressions are produced in copper and aluminum targets impacted by tungsten impactors moving at this speed? Tungsten (see Table 3.1) is one of the higher-impedance materials commonly available, and is often used as an impactor. Compare the pressure achieved when tungsten is used as an impactor with those achieved when copper or aluminum is used.

3.8.5. Work the foregoing problem for an impact velocity of 2.5 km/s, about the maximum velocity that can be achieved by a single-stage gun using chemical propellant (gun powder, hence called a *powder gun*).

3.8.6. Work the foregoing problem for an impact velocity of 1 km/s, about the maximum velocity that can be achieved by a single-stage gun using compressed air as a propellant. Such guns are called gas guns. Helium and hydrogen are more efficient propellants, and can be used to achieve projectile velocities of about 1.5 km/s. When these gases are used, the gun is usually called a (single-stage) light-gas gun.

3.8.7. In some experiments, it is necessary to affix a very thin mirror or metallic electrode to an otherwise unrestrained surface of a sample. Analyze (qualitatively) the response of the thin layer when a shock that has propagated through the sample impinges on it and reverberates within it. Consider both the case of a layer of lower and of higher shock impedance than the sample. Is an aluminum or a gold mirror or electrode to be preferred for use on an aluminum oxide sample?

3.8.8. In some experiments the stress history at a plane within a sample is measured by use of the piezoresistive effect exhibited by a manganin foil when it is compressed by a passing wave. In practice, a gauge is made by embedding the manganin foil in a thin polymeric insulator. This gauge is then placed between two plates of the sample material. Analyze (qualitatively) the response of the thin gauge as a shock propagates through the sample. It is sufficient to consider only the low-impedance polymeric insulator, as the very thin manganin foil can be shown to produce a negligible effect on the shock process.

3.8.9. When a shock is introduced into a multilayer plate assembly in which a target comprising a thin low-impedance plate backed by a thick high-impedance plate is impacted by a thick high-impedance impactor plate a sequence of shock-compressed states is produced in the low-impedance plate. Draw X–t and p–$\dot{x}$ diagrams illustrating the first few wave interactions and states produced.

3.8.10. When a shock is introduced into a multilayer plate assembly in which a target comprising a thin high-impedance plate backed by a thick low-impedance plate is impacted by a thick low-impedance impactor plate a sequence of shock-compressed states is produced in the low-impedance plate. Draw X–t and $p - \dot{x}$ diagrams illustrating the first few wave interactions and states produced.

3.8.11. The Hugoniots of several materials (including. Al alloy 2024, Cu, and U–3 wt% Mo alloy) have been determined with high accuracy. These materials are used as standards from which Hugoniots of other materials can be measured by comparison. In the experiment, a shock (usually produced by an explosive plane-wave generator) is passed through a plate of one of the standard materials and into a contacting plate of the material for which the Hugoniot is to be determined. Show how, by measuring only lengths and shock transit times, a point on the Hugoniot of the sample can be determined.

3.8.12. How can a point on the Hugoniot of a material be measured from an experiment in which an explosively generated shock is propagated through a plate of the material and allowed to reflect from its unrestrained back face?

3.8.13. Consider the effect of sudden application of a constant, uniform pressure to the surface of a halfspace. Does the work done on the boundary equal the sum of the kinetic and internal energy in the material? How do the magnitudes of these two forms of energy compare?

CHAPTER 4

Material Response I: Principles

4.1 General Remarks About Constitutive Equations

The equations of balance of mass, momentum, and energy that were presented in Chap. 2 apply to all materials that we shall consider in this book.* Since they apply equally to air, water, and steel, for example, they are obviously insufficient for complete determination of the response of any given one of these materials. We need to supplement them with a mathematical description of the differentiating characteristics of specific materials.

The materials with which we come in contact every day exhibit widely varied responses. We encounter gases, liquids, and elastic solids. Inelastic solids are also familiar. In some cases a ductile response is observed, as with soft metals, whereas brittle responses are observed in materials such as glass, rocks, and ceramics. Porous materials such as powdered-metal compacts, granular materials such as soils, and composite materials such as fiber-reinforced polymers are common. In some cases materials respond essentially instantaneously to imposed stimuli, as in elasticity, whereas evolutionary responses such as viscoplasticity and chemical reactions proceeding at finite rates are observed in other materials or in the same materials subjected to other stimuli. Shock processes can produce very large compressions and high temperatures, and do so at the highest attainable rates. This broad range of thermodynamic state and deformation rate necessitates development and use of comprehensive models of material response.

Formulation of equations describing material behavior must begin with selection of a set of variables through which this behavior can be expressed. This is an intuitive step, but one based upon experience and experimental observation. We have learned, for example, that the behavior of an elastic body can be described in terms of equations relating the force imposed on it to the deforma-

* When additional physical phenomena such as electrical interactions are to be considered, or when it is necessary to capture effects of structural complexity present in materials such as fiber-reinforced composites, the equations of Chap. 2 must be generalized.

tion produced. Specifically, we express the forces in terms of a stress tensor and the deformations in terms of a strain tensor. Similarly, gases are characterized by a relation that allows determination of pressure from specific volume and temperature. Alternatively, we can begin with an equation relating the internal energy to the deformation and entropy and then use thermodynamic principles to determine the stress and temperature relations. When the response of a material depends upon the rate at which it is deformed, it is necessary to devise an appropriate measure of this rate and to include it in the constitutive description of the material. As we shall see, there are many additional possibilities.

We shall call the set of variables required to describe the behavior of a material, or class of materials, the *constitutive variables* of the material. Constitutive variables such as stress, strain, and rates and gradients of these quantities are usually expressed in terms of their components in some coordinate system. To ensure that these variables are expressed in a form that is not essentially tied to a specific coordinate system, we formulate them as tensors. When, as in the case of strain tensors, there are various ways of expressing the concept, each is acceptable if it can be uniquely related to the others.

The mathematical description of material behavior is in the form of *constitutive equations,* which are expressions of the relationships among the constitutive variables. Except for the least comprehensive theories, several constitutive equations are required to describe a material. Truesdell's *Principle of Equipresence* holds that a dependent variable present in one constitutive equation for a given material must be assumed to be present in all of the others, except when this is specifically precluded by thermodynamic or invariance principles.

The purpose of this chapter is to provide a sketch of the principles underlying development of constitutive equations and to initiate a discussion of thermoelastic materials. As with our previous discussion of mechanical principles, we give only a brief exposition of the subject matter, referring to appropriate texts and treatises [72,98,102,103] for fuller accounts.

Constitutive equations can be inferred (in principle) directly from experimental measurements or by fitting assumed mathematical forms to measurements. Similarly, they can be inferred from the results of calculations of atomic or molecular interactions. Direct experimental measurement has proven useful in determining the equation of state of an ideal gas and descriptions of a few other simple substances, but is an impractical method of dealing with materials exhibiting complicated responses. It is also of limited use in the range of stress and rate of deformation of interest in shock-wave studies because the extent of a practical investigation and the capabilities of available instrumentation are too limited. When, as is often the case, the response of a material to sudden application of large forces is not known, experimental methods of shock physics provide a means for studying this response and the results of shock measurements can be used to establish values for material coefficients occurring in empirical constitutive equations. It is important to note that arbitrary functions fit to experimental data may not lead to properly invariant equations. It is necessary to constrain the form of the equations before attempting to match them to measurements. The foregoing comments are not intended to suggest that experimental investigations are unimportant. Indeed, they form the basis of the subject because they reveal phenomena that require explanation, indicate the path to be followed in developing a theory of the observed responses, quantify material parameters, and provide evidence as to the adequacy of the theory.

This is the first of several chapters devoted to constitutive equations. In it we address general principles; the remaining chapters in this group describe several important classes of materials.

4.2 Invariance Principles

Establishment of invariance principles governing constitutive equations is motivated by the need to rationalize the making of arbitrary choices. When there is no rational way to choose among alternatives, it is necessary to make the choice arbitrarily but to insist that the resulting theory be invariant to the choice made. Consider, for example, the choice of the spatial coordinate system in which a theory of material response is expressed. How is this frame to be selected? The space has no intrinsic features that help us select the origin of the frame, its orientation, its right- or left-handedness, or its scale.* Accordingly, we select a spatial coordinate frame arbitrarily, but require that constitutive equations describing the physical behavior of a material be invariant to this choice. We shall see how this and other requirements are implemented.

4.2.1 Transformation of Spatial Coordinates

Description of a physical event is most conveniently accomplished using a frame of reference. To explain where something is occurring, in which direction and how fast an object is moving, or where and in which direction a force is applied, for example, we use a frame of reference. If two people observe an event, each may describe it in terms of any convenient frame of reference. If the relationship between the two reference frames is known it is possible to reconcile the two reports since the event itself was completely independent of the reference frames used by the observers. Similarly, theories of material response are cast in terms of equations that involve a reference frame. Since the behavior of the material cannot depend upon the choice of a reference frame used to describe it the constitutive equations must be invariant to this choice.

---

* The equations of motion involve a preferred coordinate system called an *inertial frame*, but this concept does not play a role in developing constitutive equations.

To ensure that a theory of material response is independent of the coordinates used it must be expressed in terms of variables that transform as tensors under change of coordinates. It is sufficient to use Cartesian tensors, and that is the procedure followed in this book.

Let us consider two spatial coordinate frames in which the points are designated by the coordinates x_i and x_i^*, for $i = 1, 2, 3$. These coordinates are related by the equation

$$x_i^* = Q_{ij}(t)\, x_j + c_i(t), \tag{4.1}$$

where $\mathbf{Q}(t)$ is the orthogonal transformation required to bring the orientation and handedness of the frame x into coincidence with that of the frame x^* and $\mathbf{c}(t)$ is a vector representing the translation required to bring the origins of these frames into coincidence. As an orthogonal transformation, $\mathbf{Q}(t)$ has the property $\mathbf{Q}^{-1} = \mathbf{Q}^{\mathrm{T}}$ (or, in component notation, $Q_{ij}^{-1} = Q_{ij}^{\mathrm{T}} = Q_{ji}$). We shall use an asterisk to designate quantities referred to the frame x^*. Invariance also requires that the time be changed so that it is the same for the two observers. This will not present a problem for the constitutive equations that we shall consider because they involve time intervals but not the time itself. One must also consider the length scale of the two coordinate frames. This is taken into account when we choose the units in which coefficients entering the constitutive equations are expressed and need not be considered in connection with the coordinate transformation.

Substituting Eq. 4.1 into Eq. 2.4, gives the transformed representation of $\mathbf{F}$,

$$F_{iJ}^* = Q_{ij}\, F_{jJ}, \tag{4.2}$$

since $\mathbf{Q}(t)$ and $\mathbf{c}(t)$ are independent of $\mathbf{X}$. Similarly, the material derivative of Eq. 4.1 is

$$\dot{x}_i^* = \dot{Q}_{ij}\, x_j + Q_{ij}\, \dot{x}_j + \dot{c}_i .$$

Differentiation of this equation with respect to $\mathbf{x}^*$, and use of the relation $\partial x_i / \partial x_j^* = Q_{ij}^{-1}$ obtained from Eq. 4.1, gives the transformed representation of the velocity gradient, $\mathbf{l}$:

$$l_{ij}^* = \dot{Q}_{ik}\, \overset{-1}{Q}_{kj} + Q_{ik}\, l_{kl}\, \overset{-1}{Q}_{lj} . \tag{4.3}$$

Using these results leads to transformation laws for the symmetric and antisymmetric parts, $\mathbf{d}$ and $\mathbf{w}$, respectively, of the velocity gradient tensor $\mathbf{l}$ and of the Eulerian strain tensor $\mathbf{e}$:

$$\mathbf{d}^* = \mathbf{Q}\mathbf{d}\overset{-1}{\mathbf{Q}}, \quad \mathbf{w}^* = \mathbf{Q}\mathbf{w}\overset{-1}{\mathbf{Q}} + \dot{\mathbf{Q}}\overset{-1}{\mathbf{Q}}, \quad \mathbf{e}^* = \mathbf{Q}\mathbf{e}\overset{-1}{\mathbf{Q}} . \tag{4.4}$$

The quantities $\mathbf{F}$, $\mathbf{e}$, and $\mathbf{d}$ transform as tensors under a change of frame and can play a role in constitutive relations. The particle velocity, velocity gradient, and spin do not transform as tensors and are excluded from constitutive equations.

Lagrangian quantities such as **C** and **E** are unchanged by the change of spatial frame ($\mathbf{C}^* = \mathbf{C}$, etc.) and can appear among the constitutive variables of a theory.

As can be seen from Eq. 2.53 or directly by projecting the stress components from the frame **x** to the frame **x***, the stress transforms as a second-order tensor under a change of frame:

$$\mathbf{t}^* = \mathbf{Q}\,\mathbf{t}\,\overset{-1}{\mathbf{Q}}. \tag{4.5}$$

In some constitutive equations an expression for material stress rate is needed. Differentiation of Eq. 4.5 gives

$$\dot{\mathbf{t}}^* = \dot{\mathbf{Q}}\,\mathbf{t}\,\mathbf{Q}^{-1} + \mathbf{Q}\,\dot{\mathbf{t}}\,\mathbf{Q}^{-1} + \mathbf{Q}\,\mathbf{t}\,\dot{\mathbf{Q}}^{-1}, \tag{4.6}$$

so, as is well known, $\dot{\mathbf{t}}$ does not transform as a tensor under a change of frame. To obtain a frame-indifferent time derivative of **t**, it is necessary to calculate the rate of change in a coordinate frame that rotates with the material. There are various ways to do this; the appropriate choice is the one that lends itself most naturally to capture of the physical processes considered important. Differentiation of the elastic stress relation (see Chap. 6) leads to an equation that relates Dafalias' corodeformational rate of **t** to the deformation rate **d** through a fourth-order tensor that depends upon **F**:

$$\overset{\square}{t}_{ij} = s_{ijkl}\, d_{kl}\,, \tag{4.7}$$

where

$$\overset{\square}{t}_{ij} = \dot{t}_{ij} - l_{jm} t_{im} - l_{im} t_{mj} + (\mathrm{tr}\,\mathbf{d})\, t_{ij}\,. \tag{4.8}$$

A routine calculation based upon Eqs. 4.4_1 and 4.6 shows that $\overset{\square}{\mathbf{t}}$ transforms as a second-order tensor under a change of frame. There are several other invariant strain rates in current use, and the choice of which one is the most appropriate in specific applications is often a matter of controversy.

4.2.2 Principle of Objectivity

The *Principle of Objectivity* or *Material Frame Indifference* states that constitutive equations must be invariant under a change of spatial frame, i.e., if a constitutive equation is satisfied for a motion $\mathbf{x} = \hat{\mathbf{x}}(\mathbf{X}, t)$ it must also be satisfied for the motion $\mathbf{x}^* = \hat{\mathbf{x}}^*(\mathbf{X}, t)$ related to the first motion by Eq. 4.1.

Consider, for example, a scalar constitutive equation such as the specific internal energy function $\varepsilon = \bar{\varepsilon}(\mathbf{F}, \eta)$, where η is a scalar. When evaluated for quantities expressed in the transformed frame $\mathbf{x}^*$ we have $\varepsilon^* = \hat{\varepsilon}(\mathbf{F}^*, \eta^*)$. The invariance requirement is that $\varepsilon = \varepsilon^*$ so the internal energy function must satisfy the relation

$$\bar{\varepsilon}(\mathbf{F}, \eta) = \bar{\varepsilon}(\mathbf{F}^*, \eta^*) . \tag{4.9}$$

We have seen that $\mathbf{F}^* = \mathbf{QF}$ and, since η is a scalar, $\eta = \eta^*$, Eq. 4.9 takes the form

$$\bar{\varepsilon}(\mathbf{F}, \eta) = \bar{\varepsilon}(\mathbf{QF}, \eta) . \tag{4.10}$$

To determine the form the function $\bar{\varepsilon}(\mathbf{F}, \eta)$ must take to ensure that Eq. 4.10 is satisfied for all values of the deformation gradient and specific entropy and for all orthogonal matrices **Q**, we note that it can be written as

$$\bar{\varepsilon}(\mathbf{F}, \eta) = \bar{\varepsilon}(\mathbf{QRU}, \eta) .$$

Since this equation must hold for all orthogonal matrices **Q** it must hold, in particular, for $\mathbf{Q} = \mathbf{R}^{-1}$. Therefore, all frame-indifferent functions $\bar{\varepsilon}(\mathbf{F}, \eta)$ must be of the form

$$\varepsilon = \tilde{\varepsilon}(\mathbf{U}, \eta) . \tag{4.11}$$

To complete the proof that the internal energy function must be of the form of Eq. 4.11 we simply note that $\mathbf{U} = \mathbf{U}^*$ and $\eta = \eta^*$ so that all functions of this form are invariant to the choice of the spatial frame. Since $\mathbf{C} = \mathbf{U}^2$ and $\mathbf{E} = \frac{1}{2}(\mathbf{U}^2 - \mathbf{I})$, the function $\tilde{\varepsilon}$ can be replaced by a function of the Cauchy–Green deformation tensor, **C**, or the Lagrangian strain tensor, **E**. The form

$$\varepsilon = \hat{\varepsilon}(\mathbf{E}, \eta) \tag{4.12}$$

of the specific internal energy function often proves most convenient.

Now let us consider the stress relation $\mathbf{t} = \mathbf{t}(\mathbf{F})$. Since the stress transforms as a second-order tensor we have

$$t_{ij}^* = Q_{ik} Q_{jl} t_{kl} , \tag{4.13}$$

but $\mathbf{t}^*$ is obtained by evaluating the stress relation $\hat{\mathbf{t}}$ for the transformed deformation gradient $\mathbf{F}^* = \mathbf{QF}$. Accordingly, Eq. 4.13 takes the form

$$\bar{t}_{ij}(\mathbf{QF}) = Q_{ik} Q_{jl} \bar{t}_{kl}(\mathbf{F}) . \tag{4.14}$$

Although an objective form of $t_{ij}(\mathbf{F})$ can be derived, we shall employ thermodynamic principles to obtain $\mathbf{t}(\mathbf{F}, \eta)$ from $\bar{\varepsilon}(\mathbf{F}, \eta)$ in a way that will be seen to satisfy Eq. 4.14 automatically.

4.2.3 Material Symmetry

A final type of invariance with which we shall be concerned is invariance of material response to changes in the reference configuration of the body at the material point in question. We know that changes in reference density invariably produce a change in response, so we shall restrict our attention to volume-preserving (unimodular) changes of the reference configuration.

Fluids can be stirred about quite arbitrarily without affecting their subsequent behavior, so their constitutive equations must be invariant to all volume-preserving changes in the reference configuration. Indeed, this requirement can serve as the definition of a fluid.

The reference configuration of solids cannot be changed this much without their response being altered, but their response is usually invariant to some rotations. An unstressed solid at a specified reference temperature is said to be in its *natural configuration* and it is this configuration in which its symmetry is assessed. In particular, perfect, unstressed crystals are observed to possess symmetries of response that allow them to be rotated through certain angles about certain axes without causing any change in either their microscopic configuration or macroscopic response. Although it is only the macroscopic response with which we shall be concerned, it is useful to realize that observed symmetries in this response arise from microscopic causes, and that the symmetry of continuum-mechanical response of materials is more often inferred from knowledge of microscopic structure than from direct mechanical measurement.

Let us consider the effect of changing the reference coordinates upon various of the measures of stress, strain, etc. that we expect will enter constitutive descriptions. Suppose that

$$x_i = \hat{x}_i(\mathbf{X}, t)$$

represents the motion of a body relative to a reference configuration $\mathcal{R}^{(1)}$. Consider a different reference configuration $\mathcal{R}^{(2)}$ in which the particle positions $\boldsymbol{Y}$ are related to the original positions by the invertible function φ:

$$X_I = \varphi_I(\mathbf{Y}). \tag{4.15}$$

The deformation gradient at $\boldsymbol{X}$, i.e. relative to $\mathcal{R}^{(1)}$ is given by the usual equation

$$F_{iJ} = \frac{\partial \hat{x}_i(\mathbf{X}, t)}{\partial X_J}.$$

The deformation gradient at the same material point, undergoing the same motion, is

$$F_{iJ}^{*} = \frac{\partial \hat{x}_i(\varphi(\mathbf{Y}), t)}{\partial X_K} \frac{\partial \varphi_K(\mathbf{Y})}{\partial Y_J}$$

when related to $\mathcal{R}^{(2)}$. Writing

$$H_{KJ} = \frac{\partial \varphi_K(\mathbf{Y})}{\partial Y_J}$$

we have

$$F^*_{iJ} = F_{iK} H_{KJ} \, . \tag{4.16}$$

Scalars such as ρ, v, and ε, and spatial variables such as **c**, **d**, and **t** are invariant to changes in the reference configuration. Material variables such as **C**, **E**, and **T** transform as tensors under Eq. 4.15. Since only orthogonal transformations **H** (i.e. rotations and reflections) arise in discussion of solids, we restrict ourselves to that case. For orthogonal transformations **H**, we have $\mathbf{H}^{-1} = \mathbf{H}^{\mathrm{T}}$ and

$$\begin{aligned}
\mathbf{C}^* &= \overset{\mathrm{T}}{\mathbf{F}^*}\,\mathbf{F}^* = \overset{-1}{\mathbf{H}}\,\overset{\mathrm{T}}{\mathbf{F}}\,\mathbf{F}\,\mathbf{H} = \overset{-1}{\mathbf{H}}\,\mathbf{C}\,\mathbf{H} \\
\mathbf{E}^* &= \tfrac{1}{2}(\mathbf{C}^* - \mathbf{1}^*) = \tfrac{1}{2}(\overset{-1}{\mathbf{H}}\,\mathbf{C}\,\mathbf{H} - \overset{-1}{\mathbf{H}}\,\mathbf{I}\,\mathbf{H}) = \overset{-1}{\mathbf{H}}\,\mathbf{E}\,\mathbf{H} \\
\mathbf{T}^* &= \frac{1}{\rho^*}\overset{-1}{\mathbf{F}^*}\,\mathbf{t}^*\,\overset{-\mathrm{T}}{\mathbf{F}^*} = \frac{1}{\rho}\overset{-1}{\mathbf{F}^*}\,\mathbf{t}\,\overset{-\mathrm{T}}{\mathbf{F}^*} = \frac{1}{\rho}\overset{-1}{\mathbf{H}}\,\overset{-1}{\mathbf{F}}\,\mathbf{t}\,\overset{-\mathrm{T}}{\mathbf{F}}\,\mathbf{H} = \overset{-1}{\mathbf{H}}\,\mathbf{T}\,\mathbf{H} \, .
\end{aligned} \tag{4.17}$$

Consider the implications of the foregoing for a scalar-valued constitutive equation that depends upon a tensor variable: $\varepsilon = \hat{\varepsilon}(\mathbf{E})$. The thermomechanical response of a fluid is invariant to all volume-preserving transformations of the reference configuration. Since ε is a scalar and the only volume-preserving functions of **E** are those that depend only upon the specific volume itself, the constitutive equation for a fluid becomes $\varepsilon = \hat{\varepsilon}(v)$. A solid can be defined as a material for which the constitutive equations are invariant to some or all orthogonal transformations of an unstressed reference configuration. Since ε is a scalar, we must have

$$\hat{\varepsilon}(\mathbf{E}) = \hat{\varepsilon}(\mathbf{E}^*) = \hat{\varepsilon}(\overset{-1}{\mathbf{H}}\,\mathbf{E}\,\mathbf{H}) \, , \tag{4.18}$$

where **H** ranges over all of the orthogonal transformations for which Eq. 4.18 is satisfied for the material of interest. A *group* of transformations is a set that includes $\mathbf{H}^{(1)}\mathbf{H}^{(2)}$ if $\mathbf{H}^{(1)}$ and $\mathbf{H}^{(2)}$ are members, $\mathbf{H}^{-1}$ is a member if **H** is, and the identity is a member. It can be shown that the set of transformations to which a constitutive equation is invariant form a group, called the *isotropy group*, so results of group theory are available to help solve Eq. 4.18.

Many analyses arising in connection with shock phenomena involve *isotropic* materials, those possessing no distinguished axes in the unstressed state. The form of constitutive equations that conform to this symmetry is particularly simple. It is easy to show that the functions

$$\begin{aligned}
I_{\mathrm{E}} &= E_{KK} \\
II_{\mathrm{E}} &= \tfrac{1}{2}\left[(E_{KK})^2 - E_{IJ}\,E_{JI}\right] \\
III_{\mathrm{E}} &= \det \mathbf{E}
\end{aligned} \tag{4.19}$$

are invariant to all orthogonal transformations of **E**, so

$$\varepsilon = \varepsilon(I_{\mathrm{E}}, II_{\mathrm{E}}, III_{\mathrm{E}}) \tag{4.20}$$

is invariant to these transformations and, using the theory of isotropic functions [102, Chap. B], can be shown to serve as a scalar constitutive function of **E** for isotropic materials.

Similar results have been obtained for invariance under subsets of the orthogonal group that are appropriate to description of crystals of various classes, but results of this generality are not usually appropriate for applications of shock physics involving monocrystals. Among these applications is the use of the piezoelectric effect exhibited by crystalline quartz plates for measurement of stress-wave histories. Usually, the requirement is for a very accurate description of the thermomechanical response of the material that is valid for strains that are small (usually less than 5%), but large enough that one must take some account of nonlinear behavior to achieve the required accuracy. In this case, the internal energy density function is taken to be a polynomial in **E** and η. The material symmetry imposes constraints on the coefficient tensors of this polynomial. The forms of these tensors for the various crystal classes have been tabulated [98] and values of the coefficients measured for many materials.

Some problems involve shock propagation in laminated or fiber-reinforced materials that possess transversely isotropic symmetry, that in which distinct response is observed in one direction with identical responses being observed in all perpendicular directions.

4.3 Thermodynamic Principles

Thermodynamic considerations arise in connection with shock physics in two ways:

i) Thermodynamic principles impose restrictions upon constitutive equations, and thus help in formulating theories of the thermomechanical response of matter.

ii) Thermomechanical theories are invoked to calculate changes in temperature, phase composition, etc. produced in materials by shock compression or other processes.

The analysis of the present section is carried out in the spirit of the first application, above, but is applicable to solution of the other problem as well. The discussion follows that of Coleman and Mizel [25], Coleman and Noll [26], and Coleman and Gurtin [24].

Thermodynamic Processes. We have introduced the function $\mathbf{x} = \hat{\mathbf{x}}(\mathbf{X}, t)$ giving the motion of points of the body, the symmetric stress tensor **t** by which forces on surface elements of the body are characterized, and the body-force vector **f** describing extrinsic forces applied to interior points of the body. We also introduced the thermodynamic concepts of specific internal energy, ε, heat flux, **q**, and an extrinsic heat supply, r. To complete a minimal thermodynamic description, we must adjoin the *specific entropy* (entropy per unit mass), η, and the local absolute *temperature*, $\theta > 0$ to this list. Any choice of the functions $\hat{x}(\mathbf{X}, t)$, $\mathbf{t}(\mathbf{X}, t)$, $\mathbf{f}(\mathbf{X}, t)$, $\varepsilon(\mathbf{X}, t)$, $\mathbf{q}(\mathbf{X}, t)$, $r(\mathbf{X}, t)$, $\eta(\mathbf{X}, t)$, and $\theta(\mathbf{X}, t) > 0$ is said to comprise a *thermodynamic process* for the material in question if these functions satisfy the equations of balance of linear momentum and energy.

Our study of restrictions that thermodynamic principles impose upon the form of constitutive equations begins with the *Second Law of Thermodynamics.* This law differs from those far discussed in that it takes the form of an inequality, the *Clausius–Duhem inequality*:

$$\frac{d}{dt}\int_{p(t)} \rho\,\eta\, dv \geq \int_{p(t)} \rho \frac{r}{\theta}\, dv - \int_{s(t)} \frac{1}{\theta} q_i n_i\, ds, \tag{4.21}$$

where $p(t)$ is a material volume and $s(t)$ is the surface of $p(t)$. The left member of this inequality is the rate at which entropy increases in the body. The terms of the right member represent the rate at which entropy is introduced into the body by deposition of energy from extrinsic sources and by conduction of heat through the boundary, respectively. The inequality holds that the rate of increase of entropy is no less than that increase attributable to energy deposition and heat conduction into the body. Normally, the rate of increase will be strictly greater than the sum of these source terms as a consequence of internal dissipation in the material.

When the fields appearing in Eq. 4.21 are smooth enough (and when the equation is required to hold for all parts of $p(t)$ as well as $p(t)$ itself), we can proceed as in Sect. 2.4.1 to express this inequality in the local form

$$\dot{\eta} \geq \frac{r}{\theta} - \frac{1}{\rho}\frac{\partial}{\partial x_i}\left(\frac{q_i}{\theta}\right). \tag{4.22}$$

Using the energy-balance equation 2.83, we can write this inequality in the form

$$0 \geq \frac{1}{\theta}\dot{\varepsilon} - \dot{\eta} - \frac{1}{\rho\,\theta} t_{ij} \dot{F}_{iJ} \overset{-1}{F}_{Jj} + \frac{1}{\rho\,\theta^2} q_i \theta_{,i}\,, \tag{4.23}$$

which has the virtue that the extrinsic heat supply rate has been eliminated, leaving only quantities related to the material itself.

When the equations of balance of momentum and energy are satisfied by a thermodynamic process, the latter is said to be *admissible*. It can be shown that any set of functions $\hat{x}(\mathbf{X}, t)$, $\mathbf{t}(\mathbf{X}, t)$, $\varepsilon(\mathbf{X}, t)$, $\eta(\mathbf{X}, t)$, and $\theta(\mathbf{X}, t) > 0$ can be made to form an admissible thermodynamic process by a suitable choice of $\mathbf{f}(\mathbf{X}, t)$, and $r(\mathbf{X}, t)$. From the theoretical point of view, it is not important that some choices of f and r may be difficult to produce experimentally. The essential principle introduced by Coleman and coworkers in the papers cited is that material response functions are to be restricted so that the Clausius–Duhem inequality is satisfied for every admissible thermodynamic process.

For future developments it will be important that the values of the functions defining a thermodynamic process, and various of their derivatives, can be chosen arbitrarily and independently of one another at any value of X and t.

4.3.1 Thermoelastic Materials

In this section we begin our investigation of the process by which constitutive equations are developed. This first case, the simple thermoelastic material, provides a convenient introduction to the method.

Let us assume that a material at a point X is characterized by the response functions

$$\varepsilon = \bar{\varepsilon}(\mathbf{F}, \eta, \mathbf{g}), \ \theta = \bar{\theta}(\mathbf{F}, \eta, \mathbf{g}), \ \mathbf{t} = \bar{\mathbf{t}}(\mathbf{F}, \eta, \mathbf{g}), \text{ and } \mathbf{q} = \bar{\mathbf{q}}(\mathbf{F}, \eta, \mathbf{g}), \tag{4.24}$$

where $\mathbf{g}$ is the temperature gradient having components $g_i = \theta_{,i}$. Since these response functions do not depend upon the rate at which the deformation or entropy changes, the theory describes states of equilibrium and might better be called *thermostatics* than thermodynamics. Since materials exhibiting rate-independent responses are always in thermodynamic equilibrium, the theory under discussion can be applied to analysis of dynamic processes involving them.

If the material and the reference configuration are such that Eqs. 4.24 are independent of X, the material is said to be homogeneous. If no such configuration exists, the material is inhomogeneous. Since all of the discussion of constitutive equations is carried out for a single material point, the question of homogeneity does not enter directly, and we shall omit writing the variable X in the equations to follow. If necessary, it can be inserted in the equations when they are applied.

If we substitute Eqs. 4.24 into the inequality 4.23 we obtain

$$0 \geq \left[\frac{1}{\theta}\frac{\partial \bar{\varepsilon}}{\partial \eta} - 1\right]\dot{\eta} + \frac{1}{\hat{\theta}}\left[\frac{\partial \bar{\varepsilon}}{\partial F_{iJ}} - \frac{1}{\rho}\hat{t}_{ij}\overset{-1}{F}_{Jj}\right]\dot{F}_{iJ} + \frac{1}{\rho\theta^2}q_i g_i + \frac{1}{\theta}\frac{\partial \bar{\varepsilon}}{\partial g_i}\dot{g}_i . \tag{4.25}$$

As noted previously, the values of all of the arguments and their derivatives can be chosen arbitrarily and independently at any point. If we choose $\dot{\mathbf{F}} = \mathbf{0}$, $\mathbf{g} = \mathbf{0}$, and $\dot{\eta} = a$ (where a is not 0, but otherwise arbitrary) we conclude that

$$\bar{\theta}(\mathbf{F}, \eta, \mathbf{g}) = \frac{\partial \bar{\varepsilon}(\mathbf{F}, \eta, \mathbf{g})}{\partial \eta}. \tag{4.26}$$

Similarly, we can show that

$$\frac{\partial \bar{\varepsilon}(\mathbf{F}, \eta, \mathbf{g})}{\partial F_{iJ}} = \frac{1}{\rho} \overset{-1}{F}_{Jj} \bar{t}_{ji}(\mathbf{F}, \eta, \mathbf{g}) \tag{4.27}$$

and

$$\frac{\partial \bar{\varepsilon}(\mathbf{F}, \eta, \mathbf{g})}{\partial g_i} = 0. \tag{4.28}$$

Finally, we are left with the inequality

$$\bar{q}_i(\mathbf{F}, \eta, \mathbf{g})\, g_i \leq 0. \tag{4.29}$$

The results of this section can be summarized by saying that, under the present assumptions, particularly those of Eqs. 4.24, we find that

i) From Eq 4.28 we see that the specific internal energy function is independent of the temperature gradient:

$$\varepsilon = \bar{\varepsilon}(\mathbf{F}, \eta). \tag{4.30}$$

Equations 4.26 and 4.27 show that this function is an equation of state. We had previously called attention to Truesdell's Principle of Equipresence, which holds that all of the constitutive equations should depend on the same set of independent variables except as they may be excluded by invariance or thermodynamic requirements. The exclusion of **q** from the internal energy density equation of state is an example of such a case.

ii) The temperature-response function is obtained from this equation of state through the relation

$$\theta = \bar{\theta}(\mathbf{F}, \eta) = \partial \bar{\varepsilon}(\mathbf{F}, \eta)/\partial \eta, \tag{4.31}$$

iii) The stress-response function is obtained from Eq. 4.27, which we write in the form

$$t_{ij} = \bar{t}_{ij}(\mathbf{F}, \eta) = \rho\, F_{iJ} \frac{\partial \bar{\varepsilon}(\mathbf{F}, \eta)}{\partial F_{jJ}} = \frac{\rho}{\rho_R} F_{iJ}\, T_{jJ}. \tag{4.32}$$

iv) The assertion of the inequality 4.29 is that heat flows from regions of high temperature to those of lower temperature. In the present context, this inequality is to be interpreted as a requirement imposed upon the constitutive equation $\mathbf{q} = \bar{\mathbf{q}}(\mathbf{F}, \eta, \mathbf{g})$. In the case of Fourier's law of heat conduction, $q_i = -k_{ij}\, g_j$, for example, the requirement is simply that $\mathbf{k}$, the thermal conductivity tensor, be positive-semidefinite.

When Eqs. 4.31–4.32 are substituted into the energy equation 2.83, we obtain

$$\dot{\eta} = \frac{r}{\theta} - \frac{1}{\rho\theta} q_{i,i} , \tag{4.33}$$

showing that entropy increases solely as a result of energy deposition and heat conduction into the body. The deformation process does not enter this entropy-production equation because materials described by the constitutive equations under consideration are non-dissipative. It is important to note that the foregoing equation was developed under smoothness assumptions that omit shocks from consideration. We shall see that, even for this material the specific entropy increases when a shock passes through a material.

Differentiation of the relation $\varepsilon = \bar{\varepsilon}(\mathbf{F}, \eta)$ gives

$$\dot{\varepsilon} = \frac{\partial \bar{\varepsilon}(\mathbf{F}, \eta)}{\partial F_{iJ}} \dot{F}_{iJ} + \frac{\partial \bar{\varepsilon}(\mathbf{F}, \eta)}{\partial \eta} \dot{\eta} , \tag{4.34}$$

which can be written in the form

$$\dot{\varepsilon} = \frac{1}{\rho} t_{ij} \dot{F}_{iJ} \overset{-1}{F}_{Jj} + \theta \, \dot{\eta} = \frac{1}{\rho} t_{ij} \, d_{ij} + \theta \, \dot{\eta} , \tag{4.35}$$

where the second of these equations is obtained using Eq. 2.18 and the symmetry of **t**. Equation 4.35, which can be written in several equivalent forms, is called the *First Law of Thermodynamics*.

The foregoing expressions may be familiar from other discussions, but the particular way in which they were derived by Coleman and his colleagues lends itself to generalization to dissipative materials and materials for which the response functions are altered as the material is deformed.

The function $\bar{\varepsilon}$ of equation 4.30 is called a *thermodynamic potential*, since other response functions are derived from it by differentiation. Before these results can be applied to specific materials they must be further restricted to ensure satisfaction of the requirements for coordinate invariance and to reflect the symmetry of the material. The three other classical thermodynamic potentials, the Helmholtz free energy function, the enthalpy function, and Gibbs' function, are useful when independent variables $(\mathbf{F}, \theta)$, $(\mathbf{T}, \eta)$, or $(\mathbf{T}, \theta)$, respectively, are preferred to $(\mathbf{F}, \eta)$. The theory of thermoelastic fluids and solids is further developed in Chaps. 5 and 6, respectively.

If the foregoing equations are to be acceptable as constitutive descriptions they must satisfy the principle of objectivity, as outlined in Sect. 4.2.2. This requires that

$$\begin{aligned} \bar{\varepsilon}(\mathbf{F}, \eta) &= \bar{\varepsilon}(\mathbf{Q}\mathbf{F}, \eta) \\ \bar{\theta}(\mathbf{F}, \eta) &= \bar{\theta}(\mathbf{Q}\mathbf{F}, \eta) \\ \mathbf{Q}\,\bar{\mathbf{q}}(\mathbf{F}, \eta, \mathbf{g}) &= \bar{\mathbf{q}}(\mathbf{Q}\mathbf{F}, \eta, \mathbf{Q}\,\mathbf{g}) \\ \mathbf{Q}\,\bar{\mathbf{t}}(\mathbf{F}, \eta, X)\,\overset{-1}{\mathbf{Q}} &= \bar{\mathbf{t}}(\mathbf{Q}\mathbf{F}, \eta)\,. \end{aligned} \tag{4.36}$$

In order that Eq. 4.36_1 be satisfied **F** must appear in an invariant combination such as $C_{IJ} = F_{iI}\, F_{iJ}$ or, as we shall find convenient, $E_{IJ} = \frac{1}{2}[C_{IJ} - \delta_{IJ}]$. Accordingly, we write

$$\bar{\varepsilon}(\mathbf{F}, \eta) = \hat{\varepsilon}(\mathbf{E}, \eta)\,. \tag{4.37}$$

The response function for θ then follows in appropriate invariant form from Eq. 4.31. Substitution of Eq. 4.37 into Eq. 4.32 produces the result

$$t_{ij} = \bar{t}_{ij}(\mathbf{F}, \eta) = \rho\, F_{iI}\, F_{jJ}\, \frac{\partial \hat{\varepsilon}(\mathbf{E}, \eta)}{\partial E_{IJ}}\,, \tag{4.38}$$

also an objective form. By comparison of this result with Eq. 2.64 we see that the second Piola–Kirchhoff stress tensor is given by the constitutive relation

$$T_{IJ} = \rho_{\mathrm{R}}\, \frac{\partial \hat{\varepsilon}(\mathbf{E}, \eta)}{\partial E_{IJ}}\,. \tag{4.39}$$

Finally, the heat-flux equation takes the invariant form

$$q_i = F_{iJ}\, \hat{Q}_J(\mathbf{E}, \eta, \overset{-1}{\mathbf{F}}\mathbf{g})\,, \tag{4.40}$$

where it is important that the Lagrangian heat flux vector Q_I not be confused with the orthogonal transformation Q_{ij}.

A variety of "rate effects," of which Newtonian viscosity is the simplest example, are observed during deformation processes. The theory just outlined captures the equilibrium response of many materials of practical interest, but cannot explain viscosity or other effects of the *rate* at which deformation occurs. Because of the rapid deformation associated with shock loading, one may be led to consider theories in which the constitutive functions depend on $\dot{\mathbf{F}}$ as well as **F**, η, and **g**. This has been done by Coleman and Noll [25] as an extension of the present theory and leads to theories of viscosity.

4.3.2 Thermoelastic Materials with Internal State Variables

In the foregoing discussion we have taken the view that the only changes that occur in materials are deformation and change of temperature or entropy. This view does not recognize phenomena such as nonequilibrium excitation of molecular vibrations, chemical reaction, changes of metallurgical configuration such as dislocation density, or mesoscale changes such as cavitation or microcracking. Description of these effects is often presented in terms of *internal state variables* representing the degree of advancement of a chemical reaction, dislocation density, volume fraction of the material that is occupied by microvoids, etc. These variables are determined by evolutionary equations (usually first-order ordinary differential equations) expressing the kinetics of chemical reactions, rates of dislocation multiplication, etc. It is an important philosophical point that responses of materials in which these changes are occurring can be regarded as consequences of the history of deformation, temperature, etc. at points of the material or, as dependent only on the current state of the material, but with the description of this state involving additional variables. In the first case it is necessary to keep track of the history responsible for the responses observed and to have equations describing the effect of this history on the responses. In the second case, it is necessary to identify the additional variables required to adequately describe the current state of the material and to have equations describing the evolution of these variables. Traditional elastoplasticity theories are an example of the first case in which the history is taken into account. In this volume we shall discuss theories of viscoplasticity and fracture that fall into the second category in which *internal state variables* are used to describe the current state of the material and the response of the material in this state to subsequent stimuli. An advantage of the internal state variable theory is that these variables are usually chosen to represent specific changes of features of the material such as chemical composition, dislocation density, or porosity that are understood independently of the internal state variable theory being developed and can be measured in an experiment.

In this section we show how Coleman and Gurtin [24] extended the preceding thermodynamic analysis to include evolutionary processes described by internal state variables.

Let us consider response functions of the form

$$\begin{aligned} \varepsilon &= \bar{\varepsilon}(\mathbf{F}, \eta, \mathbf{g}, a_1, \ldots, a_N), \\ \theta &= \bar{\theta}(\mathbf{F}, \eta, \mathbf{g}, a_1, \ldots, a_N), \\ \mathbf{t} &= \bar{\mathbf{t}}(\mathbf{F}, \eta, \mathbf{g}, a_1, \ldots, a_N), \end{aligned} \tag{4.41}$$

and

$$\mathbf{q} = \bar{\mathbf{q}}(\mathbf{F}, \eta, \mathbf{g}, a_1, \ldots, a_N),$$

where $a_1, \ldots, a_N$ are N internal state variables that evolve in accordance with the N equations

$$\dot{a}_1 = f_1(\mathbf{F}, \eta, \mathbf{g}, a_1, \ldots, a_N), \ldots, \dot{a}_N = f_N(\mathbf{F}, \eta, \mathbf{g}, a_1, \ldots, a_N). \tag{4.42}$$

When these expressions are substituted into the Clausius–Duhem inequality 4.23 it takes the form

$$0 \geq \left[\frac{\partial \bar{\varepsilon}}{\partial \eta} - \bar{\theta} \right] \dot{\eta} + \left[\frac{\partial \bar{\varepsilon}}{\partial F_{iJ}} - \frac{1}{\rho} \bar{t}_{ij} \overset{-1}{F_{Jj}} \right] \dot{F}_{iJ}$$
$$+ \frac{1}{\rho \theta^2} \bar{q}_i \, g_i + \frac{\partial \bar{\varepsilon}}{\partial g_i} \dot{g}_i + \sum_{\alpha=1}^{N} \frac{\partial \bar{\varepsilon}}{\partial a_\alpha} f_\alpha . \tag{4.43}$$

Reasoning similar to that used previously produces the familiar expressions

$$\bar{t}_{ij} = \rho \, F_{jJ} \frac{\partial \bar{\varepsilon}}{\partial F_{iJ}}$$
$$\bar{\theta} = \frac{\partial \bar{\varepsilon}}{\partial \eta} \tag{4.44}$$

for the stress and temperature responses and we find that $\partial \bar{\varepsilon} / \partial g_i = 0$, i.e., the specific internal energy cannot depend on the temperature gradient. In the present case the stress and temperature responses depend not only upon the deformation and entropy but also upon the internal state variables (but not on the temperature gradient). The internal energy response function and the evolutionary equations must be so restricted as to satisfy the inequality

$$0 \geq \frac{1}{\rho \theta^2} \bar{q}_i \, g_i + \sum_{\alpha=1}^{N} \frac{\partial \bar{\varepsilon}}{\partial a_\alpha} f_\alpha . \tag{4.45}$$

Although we shall not pursue this theory further, much more can be done. In particular, Coleman and Gurtin develop equations analogous to those given above but employing different sets of independent variables. They also investigate equilibrium states to which thermodynamic processes evolve and prove stability theorems.

4.4 Exercises

4.4.1. Show that Jaumann's strain rate $\overset{\triangledown}{\mathbf{t}}$, defined by

$$\overset{\triangledown}{t}_{ij} = \dot{t}_{ij} - t_{ik} l_{jk} - t_{jk} \, l_{ik} ,$$

transforms as a second-order tensor under a change of the spatial frame.

4.4.2. Show that $\overset{\square}{\mathbf{t}}$, given by Eq. 4.8, transforms as a second-order tensor under a change of the spatial frame.

4.4.3. Consider an elastic material that has a constitutive equation of the form $T_{IJ} = C_{IJKL} \, E_{KL}$. What conditions must be satisfied by the elastic moduli C_{IJKL}

in order that the constitutive equation be invariant under the orthogonal transformation H_{IJ} of the reference coordinate frame?

4.4.4. Produce functions $\mathbf{x}(\mathbf{X},t)$, $\eta(\mathbf{X},t)$, $\theta(\mathbf{X},t)$ $\mathbf{g}(\mathbf{X},t)$ such that $\dot{\mathbf{F}}(\mathbf{X},t)$, $\dot{\eta}(\mathbf{X},t)$, $\dot{\theta}(\mathbf{X},t)$, and $\dot{\mathbf{g}}(\mathbf{X},t)$ attain arbitrary values when $\mathbf{X}=\mathbf{X}^*$ and $t=t^*$. Note that polynomials suffice.

4.4.5. For an inviscid thermoelastic fluid we have the equations of state

$$\varepsilon=\varepsilon(v,\eta,\mathbf{g}),\quad \theta=\theta(v,\eta,\mathbf{g}),\quad t_{ij}=p(v,\eta,\mathbf{g})\delta_{ij}\quad q_i=q_i(v,\eta,\mathbf{g})\,.$$

Determine the requirements imposed upon these functions in order that the Clausius–Duhem inequality be satisfied for every admissible thermodynamic processes.

4.4.6. In Chap. 11 we develop a theory of compaction of a porous material. The parent solid of which this material is made has the equation of state $\varepsilon=\varepsilon_s(v_s,\eta)$, where v_s is its specific volume and η is its specific entropy. The porous material is made by introducing voids into the parent solid so that its specific volume is $v>v_s$. We define the porosity as $\alpha=v/v_s$, which we introduce into the theory as an internal state variable. The equation of state of the porous material is taken to be $\varepsilon(v,\eta,\alpha)=\varepsilon_s(v/\alpha,\eta)$ and the compaction process is governed by the evolutionary equation $\dot{\alpha}=[\alpha-\alpha_{eq}(v,\eta)]/\tau$, where $\alpha_{eq}(v,\eta)\le\alpha$ is the equilibrium porosity at the state v, η and τ is a characteristic compaction time for the material. We assume that the material is a nonconductor of heat (i.e., $\mathbf{q}=\mathbf{0}$). Show that the Clausius–Duhem inequality is satisfied if we require that τ be positive.

CHAPTER 5

Material Response II: Inviscid Compressible Fluids

The fluids considered in this chapter are compressible materials that are incapable of supporting shear stress. In addition to the materials usually identified as fluids, theories of fluid behavior are applied to solids subjected to strong shock compression because solids behave in much the same way as fluids under these conditions. The reason for this is that, whereas a confined solid can support an arbitrarily large pressure, its resistance to shear is limited to modest values characterized by properties called shear strength, yield strength, etc. We see from Eq. 2.58 or 2.63 that the normal stress is expressible as the sum of pressure and shear terms. Because the shear stress is limited, the normal stress and the pressure approach equality when the applied stress is very large. When the relatively small difference is neglected, the solid is effectively a fluid and its response to strong shock compression is often modeled as though it *were* a fluid. The fluid model of material response is completely inappropriate for solids at low stresses (where the material is usually modeled as an elastic solid) or even at moderately higher stresses (where the material is often modeled as an elastic–plastic solid).

The physical mechanisms that limit the ability of solids to support large shear stresses need not be described in detail. The general concept is that, when the shear stress on a plane in the body exceeds the maximum value that the material can resist, some slip mechanism is activated and the deformation, which began in the continuous manner suggested by Fig. 5.1a takes the discontinuous form suggested by Fig. 5.1b. The deformed configuration in Fig. 5.1b is seen to have changed its overall shape as though sheared, but closer examination

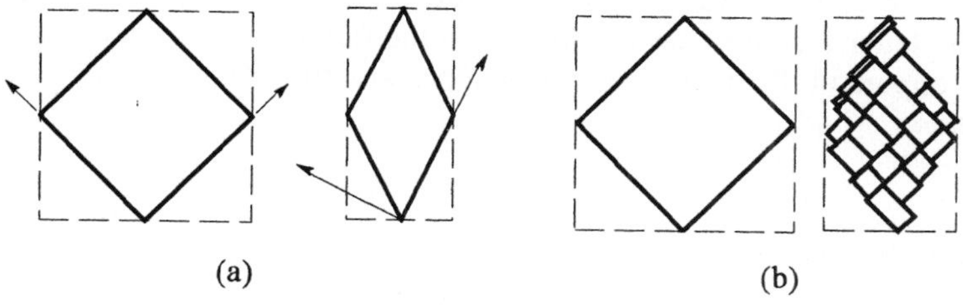

Figure 5.1. Schematic illustration of the discontinuous slip mechanism underlying the deformation that relieves shear stress and leads to a stress field in the solid that is approximately spherical, i.e. a pressure such as would be experienced in a fluid.

shows that all of the shear appears in the form of slip on discrete planes. If the shear stress is completely relieved by the slip process, the individual blocks of material thus formed are not sheared but are compressed. The scale of the individual blocks is sufficiently small that they need not be resolved in a continuum model. It is possible, of course, that relief of the shear stress is incomplete and the body is left in a compressed state involving some shear. Although the relatively small shear stress that exists in highly compressed matter can exert a disproportionate influence on some wave-propagation phenomena, it produces a negligible thermodynamic effect.

5.1 Thermodynamic Properties of Fluids

In Section 4.2.3 we noted that the response functions of a fluid should be restricted to forms that are invariant to all volume-preserving changes in the reference configuration. This means that $\varepsilon(\mathbf{F},\eta)$ depends on $\mathbf{F}$ only in the form $\varepsilon(\det\mathbf{F},\eta)$ or, equivalently, $\varepsilon(v,\eta)$. The stress relation is

$$t_{ij} = \rho F_{iJ}\frac{\partial\varepsilon(v,\eta)}{\partial F_{jJ}} = \frac{\rho}{\rho_R}\frac{\partial\varepsilon(v,\eta)}{\partial v}F_{iJ}\frac{\partial(\det\mathbf{F})}{\partial F_{jJ}}$$

$$= \frac{\partial\varepsilon(v,\eta)}{\partial v}F_{iJ}\overset{-1}{F}_{Jj} = \frac{\partial\varepsilon(v,\eta)}{\partial v}\delta_{ij},$$

so we see that the stress tensor is spherical, and

$$p = -\frac{\partial\varepsilon(v,\eta)}{\partial v}, \tag{5.1}$$

the familiar pressure equation of state for a fluid. The temperature equation of state is, as before,

$$\theta = \frac{\partial\varepsilon(v,\eta)}{\partial\eta}. \tag{5.2}$$

We see from these equations that the specific internal energy equation of state, $\varepsilon = \varepsilon(v,\eta)$, serves as a potential for $p(v,\eta)$ and $\theta(v,\eta)$ in the sense that the latter functions are obtained from the former by differentiation. Three other important thermodynamic potentials are the *Helmholtz Free Energy Function*

$$\psi(v,\theta) = \varepsilon - \theta\eta, \tag{5.3}$$

the *Enthalpy Function*

$$h(p,\eta) = \varepsilon + pv, \tag{5.4}$$

and *Gibbs' Function*

$$G(p,\theta) = h - \theta\eta. \tag{5.5}$$

Calculating the derivatives of the free energy function written in the form

$$\psi(v,\theta) = \varepsilon(v,\eta(v,\theta)) - \theta\eta(v,\theta), \tag{5.6}$$

we obtain

$$\frac{\partial\psi(v,\theta)}{\partial v} = \left.\frac{\partial\varepsilon}{\partial v}\right|_\eta + \left.\frac{\partial\varepsilon}{\partial\eta}\right|_v \left.\frac{\partial\eta}{\partial v}\right|_\theta - \theta\left.\frac{\partial\eta}{\partial v}\right|_\theta = -p, \tag{5.7}$$

and

$$\frac{\partial\psi(v,\theta)}{\partial\theta} = \left.\frac{\partial\varepsilon}{\partial\eta}\right|_v \left.\frac{\partial\eta}{\partial\theta}\right|_v - \eta - \theta\left.\frac{\partial\eta}{\partial\theta}\right|_v = -\eta, \tag{5.8}$$

where Eqs. 5.1 and 5.2 have been used to simplify the expressions. We see from these results that $\psi(v,\theta)$ serves as a potential for the equations of state $p = p(v,\theta)$ and $\eta = \eta(v,\theta)$. Similar analysis of $h(p,\eta)$ and $G(p,\theta)$ shows that these are also potentials and we have

$$\frac{\partial h(p,\eta)}{\partial p} = v, \quad \frac{\partial h(p,\eta)}{\partial\eta} = \theta, \quad \frac{\partial G(p,\theta)}{\partial p} = v, \quad \text{and} \quad \frac{\partial G(p,\theta)}{\partial\theta} = -\eta. \tag{5.9}$$

Thus far we have dealt only with the first partial derivatives of the potential functions. Second derivatives are also important because they are associated with such familiar material coefficients as bulk modulus, specific heat, and the coefficient of volumetric thermal expansion.

Before addressing material properties specifically, we note that a useful group of equations called *Maxwell relations* follows from the independence of order in which mixed second derivatives are calculated. For example, the two derivatives of the internal energy function with respect to v and η are equal:

$$\frac{\partial^2\varepsilon(v,\eta)}{\partial v\,\partial\eta} = \frac{\partial^2\varepsilon(v,\eta)}{\partial\eta\,\partial v}. \tag{5.10}$$

If we write this in the form

$$\frac{\partial}{\partial v}\left(\frac{\partial\varepsilon(v,\eta)}{\partial\eta}\right) = \frac{\partial}{\partial\eta}\left(\frac{\partial\varepsilon(v,\eta)}{\partial v}\right), \tag{5.11}$$

and use Eqs. 5.1 and 5.2, we see that

$$-\left.\frac{\partial\theta}{\partial v}\right|_\eta = \left.\frac{\partial p}{\partial\eta}\right|_v. \tag{5.12}$$

The Maxwell relations obtained from the other three potentials are

$$\left.\frac{\partial\eta}{\partial v}\right|_\theta = \left.\frac{\partial p}{\partial\theta}\right|_v, \quad \left.\frac{\partial v}{\partial\eta}\right|_p = \left.\frac{\partial\theta}{\partial p}\right|_\eta, \quad \text{and} \quad -\left.\frac{\partial\eta}{\partial p}\right|_\theta = \left.\frac{\partial v}{\partial\theta}\right|_p. \tag{5.13}$$

5.1.1 Thermodynamic Coefficients of Fluids

The *isentropic bulk modulus* and *isothermal bulk modulus*, respectively, are defined by the equations

$$B^{\eta} = -v\left.\frac{\partial p}{\partial v}\right|_{\eta} \quad \text{and} \quad B^{\theta} = -v\left.\frac{\partial p}{\partial v}\right|_{\theta}, \tag{5.14}$$

and are functions of the thermodynamic state. The *specific heats* per unit mass at constant volume and at constant pressure, respectively, are defined by the equations

$$C^{v} = \left.\frac{\partial \varepsilon}{\partial \theta}\right|_{v} \quad \text{and} \quad C^{p} = \theta\left.\frac{\partial \eta}{\partial \theta}\right|_{p}. \tag{5.15}$$

The *coefficient of volumetric thermal expansion*, β, is defined by the equation

$$\beta = \frac{1}{v}\left.\frac{\partial v}{\partial \theta}\right|_{p} \tag{5.16}$$

and the *temperature coefficient of pressure at constant volume* is defined by the equation

$$\lambda = \left.\frac{\partial p}{\partial \theta}\right|_{v}. \tag{5.17}$$

Finally, the quantity

$$\gamma = v\left.\frac{\partial p}{\partial \varepsilon}\right|_{v} \tag{5.18}$$

is called *Grüneisen's coefficient.* We shall call the functions defined by Eqs. 5.14–5.18 *material coefficients* or, more specifically, *thermodynamic coefficients.*

The reason that attention has been focused on expression of thermodynamic derivatives in terms of the various thermodynamic coefficients is that these coefficients are identified with important material responses and have been measured for many materials.

5.1.2 Relationships Among the Thermodynamic Coefficients

Since there are sixty permutations of the five variables p, v, ε, θ, and η, taken three at a time, there are many other thermodynamic derivatives of the form of those in Eqs. 5.14–5.18. These derivatives can all be expressed in terms of the material properties that have been defined and, indeed, the properties that have been defined are not all independent.

Analysis of thermodynamic derivatives and the relationships among them is an exercise in chain-rule differentiation and manipulation of derivatives of identities. We have seen that the functions $p = p(v, \eta)$ and $\theta = \theta(v, \eta)$ can be obtained from $\varepsilon = \varepsilon(v, \eta)$. To proceed further, we assume that these and similar functions are invertible for either of their dependent variables. This permits us to write, for example, $\eta = \eta(v, \varepsilon)$. Substitution leads to the identity $\eta = \eta(v, \varepsilon(v, \eta))$ from which we can form the derivatives

$$\left.\frac{\partial \eta}{\partial \eta}\right|_v \equiv 1 = \left.\frac{\partial \eta}{\partial \varepsilon}\right|_v \left.\frac{\partial \varepsilon}{\partial \eta}\right|_v , \tag{5.19}$$

so we have the reciprocal relation

$$\left.\frac{\partial \eta}{\partial \varepsilon}\right|_v = \left(\left.\frac{\partial \varepsilon}{\partial \eta}\right|_v\right)^{-1} = \frac{1}{\theta} . \tag{5.20}$$

Because reciprocal relationships of the sort represented by Eq. 5.20 hold for all permutations of the variables, the sixty possible thermodynamic derivatives occur in thirty pairs. Only one derivative in each pair need be calculated because of the reciprocal relation between the two.

Another derivative of the identity $\eta = \eta(v, \varepsilon(v, \eta))$ is

$$\left.\frac{\partial \eta}{\partial v}\right|_\eta \equiv 0 = \left.\frac{\partial \eta}{\partial v}\right|_\varepsilon + \left.\frac{\partial \eta}{\partial \varepsilon}\right|_v \left.\frac{\partial \varepsilon}{\partial v}\right|_\eta , \tag{5.21}$$

which can be written

$$\left.\frac{\partial \eta}{\partial v}\right|_\varepsilon = \frac{p}{\theta} . \tag{5.22}$$

By differentiating the expression $\varepsilon = \varepsilon(v, \eta(v, \theta))$ we obtain

$$\left.\frac{\partial \varepsilon}{\partial \theta}\right|_v = \left.\frac{\partial \varepsilon}{\partial \eta}\right|_v \left.\frac{\partial \eta}{\partial \theta}\right|_v = \theta \left.\frac{\partial \eta}{\partial \theta}\right|_v , \tag{5.23}$$

so the specific heat at constant volume can be written in the new form

$$C^v = \theta \left.\frac{\partial \eta}{\partial \theta}\right|_v . \tag{5.24}$$

Similar calculations relate other thermodynamic derivatives to the thermodynamic coefficients that have been defined. These results are given in Table 5.1.

The several thermodynamic coefficients that have been defined are not all independent and equations relating them are often useful. As an example, we note that differentiation of the function $p = p(v, \varepsilon(v, \theta))$ yields the equation

$$\left.\frac{\partial p}{\partial \theta}\right|_v = \left.\frac{\partial p}{\partial \varepsilon}\right|_v \left.\frac{\partial \varepsilon}{\partial \theta}\right|_v, \tag{5.25}$$

which, when written in terms of thermodynamic coefficients, gives the equation

$$\lambda = \frac{\gamma}{v} C^v \tag{5.26}$$

for the coefficient λ in terms of Grüneisen's coefficient; γ is usually used in place of λ in shock-physics calculations. From $v = v(p(v, \theta), \theta)$ we obtain the equation

$$\left.\frac{\partial v}{\partial \theta}\right|_v \equiv 0 = \left.\frac{\partial v}{\partial p}\right|_\theta \left.\frac{\partial p}{\partial \theta}\right|_v + \left.\frac{\partial v}{\partial \theta}\right|_p, \tag{5.27}$$

or

$$\left.\frac{\partial p}{\partial \theta}\right|_v = -\left.\frac{\partial v}{\partial \theta}\right|_p \left(\left.\frac{\partial v}{\partial p}\right|_\theta\right)^{-1} = -\left.\frac{\partial v}{\partial \theta}\right|_p \left.\frac{\partial p}{\partial v}\right|_\theta, \tag{5.28}$$

which can be written

$$\lambda = \beta B^\theta. \tag{5.29}$$

Among the many other equations relating the thermodynamic coefficients we have

$$\gamma = \frac{v\beta B^\theta}{C^v} = \frac{v\beta B^\eta}{C^p} \tag{5.30}$$

$$B^\eta - B^\theta = \theta\gamma^2 C^v / v \tag{5.31}$$

$$C^p - C^v = v\theta\beta^2 B^\theta \tag{5.32}$$

and

$$B^\eta / B^\theta = C^p / C^v. \tag{5.33}$$

Although thermodynamic analysis offers little information regarding the dependence of material properties on the state variables, analysis of the thermodynamic stability of equilibrium states produces the inequalities

$$C^v > 0, \quad B^\theta > 0 \tag{5.34}$$

constraining these coefficients [18, p. 135]. Since the right members of Eqs. 5.31 and 5.32 are clearly positive, we see that $B^\eta > B^\theta$ and $C^p > C^v$, so C^p

and B^η must also be positive. Using the relations among the various thermodynamic coefficients one can show that λ, β, and γ all have the same sign and experiments show that they are positive.

Table 5.1. Thermodynamic derivatives

$\left.\frac{\partial v}{\partial \eta}\right\vert_{\varepsilon} = \frac{\theta}{p}$	$\left.\frac{\partial \eta}{\partial \theta}\right\vert_{v} = \frac{C^v}{\theta}$	$\left.\frac{\partial \varepsilon}{\partial \theta}\right\vert_{p} = C^p - pv\beta$
$\left.\frac{\partial v}{\partial p}\right\vert_{\varepsilon} = \frac{v}{p\gamma - B^\eta}$	$\left.\frac{\partial p}{\partial \theta}\right\vert_{v} = \frac{\gamma}{v}C^v$	$\left.\frac{\partial v}{\partial \eta}\right\vert_{p} = \frac{v\beta\theta}{C^p}$
$\left.\frac{\partial v}{\partial \theta}\right\vert_{\varepsilon} = \frac{C^v}{p - \lambda\theta}$	$\left.\frac{\partial \varepsilon}{\partial v}\right\vert_{\eta} = -p$	$\left.\frac{\partial v}{\partial \theta}\right\vert_{p} = v\beta$
$\left.\frac{\partial \eta}{\partial p}\right\vert_{\varepsilon} = \frac{-pv}{\theta(B^\eta + p\gamma)}$	$\left.\frac{\partial \varepsilon}{\partial p}\right\vert_{\eta} = \frac{pv}{B^\eta}$	$\left.\frac{\partial \eta}{\partial \theta}\right\vert_{p} = \frac{C^p}{\theta}$
$\left.\frac{\partial \eta}{\partial \theta}\right\vert_{\varepsilon} = \frac{pC^v}{(p - \lambda\theta)\theta}$	$\left.\frac{\partial \varepsilon}{\partial \theta}\right\vert_{\eta} = \frac{pv}{\gamma\theta}$	$\left.\frac{\partial \varepsilon}{\partial v}\right\vert_{\theta} = -p + \frac{\gamma}{v}C^v\theta$
$\left.\frac{\partial p}{\partial \theta}\right\vert_{\varepsilon} = \frac{B^\theta(C^p - pv\beta)}{pv - v\theta\beta B^\theta}$	$\left.\frac{\partial p}{\partial v}\right\vert_{\eta} = -\frac{1}{v}B^\eta$	$\left.\frac{\partial \varepsilon}{\partial \eta}\right\vert_{\theta} = \theta - \frac{1}{\lambda}p$
$\left.\frac{\partial \varepsilon}{\partial \eta}\right\vert_{v} = \theta$	$\left.\frac{\partial \theta}{\partial v}\right\vert_{\eta} = -\frac{\gamma}{v}\theta$	$\left.\frac{\partial \varepsilon}{\partial p}\right\vert_{\theta} = \frac{pv}{B^\theta} - v\theta\beta$
$\left.\frac{\partial p}{\partial \varepsilon}\right\vert_{v} = \frac{\gamma}{v}$	$\left.\frac{\partial \theta}{\partial p}\right\vert_{\eta} = \frac{v\beta\theta}{C^p}$	$\left.\frac{\partial \eta}{\partial v}\right\vert_{\theta} = \frac{\gamma}{v}C^v$
$\left.\frac{\partial \varepsilon}{\partial \theta}\right\vert_{v} = C^v$	$\left.\frac{\partial \varepsilon}{\partial v}\right\vert_{p} = -p + \frac{C^p}{v\beta}$	$\left.\frac{\partial p}{\partial v}\right\vert_{\theta} = -\frac{1}{v}B^\theta$
$\left.\frac{\partial p}{\partial \eta}\right\vert_{v} = \frac{\gamma}{v}\theta$	$\left.\frac{\partial \varepsilon}{\partial \eta}\right\vert_{p} = \theta\left(1 - \frac{p\gamma}{B^\eta}\right)$	$\left.\frac{\partial \eta}{\partial p}\right\vert_{\theta} = -v\beta$

5.2 The Ideal Gas Equation of State

As we have seen, the thermodynamic response of a material is characterized by its equation of state. Many such equations have been developed. Some represent simple idealized views of the behavior of a class of materials in a limited range of the state variables. Others have been developed to capture the response of matter to high accuracy and over a broad range of thermodynamic states. These latter equations of state usually exist in the form of large tables stored in computers. In this and the following section we consider two conventional equations of state.

The simplest compressible fluid is the ideal gas. It is an experimental observation that the quantity $pv\mathcal{M}/\theta$ (where $\mathcal{M}$ is the molecular mass of the gas, or the average molecular mass in the case of a mixture such as air, for which $\mathcal{M} = 28.970$ kg/kg-mole) has approximately the same value for all gases under conditions not too far removed from those of the ambient atmosphere of a laboratory. The value of the foregoing ratio, denoted by the symbol $\mathcal{R}_0$, is called the *universal gas constant* and has the value $\mathcal{R}_0 = 8315$ J/(kg-mole K). The equation of state defining an *ideal gas* is

$$pv = \mathcal{R}\theta, \tag{5.35}$$

where $\mathcal{R} = \mathcal{R}_0/\mathcal{M}$. In some cases, it is convenient to introduce a *reference state* in which $p = p_{\mathrm{R}}$, $v = v_{\mathrm{R}}$, and $\theta = \theta_{\mathrm{R}}$, with values of the state variables constrained by the equation $p_{\mathrm{R}} v_{\mathrm{R}} = \mathcal{R}\theta_{\mathrm{R}}$. With this, Eq. 5.35 can be written

$$\frac{p}{p_{\mathrm{R}}}\frac{v}{v_{\mathrm{R}}} = \frac{\theta}{\theta_{\mathrm{R}}}. \tag{5.36}$$

It is important to determine the specific internal energy equation of state for of an ideal gas. To do this, let us consider the thermodynamic derivative

$$\left.\frac{\partial \varepsilon}{\partial v}\right|_{\theta} = -p + \lambda\theta = -p + \beta B^{\theta}\theta. \tag{5.37}$$

From Eq. 5.35 and the definitions of β and B^{θ} we find that

$$\beta = \left.\frac{1}{v}\frac{\partial v}{\partial \theta}\right|_{p} = \frac{\mathcal{R}}{pv} \quad \text{and} \quad B^{\theta} = -v\left.\frac{\partial p}{\partial v}\right|_{\theta} = \frac{\mathcal{R}\theta}{v} = p. \tag{5.38}$$

Substitution of these coefficients and Eq. 5.35 into Eq. 5.37 yields the result

$$\left.\frac{\partial \varepsilon}{\partial v}\right|_{\theta} = 0, \tag{5.39}$$

and we see that the specific internal energy is a function of temperature alone. Combining this result with the definition of the specific heat, C^{v}, we obtain the equation of state

$$\varepsilon = \hat{\varepsilon}(\theta) = \varepsilon_{\mathrm{R}} + \int_{\theta_{\mathrm{R}}}^{\theta} C^{v}(\theta')\,d\theta'. \tag{5.40}$$

Of course, this equation just expresses the unknown function $\hat{\varepsilon}(\theta)$ in terms of another unknown function $C^{v}(\theta)$, so some further analysis is required.

The molecular theory of an ideal gas is one in which the molecules move freely in the space available to them and interact only through elastic collisions. No cohesive forces exist. This means that all of the energy of the gas resides in the kinetic energy of the molecules and, hence, is of thermal origin. This is consistent with the form of the equation of state 5.40. The classical kinetic theory predicts that $C^{v} = f\mathcal{R}_0/(2\mathcal{M})$, where f is the number of degrees of freedom of the gas molecule, 3 for a monatomic gas, 5 for a diatomic gas, and a larger number for more complicated molecules. This equation, which agrees rather well with experimental data for small molecules at moderate temperatures, predicts that C^{v} is a constant for a given gas. Although this is not strictly true for real gases, we shall adopt the approximation that $C^{v} = C^{v}_{\mathrm{R}}$, a constant, as part of the definition of an ideal gas. With this assumption, Eq. 5.40 becomes

$$\hat{\varepsilon}(\theta) = \varepsilon_{\mathrm{R}} + C^{v}_{\mathrm{R}}(\theta - \theta_{\mathrm{R}}). \tag{5.41}$$

In this equation, ε_{R} is the specific internal energy in the reference state. If we define reference values such that

$$\varepsilon_{\mathrm{R}} = C^{v}_{\mathrm{R}}\,\theta_{\mathrm{R}}, \tag{5.42}$$

then Eq. 5.41 takes its usual form:

$$\hat{\varepsilon}(\theta) = C^{v}_{\mathrm{R}}\,\theta. \tag{5.43}$$

Substitution of Eq. 5.43 into Eq. 5.35 gives the equation of state

$$p = \frac{\mathcal{R}}{C^{v}_{\mathrm{R}}}\frac{\varepsilon}{v} \tag{5.44}$$

from which we determine Grüneisen's coefficient to be

$$\gamma = \frac{\mathcal{R}}{C^{v}_{\mathrm{R}}}. \tag{5.45}$$

Introduction of the *ratio of specific heats**,

$$\Gamma = 1 + \frac{\mathcal{R}}{C^{v}_{\mathrm{R}}}, \tag{5.46}$$

* One can show that $C^{p} - C^{v} = \mathcal{R}$ for an ideal gas, so Γ is indeed the ratio of specific heats.

simplifies the form of some equations. Note that most works on the theory of gases write γ for the ratio of specific heats rather than for Grüneisen's *coefficient* as we shall do throughout this volume.

To obtain an equation of state of the form $\varepsilon = \hat{\varepsilon}(v, \eta)$ for the ideal gas, we substitute Eqs. 5.35 and 5.43 into the First Law of Thermodynamics, $d\varepsilon = -p\,dv + \theta\, d\eta$. With some manipulation, we obtain

$$\frac{d\varepsilon}{\varepsilon} = -\frac{\mathcal{R}}{C_{\mathrm{R}}^{v}}\frac{dv}{v} + \frac{1}{C_{\mathrm{R}}^{v}}\,d\eta . \tag{5.47}$$

Introduction of reference values v_{R} and ε_{R} for specific volume and specific internal energy, respectively, allows us to write the foregoing equation in the form

$$\frac{d(\varepsilon/\varepsilon_{\mathrm{R}})}{\varepsilon/\varepsilon_{\mathrm{R}}} = -\frac{\mathcal{R}}{C_{\mathrm{R}}^{v}}\frac{d(v/v_{\mathrm{R}})}{v/v_{\mathrm{R}}} + \frac{1}{C_{\mathrm{R}}^{v}}\,d\eta , \tag{5.48}$$

which can be integrated immediately to produce the result

$$\ln\frac{\varepsilon}{\varepsilon_{\mathrm{R}}} = \ln\left[\left(\frac{v}{v_{\mathrm{R}}}\right)^{1-\Gamma}\right] + \frac{1}{C_{\mathrm{R}}^{v}}(\eta - \eta_{\mathrm{R}}) , \tag{5.49}$$

where we have designated the value of the specific entropy in the reference state by η_{R}. We can write this equation in the form

$$\frac{\varepsilon}{\varepsilon_{\mathrm{R}}} = \left(\frac{v}{v_{\mathrm{R}}}\right)^{1-\Gamma} \exp\left[\frac{\eta - \eta_{\mathrm{R}}}{C_{\mathrm{R}}^{v}}\right] . \tag{5.50}$$

Although defining an ideal gas in terms of Eqs. 5.35 and 5.43 is traditional, one could equally well adopt Eq. 5.50 as the defining relation and then derive the other two equations.

Substituting Eq. 5.50 into Eqs. 5.1 and 5.2 yields the equations

$$\begin{aligned} p &= (\Gamma - 1)\frac{\varepsilon_{\mathrm{R}}}{v_{\mathrm{R}}}\left(\frac{v}{v_{\mathrm{R}}}\right)^{-\Gamma} \exp\left[\frac{\eta - \eta_{\mathrm{R}}}{C_{\mathrm{R}}^{v}}\right] \\ \theta &= \frac{\varepsilon_{\mathrm{R}}}{C_{\mathrm{R}}^{v}}\left(\frac{v}{v_{\mathrm{R}}}\right)^{1-\Gamma} \exp\left[\frac{\eta - \eta_{\mathrm{R}}}{C_{\mathrm{R}}^{v}}\right] \end{aligned} \tag{5.51}$$

for pressure and temperature. We find that the reference values of the state variables satisfy the equations

$$p_{\mathrm{R}} = (\Gamma - 1)\,\frac{\varepsilon_{\mathrm{R}}}{v_{\mathrm{R}}} = \mathcal{R}\frac{\theta_{\mathrm{R}}}{v_{\mathrm{R}}}, \quad \text{and} \quad \theta_{\mathrm{R}} = \frac{\varepsilon_{\mathrm{R}}}{C_{\mathrm{R}}^{v}} = \frac{p_{\mathrm{R}} v_{\mathrm{R}}}{\mathcal{R}} , \tag{5.52}$$

so the pressure and temperature equations can be written in the forms

$$\frac{p}{p_{\mathrm{R}}}=\left(\frac{v}{v_{\mathrm{R}}}\right)^{-\Gamma}\exp\left[\frac{\eta-\eta_{\mathrm{R}}}{C_{\mathrm{R}}^{v}}\right]$$
$$\frac{\theta}{\theta_{\mathrm{R}}}=\left(\frac{v}{v_{\mathrm{R}}}\right)^{1-\Gamma}\exp\left[\frac{\eta-\eta_{\mathrm{R}}}{C_{\mathrm{R}}^{v}}\right]. \tag{5.53}$$

The isotherm for which $\theta=\theta^*$ is given by

$$p^{(\theta)}(v)=p_{\mathrm{R}}\frac{\theta^*}{\theta_{\mathrm{R}}}\frac{v_{\mathrm{R}}}{v}=\mathcal{R}\,\theta^*\frac{1}{v}. \tag{5.54}$$

The specific internal energy at points on this isotherm is a constant,

$$\varepsilon^{(\theta)}(v)=C_{\mathrm{R}}^{v}\,\theta^*, \tag{5.55}$$

as given by Eq. 5.43. The specific entropy at points of this isotherm is obtained from Eq. 5.53$_2$:

$$\eta^{(\theta)}(v)-\eta_{\mathrm{R}}=C_{\mathrm{R}}^{v}\ln\left[\frac{\theta^*}{\theta_{\mathrm{R}}}\left(\frac{v}{v_{\mathrm{R}}}\right)^{\Gamma-1}\right]. \tag{5.56}$$

The isentrope for which $\eta=\eta_*$, obtained directly from Eq. 5.53$_1$, takes the form

$$p^{(\eta)}(v)=p_{\mathrm{R}}\left(\frac{v}{v_{\mathrm{R}}}\right)^{-\Gamma}\exp\left[\frac{\eta^*}{C_{\mathrm{R}}^{v}}\right], \tag{5.57}$$

and the temperature on this isentrope, as obtained from Eq. 5.53$_2$, is

$$\theta^{(\eta)}(v)=\theta_{\mathrm{R}}\left(\frac{v}{v_{\mathrm{R}}}\right)^{1-\Gamma}\exp\left[\frac{\eta^*}{C_{\mathrm{R}}^{v}}\right]. \tag{5.58}$$

Substitution of Eq. 5.44, which can be written $\varepsilon=pv/(\Gamma-1)$, into the Rankine–Hugoniot equation 2.114 yields the Hugoniot

$$p^{(\mathrm{H})}(v)=p_{\mathrm{R}}\frac{(\Gamma+1)v_{\mathrm{R}}-(\Gamma-1)v}{(\Gamma+1)v-(\Gamma-1)v_{\mathrm{R}}}. \tag{5.59}$$

It remains to determine the temperature and entropy on the Hugoniot. This can be done by direct appeal to Eqs. 5.53. From Eq. 5.53$_1$ we obtain

$$\eta^{(\mathrm{H})}(v)-\eta_{\mathrm{R}}=C_{\mathrm{R}}^{(v)}\ln\left[\frac{p^{(\mathrm{H})}(v)}{p_{\mathrm{R}}}\left(\frac{v}{v_{\mathrm{R}}}\right)^{\Gamma}\right], \tag{5.60}$$

and substituting Eq. 5.59 into Eq. 5.36 yields

$$\theta^{(H)}(v) = \theta_R \frac{v}{v_R} \frac{p^{(H)}(v)}{p_R} = \theta_R \frac{v}{v_R} \frac{(\Gamma+1)v_R - (\Gamma-1)v}{(\Gamma+1)v - (\Gamma-1)v_R}. \tag{5.61}$$

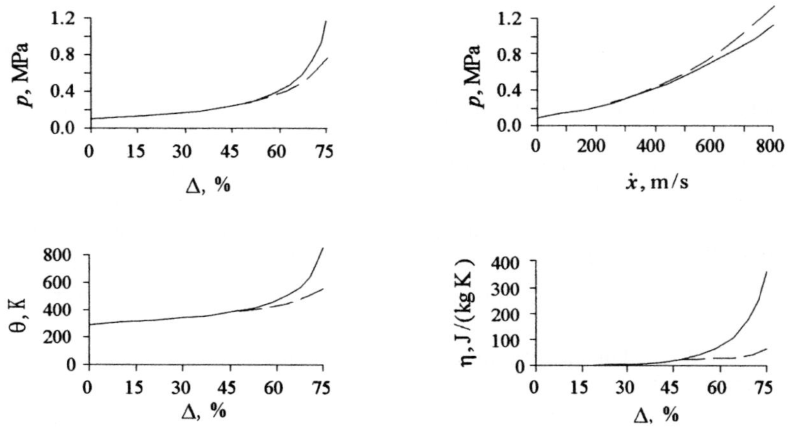

Figure 5.2. Selected Hugoniot curves for an ideal gas. The parameters used are those of air at rest at standard temperature and pressure, i.e. an initial pressure of 0.101 MPa and temperature of 293 K. The broken curves are for a second shock propagating into material that has been compressed by a first shock of strength sufficient to accelerate the gas to a particle velocity of 250 m/s.

5.3 Mie–Grüneisen Equation of State

For an ideal gas, the specific internal energy corresponds to the kinetic energy of the constituent molecules and is entirely of thermal origin. As we turn to consideration of condensed matter, we realize that the atoms are bonded together by interatomic forces and are often constrained to reside in a crystalline lattice. The internal energy derives in part from thermal excitation of the lattice but, except at very high temperatures, is dominated by the cohesive forces. This means that equations of state and thermodynamic coefficients for solids are fundamentally different from those for gases.

5.3.1 Specific Heat Coefficient for a Crystalline Solid

Lattice-dynamical calculations for solids provide useful information about their thermodynamic properties. An analysis due to Debye yields the expression

$$\frac{C^v}{3Nk} = 3\left(\frac{\theta}{\theta_D}\right)^3 \int_0^{\theta_D/\theta} \frac{\xi^4 e^\xi}{(e^\xi - 1)^2} d\xi, \tag{5.62}$$

for the specific heat of a crystalline solid (see, for example, [106], Eq. 5.33). In this expression, θ_D is a material property called the *Debye temperature*, N is the number of molecules in a unit mass of the material and k is *Boltzmann's constant*, 1.38032×10^{-23} J/K. The number of molecules in a unit mass of material is given by

$$N = \frac{N_0}{\mathcal{M}}, \tag{5.63}$$

where N_0 is a constant called *Avogadro's number* (it is equal to 6.0251×10^{26} molecules/kg-mole) and $\mathcal{M}$ is the average molecular weight of the material. The integral in Eq. 5.62 must be evaluated numerically and, because of the nature of the integrand, the integration must be initiated using the high-temperature expansion

$$\frac{C^v}{3Nk} = 1 - \frac{1}{20}\left(\frac{\theta_D}{\theta}\right)^2 + \cdots.$$

A graph illustrating the dependence of C^v on temperature is shown in Fig. 5.3.

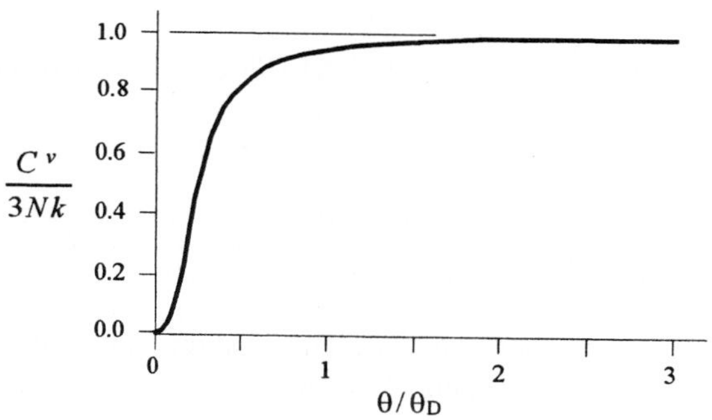

Figure 5.3. The specific heat C^v, as given by Eq. 5.62, is plotted as a function of θ/θ_D.

The Debye theory leading to the foregoing expression for the specific heat shows that Grüneisen's coefficient is a function of v alone and also provides the equation

$$\theta_D(v) = \theta_{DR} \exp\left[-\int_{v_R}^{v} \frac{\gamma(v')}{v'}\, dv'\right] \tag{5.64}$$

relating this coefficient and the Debye temperature. The constant θ_{DR} is the Debye temperature at the reference specific volume. Equations 5.62 and 5.64

express C^v as a function of v and θ. We shall see in the next section that C^v can also be expressed as a function of the single variable η.

Grüneisen's coefficient plays an important role in determination and application of equations of state of crystalline materials. Equations 5.30 are useful for determining γ in a reference state corresponding to the laboratory ambient. Determination of $\gamma(v)$ is more difficult, although a great deal of theoretical and experimental effort has been expended on the problem. Of the several equations for the dependence $\gamma(v)$ that are in use in shockwave calculations [48,90] we shall often use the simplest empirical relation,

$$\gamma(v) = \frac{\gamma_R}{v_R} v \,. \tag{5.65}$$

The equation

$$\gamma(v) = \frac{\gamma_R}{v_R} v + \frac{2}{3}\left(1 - \frac{v}{v_R}\right) \tag{5.66}$$

is used in the SESAME library [70] and the equation

$$\gamma(v) = \frac{\gamma_R}{v_R} v + \frac{2}{3}\left(1 - \frac{v}{v_R}\right)^2 \tag{5.67}$$

is used for the ANEOS equation of state [96]. Three theoretical models based on the $\theta = 0\,\mathrm{K}$ pressure isotherm, $p^{(\mathrm{K})}(v)$, (often called the *cold compression curve*) are represented by the equation

$$\gamma(v) = \frac{t-2}{3} - \frac{v}{2} \frac{d^2[v^{2t/3} p^{(\mathrm{K})}(v)]}{dv^2} \left(\frac{d[v^{2t/3} p^{(\mathrm{K})}(v)]}{dv}\right)^{-1}, \tag{5.68}$$

where $t = 0$, 1, and 2 correspond to the Slater–Landau, Dugdale–McDonald, and Vaschenko–Zubarev (free-volume) theory, respectively. These equations are useful when a theoretical expression for the cold compression curve of a particular material is available or when a semi-empirical equation for this curve (see [48, p. 133]) is adopted. All of the foregoing equations capture the decrease in $\gamma(v)$ that is observed as the material is compressed, but the predictions differ. For copper at 30% compression γ from Eq. 5.67 exceeds the value from Eq. 5.65 by 14.3%. The value of γ/v from Eq. 5.67 is less than the value from Eq. 5.65 by 14.2%. In each case the value given by Eq. 5.66 falls between the other values.

When Grüneisen's coefficient is given by Eq. 5.65, the expression for the Debye temperature becomes

$$\theta_D(v) = \theta_{DR} \exp\left(\frac{\gamma_R}{v_R}(v_R - v)\right). \tag{5.69}$$

Either this result or Eq. 5.64 shows that θ_D increases with increasing compression.

The value of C^v given by Eq. 5.62 approaches $3Nk$ in the high-temperature limit. This result,

$$C^v = 3Nk \ , \tag{5.70}$$

is called the Dulong–Petit value for the specific heat. An important observation to be drawn from Fig. 5.3 is that this limit is approached within 5% for $\theta > \theta_D$. Since the Debye temperature is near room temperature for many materials, and increases with increasing compression, use of the Dulong–Petit specific heat is justified for many practical applications and this approximation will often be adopted in this book.

Equation 5.70 should be used with caution when analyzing materials for which θ_D is significantly higher than that for which a value of C^v is sought. It is also important to recognize that the subject of shock physics includes a large body of work devoted to very strong shocks that compress metals to specific volumes as small as one-third of their normal value and temperatures of many thousands of Kelvin. Under these conditions materials are in far different thermodynamic states than were contemplated when the foregoing theory was developed, and much more refined equations of state must be used [62,70].

5.3.2. Complete Mie–Grüneisen Equation of State

Grüneisen's coefficient can be defined by any one of the equivalent thermodynamic derivatives

$$\gamma = v\frac{\partial p}{\partial \varepsilon}\bigg|_v = -\frac{v}{\theta}\frac{\partial \theta}{\partial v}\bigg|_\eta = -v\frac{\partial[\ln(\theta/\theta_R)]}{\partial v}\bigg|_\eta . \tag{5.71}$$

A material is called a Mie–Grüneisen material if γ is a function of v alone, i.e., $\partial\gamma(v,\eta)/\partial\eta = 0$. Using Eq. 5.71_3, this condition can be written

$$\frac{\partial^2}{\partial\eta\partial v}\left[\ln\left(\frac{1}{\theta_R}\frac{\partial\varepsilon(v,\eta)}{\partial\eta}\right)\right] = 0 , \tag{5.72}$$

where we have introduced the thermodynamic derivative $\theta = \partial\varepsilon(v,\eta)/\partial\eta$.

Integration of Eq. 5.72 with respect to η and then v yields the result

$$\frac{\partial\varepsilon(v,\eta)}{\partial\eta} = \theta_R\, \chi(v)\, \varpi(\eta) , \tag{5.73}$$

where we shall write $\chi(v)$ in the form

$$\chi(v) = \exp\left[\int_{v_R}^{v} \varphi(v')\,dv'\right]. \tag{5.74}$$

Both φ and ϖ are functions that remain to be determined.

A further integration yields a specific internal energy equation of state of the form

$$\varepsilon(v,\eta) = \varepsilon_R + \theta_R\,\chi(v)\,\vartheta(\eta) + \theta_R\,\zeta(v), \tag{5.75}$$

with both ϑ and ζ as yet undetermined functions. The pressure and temperature equations of state following from Eq. 5.75 are

$$p(v,\eta) = -\frac{\partial\varepsilon(v,\eta)}{\partial v} = -\theta_R\,\chi'(v)\,\vartheta(\eta) - \theta_R\,\zeta'(v) \tag{5.76}$$

and

$$\theta(v,\eta) = \frac{\partial\varepsilon(v,\eta)}{\partial\eta} = \theta_R\,\chi(v)\,\vartheta'(\eta), \tag{5.77}$$

where the primes denote differentiation with respect to the indicated argument.

Grüneisen's coefficient for this material, calculated using Eq. 5.71_2, is

$$\gamma = -v\,\varphi(v). \tag{5.78}$$

From this result, we recover the initial premise that γ depends on v alone, identify the function φ as

$$\varphi(v) = -\frac{\gamma(v)}{v}, \tag{5.79}$$

and express $\chi(v)$ in the form

$$\chi(v) = \exp\left[-\int_{v_R}^{v} \frac{\gamma(v')}{v'}\,dv'\right]. \tag{5.80}$$

The function $\chi(v)$ occurs frequently in thermodynamic calculations.

The specific heat at constant volume for the material described by Eq. 5.75 is

$$C^v = \theta\left.\frac{\partial\eta}{\partial\theta}\right|_v = \frac{\theta}{(\partial\theta/\partial\eta)|_v} = \frac{\vartheta'(\eta)}{\vartheta''(\eta)}, \tag{5.81}$$

which shows that C^v is a function of η alone for the Mie–Grüneisen material. To express the function $\vartheta(\eta)$ in terms of C^v we integrate Eq. 5.81, which we write in the form

$$\frac{1}{C^{v}(\eta)} = \frac{d}{d\eta}[\ln \vartheta'(\eta)], \tag{5.82}$$

so that

$$\vartheta'(\eta) = \exp\left[\int_{\eta_R}^{\eta} \frac{d\eta'}{C^{v}(\eta')}\right]. \tag{5.83}$$

Integration of this result gives the equation

$$\vartheta(\eta) - \vartheta(\eta_R) = \int_{\eta_R}^{\eta} \exp\left[\int_{\eta_R}^{\eta'} \frac{d\eta''}{C^{v}(\eta'')}\right] d\eta' = \int_{\eta_R}^{\eta} \omega(\eta') d\eta', \tag{5.84}$$

where

$$\omega(\eta) = \exp\left[\int_{\eta_R}^{\eta} \frac{d\eta'}{C^{v}(\eta')}\right], \tag{5.85}$$

expressing $\vartheta(\eta)$ in terms of $C^{v}(\eta)$. In forming this equation, we have chosen the constant of integration so that $\theta(v_R, \eta_R) = \theta_R$. From these results and Eq. 5.77 we find that $\theta_R\, \chi(v)$ is the temperature on the isentrope for which $\eta = \eta_R$.

Now let us return to Eq. 5.75. On the isentrope for which $\eta = \eta_R$, it becomes

$$\varepsilon^{(\eta)}(v; \eta_R) = \varepsilon_R + \theta_R\, \chi(v)\, \vartheta(\eta_R) + \theta_R\, \zeta(v), \tag{5.86}$$

so we identify the function $\zeta(v)$ as

$$\zeta(v) = \frac{1}{\theta_R}[\varepsilon^{(\eta)}(v; \eta_R) - \varepsilon_R] - \chi(v)\, \vartheta(\eta_R), \tag{5.87}$$

and write Eq. 5.75 in the form

$$\varepsilon(v, \eta) = \varepsilon^{(\eta)}(v; \eta_R) + \theta_R\, \chi(v) \int_{\eta_R}^{\eta} \omega(\eta') d\eta'. \tag{5.88}$$

Equation 5.88 is the *complete Mie–Grüneisen equation of state*. The pressure and temperature equations of state following from Eq. 5.88 are

$$p(v, \eta) = p^{(\eta)}(v; \eta_R) + \theta_R \frac{\gamma(v)}{v} \chi(v) \int_{\eta_R}^{\eta} \omega(\eta') d\eta', \tag{5.89}$$

and

$$\theta(v, \eta) = \theta_R\, \chi(v)\, \omega(\eta). \tag{5.90}$$

If we choose the isentrope so that $\varepsilon^{(\eta)}(v_{\mathrm{R}};\eta_{\mathrm{R}})=\varepsilon_{\mathrm{R}}$ and $p^{(\eta)}(v_{\mathrm{R}};\eta_{\mathrm{R}})=p_{\mathrm{R}}$, then the reference state is $\varepsilon=\varepsilon_{\mathrm{R}}$, $v=v_{\mathrm{R}}$, $\eta=\eta_{\mathrm{R}}$ $p=p_{\mathrm{R}}$, and $\theta=\theta_{\mathrm{R}}$.

This equation of state is special in view of the restriction that γ depends only on v or, equivalently, that C^v depends only on η. Its generality is associated with the allowed variability of these functions and of the isentrope $\varepsilon^{(\eta)}(v;\eta_{\mathrm{R}})$. The isentrope appearing in this equation is a reference curve that conveys information about the elastic contribution to the specific internal energy (in a form that also includes a thermal contribution unless the isentrope chosen happens to be the cold compression curve.)

In the analysis of this section we have introduced the parameter η_{R}, the value of the entropy for the material in a particular state. For each material this quantity has a unique value for given values of v_{R} and θ_{R}. The basis for this assertion lies in a postulate called the *Third Law of Thermodynamics* or *Nernst's Theorem*, which asserts that the entropy vanishes when $\theta=0$ and on the experimental observation that $C^v\to 0$ as $\theta\to 0$, and does so rapidly enough that we also have $C^v/\theta\to 0$ as $\theta\to 0$. Integration of the thermodynamic derivative

$$\frac{\partial\eta(v,\theta)}{\partial\theta}=\frac{C^v(v,\theta)}{\theta} \tag{5.91}$$

along the line $v=v_{\mathrm{R}}$ gives

$$\eta_{\mathrm{R}}=\eta(v_{\mathrm{R}},\theta_{\mathrm{R}})=\int_0^{\theta_{\mathrm{R}}}\frac{C^v(v_{\mathrm{R}},\theta')}{\theta'}d\theta'\,. \tag{5.92}$$

In many cases this integral can be evaluated using the Debye specific heat. The calculation may not be this simple when phase transformations or other complications occur, but it is clear that a specific, positive, value of the entropy corresponding to each state $v_{\mathrm{R}},\theta_{\mathrm{R}}$ still exists. For our work this knowledge is usually sufficient since we shall be interested only in entropy differences between states. Adoption of the same philosophy regarding ε_{R} leads to the value

$$\varepsilon_{\mathrm{R}}=\varepsilon(v_{\mathrm{R}},\theta_{\mathrm{R}})=\int_0^{\theta_{\mathrm{R}}}C^v(v_{\mathrm{R}},\theta')d\theta'\,. \tag{5.93}$$

Again, we are usually interested only in the difference in the value of ε for different states, in which case ε_{R} can also be chosen arbitrarily.

It is frequently convenient to specialize the foregoing equations of state to the case that C^v is constant, $C^v=C^v_{\mathrm{R}}$, and $\gamma(v)/v=\gamma_{\mathrm{R}}/v_{\mathrm{R}}$, also constant. They then take the simpler forms

$$
\begin{aligned}
\varepsilon(v,\eta) &= \varepsilon^{(\eta)}(v;\eta_{\mathrm{R}}) + C_{\mathrm{R}}^{v}\,\theta_{\mathrm{R}}\,\chi_{\mathrm{c}}(v)\big[\omega_{\mathrm{c}}(\eta)-1\big] \\
p(v,\eta) &= p^{(\eta)}(v;\eta_{\mathrm{R}}) + \frac{\gamma_{\mathrm{R}}}{v_{\mathrm{R}}} C_{\mathrm{R}}^{v}\,\theta_{\mathrm{R}}\,\chi_{\mathrm{c}}(v)\big[\omega_{\mathrm{c}}(\eta)-1\big] \\
\theta(v,\eta) &= \theta_{\mathrm{R}}\,\chi_{\mathrm{c}}(v)\,\omega_{\mathrm{c}}(\eta),
\end{aligned}
\tag{5.94}
$$

where

$$
\chi_{\mathrm{c}}(v) = \exp\left[\frac{\gamma_{\mathrm{R}}}{v_{\mathrm{R}}}(v_{\mathrm{R}}-v)\right] \quad \text{and} \quad \omega_{\mathrm{c}}(\eta) = \exp\left[\frac{\eta-\eta_{\mathrm{R}}}{C_{\mathrm{R}}^{v}}\right]. \tag{5.95}
$$

In these and various subsequent equations the subscript c is used as a reminder that $\gamma(v)/v$ and C^{v} are taken to be material constants.

It is often useful to express these equations of state in terms of temperature instead of specific entropy. This can be done by eliminating $\omega_{\mathrm{c}}(\eta)$ in favor of θ in Eqs. 5.94, but a better approach is to proceed by developing the Helmholtz free energy function equivalent to Eq. 5.94_1. The Helmholtz free energy function $\psi(v,\theta)$ given by Eq. 5.3 is obtained as follows: Use Eq. 5.94_3 to eliminate $\omega_{\mathrm{c}}(\eta)$ from Eq. 5.94_1. Then substitute Eq. 5.95_2 into Eq. 5.94_3 and solve the resulting equation for η. Finally, substitute these results into Eq. 5.3 to obtain the equation

$$
\psi(v,\theta) = \varepsilon^{(\eta)}(v;\eta_{\mathrm{R}}) + C_{\mathrm{R}}^{v}[\theta-\theta_{\mathrm{R}}\,\chi_{\mathrm{c}}(v)] - C_{\mathrm{R}}^{v}\,\theta \ln\left(\frac{\theta}{\theta_{\mathrm{R}}\,\chi_{\mathrm{c}}(v)}\right) - \theta\eta_{\mathrm{R}}. \tag{5.96}
$$

A somewhat more appealing result is obtained if we replace the reference isentrope with an isotherm. An equation relating these curves is obtained if Eq. 5.94_3 is used to eliminate $\omega_{\mathrm{c}}(\eta)$ from Eq. 5.94_1 and the resulting equation evaluated for $\theta=\theta_{\mathrm{R}}$. The result is

$$
\varepsilon^{(\theta)}(v;\theta_{\mathrm{R}}) = \varepsilon^{(\eta)}(v;\eta_{\mathrm{R}}) - C_{\mathrm{R}}^{v}\,\theta_{\mathrm{R}}[\chi_{\mathrm{c}}(v)-1], \tag{5.97}
$$

so the Helmholtz free energy function can be expressed in the form

$$
\psi(v,\theta) = \varepsilon^{(\theta)}(v;\theta_{\mathrm{R}}) + C_{\mathrm{R}}^{v}(\theta-\theta_{\mathrm{R}}) - C_{\mathrm{R}}^{v}\,\theta \ln\left(\frac{\theta}{\theta_{\mathrm{R}}\,\chi_{\mathrm{c}}(v)}\right) - \theta\,\eta_{\mathrm{R}}. \tag{5.98}
$$

The pressure, specific entropy, and specific internal energy equations of state that follow from this function are

$$
\begin{aligned}
p(v,\theta) &= p^{(\theta)}(v;\theta_{\mathrm{R}}) + \frac{\gamma_{\mathrm{R}}}{v_{\mathrm{R}}} C_{\mathrm{R}}^{v}(\theta-\theta_{\mathrm{R}}) \\
\eta(v,\theta) &= \eta_{\mathrm{R}} + C_{\mathrm{R}}^{v} \ln\left(\frac{\theta}{\theta_{\mathrm{R}}\,\chi_{\mathrm{c}}(v)}\right) \\
\varepsilon(v,\theta) &= \varepsilon_{\mathrm{R}} + \varepsilon^{(\theta)}(v;\theta_{\mathrm{R}}) + C_{\mathrm{R}}^{v}(\theta-\theta_{\mathrm{R}}).
\end{aligned}
\tag{5.99}
$$

Mie–Grüneisen p–v–ε Equation of State. Evaluation of Eq. 5.76 for $\eta = \eta_R$ allows us to write

$$\theta_R\, \zeta'(v) = -p^{(\eta)}(v;\eta_R) - \theta_R\, \chi'(v)\vartheta(\eta_R), \tag{5.100}$$

and solution of Eq. 5.88 for $\vartheta(\eta) - \vartheta(\eta_R)$ gives

$$\vartheta(\eta) - \vartheta(\eta_R) = \frac{1}{\theta_R\, \chi(v)}\left[\varepsilon(v,\eta) - \varepsilon^{(\eta)}(v;\eta_R)\right]. \tag{5.101}$$

Substitution of these results into Eq. 5.76 leads to the equation

$$p(v,\eta) = p^{(\eta)}(v;\eta_R) + \frac{\gamma(v)}{v}\left[\varepsilon(v,\eta) - \varepsilon^{(\eta)}(v;\eta_R)\right]. \tag{5.102}$$

Usually, this equation is applied to the case in which $p(v,\eta)$ and $\varepsilon(v,\eta)$ are associated values of pressure and specific internal energy on a curve such as a Hugoniot or an isotherm. In this case, Eq. 5.102 is written

$$p(v) = p^{(\eta)}(v) + \frac{\gamma(v)}{v}\left[\varepsilon(v) - \varepsilon^{(\eta)}(v)\right], \tag{5.103}$$

which is the usual expression of the Mie–Grüneisen p–v–ε equation of state with an isentrope as the reference curve.

When the pressure and specific internal energy on some curve other than the isentrope, say a Hugoniot $p^{(H)}(v)$, are substituted into Eq. 5.103 we obtain

$$p^{(H)}(v) = p^{(\eta)}(v) + \frac{\gamma(v)}{v}\left[\varepsilon^{(H)}(v) - \varepsilon^{(\eta)}(v)\right]. \tag{5.104}$$

Subtracting this equation from Eq. 5.103 yields the result

$$p(v) - p^{(H)}(v) = \frac{\gamma(v)}{v}\left[\varepsilon(v) - \varepsilon^{(H)}(v)\right], \tag{5.105}$$

an equation of the same form as Eq. 5.103, but referred to a Hugoniot instead of an isentrope. Similar equations are obtained when the Hugoniot is replaced by an isotherm or any other reference curve, and we refer to any one of these equations as a *Mie–Grüneisen p–v–ε equation of state*. Mie–Grüneisen equations are widely used in analysis of shock compression in the range of pressures up to a few hundred GPa in many common solids. Most often, the phrase "Mie–Grüneisen equation" refers to Eq. 5.105 rather than the complete Mie–Grüneisen equation of state.

Adjoining a Mie–Grüneisen p–v–ε equation of state to the three jump conditions as represented, for example, by Eqs. 2.113, gives four equations in the five variables p^+, v^+, $\dot{x}^+$, ε^+, and U_S. When one of these variables is given as a measure of the stimulus driving the shock, the others can be determined.

5.3.3 Thermodynamic Response Curves

If a complete equation of state such as $\varepsilon = \varepsilon(v, \eta)$ is available, determination of isentropes, isotherms, and Hugoniots is, in principle, both simple and direct. We have

$$p(v, \eta) = -\frac{\partial \varepsilon(v, \eta)}{\partial v} \tag{5.106}$$

and the curves

$$p = p(v, \eta^*) \tag{5.107}$$

for various values of entropy, η^*, are the isentropes.

Equation 5.2,

$$\theta(v, \eta) = \frac{\partial \varepsilon(v, \eta)}{\partial \eta}, \tag{5.108}$$

can be inverted to give $\eta = \eta(v, \theta)$ and the result substituted into the function $p = p(v, \eta)$ obtained previously to give an expression $p = p(v, \theta)$. The curves

$$p = p(v, \theta^*) \tag{5.109}$$

for various values of temperature, θ^*, are the isotherms.

The Hugoniot is obtained by substituting the state function $\hat{\varepsilon}(p, v)$ into the Rankine–Hugoniot equation, giving the implicit equation

$$\varepsilon(p, v) = \tfrac{1}{2}(p + p^*)(v^* - v) + \hat{\varepsilon}(p^*, v^*) \tag{5.110}$$

for the p–v Hugoniot curve centered on the point $p = p^*$, $v = v^*$.

The foregoing analysis is difficult to apply because the equation of state, $\varepsilon = \varepsilon(v, \eta)$, is not usually known. However, several empirical expressions for isotherms are available and Hugoniot curves have been measured for many materials. The 0K isotherm for a material, called the *cold compression curve*, $p = p^{(K)}(v)$, has been calculated for a number of materials using theories of the physics of solids. Several empirical expressions for this curve are available [48, p. 133]. What is needed is a way to evaluate the complete thermodynamic state at points on isotherms, isentropes, and Hugoniots, and to extrapolate these data to cover regions of the p–v plane in the neighborhood of the known curve.

Mie–Grüneisen Hugoniot. One case in which the analysis just described can be carried out is that of the complete Mie–Grüneisen equation of state. By using Eqs. 5.88 and 5.89, along with the definition 5.80, we can write the specific internal energy as a function of specific volume and pressure:

$$\varepsilon(v, p) = \frac{v}{\gamma(v)}\left[p - p^{(\eta)}(v; \eta_R)\right] + \varepsilon^{(\eta)}(v; \eta_R). \tag{5.111}$$

The pressure and specific internal energy at points on the Hugoniot centered on $p = p_R$, $v = v_R$, and $\varepsilon = \varepsilon_R$ are related by the Rankine–Hugoniot equation, so we have

$$\begin{aligned} &\frac{v}{\gamma(v)}\left[p^{(H)}(v) - p^{(\eta)}(v; \eta_R)\right] + \varepsilon^{(\eta)}(v; \eta_R) \\ &\quad = \frac{1}{2}\left[p^{(H)}(v) + p_R\right](v_R - v) + \varepsilon_R, \end{aligned} \tag{5.112}$$

which can be solved for the Mie–Grüneisen Hugoniot,

$$p^{(H)}(v) = \frac{p^{(\eta)}(v; \eta_R) + \dfrac{\gamma(v)}{2v} p_R (v_R - v) + \dfrac{\gamma(v)}{v}\left[\varepsilon_R - \varepsilon^{(\eta)}(v; \eta_R)\right]}{1 - \dfrac{\gamma(v)}{2v}(v_R - v)}. \tag{5.113}$$

The principal Hugoniot corresponding to the reference state $p_R = 0$, $v = v_R$, $\varepsilon_R = 0$ takes the simpler form

$$p^{(H)}(v) = \frac{p^{(\eta)}(v; \eta_R) - \dfrac{\gamma(v)}{v}\varepsilon^{(\eta)}(v; \eta_R)}{1 - \dfrac{\gamma(v)}{2v}(v_R - v)}. \tag{5.114}$$

As we shall see, isentropes can be related to isotherms, so the Mie–Grüneisen Hugoniot can be expressed in terms of either of these curves.

5.3.3.1 Isotherm

When a p–v isotherm is available it can be used to calculate an isentrope or Hugoniot. In cases in which an isotherm is not available, one can construct it from a measured Hugoniot.

Isotherms for some materials are available as the result of direct measurement or extrapolation of low-pressure data. Isotherms derived by extrapolating low-pressure data are usually obtained by substituting measured values of the isothermal bulk modulus, B^{θ} and its first pressure derivative, $B'^{\theta} \equiv \partial B^{\theta}(p, \theta)/\partial p$, into an empirical equation for the curve. Two such equations are the Birch–Murnaghan equation,

$$p^{(\Theta)}(v)=\frac{3}{2}B^{\Theta}\left(\frac{v}{v_R}\right)^{-7/3}\left[1-\left(\frac{v}{v_R}\right)^{2/3}\right]\left\{1+\frac{3}{4}(B'^{\Theta}-4)\left[\left(\frac{v}{v_R}\right)^{-2/3}-1\right]\right\},$$

(5.115)

and one proposed by Rose et al. (see, for example, [90, Eqs. 1.34 and 1.35]),

$$p^{(\Theta)}(v)=3B^{\Theta}\left(\frac{v}{v_R}\right)^{-2/3}\left[1-\left(\frac{v}{v_R}\right)^{1/3}\right]\exp\left\{\frac{3}{2}(B'^{\Theta}-1)\left[1-\left(\frac{v}{v_R}\right)^{1/3}\right]\right\},$$

(5.116)

for the temperature at which the parameters v_R, B^{Θ}, and B'^{Θ} were determined at zero pressure.

The entropy at points on an isotherm can be calculated from the thermodynamic derivative

$$\left.\frac{\partial\eta}{\partial v}\right|_{\Theta}=\frac{\gamma(v)}{v}C^{v}(v,\Theta). \tag{5.117}$$

On the isotherm $\Theta=\Theta^*$ this is the ordinary differential equation

$$\frac{d\eta(v,\Theta^*)}{dv}=\frac{\gamma(v)}{v}C^{v}(v;\Theta^*), \tag{5.118}$$

which has the solution

$$\eta^{(\Theta)}(v,\Theta^*)=\eta^*+\int_{v^*}^{v}\frac{\gamma(v')}{v'}C^{v}(v;\Theta^*)\,dv'. \tag{5.119}$$

If we assume that $\gamma(v)/v=\gamma_R/v_R$ and $C^v=C^v_R$, we get the more explicit result

$$\eta^{(\Theta)}(v;\Theta^*)=\eta^*+\frac{\gamma_R}{v_R}C^v_R(v-v^*). \tag{5.120}$$

The specific internal energy on this isotherm can be obtained by integrating the thermodynamic derivative

$$\left.\frac{\partial\varepsilon}{\partial v}\right|_{\Theta=\Theta^*}=-p(v;\Theta^*)+\frac{\gamma(v)}{v}C^{v}(v;\Theta^*)\Theta^*, \tag{5.121}$$

giving

$$\varepsilon^{(\Theta)}(v;\Theta^*)=\varepsilon^*_R-\int_{v^*_R}^{v}p^{(\Theta)}(v';\Theta^*)\,dv'+\Theta^*\int_{v^*_R}^{v}\frac{\gamma(v')}{v'}C^{v}(v';\Theta^*)\,dv', \tag{5.122}$$

where ε_R^* is the internal energy when $v = v_R^*$ and $\theta = \theta^*$. If the material properties $\gamma(v)/v$ and C^v are constant this equation takes the simpler form

$$\varepsilon^{(\theta)}(v, \theta^*) = \varepsilon_R^* - \int_{v_R^*}^{v} p^{(\theta)}(v'; \theta^*)\,dv' + \frac{\gamma_R}{v_R} C_R^v \theta^* (v - v_R^*)\,. \tag{5.123}$$

Transforming an Isotherm to a Different Temperature. Isotherms are most often measured at room temperature, approximately 300 K. When $\gamma(v)$ and $C^v(v, \theta)$ are known it is possible to transform a known isotherm to one for a different temperature.

Integration of the thermodynamic derivative

$$\left.\frac{\partial p}{\partial \theta}\right|_v = \frac{\gamma(v)}{v} C^v(v, \theta)$$

at a given value of v yields the equation

$$p^{(\theta)}(v; \theta^{**}) = p^{(\theta)}(v; \theta^*) + \frac{\gamma(v)}{v} \int_{\theta^*}^{\theta^{**}} C^v(v, \theta')\,d\theta'\,, \tag{5.124}$$

relating the θ^* and θ^{**} isotherms. Similarly, integration of the thermodynamic derivative

$$\left.\frac{\partial \varepsilon}{\partial \theta}\right|_v = C^v(v, \theta)$$

at a given value of v yields the equation

$$\varepsilon^{(\theta)}(v; \theta^{**}) = \varepsilon^{(\theta)}(v; \theta^*) + \int_{\theta^*}^{\theta^{**}} C^v(v, \theta')\,d\theta'\,. \tag{5.125}$$

In the special case that $\gamma(v)/v$ and $C^v(v, \theta)$ are constants, Eqs. 5.124 and 5.125 take the simpler forms

$$\begin{aligned} \varepsilon^{(\theta)}(v; \theta^{**}) &= \varepsilon^{(\theta)}(v; \theta^*) + C_R^v(\theta^{**} - \theta^*) \\ p^{(\theta)}(v; \theta^{**}) &= p^{(\theta)}(v; \theta^*) + \frac{\gamma_R}{v_R} C_R^v(\theta^{**} - \theta^*)\,, \end{aligned} \tag{5.126}$$

and we see that, in this special case, the two isotherms have the same shape and differ only by a translation in specific internal energy or pressure.

Figure 5.4 shows the result of applying the foregoing analysis to the transformation of a room-temperature pressure isotherm represented by Eq. 5.116 to two higher temperatures.

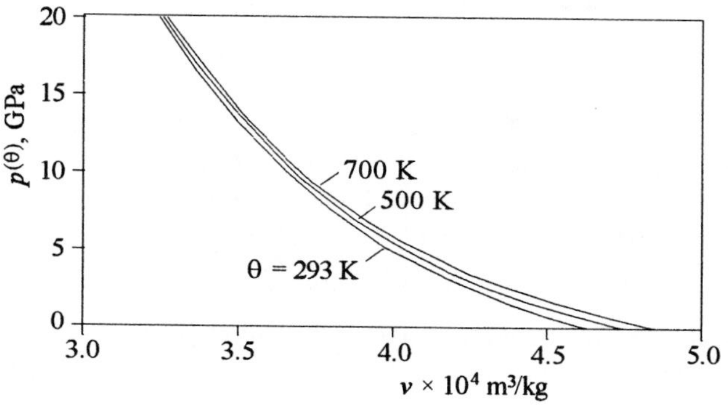

Figure 5.4. A 293 K isotherm for NaCl represented by Eq. 5.116 has been transformed to 500 and 700 K, respectively, using Eq. 5.126.

5.3.3.2 Isentrope

Isentropes, curves on which η is constant, are also of interest in connection with shock wave studies. In particular, decompression processes are usually isentropic (see Sect. 9.1). One is not usually presented with an isentrope derived by the methods of atomic physics or measured experimentally, but the results of ultrasonic experiments and some experiments with weak shock waves can be interpreted to yield isentropic bulk moduli and, therefore, isentropes valid over a range of modest compressions. The analysis of these isentropes proceeds in a manner analogous to the analysis of isotherms presented in the previous section.

The specific entropy of the material at some reference specific volume and temperature can be determined using Eq. 5.92.

The temperature on an isentrope is calculated using the thermodynamic derivative

$$\gamma = -\frac{v}{\theta}\frac{\partial \theta}{\partial v}\bigg|_{\eta} . \tag{5.127}$$

On the isentrope corresponding to $\eta = \eta*$, this expression becomes an ordinary differential equation for $\theta(v; \eta^*)$:

$$\frac{d\theta}{dv} = -\frac{\gamma(v)}{v}\theta, \tag{5.128}$$

which can be written

$$\frac{d\theta^{(\eta)}(v,\eta^*)}{\theta} = -\frac{\gamma(v)}{v}dv. \tag{5.129}$$

The solution of this equation is

$$\theta^{(\eta)}(v,\eta^*) = \theta^*\exp\left[-\int_{v^*}^{v}\frac{\gamma(v')}{v'}dv'\right] = \theta^*\chi(v), \tag{5.130}$$

where the constant of integration has been chosen so that $\theta = \theta^*$ is the temperature when $v = v^*$ and $\eta = \eta^*$. When $\gamma(v)$ is given by Eq. 5.65, this takes the simpler form

$$\theta^{(\eta)}(v,\eta^*) = \theta^*\chi_c(v). \tag{5.131}$$

Proceeding as before, we obtain the equation

$$\varepsilon^{(\eta)}(v,\eta^*) = \varepsilon^* - \int_{v^*}^{v} p^{(\eta)}(v',\eta^*)dv' \tag{5.132}$$

for the internal energy at points on the isentrope for which $\varepsilon = \varepsilon^*$ and $\eta = \eta^*$ when $v = v^*$.

5.3.3.3 Hugoniot

The Hugoniot curve is defined by the constraint on the equation of state that the Rankine–Hugoniot equation be satisfied. To calculate the temperature and entropy at points on a Hugoniot, we begin with the Rankine–Hugoniot equation, which we write in the form

$$\varepsilon^{(H)}(v) = \varepsilon^- + \frac{1}{2}\left[p^{(H)}(v) + p^-\right](v^- - v), \tag{5.133}$$

where $\varepsilon = \varepsilon^{(H)}(v)$ and $p = p^{(H)}(v)$ are the ε–v and p–v Hugoniots centered on $p = p^-$ and $v = v^-$. Differentiating Eq. 5.133 yields the equation

$$\frac{d\varepsilon^{(H)}(v)}{dv} = \frac{1}{2}\left[(v^- - v)\frac{dp^{(H)}(v)}{dv} - p^{(H)}(v) - p^-\right]. \tag{5.134}$$

Along the Hugoniot, the differential of the specific internal energy function, $d\varepsilon = \theta d\eta - p dv$, can be written in the form

$$\frac{d\varepsilon^{(\mathrm{H})}(v)}{dv} = \theta^{(\mathrm{H})}(v)\,\frac{d\eta^{(\mathrm{H})}(v)}{dv} - p^{(\mathrm{H})}(v)\,. \tag{5.135}$$

Combining the foregoing two equations, we obtain

$$\frac{d\eta^{(\mathrm{H})}(v)}{dv} = \frac{\kappa(v)}{2\theta^{(\mathrm{H})}(v)}\,, \tag{5.136}$$

where

$$\kappa(v) = p^{(\mathrm{H})}(v) - p^- + (v^- - v)\frac{dp^{(\mathrm{H})}(v)}{dv}\,. \tag{5.137}$$

Before this equation can be solved we must determine the temperature $\theta^{(\mathrm{H})}(v)$. To do this, we note that the differential of the equation $\theta = \theta(v, \eta)$ is

$$d\theta = \left.\frac{\partial\theta}{\partial v}\right|_{\eta} dv + \left.\frac{\partial\theta}{\partial\eta}\right|_{v} d\eta\,, \tag{5.138}$$

which can be written

$$d\theta = -\,\theta\,\frac{\gamma(v)}{v}\,dv + \frac{\theta}{C^{v}(\eta)}\,d\eta\,. \tag{5.139}$$

On the Hugoniot, this equation takes the form

$$\frac{d\theta^{(\mathrm{H})}(v)}{dv} = -\frac{\gamma(v)}{v}\,\theta^{(\mathrm{H})}(v) + \frac{1}{C^{v}(\eta^{(\mathrm{H})}(v))}\,\theta^{(\mathrm{H})}(v)\,\frac{d\eta^{(\mathrm{H})}(v)}{dv}\,. \tag{5.140}$$

When this equation is solved for $d\eta^{(\mathrm{H})}/dv$ and the result substituted into Eq. 5.136 we obtain

$$\frac{d\theta^{(\mathrm{H})}(v)}{dv} + \frac{\gamma(v)}{v}\,\theta^{(\mathrm{H})}(v) = \frac{\kappa(v)}{2C^{v}(\eta)}\,. \tag{5.141}$$

When C^{v} has the constant value C^{v}_{R}, Eq. 5.141 takes the simpler form

$$\frac{d\theta^{(\mathrm{H})}(v)}{dv} + \frac{\gamma(v)}{v}\,\theta^{(\mathrm{H})}(v) = \frac{\kappa(v)}{2C^{v}_{\mathrm{R}}}\,. \tag{5.142}$$

In this case, the solution can be reduced to quadrature by well-known means and the result is

$$\theta^{(\mathrm{H})}(v) = \chi(v)\left\{\theta^- + \frac{1}{2C^{v}_{\mathrm{R}}}\int_{v^-}^{v}\frac{\kappa(v')}{\chi(v')}\,dv'\right\}, \tag{5.143}$$

and $\eta^{(\mathrm{H})}(v)$ is given by

$$\eta^{(H)}(v) = \eta^{-} + \frac{1}{2}\int_{v^{-}}^{v} \frac{\kappa(v')}{\theta^{(H)}(v')} dv', \tag{5.144}$$

where the constants of integration have been chosen so that $\theta = \theta^{-}$ and $\eta = \eta^{-}$ when $v = v^{-}$. Recourse to numerical integration is usually required to evaluate the temperature and specific entropy Hugoniots. In the final portion of this section we pursue the analysis further under the most common assumptions on $\gamma(v)$ and $p^{(H)}(v)$.

Linear $U_S - \dot{x}$ Hugoniot. When the p–v Hugoniot is given by Eq. 3.12 the function $\kappa(v)$ of Eq. 5.137 becomes

$$\kappa(v) = -2\rho_R^3 S C_B^2 \frac{(v_R - v)^2}{[1 - \rho_R S(v_R - v)]^3}, \tag{5.145}$$

so

$$\theta^{(H)}(v) = \chi(v)\left\{\theta^{-} - \frac{\rho_R^3 S C_B^2}{C_R^v} \int_{v^{-}}^{v} \frac{1}{\chi(v')} \frac{(v_R - v')^2}{[1 - \rho_R S(v_R - v')]^3} dv'\right\}, \tag{5.146}$$

and, with these results, the temperature and entropy Hugoniots can easily be evaluated numerically. Example results of the determination of temperature and specific entropy on the Hugoniots of copper and aluminum alloy 2024 using these equations are given in Fig. 5.5.

Admissibility of Solutions to the Jump Equations. In the discussions of Chaps. 2 and 3, it was assumed that stable shocks corresponding to solutions of the jump equations actually exist. This is not true for all possible solutions of the jump conditions. Although this issue might usefully have been addressed earlier, it was necessary to defer the discussion until the equations for calculating the specific entropy of shock-compressed material were developed. The entropy-production inequality holds that a solution of the jump equations that results in a decrease in entropy is inadmissible. Examination of Eq. 5.136 (a sketch is helpful) shows that the specific entropy for states on the Hugoniot increases with increasing compression when the Hugoniot is concave upward and decreases with increasing compression when it is concave downward. Since a process that results in a decrease in entropy is inadmissible, we conclude that compression shocks cannot exist when the Hugoniot is concave downward. The same analysis of decompression shocks shows that they cannot exist for materials having a Hugoniot that is concave upward. As we shall see in Chap. 9, the cases in which shocks cannot exist are exactly those that admit smooth waves as solutions.

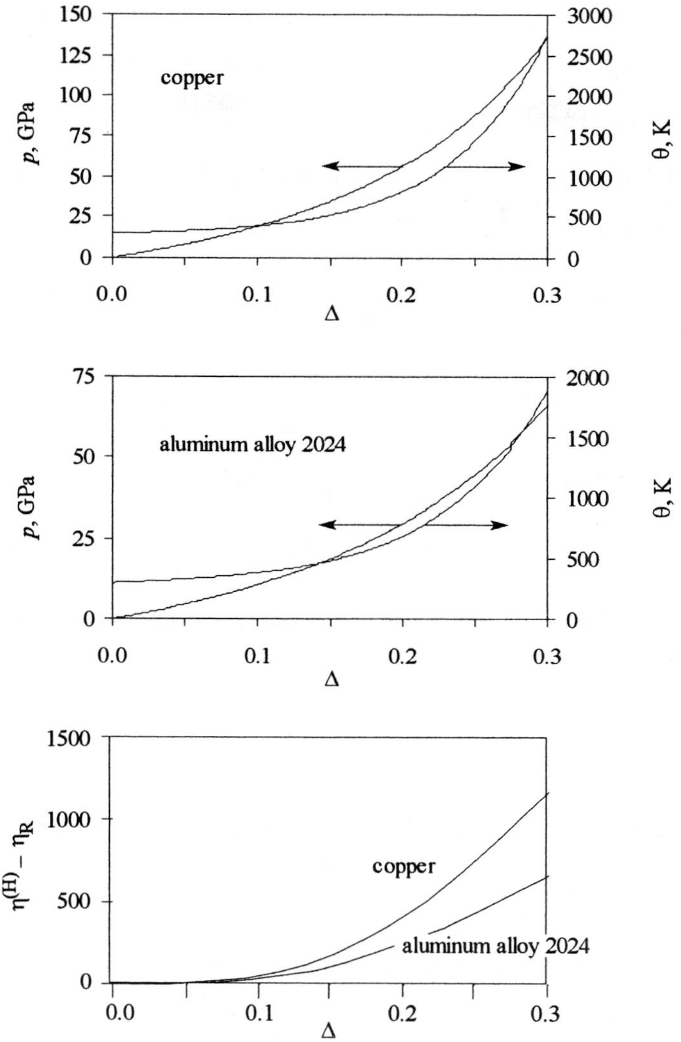

Figure 5.5. Temperature, pressure, and specific entropy Hugoniots as functions of compression for copper and aluminum alloy 2024 calculated from the Hugoniot data of Table 3.1.

Entropy Jump at a Weak Shock. Two stress–deformation curves that are important in analysis of longitudinal wave propagation are the Hugoniot, the locus of endstates of shock transitions, and the isentrope, the path followed by the state point as a smooth elastic wave passes a given material particle. The specific issue addressed in this section is the degree to which these curves differ and, in particular, the jump in entropy that occurs across a weak shock.

The Rankine–Hugoniot jump condition shows that the equation

$$H(v) = \varepsilon^{(\mathrm{H})}(v) - \varepsilon^- + \tfrac{1}{2}\left[p^{(\mathrm{H})}(v) + p^- \right](v - v^-) = 0 \tag{5.147}$$

must be satisfied along a Hugoniot curve centered on $\mathcal{S}^- = \{v^-, p^-, \varepsilon^-, \dot{x}^-\}$.

Differentiating the function $H(v)$ and substituting from Eq. 5.135, we obtain

$$\begin{aligned} 0 = \frac{dH}{dv} &= \frac{d\varepsilon^{(\mathrm{H})}}{dv} + \frac{1}{2}(v - v^-)\frac{dp^{(\mathrm{H})}}{dv} + \frac{1}{2}(p^{(\mathrm{H})} + p^-) \\ &= \theta\frac{d\eta^{(\mathrm{H})}}{dv} - \frac{1}{2}(p^{(\mathrm{H})} - p^-) + \frac{1}{2}(v - v^-)\frac{dp^{(\mathrm{H})}}{dv}. \end{aligned} \tag{5.148}$$

Evaluating this derivative at the center point $\mathcal{S}^-$, we find that

$$\left.\frac{d\eta^{(\mathrm{H})}}{dv}\right|_{\mathcal{S}^-} = 0, \tag{5.149}$$

i.e., the rate of change of entropy with respect to specific volume along the Hugoniot at the center point is zero.

Differentiating Eq. 5.148, we obtain

$$0 = \frac{d\theta^{(\mathrm{H})}}{dv}\frac{d\eta^{(\mathrm{H})}}{dv} + \theta^{(\mathrm{H})}\frac{d^2\eta^{(\mathrm{H})}}{dv^2} + \frac{1}{2}(v - v^-)\frac{d^2p^{(\mathrm{H})}}{dv^2}. \tag{5.150}$$

Evaluation of this result at the center point and using Eq. 5.149 yields

$$\left.\frac{d^2\eta^{(\mathrm{H})}}{dv^2}\right|_{\mathcal{S}^-} = 0. \tag{5.151}$$

Finally, let us take a third derivative. From Eq. 5.150 we obtain

$$\begin{aligned} 0 = &\frac{d^2\theta^{(\mathrm{H})}}{dv^2}\frac{d\eta^{(\mathrm{H})}}{dv} + 2\frac{d\theta^{(\mathrm{H})}}{dv}\frac{d^2\eta^{(\mathrm{H})}}{dv^2} \\ &+ \theta\frac{d^3\eta^{(\mathrm{H})}}{dv^3} + \frac{1}{2}\frac{d^2p^{(\mathrm{H})}}{dv^2} + \frac{1}{2}(v - v^-)\frac{d^3p^{(\mathrm{H})}}{dv^3}, \end{aligned}$$

which yields

$$\left.\frac{d^3\eta^{(\mathrm{H})}}{dv^3}\right|_{\mathcal{S}^-} = -\frac{1}{2\theta^-}\left.\frac{d^2p^{(\mathrm{H})}}{dv^2}\right|_{\mathcal{S}^-}. \tag{5.152}$$

The expanded form of the function $\eta^{(\mathrm{H})}(v)$ giving the value of the entropy along the Hugoniot is

$$\eta^{(\mathrm{H})}(v) = \eta^{(\mathrm{H})}(v^-) + \left.\frac{d\eta^{(\mathrm{H})}}{dv}\right|_{S^-} [\![v]\!] + \frac{1}{2}\left.\frac{d^2\eta^{(\mathrm{H})}}{dv^2}\right|_{S^-} [\![v]\!]^2 + \frac{1}{6}\left.\frac{d^3\eta^{(\mathrm{H})}}{dv^3}\right|_{S^-} [\![v]\!]^3 + \cdots,$$

but substitution of Eqs. 5.149, 5.151, and 5.152 into this expression shows that the jump in entropy across a shock is given by

$$[\![\eta]\!] = -\frac{1}{12\theta^-}\left.\frac{d^2 p^{(\mathrm{H})}(v)}{dv^2}\right|_{S^-} [\![v]\!]^3 + \cdots, \tag{5.153}$$

to within terms of fourth order in $[\![v]\!]$. This equation shows that the entropy jump across a weak shock is very small so that the shock transition is almost isentropic.

We also record a result, similar to the one above, that was derived [109, Eq. 1.89] starting with the enthalpy response function:

$$[\![\eta]\!] = \frac{1}{12\theta^-}\left.\frac{d^2 v^{(\mathrm{H})}(p)}{dp^2}\right|_{S^-} [\![p]\!]^3 + \cdots. \tag{5.154}$$

5.3.4 Relationships Among Thermodynamic Response Curves

In this section, we derive equations relating the various thermodynamic curves. The graphical depictions of these results that are presented are calculated using the Dulong–Petit specific heat, the equation $\gamma(v) = \gamma_{\mathrm{R}}\, v/v_{\mathrm{R}}$, and the linear $U_{\mathrm{S}} - \dot{x}$ Hugoniot of Eq. 3.10. In the foregoing equation for $\gamma(v)$, the specific volume v_{R} is the value at which $\gamma(v_{\mathrm{R}}) = \gamma_{\mathrm{R}}$, and is not to be confused with other reference values of the specific volume to be introduced. Results obtained on this basis are not the most refined available, but are easily obtained, illustrative, and acceptable for many purposes. Applications often require that a Hugoniot be centered on a specific reference state or that an isotherm or isentrope pass through a given state. The equations presented in this section have been developed to meet these requirements. A distinct reference state is adopted for each of the several curves considered in this section. This complicates the notation and appearance of the equations developed, but adds to their generality and is necessary for some applications.

5.3.4.1 Relationships Between Isotherms and Isentropes

Isotherms and isentropes are related to one another by the Mie–Grüneisen equation

$$p^{(\eta)}(v;\eta_{\mathrm{R}}^{(\eta)}) - p^{(\theta)}(v;\theta_{\mathrm{R}}^{(\theta)}) = \frac{\gamma(v)}{v}\left[\varepsilon^{(\eta)}(v;\eta_{\mathrm{R}}^{(\eta)}) - \varepsilon^{(\theta)}(v;\theta_{\mathrm{R}}^{(\theta)})\right]. \tag{5.155}$$

We also have Eqs. 5.122 and 5.132, which we write in the forms

$$\varepsilon^{(\theta)}(v;\theta_{\mathrm{R}}^{(\theta)}) = \varepsilon_{\mathrm{R}}^{(\theta)} + \theta_{\mathrm{R}}^{(\theta)}\int_{v_{\mathrm{R}}^{(\theta)}}^{v}\frac{\gamma(v')}{v'}C^{(v)}(v';\theta_{\mathrm{R}}^{(\theta)})\,dv' - \int_{v_{\mathrm{R}}^{(\theta)}}^{v}p^{(\theta)}(v';\theta_{\mathrm{R}}^{(\theta)})\,dv' \tag{5.156}$$

and

$$\varepsilon^{(\eta)}(v;\eta_{\mathrm{R}}^{(\eta)}) = \varepsilon_{\mathrm{R}}^{(\eta)} - \int_{v_{\mathrm{R}}^{(\eta)}}^{v}p^{(\eta)}(v';\eta_{\mathrm{R}}^{(\eta)})\,dv', \tag{5.157}$$

relating the specific internal energy to the pressure on each of these curves. To obtain an equation relating the isotherm and isentrope, we substitute Eqs. 5.156 and 5.157 into Eq. 5.155 and differentiate the result, giving

$$\begin{aligned}\frac{dp^{(\eta)}(v;\eta_{\mathrm{R}}^{(\eta)})}{dv} &+ \left[\frac{\gamma(v)}{v} - \frac{v}{\gamma(v)}\frac{d}{dv}\left(\frac{\gamma(v)}{v}\right)\right]p^{(\eta)}(v;\eta_{\mathrm{R}}^{(\eta)}) \\ &= \frac{dp^{(\theta)}(v;\theta_{\mathrm{R}}^{(\theta)})}{dv} + \left[\frac{\gamma(v)}{v} - \frac{v}{\gamma(v)}\frac{d}{dv}\left(\frac{\gamma(v)}{v}\right)\right]p^{(\theta)}(v;\theta_{\mathrm{R}}^{(\theta)}) \\ &\quad - \left(\frac{\gamma(v)}{v}\right)^2 C^{v}\theta_{\mathrm{R}}^{(\theta)}.\end{aligned} \tag{5.158}$$

We see that this is the familiar linear, first-order ordinary differential equation. If the pressure isotherm is known it is an equation for the pressure isentrope and if the pressure isentrope is known, it is an equation for the pressure isotherm. When the isotherm is sought, the specific heat is expressed as a function of specific volume and the constant temperature on the isotherm. When an isentrope is sought, the specific heat is expressed as a function of the constant specific entropy on the isentrope. When Eq. 5.158 is solved, the specific internal energy isotherm or isentrope is obtained from Eq. 5.156 or Eq. 5.157, as appropriate.

If we adopt the approximations that C^{v} and $\gamma(v)/v$ are constants, equation 5.158 takes a simpler form, but the isotherms and isentropes are more easily determined directly from the complete Mie–Grüneisen equation of state. When seeking equations for the isotherms in terms of the isentropes, we use Eqs. 5.94. When seeking equations for the isentropes in terms of the isotherms, we use Eqs. 5.99. In writing equations for these curves, we have allowed the isotherms and isentropes to have different reference states. For the pressure and specific internal energy isotherms, we obtain the equations

$$p^{(\theta)}(v;\theta_{\rm R}^{(\theta)}) = p^{(\eta)}(v;\eta_{\rm R}^{(\eta)}) + \chi_{\rm c}^{(\theta)}(v)\left[p_{\rm R}^{(\theta)} - p^{(\eta)}(v_{\rm R}^{(\theta)};\eta_{\rm R}^{(\eta)})\right] - \frac{\gamma_{\rm R}}{v_{\rm R}} C_{\rm R}^{v}\theta_{\rm R}^{(\theta)}\left[\chi_{\rm c}^{(\theta)}(v) - 1\right], \tag{5.159}$$

and

$$\varepsilon^{(\theta)}(v;\theta_{\rm R}^{(\theta)}) = \varepsilon_{\rm R}^{(\theta)} - \varepsilon_{\rm R}^{(\eta)} + \frac{v_{\rm R}}{\gamma_{\rm R}}\left[\chi_{\rm c}^{(\theta)}(v) - 1\right]\left[p_{\rm R}^{(\theta)} - p^{(\eta)}(v_{\rm R}^{(\theta)};\eta_{\rm R}^{(\eta)})\right] + \varepsilon^{(\eta)}(v;\eta_{\rm R}^{(\eta)}) - C_{\rm R}^{v}\theta_{\rm R}^{(\theta)}\left[\chi_{\rm c}^{(\theta)}(v) - 1\right]. \tag{5.160}$$

For the pressure and specific internal energy isentropes through the point $p = p_{\rm R}^{(\eta)}$, $v = v_{\rm R}^{(\eta)}$, we obtain the equations

$$p^{(\eta)}(v;\eta_{\rm R}^{(\eta)}) = p^{(\theta)}(v;\theta_{\rm R}^{(\theta)}) + \frac{\gamma_{\rm R}}{v_{\rm R}} C_{\rm R}^{v}\theta_{\rm R}^{(\theta)}\left[\chi_{\rm c}^{(\eta)}(v) - 1\right] + \chi_{\rm c}^{(\eta)}(v)\left[p_{\rm R}^{(\eta)} - p^{(\theta)}(v_{\rm R}^{(\eta)};\theta_{\rm R}^{(\theta)})\right] \tag{5.161}$$

and

$$\varepsilon^{(\eta)}(v;\eta_{\rm R}^{(\eta)}) = \varepsilon^{(\theta)}(v;\theta_{\rm R}^{(\theta)}) + C_{\rm R}^{(v)}\theta_{\rm R}^{(\theta)}\left[\chi_{\rm c}^{(\eta)}(v) - 1\right] + \varepsilon_{\rm R}^{(\eta)} - \varepsilon^{(\theta)}(v_{\rm R}^{(\eta)};\theta_{\rm R}^{(\theta)}) + \frac{v_{\rm R}}{\gamma_{\rm R}}\chi_{\rm c}^{(\eta)}(v)\left[p_{\rm R}^{(\eta)} - p^{(\theta)}(v_{\rm R}^{(\eta)};\theta_{\rm R}^{(\theta)})\right]. \tag{5.162}$$

5.3.4.2 Relationships Between Hugoniots and Isotherms

As we have mentioned, atomic-interaction calculations or extrapolation equations such as 5.116 can be used to obtain isotherms for various materials. When a Hugoniot is needed, it can be obtained from the isotherm. In principle, this can be done by calculation of the specific-volume difference as a function of pressure or the pressure difference as a function of specific volume (see Fig. 5.6). In practice, the better choice is calculation of the pressure offset because this can be done in terms of more convenient thermodynamic properties.

Let us assume that the isotherm $p = p^{(\theta)}(v, \theta_0)$ is known. This isotherm is related to the Hugoniot that we seek by the Mie–Grüneisen equation of state,

$$p^{(\rm H)}(v) = p^{(\theta)}(v, \theta_0) + \frac{\gamma(v)}{v}\left[\varepsilon^{(\rm H)}(v) - \varepsilon^{(\theta)}(v, \theta_0)\right], \tag{5.163}$$

and the pressure and internal energy on the Hugoniot are related by the Rankine–Hugoniot equation

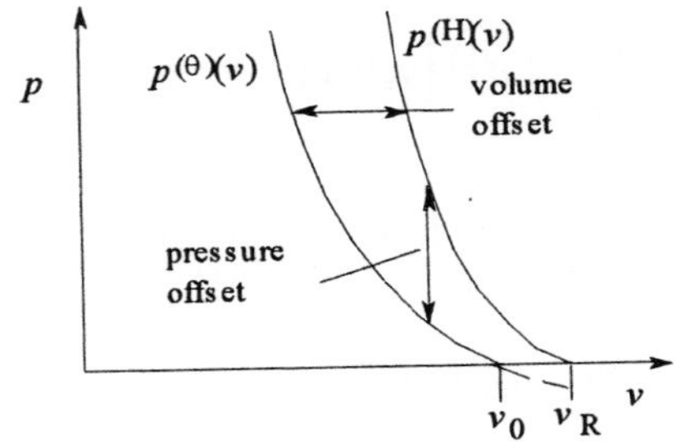

Figure 5.6. Illustration of the relationship between a p–v Hugoniot and an isotherm

$$\varepsilon^{(H)}(v) = \varepsilon_R + \frac{1}{2} p^{(H)}(v)\,(v_R - v)\,, \tag{5.164}$$

where the reference state has been chosen to be that for which $p = 0$ when $v = v_R$.

Substituting Eq. 5.123 and Eq. 5.164 into Eq. 5.163, and solving for $p^{(H)}(v)$ gives

$$p^{(H)}(v) = \frac{p^{(\theta)}(v,\theta_0) + \dfrac{\gamma(v)}{v}\left[\varepsilon_R - \varepsilon_0 + \displaystyle\int_{v_0}^{v} p^{(\theta)}(v',\theta_0)\,dv' - \theta_0 \int_{v_0}^{v} \frac{\gamma(v')}{v'} C^v\,dv'\right]}{1 - \dfrac{\gamma(v)}{2v}(v_0 - v)}\,, \tag{5.165}$$

where v_0, ε_0, and θ_0 are values on the isotherm at the pressure $p = 0$. The specific volume and specific internal energy offsets at $p = 0$ are given by

$$v_0 = v_R \exp\left[\beta(\theta_0 - \theta_R)\right] \quad \text{and} \quad \varepsilon_R - \varepsilon_0 = \int_{\theta_0}^{\theta_R} C^p\,d\theta\,, \tag{5.166}$$

where β is the coefficient of volumetric thermal expansion and C^p is the specific heat at the constant pressure $p = 0$.

In the case that $\gamma(v)$ is given by Eq. 5.65, Eq. 5.165 takes the simpler form

$$p^{(H)}(v) = \frac{p^{(\theta)}(v,\theta_0) + \dfrac{\gamma_R}{v_R}\left[\varepsilon_R - \varepsilon_0 + \displaystyle\int_{v_0}^{v} p^{(\theta)}(v',\theta_0)\,dv' - \theta_0 \frac{\gamma_R}{v_R} C_R^v\,(v - v_0)\right]}{1 - \dfrac{\gamma_R}{2v_R}(v_0 - v)}\,. \tag{5.167}$$

Examination of Fig. 5.6 shows that calculation of the Hugoniot for specific volumes greater than v_0 requires extrapolation of the isotherm into the tensile region or calculation of a pressure offset from the line $p = 0$. Because it is easier, we shall adopt the former procedure. Usually, errors in the Hugoniot at low pressure are unimportant because only the higher-pressure portion of the curve is used. An example calculation of a Hugoniot from an isotherm is shown in Fig. 5.7.

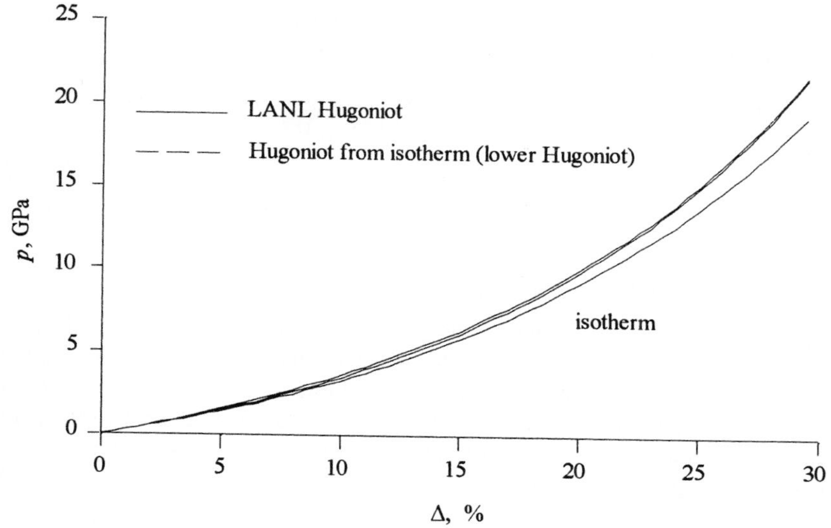

Figure 5.7. An isotherm and a derived Hugoniot for sodium chloride (NaCl). The isotherm is obtained from Eq. 5.116 using the modulus values $B^{\theta} = 23.97\,\text{GPa}$ and $B'^{\theta} = 5.05$ and the offset to the Hugoniot is calculated using $C^{v} = 854\,\text{J/(kg K)}$, $\rho_R = 2165\,\text{kg/m}^3$ $\theta_R = 293\,\text{K}$, and $\gamma_R = 1.60$. The Hugoniot is in sufficiently good agreement with a measured Hugoniot [46] that the difference is not easily seen on a graph of this scale.

Either Eq. 5.165 or the simpler equation 5.167 can be used to calculate an isotherm through the point at which a known Hugoniot is centered. Taking the simpler case as an example, we have, after differentiating,

$$\frac{dp^{(\theta)}(v;\theta_0)}{dv} + \frac{\gamma_R}{v_R} p^{(\theta)}(v;\theta_0) = \hat{\kappa}_c(v,\theta_0) + \left(\frac{\gamma_R}{v_R}\right)^2 C_R^v \theta_0, \qquad (5.168)$$

where

$$\hat{\kappa}_c(v) = \left[1 - \frac{\gamma_R}{2v_R}(v_R - v)\right]\frac{dp^{(H)}(v)}{dv} + \frac{\gamma_R}{2v_R} p^{(H)}(v). \qquad (5.169)$$

This is in the form of Eq. 5.142 and can be reduced to quadrature in the same way, thus yielding the isotherm

$$p^{(\theta)}(v,\theta_0)=\chi_c(v)\left[p_R+\int_{v_0}^{v}\frac{\hat{\kappa}_c(v')}{\chi_c(v')}dv'\right]-\frac{\gamma_R}{v_R}C_R^v\theta_0[1-\chi_c(v)] \tag{5.170}$$

that passes through the state $p=0$, $v=v_0$ and corresponds to the temperature $\theta=\theta_0$.

Once the pressure isotherm has been determined, Eq. 5.163 can be solved for the specific internal energy isotherm

$$\begin{aligned}\varepsilon^{(\theta)}(v;\theta_0)&=\frac{v_R}{\gamma_R}\left[p^{(\theta)}(v;\theta_0)-p^{(H)}(v)\right]+\varepsilon^{(H)}(v)\\&=\varepsilon_R+\frac{v_R}{\gamma_R}p^{(\theta)}(v;\theta_0)+\left[\frac{1}{2}(v_R-v)-\frac{v_R}{\gamma_R}\right]p^{(H)}(v),\end{aligned} \tag{5.171}$$

where the second of these equations is obtained using Eq. 5.123.

Cold Isotherm. The 0K isotherm (also called the cold isotherm or cold compression curve) plays an important role in determination of equations of state for condensed matter. Often the specific internal energy is obtained by adding a thermal energy term to the elastic energy at 0K, as represented by the cold isotherm. A number of quantum-mechanical and semi-classical methods have been used to calculate the cold isotherm. The thermal term can be obtained from the Debye theory or any of several more refined models. This subject area forms a major part of modern investigations of the equation of state of matter, but the analyses are based on methods of atomic physics that are beyond the scope of this book. In this section we restrict attention to the relationship between the cold isotherm and the Hugoniot.

These two curves are related by Eq. 5.165 which, when $\theta=0$, can be written

$$\left[1-\frac{\gamma(v)}{v}(v_0-v)\right]p^{(H)}(v)=p^{(K)}(v)+\frac{\gamma(v)}{v}\left[\varepsilon_R-\varepsilon_0+\int_{v_0}^{v}p^{(K)}(v')dv'\right], \tag{5.172}$$

where ε_R is the specific internal energy at $p=0$ on the Hugoniot and ε_0 is the specific internal energy at $p=0$ and $v=v_0$ on the cold isotherm. The equation $\gamma(v)/v=\gamma_R/v_R$ can be used, but the reference specific internal energy terms require attention because of the behavior of the specific heat at low temperatures. We can choose $\varepsilon_0=0$ and calculate ε_R relative to this value using Eq. 5.93 with C^v given by Eq. 5.62. The specific volume v_0 can be determined by evaluating Eq. 5.172 at $v=v_0$ and solving the resulting equation,

$$p^{(H)}(v_0) = \frac{\gamma_R}{v_R} \varepsilon_R \tag{5.173}$$

for v_0. With these results, the Hugoniot $p^{(H)}(v)$ is given by Eq. 5.172 when the cold isotherm is known. If the Hugoniot is known, the cold isotherm can be obtained by solving this equation for $p^{(K)}(v)$.

5.3.4.3 Relationship Between a Hugoniot and an Isentrope

An isentrope can be determined from a Hugoniot by calculation of the pressure offset of the two curves in much the same way as was done in the preceding analysis. A Hugoniot and an associated isentrope are shown schematically in Fig. 5.8.

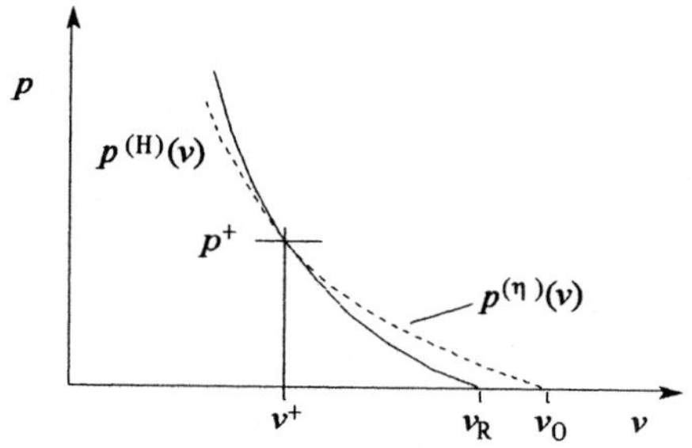

Figure 5.8. Illustration of the relationship of a Hugoniot curve and the isentrope through the point (p^+, v^+) on the Hugoniot centered on the state $p = 0$, $v = v_R$.

The pressure-offset calculation made using the Mie–Grüneisen and Rankine–Hugoniot equations and Eq. 5.132 relating pressure and specific internal energy on an isentrope gives the equation

$$\begin{aligned} p^{(\eta)}(v;\eta^+) = {} & \frac{\gamma(v)}{v}\left[-\int_{v^+}^{v} p^{(\eta)}(v';\eta^+)\,dv' + \frac{1}{2}p^+(v_R - v^+) \right] \\ & + \left[1 - \frac{1}{2}\frac{\gamma(v)}{v}(v_R - v) \right] p^{(H)}(v), \end{aligned} \tag{5.174}$$

relating the Hugoniot and the isentrope. To solve this integral equation for the isentrope in terms of a known Hugoniot we begin by differentiating it to convert it to a linear first-order ordinary differential equation for which the solution can be reduced to quadrature. When $\gamma/v = \gamma_R/v_R$ this latter calculation leads to the particularly simple result

$$\frac{dp^{(\eta)}(v;\eta^+)}{dv} + \frac{\gamma_R}{v_R} p^{(\eta)}(v;\eta^+) = \kappa_c(v), \tag{5.175}$$

where

$$\kappa_c(v) = \frac{1}{2}\frac{\gamma_R}{v_R} p^{(H)}(v) + \left[1 - \frac{1}{2}\frac{\gamma_R}{v_R}(v_R - v)\right]\frac{dp^{(H)}(v)}{dv}.$$

The solution of Eq. 5.175 is

$$p^{(\eta)}(v;\eta^+) = \chi_c(v)\left[p^+ + \int_{v^+}^{v} \frac{\kappa_c(v')}{\chi_c(v')}dv'\right], \tag{5.176}$$

where

$$\chi_c(v) = \exp\left[\frac{\gamma_R}{v_R}(v^+ - v)\right]. \tag{5.177}$$

Equation 5.176 is easily integrated by numerical means, and an example result is shown in Fig. 5.9.

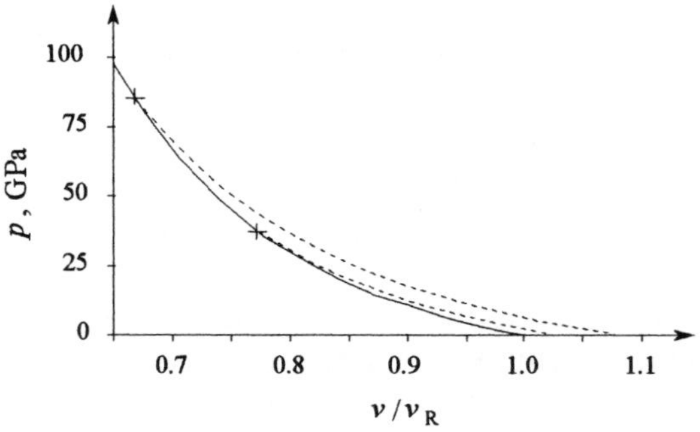

Figure 5.9. Pressure–specific-volume Hugoniot and decompression isentropes for aluminum. The higher-pressure case is for release from $v = 0.67\,v_R$ (85 GPa) and the lower-pressure case is for release from $v = 0.77\,v_R$ (38 GPa). It is easy to see that the decompression isentrope differs little from the Hugoniot at low pressures, thus justifying the approximation in which the former is replaced with the latter when calculations are made in this range.

The temperature at points on the decompression isentrope is given by Eq. 5.130, where v^* characterizes the Hugoniot state at which decompression begins and θ^* is the temperature of the material at this Hugoniot state. The result of making this calculation for the isentropes shown in Fig. 5.9 is given in Fig 5.10.

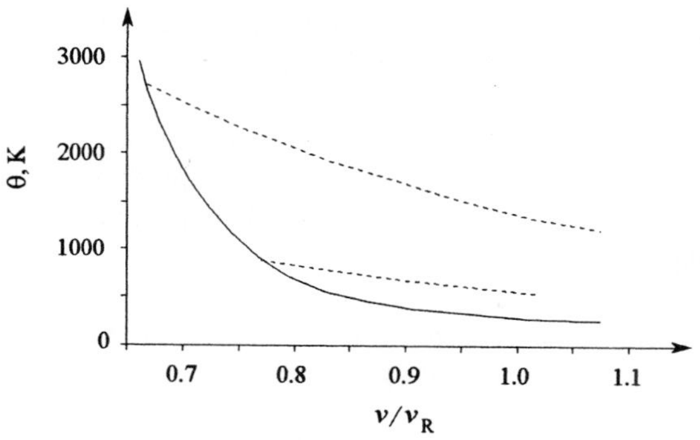

Figure 5.10. Temperature–specific volume Hugoniot (solid line) and decompression isentropes (broken lines) for aluminum. The higher-pressure case is for release from $v = 0.67\, v_R$ (2700 K) and the lower-pressure case is for release from $v = 0.77\, v_R$ (893 K).

5.4 Exercises

5.4.1. Derive Eq. 5.30.

5.4.2. Derive Eq. 5.31. Hint: Consider the function $p = p(v, \theta(v, \eta))$.

5.4.3. Derive Eq. 5.32. Hint: Consider the function $\eta = \eta(\theta, v(p, \theta))$.

5.4.4. Derive Eq. 5.33.

5.4.5. Work out a specific form for the p–$\dot{x}$ Hugoniot for an ideal gas.

5.4.6. Work out a specific form for the p–v second-shock Hugoniot for an ideal gas.

5.4.7. Can an ideal gas be compressed to arbitrarily large density by a shock? If this gas is compressed by two shocks passing sequentially through it (a two-step compression process), and by a single shock resulting in the same final pressure, How do the final densities differ?

5.4.8. Calculate $B^{(\theta)}$, $B^{(\eta)}$, $C^{(p)}$, $C^{(v)}$, β, and γ for an ideal gas.

5.4.9. Using Eq. 5.136, develop, in detail, the argument for entropy change at a shock. In particular, show that the entropy density

i. increases upon passage of a decompression shock when the $p-v$ Hugoniot is concave upward and

ii. decreases upon passage of a compression shock when the $p-v$ Hugoniot is concave downward.

5.4.10. Show that Eq. 5.143 is a solution of the linear first-order ordinary differential equation 5.142. Use a personal computer and a spreadsheet program to calculate $\theta^{(\mathrm{H})}(v)$ from Eq. 5.146 when $v^{-} = v_{\mathrm{R}}$. It is sufficient to use the trapeziodal rule for the integration.

5.4.11. Derive the equations for transforming a known isentrope of a Mie–Grüneisen material to a different specific entropy.

5.4.12. Derive the equation for a second-shock Hugoniot of a Mie–Grüneisen material having a known principal Hugoniot.

CHAPTER 6

Material Response III: Elastic Solids

The stress response of finitely deformed elastic solids has been a subject of investigation for more than three centuries. Its importance to shock physics derives in part from the need to very accurately describe the small nonlinearities observed when certain strong solids are compressed by a few per cent and in part by the need to describe the elastic contribution to finite elastic–plastic deformations. In uniaxial deformations neither the shear strain nor the compression dominates the process but states of large compression combined with small shear do arise in more general elastic deformations and can also arise in connection with the elastic part of the deformation in problems of uniaxial elastic–plastic strain to be discussed in Chaps. 7 and 10. This suggests the need for a theory in which large compression is taken into account but in which simplifications made possible because the shear strain is small have been introduced.

6.1 Objective Stress Relation

In Sect. 4.3.1 we described an elastic material in terms of a specific internal energy equation of state $\varepsilon = \bar{\varepsilon}(\mathbf{F}, \eta)$. For this to be an acceptable constitutive equation, ε must transform as a scalar under the change of the spatial frame represented by Eq. 4.1. It must also capture observed symmetries of response by the group of changes of reference frame to which it is invariant. These requirements restrict the way in which ε depends on **F**. Both the specific internal energy ε and the specific entropy η transform as scalars under the change of spatial frame, i.e. $\varepsilon^* = \varepsilon$ and $\eta^* = \eta$. We have seen in Eq. 4.2 that **F** transforms according to the law $\mathbf{F}^* = \mathbf{QF}$, so satisfaction of the equation $\varepsilon^* = \varepsilon$ requires that the response function satisfy the equation

$$\bar{\varepsilon}(\mathbf{F}, \eta) = \bar{\varepsilon}(\mathbf{QF}, \eta) .$$

This means that **F** must appear in $\bar{\varepsilon}$ in an invariant combination. The simplest possibility is in terms of $\mathbf{C} = \mathbf{F}^{\mathrm{T}}\mathbf{F}$, but the Lagrangian strain tensor $\mathbf{E} = (\mathbf{C} - \mathbf{I})/2$ is equally acceptable, and will prove more convenient for application of the theory. Accordingly, we take

$$\varepsilon = \hat{\varepsilon}(\mathbf{E}, \eta) . \tag{6.1}$$

Substitution of this result into Eq. 4.31 gives the temperature-response function

$$\hat{\theta}(\mathbf{E}, \eta) = \frac{\partial \hat{\varepsilon}(\mathbf{E}, \eta)}{\partial \eta}, \tag{6.2}$$

which is properly invariant. Similarly, we find the objective form

$$T_{IJ} = \rho_R \frac{\partial \hat{\varepsilon}(\mathbf{E}, \eta)}{\partial E_{IJ}} \tag{6.3}$$

for the stress response function giving the second Piola–Kirchhoff tensor, $\mathbf{T}$. When Eq. 6.3 is written in terms of the Cauchy stress we obtain the result

$$\hat{t}_{ij}(\mathbf{E}, \eta) = \rho F_{iJ} F_{jI} \frac{\partial \hat{\varepsilon}(\mathbf{E}, \eta)}{\partial E_{IJ}}. \tag{6.4}$$

6.2 Third-order Stress and Temperature Equations of State

For practical applications it is often appropriate to approximate the internal energy response function by an expansion in powers of $\mathbf{E}$ and $\eta - \eta_R$, i.e., about the unstressed reference state $\mathbf{T} = \mathbf{0}$, $\mathbf{E} = \mathbf{0}$, and $\eta = \eta_R$. The expansion is most useful for description of strong solids such as quartz and aluminum oxide that can be elastically compressed by a few per cent, and for which a very accurate description of the slightly nonlinear response is needed. The expansions of $\hat{\varepsilon}(\mathbf{E}, \eta)$ are truncated at the cubic, or occasionally the quartic, term. This results in polynomial expressions that can very accurately represent small nonlinearities, but do not accurately approximate rational functions such as the pressure Hugoniot associated with the linear $U_S - \dot{x}$ response that is usually the best representation of the material behavior at large compressions.

When the expansion is carried out to include third-order terms we have

$$\begin{aligned}
\hat{\varepsilon}(\mathbf{E}, \eta) &= \varepsilon_R + \theta_R(\eta - \eta_R) \\
&\quad + \frac{1}{2\rho_R} C^{\eta}_{IJKL} E_{IJ} E_{KL} - \theta_R(\eta - \eta_R)\gamma_{IJ} E_{IJ} + \frac{\theta_R}{2C^{\mathbf{E}}_R}(\eta - \eta_R)^2 \\
&\quad + \frac{1}{6\rho_R} C^{\eta}_{IJKLMN} E_{IJ} E_{KL} E_{MN} + \frac{1}{2\rho_R} \left.\frac{\partial C^{\eta}_{IJKL}}{\partial \eta}\right|_{0,\eta_R} (\eta - \eta_R) E_{IJ} E_{KL} \\
&\quad - \frac{1}{2}\frac{\theta_R}{C^{\mathbf{E}}_R}\gamma_{IJ}(\eta - \eta_R)^2 E_{IJ} + \frac{1}{6}\frac{\theta_R}{(C^{\mathbf{E}}_R)^2}\left(1 - \left.\frac{dC^{\mathbf{E}}(\eta)}{d\eta}\right|_{\eta_R}\right)(\eta - \eta_R)^3.
\end{aligned} \tag{6.5}$$

The coefficients in this equation are defined as follows:

$$\varepsilon_R = \hat{\varepsilon}(\mathbf{0}, \eta_R), \quad \theta_R = \left.\frac{\partial \hat{\varepsilon}}{\partial \eta}\right|_{0,\eta_R}, \quad \frac{\theta_R}{C^{\mathbf{E}}_R} = \left.\frac{\partial^2 \hat{\varepsilon}(\mathbf{E}, \eta)}{\partial \eta^2}\right|_{0,\eta_R} = \left.\frac{\partial \hat{\theta}(\mathbf{E}, \eta)}{\partial \eta}\right|_{0,\eta_R}, \tag{6.6a}$$

$$\gamma_{IJ} = -\frac{1}{\theta_R}\frac{\partial^2 \hat{\varepsilon}(\mathbf{E},\eta)}{\partial\eta\,\partial E_{IJ}}\bigg|_{\mathbf{0},\eta_R} = -\frac{1}{\theta_R}\frac{\partial\hat{\theta}(\mathbf{E},\eta)}{\partial E_{IJ}}\bigg|_{\mathbf{0},\eta_R} = -\frac{1}{\rho_R\,\theta_R}\frac{\partial T_{IJ}}{\partial\eta}\bigg|_{\mathbf{E}},$$

$$C^{\eta}_{IJKL} = \rho_R \frac{\partial^2 \hat{\varepsilon}(\mathbf{E},\eta)}{\partial E_{IJ}\,\partial E_{KL}}\bigg|_{\mathbf{0},\eta_R}, \quad C^{\eta}_{IJKLMN} = \rho_R \frac{\partial^3 \hat{\varepsilon}(\mathbf{E},\eta)}{\partial E_{IJ}\,\partial E_{KL}\,\partial E_{MN}}\bigg|_{\mathbf{0},\eta_R}. \tag{6.6b}$$

In writing these equations we have assumed that $C^{\mathbf{E}}$, the specific heat at constant strain, is a function of η alone and γ, the Grüneisen coefficient tensor, is a function of $\mathbf{E}$ alone. These assumptions are made for consistency with the Mie–Grüneisen theory discussed in Chap. 5. The coefficients $\mathbf{C}^{\eta}$ are called *isentropic elastic stiffness moduli* of second and third order, respectively, according to the definitions of Brugger [15]. One can, of course, extend the expansion to higher order.

When this equation of state is substituted into Eq. 6.3, we obtain the stress equation of state

$$\begin{aligned} T_{IJ} = {} & \left[C^{\eta}_{IJKL} + \frac{\partial C^{\eta}_{IJKL}}{\partial\eta}\bigg|_{\mathbf{0},\eta_R} (\eta-\eta_R) \right] E_{KL} + \frac{1}{2} C^{\eta}_{IJKLMN}\, E_{KL}\, E_{MN} \\ & - \rho_R\,\theta_R\,(\eta-\eta_R) \left[1 - \frac{1}{2}\frac{\eta-\eta_R}{C^{\mathbf{E}}_R} \right] \gamma_{IJ}\,. \end{aligned} \tag{6.7}$$

Substitution of Eq. 6.5 into Eq. 6.2 yields the temperature equation of state

$$\begin{aligned} \theta(\mathbf{E},\eta) = \theta_R \Bigg[& 1 - \gamma_{IJ}\,E_{IJ} + \frac{\eta-\eta_R}{C^{\mathbf{E}}_R} + \frac{1}{2\rho_R\,\theta_R}\frac{\partial C^{\eta}_{IJKL}}{\partial\eta}\bigg|_{\mathbf{0},\eta_R} E_{IJ}\,E_{KL} \\ & - \frac{1}{C^{\mathbf{E}}_R}\gamma_{IJ}\,(\eta-\eta_R)\,E_{IJ} + \frac{1}{2}\left(1 - \frac{dC^{\mathrm{E}}(\eta)}{d\eta}\bigg|_{\mathbf{0},\eta_R} \right) \frac{(\eta-\eta_R)^2}{(C^{\mathbf{E}}_R)^2} \Bigg]. \end{aligned} \tag{6.8}$$

When the similar expansion is carried out in terms of the Helmholtz free energy function, one obtains isothermal elastic stiffness moduli. Expansions of the enthalpy or Gibbs functions give the strain in terms of stress and either entropy or temperature, with the associated coefficients being called isentropic and isothermal *elastic compliance moduli*, respectively.

In Chap. 5 we showed that the entropy jump at a shock propagating in an inviscid fluid was proportional to the cube of the jump of the pressure or specific volume. This result is also obtained for a thermoelastic solid, which means that, when the specific entropy is held constant, Eq. 6.7 can be interpreted as a Hugoniot as well as an isentrope.

Because of the symmetry of **T** and **E**, the stiffness tensor C^{η}_{IJKL} is invariant to interchange of indices of the first pair, the second pair, or both and, because E_{IJ} and E_{KL} can be interchanged, the first and second pairs of indices can be interchanged. This motivates introduction of the Voigt notation in which the stresses are renamed in accordance with the scheme $T_{11} \to T_1$, $T_{22} \to T_2$, $T_{33} \to T_3$, $T_{23} \to T_4$, $T_{13} \to T_5$, and $T_{12} \to T_6$, and the strains are renamed in accordance with the scheme $E_{11} \to E_1$, $E_{22} \to E_2$, $E_{33} \to E_3$, $2E_{23} \to E_4$, $2E_{13} \to E_5$, and $2E_{12} \to E_6$. It is then possible to express the stiffness tensor in the simpler form $C^{\eta}_{\alpha\beta}$, where α and β range over the values 1, 2, …, 6 according to the scheme $11 \to 1$, $22 \to 2$, $33 \to 3$, $23 \to 4$, $13 \to 5$, and $12 \to 6$, so Eq. 6.7 can be written

$$T_{\alpha} = \left[C^{\eta}_{\alpha\beta} + \left.\frac{\partial C^{\eta}_{\alpha\beta}}{\partial \eta}\right|_{0,\eta_R} (\eta - \eta_R) \right] E_{\beta} + \frac{1}{2} C^{\eta}_{\alpha\beta\gamma} E_{\beta} E_{\gamma}$$

$$- \rho_R \, \theta_R \, (\eta - \eta_R) \left[1 - \frac{\eta - \eta_R}{2 C^{\mathbf{E}}_{\mathrm{R}}} \right] \gamma_{\alpha} . \tag{6.9}$$

The Voigt notation and Eq. 6.9 have been introduced because many higher-order elastic coefficients have been measured and the results are usually reported in this notation, which is further explained in reference [98, p.124]. It is important to realize that the various quantities are not tensors when in Voigt notation.

In many practical applications, including those of shock physics, it is often appropriate to approximate the energy equation of state by an expansion such as that of Eq. 6.7. In this case, invariance of the response to certain transformations of the reference coordinates is manifest in the structure of the tensor-valued coefficients $\mathbf{C}^{\eta}$. These tensors have been developed to various orders for all crystal classes and tables of the results are given in [14,15,98].

6.3 Stress and Temperature Relations for Isotropic Materials

In dealing with practical problems of shock physics, we are often interested in isotropic materials. To develop specific results for this case, let us consider Eq. 6.7 using the representation for ε given by Eq. 4.20. We have

$$\frac{1}{\rho_R} T_{IJ} = \frac{\partial \varepsilon (I_{\mathbf{E}}, II_{\mathbf{E}}, III_{\mathbf{E}}, \eta)}{\partial E_{IJ}} = \frac{\partial \varepsilon}{\partial I_{\mathbf{E}}} \frac{\partial I_{\mathbf{E}}}{\partial E_{IJ}} + \frac{\partial \varepsilon}{\partial II_{\mathbf{E}}} \frac{\partial II_{\mathbf{E}}}{\partial E_{IJ}} + \frac{\partial \varepsilon}{\partial III_{\mathbf{E}}} \frac{\partial III_{\mathbf{E}}}{\partial E_{IJ}} . \tag{6.10}$$

Differentiating Eqs. 4.19 gives

$$\frac{\partial I_{\mathbf{E}}}{\partial E_{IJ}} = \delta_{IJ}, \quad \frac{\partial II_{\mathbf{E}}}{\partial E_{IJ}} = I_{\mathbf{E}} \delta_{IJ} - E_{IJ}, \quad \frac{\partial III_{\mathbf{E}}}{\partial E_{IJ}} = E_{IK} E_{KJ} - I_{\mathbf{E}} E_{IJ} + II_{\mathbf{E}} \delta_{IJ} , \tag{6.11}$$

which, when substituted into the preceding result leads to the exact representation of $\mathbf{T}$,

$$\frac{1}{\rho_R} T_{IJ} = \left(\frac{\partial \varepsilon}{\partial I_{\mathbf{E}}} + I_{\mathbf{E}} \frac{\partial \varepsilon}{\partial II_{\mathbf{E}}} + II_{\mathbf{E}} \frac{\partial \varepsilon}{\partial III_{\mathbf{E}}} \right) \delta_{IJ}$$
$$- \left(\frac{\partial \varepsilon}{\partial II_{\mathbf{E}}} + I_{\mathbf{E}} \frac{\partial \varepsilon}{\partial III_{\mathbf{E}}} \right) E_{IJ} + \frac{\partial \varepsilon}{\partial III_{\mathbf{E}}} E_{IK} E_{KJ} , \qquad (6.12)$$

for the case of an isotropic elastic solid described by an arbitrary specific internal energy equation of state.

Expansion of this result leads to an expression of the form of Eq. 6.7 that is appropriate to isotropic materials. An expression for ε that is accurate to cubic terms in $\mathbf{E}$ and $\eta - \eta_R$ is

$$\begin{aligned}
\rho_R \varepsilon &= \rho_R \varepsilon_R + \rho_R \theta_R (\eta - \eta_R) \\
&+ \tfrac{1}{2}(\lambda_R + 2\mu_R) I_{\mathbf{E}}^2 - 2\mu_R II_{\mathbf{E}} - \rho_R \theta_R \gamma_R (\eta - \eta_R) I_{\mathbf{E}} \\
&+ \frac{\rho_R \theta_R}{2C_R^{\mathbf{E}}} (\eta - \eta_R)^2 + \tfrac{1}{6}(\nu_1 + 6\nu_2 + 8\nu_3) I_{\mathbf{E}}^3 - 2(\nu_2 + 2\nu_3) I_{\mathbf{E}} II_{\mathbf{E}} \\
&+ 4\nu_3 III_{\mathbf{E}} + \frac{1}{2} \left[\left.\frac{\partial \lambda(\mathbf{E}, \eta)}{\partial \eta}\right|_{0,\eta_R} I_{\mathbf{E}}^2 + 2 \left.\frac{\partial \mu(\mathbf{E}, \eta)}{\partial \eta}\right|_{0,\eta_R} (I_{\mathbf{E}}^2 - 2II_{\mathbf{E}}) \right] (\eta - \eta_R) \\
&+ \frac{1}{6} \frac{\rho_R \theta_R}{(C_R^{\mathbf{E}})^2} \left[1 - \left.\frac{dC^{\mathbf{E}}(\eta)}{d\eta}\right|_{0,\eta_R} \right] (\eta - \eta_R)^3 - \frac{\rho_R \theta_R}{2C_R^{\mathbf{E}}} \gamma_R (\eta - \eta_R)^2 I_{\mathbf{E}} .
\end{aligned} \qquad (6.13)$$

The stress equation of state that follows from Eq. 6.13 is

$$\begin{aligned}
T_{IJ} &= \lambda_R I_{\mathbf{E}} \delta_{IJ} + 2\mu_R E_{IJ} - \rho_R \theta_R \gamma_R (\eta - \eta_R) \delta_{IJ} \\
&+ \tfrac{1}{2}(\nu_1 + 6\nu_2 + 8\nu_3) I_{\mathbf{E}}^2 \delta_{IJ} - 2(\nu_2 + 2\nu_3) II_{\mathbf{E}} \delta_{IJ} \\
&- 2(\nu_2 + 2\nu_3) I_{\mathbf{E}} (I_{\mathbf{E}} \delta_{IJ} - E_{IJ}) + 4\nu_3 (E_{IK} E_{KJ} - I_{\mathbf{E}} E_{IJ} + II_{\mathbf{E}} \delta_{IJ}) \qquad (6.14) \\
&+ \left[\left.\frac{\partial \lambda(\mathbf{E}, \eta)}{\partial \eta}\right|_{0,\eta_R} + \left.\frac{\partial \mu(\mathbf{E}, \eta)}{\partial \eta}\right|_{0,\eta_R} \right] I_{\mathbf{E}} \delta_{IJ} (\eta - \eta_R) \\
&+ \left.\frac{\partial \mu(\mathbf{E}, \eta)}{\partial \eta}\right|_{0,\eta_R} E_{IJ} (\eta - \eta_R) - \frac{\rho_R \theta_R}{2C_R^{\mathbf{E}}} \gamma_R (\eta - \eta_R)^2 \delta_{IJ} ,
\end{aligned}$$

and the temperature equation of state is

$$\theta(\mathbf{E},\eta)=\theta_R\left[1-\gamma_R\,I_{\mathbf{E}}+\frac{\eta-\eta_R}{C_R^{\mathbf{E}}}-\gamma_R\frac{\eta-\eta_R}{C_R^{\mathbf{E}}}I_{\mathbf{E}}\right.$$
$$+\frac{1}{2\rho_R\,\theta_R}\left(\left.\frac{\partial\lambda(\mathbf{E},\eta)}{\partial\eta}\right|_{\mathbf{0},\eta_R}I_{\mathbf{E}}^2+2\left.\frac{\partial\mu(\mathbf{E},\eta)}{\partial\eta}\right|_{\mathbf{0},\eta_R}(I_{\mathbf{E}}^2-II_{\mathbf{E}})\right) \tag{6.15}$$
$$\left.+\frac{1}{2}\left(1-\left.\frac{dC^{\mathrm{E}}(\eta)}{d\eta}\right|_{\mathrm{R}}\right)\frac{(\eta-\eta_R)^2}{(C_R^{\mathbf{E}})^2}\right].$$

The coefficients in these equations have been written in a form that identifies those of the quadratic energy terms with the usual Lamé coefficients λ_R and μ_R and the coefficients ν_1, ν_2 and ν_3 with the third-order Lamé coefficients defined by Toupin and Bernstein [101]. It is important to note that the lowest-order terms of these equations do not correspond exactly to the expressions used in the linear theory of thermoelasticity because **E** is not linear in the displacement gradients. To recover the linear theory one must replace **E** by its linear approximation

$$\widetilde{E}_{ij}=\frac{1}{2}\left(\frac{\partial u_i}{\partial x_j}+\frac{\partial u_j}{\partial x_i}\right), \tag{6.16}$$

where we have identified the material and spatial coordinates since the distinction disappears in the approximation of small deformation, and where we also have $\mathbf{t}=\mathbf{T}$ in this approximation, leaving us with the resulting linear expression

$$t_{ij}=\lambda_R\,\widetilde{E}_{kk}\,\delta_{ij}+2\mu_R\,\widetilde{E}_{ij}-\rho_R\,\theta_R\,\gamma_R(\eta-\eta_R)\delta_{ij}$$
$$\theta=\theta_R\left[1-\gamma_R\,I_E+\frac{\eta-\eta_R}{C_R^{\mathbf{E}}}\right]. \tag{6.17}$$

Adopting the definition of the deviatoric strain,

$$\widetilde{E}'_{ij}=\widetilde{E}_{ij}-\tfrac{1}{3}\widetilde{E}_{kk}\,\delta_{ij}\,, \tag{6.18}$$

we can write Eq. 6.17_1 in the form

$$t_{ij}=\left[\left(\lambda_R+\tfrac{2}{3}\mu_R\right)\widetilde{E}_{kk}-\rho_R\,\theta_R\,\gamma_R(\eta-\eta_R)\right]\delta_{ij}+2\mu_R\,\widetilde{E}'_{ij}\,\delta_{ij}\,, \tag{6.19}$$

in which the first term gives the pressure (the coefficient $\lambda_R+(2\mu_R/3)$ is the isentropic bulk modulus) and the second gives the shear stress.

Uniaxial Motions. Uniaxial motions have been discussed in Sect. 2.2. In some cases the principal stress components associated with these motions are along and transverse to the axis of motion. This occurs in isotropic materials and in some materials of lower symmetry if the axis of motion is suitably chosen. It does not occur for all orientations or all materials, but Brugger [16] has determined the cases where pure longitudinal motion is possible. When such motions are possible, Eq. 6.9 yields the isentropic stress relation

$$T_{11} = C_{11}^{\eta} E_{11} + \tfrac{1}{2} C_{111}^{\eta} (E_{11})^2 - \rho_R \theta_R \gamma_R (\eta - \eta_R)$$

$$+ \left.\frac{\partial C_{11}^{\eta}}{\partial \eta}\right|_R E_{11} (\eta - \eta_R) + \frac{\rho_R \theta_R}{2 C_R^E} \gamma_R (\eta - \eta_R)^2, \tag{6.20}$$

where the elastic moduli are in Voigt notation. Substituting this result into Eq. 2.64, and using the kinematical relations for uniaxial motions, gives

$$t_{11} = T_{11} F_{11} = C_{11}^{\eta} E_{11} F_{11} + \tfrac{1}{2} C_{111}^{\eta} (E_{11})^2 F_{11} - \rho_R \theta_R \gamma_R F_{11} (\eta - \eta_R)$$

$$+ \left.\frac{\partial C_{11}}{\partial \eta}\right|_{0,\eta R} E_{11} F_{11} (\eta - \eta_R) + \frac{\rho_R \theta_R}{2 C_R^E} \gamma_R F_{11} (\eta - \eta_R)^2$$

$$= C_{11}^{\eta} U_X - \rho_R \theta_R \gamma_R (\eta - \eta_R) + \tfrac{1}{2} (3 C_{11}^{\eta} + C_{111}^{\eta}) (U_X)^2 \tag{6.21}$$

$$- \rho_R \theta_R \gamma_R U_X (\eta - \eta_R) + \left.\frac{\partial C_{11}^{\eta}}{\partial \eta}\right|_R U_X (\eta - \eta_R)$$

$$+ \frac{\rho_R \theta_R}{2 C_R^E} \gamma_R (\eta - \eta_R)^2 .$$

It is important to recognize that, even though the motion is entirely along one axis, there are nonvanishing transverse stresses. These stresses play no role in the analysis of longitudinal elastic waves, but are important for analyzing the yield phenomena that are the focus of analyses of elastic–plastic waves.

6.3.1 Separation of Dilatation and Distortion

In the preceding chapter we pointed out that a solid body can support a large pressure but fails by some mechanism of inelastic flow when subjected to a relatively small shear stress. This suggests that we separate the deformation into dilatational and distortional parts and simplify the distortional part by introduction of the approximations that can be made when it is small. Preparatory to this, we perform the separation, but without incorporating the simplifications that result when the distortion is small. We proceed by analyzing the deformation as though it resulted from a dilatation followed by a distortion. This process is illustrated in Fig. 6.1.

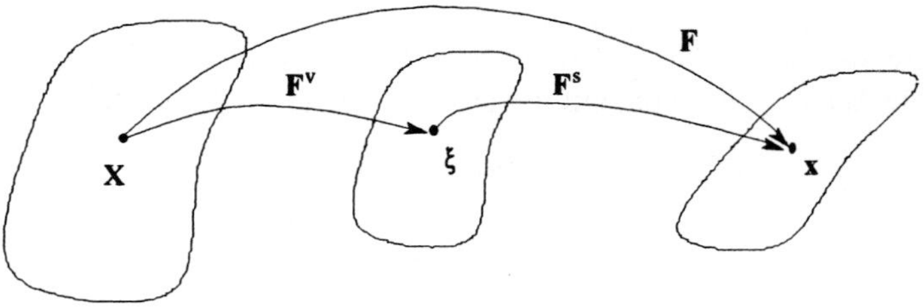

Figure 6.1. Schematic illustration of a deformation resulting when a distortion, $\mathbf{F}^{s}$, is superimposed on a dilatation, $\mathbf{F}^{v}$.

The isotropic dilatation from the reference state to the intermediate state is described by the equation

$$\xi_{\alpha} = \hat{\xi}_{\alpha}(\mathbf{X}), \tag{6.22}$$

and the isochoric distortion from the intermediate state to the current state is described by the equation

$$x_i = \hat{x}_i(\xi), \tag{6.23}$$

so the total deformation is given by

$$x_i = \hat{x}_i(\xi(\mathbf{X})). \tag{6.24}$$

Calculation of the deformation gradient using the chain rule gives

$$F_{iI} = F^{s}_{i\alpha} F^{v}_{\alpha I}, \tag{6.25}$$

where

$$F^{s}_{i\alpha} = \frac{\partial \hat{x}_i(\xi)}{\partial \xi_{\alpha}} \quad \text{and} \quad F^{v}_{\alpha I} = \frac{\partial \hat{\xi}_{\alpha}(\mathbf{X})}{\partial X_I}. \tag{6.26}$$

Since $\hat{\xi}(\mathbf{X})$ is to represent an isotropic dilatation, we have

$$\xi_{\alpha} = (v/v_{\mathrm{R}})^{1/3} \delta_{\alpha I} X_I, \tag{6.27}$$

and, therefore,

$$F^{v}_{\alpha I} = (v/v_{\mathrm{R}})^{1/3} \delta_{\alpha I}. \tag{6.28}$$

The intent of the decomposition of Eq. 6.25 is to form the parts so that $\mathbf{F}^{v}$ captures all of the dilatation and that $\mathbf{F}^{s}$ is a pure distortion. Since, by Eq. 2.23, the determinant of $\mathbf{F}$ is the ratio of current to reference specific volumes, we have

$$v/v_R = \det(\mathbf{F}^s\mathbf{F}^v) = \det\mathbf{F}^s \det\mathbf{F}^v = (v/v_R)\det\mathbf{F}^s . \qquad (6.29)$$

We see that $\det\mathbf{F}^s = 1$ and, therefore, $\mathbf{F}^s$ represents a pure distortion.

The components of the Lagrangian strain tensor $\mathbf{E}$ can be written

$$E_{IJ} \equiv \tfrac{1}{2}(F_{iI}F_{iJ} - \delta_{IJ}) = \tfrac{1}{2}[(v/v_R)^{2/3}F^s_{i\alpha}F^s_{i\beta}\delta_{\alpha I}\delta_{\beta J} - \delta_{IJ}] . \qquad (6.30)$$

If we define strain tensors associated with $\mathbf{F}^v$ and $\mathbf{F}^s$ in the same way, we have

$$E^v_{IJ} \equiv \tfrac{1}{2}(F^v_{\alpha I}F^v_{\alpha J} - \delta_{IJ}) = \tfrac{1}{2}[(v/v_R)^{2/3} - 1]\delta_{IJ} , \qquad (6.31)$$

and

$$E^s_{\alpha\beta} \equiv \tfrac{1}{2}(F^s_{i\alpha}F^s_{i\beta} - \delta_{\alpha\beta}) . \qquad (6.32)$$

Equations 6.30 and 6.32 can be combined to give

$$E_{IJ} = \tfrac{1}{2}[(v/v_R)^{2/3} - 1]\delta_{IJ} + (v/v_R)^{2/3}E^s_{\alpha\beta}\delta_{\alpha I}\delta_{\beta J} . \qquad (6.33)$$

Evaluation of the invariants of $\mathbf{E}$ according to Eqs. 4.19 yields the result

$$\begin{aligned} I_\mathbf{E} &= \tfrac{3}{2}(A-1) + AI_S \\ II_\mathbf{E} &= \tfrac{3}{4}(A-1)^2 + A(A-1)I_S + A^2 II_S \\ III_\mathbf{E} &= \tfrac{1}{8}(A-1)^3 - \tfrac{1}{4}A(2A-1)I_S - \tfrac{1}{2}A^2 II_S , \end{aligned} \qquad (6.34)$$

where

$$A = (v/v_R)^{2/3}, \quad I_S = E^s_{\alpha\alpha}, \quad \text{and} \quad II_S = \tfrac{1}{2}(I_S^2 - E^s_{\alpha\beta}E^s_{\beta\alpha}) , \qquad (6.35)$$

with I_S and II_S being the first and second invariants of $\mathbf{E}^s$. The third invariant of $\mathbf{E}$, Eq. 6.34_3, was obtained from the equation

$$(v/v_R)^2 = 1 + 2I_\mathbf{E} + 4II_\mathbf{E} + 8III_\mathbf{E} , \qquad (6.36)$$

which is easily proven for the case that $\mathbf{E}$ is diagonal, with the general result following from the invariance of the equation.

For the analysis of a deformation decomposed into dilatational and distortional parts, as has been done here, it is appropriate to replace the invariants $I_\mathbf{E}$, $II_\mathbf{E}$, and $III_\mathbf{E}$ by the equivalent set of invariants, A, I_S, and II_S that are related to the first set by the equations

$$A = \left[1 + 2I_\mathbf{E} + 4II_\mathbf{E} + 8III_\mathbf{E}\right]^{1/3} \qquad (6.37a)$$

$$I_S = -\frac{3}{2}\left(\frac{A-1}{A}\right) + \frac{1}{A} I_E I_S = -\frac{3}{2}\left(\frac{A-1}{A}\right) + \frac{1}{A} I_E$$

$$II_S = \frac{3}{4}\left(\frac{A-1}{A}\right)^2 - \frac{A-1}{A^2} I_E + \frac{1}{A^2} II_E . \tag{6.37b}$$

Stress and Temperature Equations of State. The stress equation analogous to Eq. 6.10 can be written

$$\frac{1}{\rho_R} T_{IJ} = \frac{\partial \varepsilon(A, I_S, II_S, \eta)}{\partial E_{IJ}} = \frac{\partial \varepsilon}{\partial A}\frac{\partial A}{\partial E_{IJ}} + \frac{\partial \varepsilon}{\partial I_S}\frac{\partial I_S}{\partial E_{IJ}} + \frac{\partial \varepsilon}{\partial II_S}\frac{\partial II_S}{\partial E_{IJ}}, \tag{6.38}$$

and the Cauchy stress components are given by

$$t_{ij} = \frac{\rho}{\rho_R} F_{iI} F_{jJ} T_{IJ} = \left(\frac{v}{v_R}\right)^{-1/3} F^s_{i\alpha} F^s_{j\beta} \delta_{\alpha I} \delta_{\beta J} T_{IJ} . \tag{6.39}$$

With a view toward eventual restriction to small shear strains, we shall consider the case in which $\varepsilon(A, I_S, II_S, \eta)$ is quadratic in I_S, and II_S, leading to derivatives that are linear in these invariants. Accordingly, we have

$$\begin{aligned}\varepsilon(A, I_S, II_S, \eta) &= \varepsilon_1(A) + \varepsilon_2(A) I_S + \varepsilon_3(A) I_S^2 + \varepsilon_4(A) II_S \\ &+ \varepsilon_5(A)(\eta - \eta_R) + \varepsilon_6(A) I_S(\eta - \eta_R) + \varepsilon_7(A)(\eta - \eta_R)^2,\end{aligned} \tag{6.40}$$

and the stress equation of state is

$$\begin{aligned}\frac{1}{\rho_R} T_{IJ} = \frac{\partial \varepsilon}{\partial E_{IJ}} &= \big[\varepsilon_1'(A) + \varepsilon_2'(A) I_S + \varepsilon_3'(A) I_S^2 + \varepsilon_4'(A) II_S \\ &+ \varepsilon_5'(A)(\eta - \eta_R) + \varepsilon_6'(A) I_S(\eta - \eta_R) + \varepsilon_7'(A)(\eta - \eta_R)^2\big]\frac{\partial A}{\partial E_{IJ}} \\ &+ \big[\varepsilon_2(A) + 2\,\varepsilon_3(A) I_S\big]\frac{\partial I_S}{\partial E_{IJ}} + \varepsilon_4(A)\frac{\partial II_S}{\partial E_{IJ}} + \varepsilon_6(A)\frac{\partial I_S}{\partial E_{IJ}}(\eta - \eta_R),\end{aligned} \tag{6.41}$$

where the primes designate differentiation with respect to A and where

$$\frac{\partial A}{\partial E_{IJ}} = \frac{2}{3}(1 + 2I_S + 4II_S)\delta_{IJ} - \frac{4}{3}(1 + 2I_S)E^s_{\alpha\beta}\delta_{\alpha I}\delta_{\beta J} + \frac{8}{3}E^s_{\alpha\gamma}E^s_{\gamma\beta}\delta_{\alpha I}\delta_{\beta J}$$

$$\begin{aligned}\frac{\partial I_S}{\partial E_{IJ}} = \frac{2}{3A}\big[&-2(2I_S + I_S^2 + 3II_S + 2I_S II_S)\delta_{IJ} \\ &+ (3 + 8I_S + 4I_S^2)E^s_{\alpha\beta}\delta_{\alpha I}\delta_{\beta J} - 2(3 + 2I_S)E^s_{\alpha\gamma}E^s_{\gamma\beta}\delta_{\alpha I}\delta_{\beta J}\big]\end{aligned} \tag{6.42a}$$

$$\frac{\partial II_S}{\partial E_{IJ}} = -\frac{1}{A}\Big[(2I_S^2 + 8I_S II_S + 2II_S + 8II_S^2)\delta_{IJ}$$

$$+ (1 - 2I_S - 4I_S^2 - 4II_S - 8I_S II_S)E^s_{\alpha\beta}\,\delta_{\alpha I}\,\delta_{\beta J} \tag{6.42b}$$

$$+ 4(I_S + 2II_S)E^s_{\alpha\gamma}\,E^s_{\gamma\beta}\delta_{\alpha I}\,\delta_{\beta J}\Big].$$

The temperature equation of state is

$$\theta = \varepsilon_5(A) + \varepsilon_6(A)I_S + 2\varepsilon_7(A)(\eta - \eta_R)\,. \tag{6.43}$$

Uniaxial Deformation. The foregoing analysis is easily specialized to the case of uniaxial strain, for which only the diagonal components of the deformation gradient are nonzero,

$$\mathbf{F} = \mathrm{diag}\left\| v/v_R,\ 1,\ 1 \right\|, \tag{6.44}$$

and the only nonzero component of the Lagrangian strain is

$$E_{11} = \tfrac{1}{2}(F_L^2 - 1) = \tfrac{1}{2}[(v/v_R)^2 - 1]\ . \tag{6.45}$$

As in the general case, this deformation can be decomposed into the dilatation of Eq. 6.28 and the isochoric distortion

$$\mathbf{F}^s = \mathrm{diag}\left\| (v/v_R)^{2/3},\ (v/v_R)^{-1/3},\ (v/v_R)^{-1/3} \right\|. \tag{6.46}$$

The Lagrangian strain associated with this deformation gradient is

$$\mathbf{E}^s = \mathrm{diag}\left\| \tfrac{1}{2}[(v/v_R)^{4/3} - 1],\ \tfrac{1}{2}[(v/v_R)^{-2/3} - 1],\ \tfrac{1}{2}[(v/v_R)^{-2/3} - 1] \right\|. \tag{6.47}$$

When $|(v/v_R) - 1| \ll 1$ Eqs. 6.46 and 6. 47 take the approximate forms

$$\mathbf{F}^s = \mathrm{diag}\left\| 1 - \tfrac{2}{3}(v/v_R),\ 1 + \tfrac{1}{3}(v/v_R),\ 1 + \tfrac{1}{3}(v/v_R) \right\|, \tag{6.48}$$

and

$$\widetilde{\mathbf{E}}^s = \mathrm{diag}\left\| -\tfrac{2}{3}[1 - (v/v_R)],\ \tfrac{1}{3}[1 - (v/v_R)],\ \tfrac{1}{3}[1 - (v/v_R)] \right\|. \tag{6.49}$$

6.3.2 Finite Dilatation Combined with Small Distortion

In the previous section we addressed decomposition of the deformation into dilatational and deviatoric parts, but we have not yet introduced the simplifications that can be made when we omit all but the lowest-order terms in $\mathbf{E}^s$. We shall now proceed with this process. Note that states of finite dilatation combined with small distortion do not arise in cases of uniaxial elastic strain because the dilatation and distortion are related almost linearly.

When the shear strain is small we can write $\mathbf{E}^s = \widetilde{\mathbf{E}}^s$ and neglect quadratic terms in $\widetilde{\mathbf{E}}^s$. The criterion that $\mathbf{E}^s$ be a pure distortion is not $\mathrm{tr}\,\mathbf{E}^s = 0$ as in the linear theory but we recover this condition when quadratic terms are neglected. When $\mathbf{E}^s$ is replaced by $\widetilde{\mathbf{E}}^s$ and the quadratic terms in II_S are neglected, the invariants I_S and II_S both vanish. Introduction of these approximations into Eqs. 6.42 yields the simpler equations

$$\begin{aligned} \frac{\partial A}{\partial E_{IJ}} &= \frac{2}{3}(\delta_{IJ} - 2\widetilde{E}^s_{\alpha\beta}\,\delta_{\alpha I}\,\delta_{\beta J}) \\ \frac{\partial I_S}{\partial E_{IJ}} &= \frac{2}{A}\widetilde{E}^s_{\alpha\beta}\,\delta_{\alpha I}\,\delta_{\beta J} \\ \frac{\partial II_S}{\partial E_{IJ}} &= -\frac{1}{A}\widetilde{E}^s_{\alpha\beta}\,\delta_{\alpha I}\,\delta_{\beta J}. \end{aligned} \tag{6.50}$$

When Eqs. 6.50 are substituted into Eq. 6.41 and quadratic terms are neglected, we obtain the simplified stress and temperature equations of state

$$\begin{aligned} \frac{1}{\rho_R}T_{IJ} &= \varepsilon_1^*(A,\eta)\,\delta_{IJ} + \varepsilon_2^*(A,\eta)\,\widetilde{E}^s_{\alpha\beta}\,\delta_{\alpha I}\,\delta_{\beta J} \\ \theta &= \varepsilon_5(A) + 2\varepsilon_7(A)(\eta - \eta_R), \end{aligned} \tag{6.51}$$

where

$$\begin{aligned} \varepsilon_1^*(A,\eta) &= \frac{2}{3}\left[\varepsilon_1'(A) + \varepsilon_5'(A)(\eta-\eta_R) + \varepsilon_7'(A)(\eta-\eta_R)^2\right] \\ \varepsilon_2^*(A,\eta) &= -\frac{4}{3}\left[\varepsilon_1'(A) + \varepsilon_5'(A)(\eta-\eta_R) + \varepsilon_7'(A)(\eta-\eta_R)^2\right] \\ &\quad + \frac{1}{A}\left[2\varepsilon_2(A) - \varepsilon_4(A) + 2\varepsilon_6(A)(\eta-\eta_R)\right]. \end{aligned} \tag{6.52}$$

The Piola–Kirchhoff and Cauchy stress tensors are related by Eq. 6.39. Since this equation involves the deformation gradient $\mathbf{F}^s$, we need to determine the form of this deformation gradient when $\mathbf{E}^s$ is small. The complication is that, even though $\mathbf{E}^s$ is small, $\mathbf{F}^s$ may involve a finite rotation. Equation 2.9, which we write $F^s_{i\alpha} = R_{i\beta}\,U^s_{\beta\alpha}$, leads to the equations

$$E^s_{\alpha\beta} = \tfrac{1}{2}(F^s_{i\alpha}F^s_{i\beta} - \delta_{\alpha\beta}) = \tfrac{1}{2}(R_{i\gamma}U^s_{\gamma\alpha}R_{i\delta}U^s_{\delta\beta} - \delta_{\alpha\beta}) = \tfrac{1}{2}(U^s_{\gamma\alpha}U^s_{\gamma\beta} - \delta_{\alpha\beta}) \tag{6.53}$$

for $\mathbf{E}^s$. When we restrict attention to small strains, we have

$$\widetilde{U}^s_{\alpha\beta} = \delta_{\alpha\beta} + \widetilde{E}^s_{\alpha\beta} + \cdots, \tag{6.54}$$

as can be verified by substituting this expression into Eq. 6.53 and discarding the quadratic term in $\widetilde{\mathbf{E}}^{s}$. With this, we have

$$F^{s}_{i\alpha} = R_{i\gamma}\,\widetilde{U}^{s}_{\gamma\alpha} = R_{i\alpha} + R_{i\gamma}\,\widetilde{E}^{s}_{\gamma\alpha} + \cdots . \tag{6.55}$$

Substitution of Eqs. 6.51 and 6.55 into Eq. 6.39_2 yields the equation

$$t_{ij} = \rho_R\,(v/v_R)^{-1/3}\left[\varepsilon^*_1(A,\eta)\,\delta_{ij} + [2\varepsilon^*_1(A,\eta) + \varepsilon^*_2(A,\eta)]R_{i\alpha}\widetilde{E}^{s}_{\alpha\beta}\overset{-1}{R}_{\beta j}\right] \tag{6.56}$$

for the Cauchy stress accurate to first order in $\widetilde{\mathbf{E}}^{s}$. This equation is of the form

$$t_{ij} = t'_{ij} - p\,\delta_{ij}\,, \tag{6.57}$$

where $\mathbf{t}'$ and p are the pressure and the deviatoric stress, respectively.

The pressure can be calculated from a suitable equation of state as discussed in Chap. 5. The coefficient of the stress deviator,

$$t'_{ij} = \rho_R\,(v/v_R)^{-1/3}\,[2\varepsilon^*_1(A,\eta) + \varepsilon^*_2(A,\eta)]R_{i\alpha}\widetilde{E}^{s}_{\alpha\beta}\overset{-1}{R}_{\beta j}\,, \tag{6.58}$$

is usually designated 2μ, so we have

$$\rho_R\,(v/v_R)^{-1/3}\,[2\varepsilon^*_1(A,\eta) + \varepsilon^*_2(A,\eta)] = 2\mu(v,\eta)\,. \tag{6.59}$$

In states of uniaxial strain Eq. 6.58 becomes

$$t'_{ij} = 2\mu(v,\eta)\,\widetilde{E}^{s}_{ij}\,, \tag{6.60}$$

when Eq. 6.59 is used. When we also use Eq. 6.49 this becomes

$$\mathbf{t}' = 2\mu(v,\eta)\,\mathrm{diag}\left\| -\tfrac{2}{3}[1-(v/v_R)],\ \tfrac{1}{3}[1-(v/v_R)],\ \tfrac{1}{3}[1-(v/v_R)] \right\|. \tag{6.61}$$

Steinberg et al. [91] have developed the equation

$$\mu(v,\theta) = \mu_R\,[1 + \mu_1\,(v/v_R)^{1/3}\,p(v) + \mu_2\,(\theta-\theta_R)]\,, \tag{6.62}$$

for the shear modulus as a function of specific volume and temperature, and have provided values of the coefficients μ_1 and μ_2 for several materials. The temperature in this equation can be replaced by a function of specific volume and specific entropy by use of Eq. 6.51_2.

It is useful to note that the internal energy equation of state used in developing the foregoing stress relations can be replaced by the specific Helmholtz free energy equation of state, in which case the equations all have the same form except that the elastic moduli become isothermal moduli, the various coefficients depend on the temperature instead of the specific entropy, and the thermal terms are somewhat altered.

6.4 Exercises

6.4.1 Show that $(v/v_{\mathrm{R}})^{1/3} = 1 - \frac{1}{3}[1-(v/v_{\mathrm{R}})] + \mathrm{O}[1-(v/v_{\mathrm{R}})]^2$.

6.4.2 Write out the third-order expansion of $\hat{\varepsilon}(\mathbf{E}, \eta)$ and show that it can be placed in the form of Eq. 6.5.

6.4.3 Two of the coefficients of terms in Eq. 6.5 involve derivatives with respect to specific entropy. How can these coefficients be replaced by coefficients that involve derivatives with respect to temperature?

6.4.4 Show that $R_{iI} = \delta_{iI}$ for uniaxial deformations.

6.4.5 Derive Eq. 6.56.

6.4.6 Show that the Hugoniot and isentrope differ by cubic and higher terms in the volume jump at a weak shock propagating in a thermoelastic solid.

CHAPTER 7

Material Response IV: Elastic–Plastic and Elastic–Viscoplastic Solids

Most solids exhibit elastic response only within a narrow range of stress or strain. Materials respond elastically to large compressive forces if they take the form of a uniform pressure, but their resistance to shear stress is limited. When ductile solids such as soft metals are subjected to shear stress, they can be severely deformed without fracturing, but they do not recover their original shape when the applied stress is removed. It is this response, plastic deformation, that is the subject of this chapter. Its investigation necessarily involves consideration of elastic response, and the theory is more complicated than that of elastic response alone. The observed phenomena are more varied than elastic responses and a range of theories is used to capture these phenomena. The simpler theories capture the observed phenomena in only the most rudimentary way. The more comprehensive theories capture a broader range of phenomena, and/or capture the more basic observations with a higher degree of fidelity, but this is accomplished only at a cost in complexity. The simpler theories are based upon the premise that the material response is independent of the rate of deformation, but rate effects are commonly observed and are captured by the more comprehensive theories.

The subject matter of this chapter differs from that of those preceding because theories of plasticity are not in the settled state of the theories of elastic deformation or fluid flow. As a result, there are many theories of plasticity and none of them captures all of the responses that are observed when metals are deformed beyond their range of elastic response. It is difficult to conduct shock-wave experiments to measure shear stress, which is the quantity controlling most aspects of plasticity. Examination of ductile metals recovered after having been subjected to a shock-compression/decompression process reveals that the process is very chaotic, which indicates that continuum theories based upon traditional microscopic descriptions of plastic deformation omit consideration of many mechanisms of inelastic deformation.

The setting in which elastic–plastic response is most easily observed is that in which a ductile metal rod is subjected to tension along its axis. When the applied stress is plotted against the associated strain, a curve such as that of

Fig. 7.1b is obtained. If the maximum strain imposed in a given test is sufficiently small, the elongation process is reversible (i.e., the curve is retraced upon unloading) and is described by elasticity theory. When the bar is stretched beyond some limit (usually less than 0.5 % increase in length) much smaller stress increments are sufficient to produce a given strain increment, and the loading curve is no longer retraced when the applied tension is reduced. When the applied tension is completely removed, some residual elongation, called *plastic deformation*, is observed, as shown in the figure. The unloading is accompanied by an axial contraction of the rod that is described by elasticity theory based upon strain measured relative to the residual configuration. The simplest theory of the phenomena described, that of an ideal elastic–plastic material, corresponds to the stress–strain curve of Fig. 7.1c, i.e., once the bar has been deformed beyond the *elastic limit* or *yield point* of the material it can be further extended without application of additional stress. When additional stress must be applied to produce additional strain, as in Fig. 7.1b, the material is said to exhibit hardening behavior.

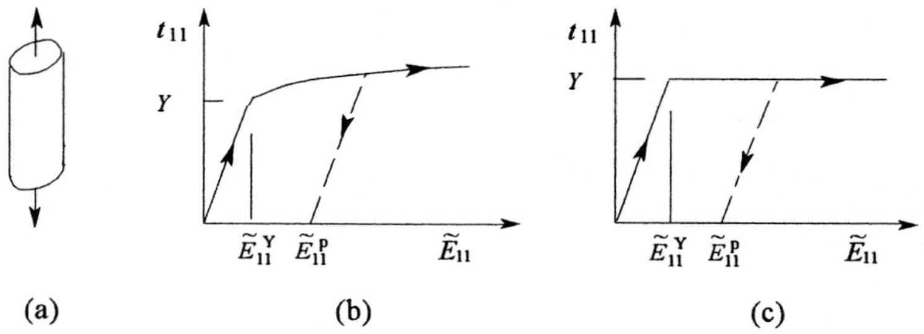

Figure 7.1. Stress–strain response of a rod in states of uniaxial stress.

Although we have been discussing plasticity in terms of tensile response, it is also observed in compression. Both experimental observation and our understanding of the underlying mechanism of plastic deformation indicate that yielding is a shear phenomenon that occurs independently of any superimposed dilatation to which the material may be subjected. A stress–strain curve is normally measured in a test conducted at a low rate of deformation ($\dot{l}/L \approx 0.001\,\mathrm{s}^{-1}$). When the rate of deformation is increased it is often observed that a higher value of stress is required to produce a given strain. This response, called *rate dependence* or *viscoplasticity*, can be important in understanding shock phenomena because of the very rapid deformation encountered in this context ($\dot{l}/L$ often exceeds $10^5\,\mathrm{s}^{-1}$).

Mechanism of Plastic Deformation. Plastic deformation arises through the operation of many mechanisms of rearrangement of the microscopic configuration of the material. Most important among them is the motion of dislocations in a crystal lattice, although adiabatic shear bands, twins, and other defect structures also play a role. Even the most careful consideration of the microscopic mechanisms of inelastic deformation has not yet yielded a theory capable of producing predictions in detailed quantitative agreement with observations of the operation of these mechanisms. Nevertheless, knowledge of deformation mechanisms has motivated development of continuum theories that have been very effective in modeling observed responses.

Sliding the planes of a perfect crystal over one another requires a very large shear stress—much larger than a typical yield stress. When dislocations are present in the lattice, as they always are in all but the smallest and most perfectly prepared monocrystals (There are many dislocations—10^4 to 10^{10} intersect a typical mm^2 area.), shear occurs at much lower stress because the dislocations can move rather easily. The situation is illustrated schematically in Fig. 7.2, which shows how one part of a sheared crystal slips over the other in small steps as a dislocation jumps from one site to the next. It is important to note that deformation attributable to dislocation motion accumulates gradually in response to an applied shear stress. The stress application leads to a deformation rate as opposed to a deformation itself as in the case of elasticity. As a result, application of what is known of dislocation behavior to development of a theory of deformation leads to an elastic–viscoplastic theory. The physical processes underlying plastic deformation lead to microstructural changes in the material that result in altered yield thresholds and permanent metallurgical changes.

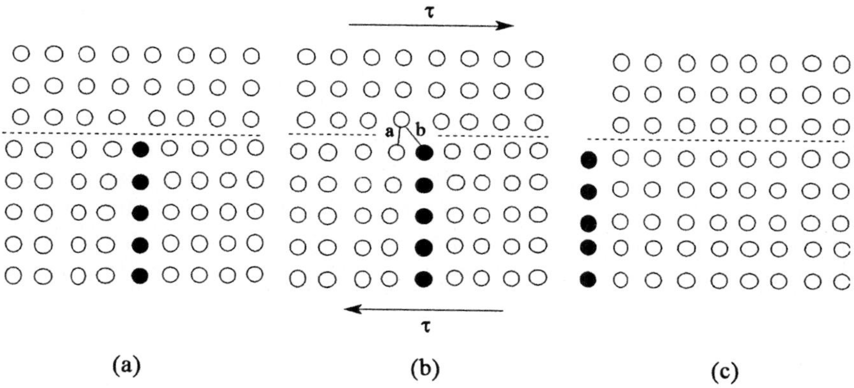

Figure 7.2. Illustration of slip of a crystal due to the motion of an edge dislocation. The dislocation occurs at the edge of the extra half-plane of atoms shown as filled circles. When shear stress is applied, as shown in part (b) of the figure, the elastic deformation causes the bond designated *a* to transfer to the bond designated *b*. Through a succession of such steps, we finally arrive at the configuration shown in part (c) of the figure. In this configuration the crystal is unstressed, but a plastic shear has accumulated. The slip plane is indicated by the dotted line and the Burgers vector corresponds to the offset shown in part (c) of the figure.

7.1 Elastic–Plastic Response to Small Strain

Before undertaking development of a theory based upon dislocation mechanics, it is appropriate to consider a simple rate-independent continuum theory that lends itself to detailed analysis of many practical problems [47]. Examination of solutions based upon this theory offers insight into the effects of wave interaction that are not apparent from either experimental measurements or typical finite-difference calculations. Waveforms observed in experiments are smoothed by both dispersion in the material and the temporal resolution of the instrumentation used. Numerical simulations usually involve an artificial viscosity that is included to smooth the wave so the partial differential equations that describe the process are meaningful.

Ideal elastoplasticity at small strain incorporates a linear elastic stress–strain relation and the constraint that the plastic response does not produce any change in specific volume. It is the simplest model of elastic–plastic response that is useful for analysis of shock phenomena. The restriction to small strains does not prove as limiting as one might suppose because elastic–plastic effects are most pronounced when the strength of a shock is only moderately in excess of the limit of elastic response. Imposition of this restriction allows us to use the Reuss theory in which the components of the small-deformation approximation to the strain (given by Eq. 6.16) are additively decomposed into parts $\widetilde{E}^{\mathrm{e}}_{ij}$ and $\widetilde{E}^{\mathrm{p}}_{ij}$ representing the elastic and plastic contributions, respectively, to the deformation:

$$\widetilde{E}_{ij} = \widetilde{E}^{\mathrm{e}}_{ij} + \widetilde{E}^{\mathrm{p}}_{ij} . \tag{7.1}$$

The stress is determined solely by the elastic part of the strain and, since this strain is small, the linear stress–strain relations are appropriate. For the simple case of an isotropic solid to which we restrict attention here, the stress and strain are related by Eq. 6.17 with $\widetilde{E}_{ij}$ replaced by $\widetilde{E}^{\mathrm{e}}_{ij}$. Finally, both experimental observation and microscopic models of the underlying cause of plastic deformation of metals justifies the approximation that there is no volume change that is attributable to the plastic part of the strain, a condition that can be written in the form

$$\widetilde{E}^{\mathrm{p}}_{kk} = 0 . \tag{7.2}$$

We postulate that yielding occurs when some critical state of stress is reached. For example, when a rod is subjected to increasing tension along its axis, a point at which yielding occurs is eventually reached.* This stress is a material property called the *yield stress*, Y, and the criterion for yielding in this configuration is simply that the applied stress reach the yield value. To develop

* Brittle materials that fracture before yielding are not covered by the theory of plasticity.

a theory of plasticity, we begin by generalizing the foregoing yield condition to arbitrary stress states.

In the example of stretching a rod along the x_1 axis, the only nonzero component of the stress field is t_{11} and the yield condition is

$$t_{11} = Y \,. \tag{7.3}$$

In general, we consider a yield condition in the form

$$f(\mathbf{t}) = 0 \,. \tag{7.4}$$

Yielding is a shear phenomenon and, as a measure of shear stress, we adopt the *stress deviator* tensor having components

$$t'_{ij} = t_{ij} + p\delta_{ij} \tag{7.5}$$

obtained by eliminating the mechanical pressure *p,* given by Eq. 2.57, from the total stress. Yielding is assumed to depend only upon $\mathbf{t}'$; the pressure plays no role. Accordingly, the yield function *f* depends only upon $\mathbf{t}'$, and we can write the yield condition in the form

$$\hat{f}(\mathbf{t}') = 0 \,. \tag{7.6}$$

Since we have restricted our considerations to isotropic materials, $\hat{f}$ must be a function only of the invariants of $\mathbf{t}'$ under orthogonal transformations. It is customary to select as invariants

$$J'_1 = t'_{kk}, \quad J'_2 = \tfrac{1}{2} t'_{ij} t'_{ij}, \quad J'_3 = \tfrac{1}{3} t'_{ij} t'_{jk} t'_{ki} \,. \tag{7.7}$$

By definition of $\mathbf{t}'$, we have $J'_1 = 0$, so the yield function takes the form

$$\tilde{f}(J'_2, J'_3) = 0 \,. \tag{7.8}$$

When we discussed extension of a rod, we saw that yielding occurred when the tension reached a critical value, *Y*. When yielding occurs under compression at the same stress magnitude, the yield criterion can be generalized to $|t_{11}| = Y$. We assume, more generally, that the stress magnitude at which yielding occurs is unchanged as $\mathbf{t}' \to -\mathbf{t}'$. This means that $\tilde{f}(J'_2, J'_3)$ must be an even function of J'_3. Two yield functions that we shall consider are the *Tresca criterion* and the *von Mises criterion.*

According to the Tresca criterion, yielding occurs at a material point when the maximum shear stress at that point reaches a critical value (the shear stress is evaluated on a plane through the point, and the maximum is taken over all directions on all planes through the point). This criterion is the continuum analog of the yield criterion that arises from the interpretation of yielding in terms of the motion of lattice dislocations. When written in terms of the principal deviatoric stress components, the magnitudes of the principal shear stresses are $|t'_1 - t'_2|/2$, $|t'_1 - t'_3|/2$, and $|t'_2 - t'_3|/2$, and the maximum shear stress is

$$\tau_{max} = \tfrac{1}{2}\max\left[\,|t'_1 - t'_2|, |t'_1 - t'_3|, |t'_2 - t'_3|\,\right]. \tag{7.9}$$

For the bar, we have

$$\tau_{max} = \tfrac{1}{2}\,|t_{11}|\,. \tag{7.10}$$

At the yield point,

$$\tau_{max} = Y/2, \tag{7.11}$$

so the yield condition becomes

$$\max\left[\,|t'_1 - t'_2|, |t'_1 - t'_3|, |t'_2 - t'_3|\,\right] - Y = 0, \tag{7.12}$$

or, equivalently,

$$\max\left[\,|t_1 - t_2|, |t_1 - t_3|, |t_2 - t_3|\,\right] - Y = 0. \tag{7.13}$$

The Tresca criterion can be written in terms of J'_2 and J'_3, but the result is complicated and is not needed for our work.

The von Mises yield criterion,

$$J'_2 - \tfrac{1}{3}Y^2 = 0, \tag{7.14}$$

is the simplest acceptable function of the invariants of $\mathbf{t}'$, is in reasonable accord with measured responses, and closely approximates the Tresca condition. Since yielding is a consequence of shear rather than tension or compression it may be of interest to express Eq. 7.14 in terms of the maximum shear stress corresponding to the tensile yield stress Y measured in a uniaxial stress test. In this case, we have $\tau_{max} = Y/2$ so the von Mises yield criterion can be written

$$J'_2 - \tfrac{4}{3}\tau_{max}^2 = 0\,.$$

The yield stress is usually observed to increase as the material is deformed. This increase can be correlated with the plastic strain, in which case it is called *strain hardening*, or with the work done in producing the plastic strain, in which case it is called *work hardening*.

The remaining ingredient required to make a theory of plasticity is a *flow rule*, an equation relating the part of the strain rate attributable to plastic deformation to the stress. A flow rule is not needed for analysis of uniaxial shocks using the theory of this section, but a brief account is included for completeness. Because the plastic part of the deformation is isochoric, $\widetilde{\mathbf{E}}^{\mathrm{p}}$ is a deviator tensor so the strain rate $\partial\widetilde{\mathbf{E}}^{\mathrm{p}}/\partial t$ (which is the same as $\mathbf{d}^{\mathrm{p}}$ in the limit of small deformations) is also a deviator tensor. Therefore, we are led to adopt a flow rule of the form

$$\frac{\partial \widetilde{E}_{ij}^{\,\mathrm{p}}}{\partial t} = \Lambda\, t'_{ij}\,, \tag{7.15}$$

where Λ is a function to be determined. If we multiply each member of Eq. 7.15 by t'_{jk} and contract, we obtain the equation

$$\frac{\partial \widetilde{E}_{ij}^{\,\mathrm{p}}}{\partial t} t'_{ij} = \Lambda\, t'_{ij}\, t'_{ji} = 2\Lambda J'_2\,. \tag{7.16}$$

The stress power, $\operatorname{tr}(\mathbf{td})$, associated with this deformation is

$$\rho_{\mathrm{R}}\, \dot{W} = \frac{\partial \widetilde{E}_{ij}}{\partial t} t_{ij} = \frac{\partial \widetilde{E}_{ij}^{\,\mathrm{e}}}{\partial t} t_{ij} + \frac{\partial \widetilde{E}_{ij}^{\,\mathrm{p}}}{\partial t} t_{ij} = \rho_{\mathrm{R}}\, \dot{W}^{\,\mathrm{e}} + \rho_{\mathrm{R}}\, \dot{W}^{\,\mathrm{p}}\,, \tag{7.17}$$

where

$$\rho_{\mathrm{R}}\, \dot{W}^{\,\mathrm{e}} = \frac{\partial \widetilde{E}_{ij}^{\,\mathrm{e}}}{\partial t} t_{ij} \quad \text{and} \quad \rho_{\mathrm{R}}\, \dot{W}^{\,\mathrm{p}} = \frac{\partial \widetilde{E}_{ij}^{\,\mathrm{p}}}{\partial t} t_{ij} = \frac{\partial \widetilde{E}'_{ij}}{\partial t} t'_{ij} \tag{7.18}$$

are the contributions of the elastic and plastic parts of the deformation. The latter is often called the *plastic working*. Since $t_{ij} = t'_{ij} - p\,\delta_{ij}$ and $\widetilde{E}_{kk}^{\,\mathrm{p}} = 0$ we have

$$\rho_{\mathrm{R}}\, \dot{W}^{\,\mathrm{p}} = \frac{\partial \widetilde{E}_{ij}^{\,\mathrm{p}}}{\partial t} t_{ij} = \frac{\partial \widetilde{E}_{ij}^{\,\mathrm{p}}}{\partial t} (t'_{ij} - p\,\delta_{ij}) = \frac{\partial \widetilde{E}_{ij}^{\,\mathrm{p}}}{\partial t} t'_{ij}\,. \tag{7.19}$$

When the last of Eqs. 7.19 is substituted into Eq. 7.16 and the result solved for the function Λ, we have

$$\Lambda = \frac{1}{2J'_2} \frac{\partial \widetilde{E}_{ij}^{\,\mathrm{p}}}{\partial t} t'_{ij} = \frac{\rho_{\mathrm{R}}\, \dot{W}^{\,\mathrm{p}}}{2J'_2}\,, \tag{7.20}$$

so the flow rule, Eq. 7.15, becomes

$$\frac{\partial \widetilde{E}_{ij}^{\,\mathrm{p}}}{\partial t} = \frac{\rho_{\mathrm{R}}\, \dot{W}^{\,\mathrm{p}}}{2J'_2} t'_{ij}\,. \tag{7.21}$$

The linear elastic stress relation can be written

$$\widetilde{E}_{ij}^{\,\mathrm{e}} = -\frac{1}{3\lambda_{\mathrm{R}} + 2\mu_{\mathrm{R}}}\, p\,\delta_{ij} + \frac{1}{2\mu_{\mathrm{R}}} t'_{ij}\,, \tag{7.22}$$

so the elastic part of the strain rate is given by

$$\frac{\partial \widetilde{E}_{ij}^{\,\mathrm{e}}}{\partial t} = -\frac{1}{3\lambda_{\mathrm{R}} + 2\mu_{\mathrm{R}}} \frac{\partial p}{\partial t} \delta_{ij} + \frac{1}{2\mu_{\mathrm{R}}} \frac{\partial t'_{ij}}{\partial t}\,. \tag{7.23}$$

When the spherical and deviator parts of this equation are separated, we have

$$\vartheta = -\frac{3}{3\lambda_R + 2\mu_R} p \tag{7.24}$$

and

$$\frac{\partial \widetilde{E}'^{e}_{ij}}{\partial t} = \frac{1}{2\mu_R} \frac{\partial t'_{ij}}{\partial t}, \tag{7.25}$$

where

$$E'^{e}_{ij} = E^{e}_{ij} - \tfrac{1}{3}\vartheta \delta_{ij},$$

and $\vartheta = \widetilde{E}^{e}_{kk}$ is the small-strain approximation to the dilatation. Adding Eqs 7.21 and 7.25 yields the result

$$\frac{\partial \widetilde{E}'_{ij}}{\partial t} = \frac{1}{2\mu_R} \frac{\partial t'_{ij}}{\partial t} + \frac{\rho_R \dot{W}^{p}}{2J'_2} t'_{ij}, \tag{7.26}$$

and Eqs. 7.24 and 7.26, together, form the constitutive description of the ideal elastic–plastic material at small strain. It is important to note that, despite the appearance of time derivatives in this description, Eq. 7.26 is homogeneous in t so the response is actually rate independent.

Small Uniaxial Deformation. Let us now consider the uniaxial deformation discussed in Sect. 2.2. Since the problem is set in principal coordinates, the off-diagonal strain components vanish. The definition of uniaxial deformation implies that the lateral components of strain also vanish:

$$\widetilde{E}_{22} = \widetilde{E}_{33} = 0. \tag{7.27}$$

By symmetry, the two in-plane components of each of the variables are equal, so we need only give results for the x_2 components.

It is the purpose of this section to specialize the foregoing constitutive equations to this deformation so they will be available for analysis of propagation of a plane shock in an ideal elastic–plastic solid.

Range of Elastic Response. In the elastic region the total strain $\widetilde{E}_{ij}$ and its elastic part $\widetilde{E}^{e}_{ij}$ are the same. According to Eq. 6.17, the nonvanishing stress components are

$$t_{11} = (\lambda_R + 2\mu_R)\widetilde{E}^{e}_{11}, \quad t_{22} = t_{33} = \lambda_R \widetilde{E}^{e}_{11} = \frac{\lambda_R}{\lambda_R + 2\mu_R} t_{11}. \tag{7.28}$$

The pressure is

$$p = -\tfrac{1}{3} t_{kk} = -(\lambda_R + \tfrac{2}{3}\mu_R)\widetilde{E}^{e}_{11} = -\frac{3\lambda_R + 2\mu_R}{3(\lambda_R + 2\mu_R)} t_{11}, \tag{7.29}$$

and the nonvanishing stress deviator components are

$$t'_{11} = \frac{4\mu_R}{3(\lambda_R + 2\mu_R)} t_{11}, \quad \text{and} \quad t'_{22} = t'_{33} = -\frac{2\mu_R}{3(\lambda_R + 2\mu_R)} t_{11}. \tag{7.30}$$

Finally, the maximum shear stress is

$$-\tau_{45°} = -\tfrac{1}{2}(t_{11} - t_{22}) = -\frac{\mu_R}{\lambda_R + 2\mu_R} t_{11} = -\mu_R \tilde{E}^e_{11}, \tag{7.31}$$

and is, as the notation suggests, the shear stress present on planes lying at 45° to the x axis.

The Tresca criterion gives the yield stress as

$$|t_{11}| = t_{11}^{\mathrm{HEL}}, \qquad |t_{22}| = t_{22}^{\mathrm{HEL}}, \tag{7.32}$$

where the definitions

$$t_{11}^{\mathrm{HEL}} \equiv \frac{\lambda_R + 2\mu_R}{2\mu_R} Y, \quad t_{22}^{\mathrm{HEL}} \equiv \frac{\lambda_R}{2\mu_R} Y, \tag{7.33}$$

have been introduced in anticipation of application of these results to the analysis of shock-propagation problems. The superscript HEL stands for *Hugoniot Elastic Limit*, and is the state on the Hugoniot curve at which yielding begins. In this case, as in the case of the bar, a simple calculation shows that the von Mises and Tresca yield criteria give the same result (although this is not true for all stress fields).

Range of Elastic–Plastic Response. In the plastic regime we have the usual elastic stress relations but, in contrast to the analysis of the elastic regime, $\tilde{\mathbf{E}}^e$ is no longer the total strain. The stress relations are

$$t_{11} = (\lambda_R + 2\mu_R)\tilde{E}^e_{11} + 2\lambda_R \tilde{E}^e_{22}, \quad t_{22} = \lambda_R \tilde{E}^e_{11} + 2(\lambda_R + \mu_R)\tilde{E}^e_{22}, \tag{7.34}$$

which we use in the equivalent form

$$\tilde{E}^e_{11} = \frac{(\lambda_R + \mu_R)t_{11} - \lambda_R t_{22}}{\mu_R(3\lambda_R + 2\mu_R)}, \quad \tilde{E}^e_{22} = \frac{-\lambda_R t_{11} + (\lambda_R + 2\mu_R)t_{22}}{2\mu_R(3\lambda_R + 2\mu_R)}. \tag{7.35}$$

The axial and lateral strains are each made up of both elastic and plastic parts, as given by Eq. 7.1. The lateral constraint takes the form

$$\tilde{E}_{22} = \tilde{E}^e_{22} + \tilde{E}^p_{22} = 0, \tag{7.36}$$

and the condition that the plastic part of the deformation be isochoric is expressed by the equation

$$\tilde{E}^p_{11} + 2\tilde{E}^p_{22} = 0. \tag{7.37}$$

From Eqs. 7.35_2, 7.36, and 7.37 we obtain

$$\tilde{E}^{\mathrm{p}}_{11} = \frac{-\lambda_{\mathrm{R}} t_{11} + (\lambda_{\mathrm{R}} + 2\mu_{\mathrm{R}}) t_{22}}{\mu_{\mathrm{R}} (3\lambda_{\mathrm{R}} + 2\mu_{\mathrm{R}})}, \quad \tilde{E}^{\mathrm{p}}_{22} = \frac{\lambda_{\mathrm{R}} t_{11} - (\lambda_{\mathrm{R}} + 2\mu_{\mathrm{R}}) t_{22}}{2\mu_{\mathrm{R}} (3\lambda_{\mathrm{R}} + 2\mu_{\mathrm{R}})} . \tag{7.38}$$

Finally, the yield condition

$$| t_{11} - t_{22} | = Y \tag{7.39}$$

must be satisfied in the plastic regime.

The five equations 7.35–7.37 and 7.39 can be solved for any five of the unknown quantities t_{11}, t_{22}, $\tilde{E}^{\mathrm{e}}_{11}$, $\tilde{E}^{\mathrm{p}}_{11}$, $\tilde{E}^{\mathrm{e}}_{22}$, and $\tilde{E}^{\mathrm{p}}_{22}$ once the value of the remaining unknown is given as a measure of the stimulus producing the deformation. Let us assume that t_{11} is given. There are several cases to consider, depending upon the history of deformation that the material has experienced.

Let us suppose first that a stress, t_{11}, is applied to material that is unstrained and at zero stress. To determine the sign of $t_{11} - t_{22}$ in the plastic range it is only necessary to evaluate Eqs. 7.34 at the yield point, at which $\tilde{E}^{\mathrm{e}}_{22} = 0$. This gives

$$t_{11} = (\lambda_{\mathrm{R}} + 2\mu_{\mathrm{R}}) \tilde{E}^{\mathrm{e}}_{11}, \quad t_{22} = \lambda_{\mathrm{R}} \tilde{E}^{\mathrm{e}}_{11} . \tag{7.40}$$

In tension, where $\tilde{E}^{\mathrm{e}}_{11} > 0$, we have $t_{11} > t_{22} > 0$ so that $t_{11} - t_{22} > 0$ and the yield condition is

$$t_{11} - t_{22} = +Y . \tag{7.41}$$

In compression, where $\tilde{E}^{\mathrm{e}}_{11} < 0$, we have $t_{11} < t_{22} < 0$ so that $t_{11} - t_{22} < 0$ and the yield condition is

$$t_{11} - t_{22} = -Y . \tag{7.42}$$

These cases can be combined by writing Eqs. 7.41 and 7.42 in the form

$$t_{11} - t_{22} = \chi Y , \tag{7.43}$$

where

$$\chi = \frac{t_{11}}{| t_{11} |} = \begin{cases} +1 & \text{in tension} \\ -1 & \text{in compression.} \end{cases} \tag{7.44}$$

We can now use the yield condition 7.43 to express t_{22} in terms of t_{11}. We obtain, from Eqs. 7.35 and 7.38,

$$\begin{aligned} \tilde{E}^{\mathrm{e}}_{11} &= \frac{\mu_{\mathrm{R}} t_{11} + \chi \lambda_{\mathrm{R}} Y}{\mu_{\mathrm{R}} (3\lambda_{\mathrm{R}} + 2\mu_{\mathrm{R}})}, \quad & \tilde{E}^{\mathrm{e}}_{22} &= \frac{2\mu_{\mathrm{R}} t_{11} - \chi (\lambda_{\mathrm{R}} + 2\mu_{\mathrm{R}}) Y}{2\mu_{\mathrm{R}} (3\lambda_{\mathrm{R}} + 2\mu_{\mathrm{R}})} \\ \tilde{E}^{\mathrm{p}}_{11} &= \frac{2\mu_{\mathrm{R}} t_{11} - \chi (\lambda_{\mathrm{R}} + 2\mu_{\mathrm{R}}) Y}{\mu_{\mathrm{R}} (3\lambda_{\mathrm{R}} + 2\mu_{\mathrm{R}})}, \quad & \tilde{E}^{\mathrm{p}}_{22} &= \frac{-2\mu_{\mathrm{R}} t_{11} + \chi (\lambda_{\mathrm{R}} + 2\mu_{\mathrm{R}}) Y}{2\mu_{\mathrm{R}} (3\lambda_{\mathrm{R}} + 2\mu_{\mathrm{R}})} . \end{aligned} \tag{7.45}$$

The total uniaxial strain is

$$\widetilde{E}_{11} = \frac{1}{B_R}\left(t_{11} + \frac{2}{3}\chi Y \right), \tag{7.46}$$

where

$$B_R = \lambda_R + \tfrac{2}{3}\mu_R \tag{7.47}$$

is the bulk modulus of elasticity. Equation 7.46 can be solved for t_{11} to yield

$$t_{11} = B_R \widetilde{E}_{11} + \tfrac{2}{3}\chi Y . \tag{7.48}$$

Substitution of this result into Eq. 7.43 gives

$$t_{22} = B_R \widetilde{E}_{11} - \tfrac{1}{3}\chi Y . \tag{7.49}$$

The specific volume is given by $\mathrm{tr}\,\widetilde{\mathbf{E}} = (v/v_R) - 1$ but, since $\widetilde{E}_{22} = \widetilde{E}_{33} = 0$, we have $\widetilde{E}_{11} = (v/v_R) - 1$ and Eq. 7.48 can be written in another common and useful form

$$t_{11} = B_R\,[(v/v_R) - 1] + \tfrac{2}{3}\chi Y . \tag{7.50}$$

We recall that all of these relationships are valid only in the plastic range, which is defined by either of the relations

$$\chi \widetilde{E}_{11} > \widetilde{E}_{11}^{\mathrm{HEL}}, \quad \chi\, t_{11} > t_{11}^{\mathrm{HEL}}, \tag{7.51}$$

where we adjoin the definition

$$\widetilde{E}_{11}^{\mathrm{HEL}} = \frac{Y}{2\mu_R} \tag{7.52}$$

to the yield-point definitions of Eqs. 7.33.

Using Eqs. 7.34–7.37, and 7.1, we can show that

$$t_{22} = \tfrac{1}{2}\left[(3\lambda_R + 2\mu_R)\widetilde{E}_{11} - t_{11} \right], \tag{7.53}$$

a useful result that holds in both the elastic and plastic ranges. Substitution of these uniaxial fields into Eq. 7.26 shows that it is satisfied in the range of plastic deformation.

Figure 7.3 shows stress–strain paths for an ideal elastic–plastic solid subject to uniaxial deformation from a state of zero stress and strain. The dotted line marked p is given by the equation $p = [\,\lambda_R + (2\mu_R/3)]\widetilde{E}_{11}$ and represents the contribution of pressure to the total stress. The solid line represents the longitudinal stress component, given by the equation $t_{11} = (\lambda_R + 2\mu_R)\widetilde{E}_{11}$ in the elastic range and by Eq. 7.48 in the plastic range. The broken line represents the lateral component of stress, and is given by $t_{22} = \lambda_R \widetilde{E}_{11}$ in the elastic range and by Eq. 7.49 in the plastic range. When the material is loaded beyond yield in this way it is said to be in a state of *forward yield.*

At points below the yield stress, all of the strain is attributable to the elastic response of the material. Beyond the yield point, the elastic contribution to the total strain is the sum of the total strain at the yield point and the strain attributable to the increment of pressure beyond the yield-point value of this quantity. The remaining strain is attributable to plastic flow of the material.

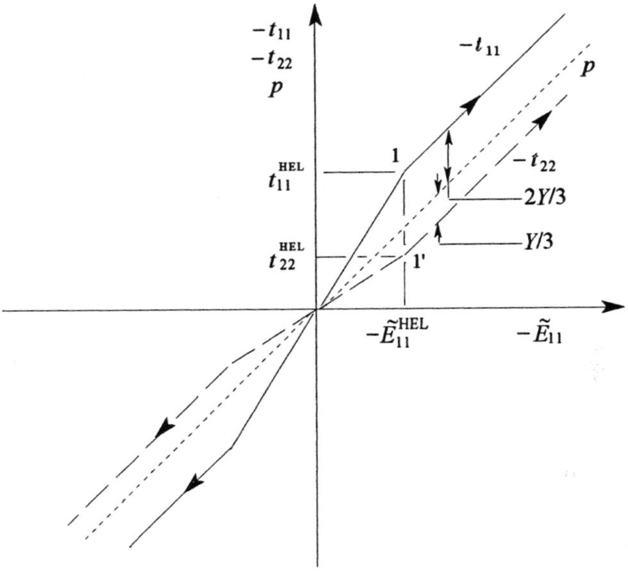

Figure 7.3. Stress–strain curves for an ideal elastic–plastic solid subject to uniaxial deformation from a state of zero stress and strain.

It is also important to determine the stress–strain paths followed upon unloading, reloading, etc. All of these responses play a role in the analysis of elastic–plastic wave propagation. Let us consider unloading from a stress state $t_{11}^{(2)}$ to the state $t_{11} = 0$. The initial part of each stress–strain path is an elastic load release. This process is described by the elasticity theory that we have been using, with all of the deformation being associated with the elastic part of the strain. In this simple one-dimensional example it is easiest to follow the process graphically.

The state point defined by t_{11} moves toward zero stress and strain along a line having the same slope, $\lambda_R + 2\mu_R$, as the elastic loading process. As the state point crosses the pressure line there is a reversal of shear stress direction but the point continues to move along the same path until this reversed shear stress reaches the yield value. At this transition, the state point moves along the yield surface, i.e., along a line having slope $\lambda_R + (2\mu_R/3)$ in the stress–strain plane. This line intercepts the E_{11} axis at the appropriate one of the points $\widetilde{E}_{11} = \chi\, \widetilde{E}_{11}^{R} \equiv 2\chi\, Y/(3B_R)$ defining the states of residual plastic strain. The stress component t_{22} also changes during the decompression process, following the path indicated by the arrow on the broken line in Fig. 7.4. It is important to note that t_{22} does not vanish when the unloading of the faces of the slab is completed. A residual in-plane stress is required to maintain the constraint of uniaxial deformation.

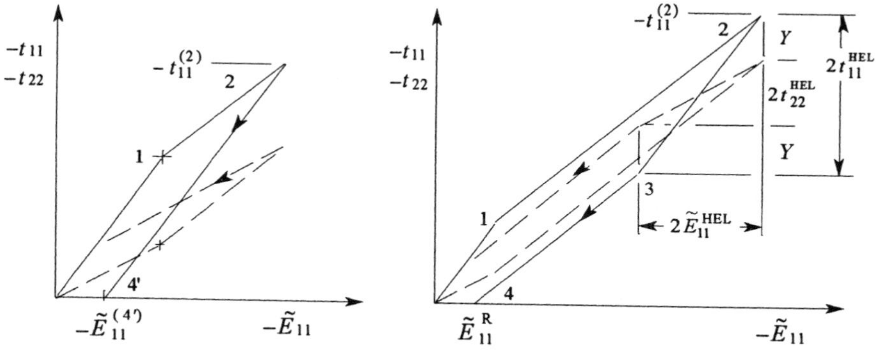

(a) (b)

Figure 7.4. Stress–strain paths for decompression (from state 2) of a laterally-restrained slab that is subject to a uniform normal compression. Part (a) is for the range, $t_{11}^{HEL} < -t_{11}^{(2)} \le 2\,t_{11}^{HEL}$, in which the decompression is entirely elastic. Part (b) is for $-t_{11}^{(2)} > 2\,t_{11}^{HEL}$, the range in which decompression involves both elastic and plastic strains. Solid lines refer to $-t_{11}$ and broken lines refer to $-t_{22}$. The quantity $\widetilde{E}_{11}^{R} = 2Y/(3B)$ is a material property but $\widetilde{E}_{11}^{(4')}$ is a function of $t_{11}^{(2)}$.

We shall call a yield condition reached by loading from zero stress and strain a *forward yield point* and the material states reached by continuation of the deformation beyond this point states of forward yield. A yield point reached by an unloading process is called a *reverse yield point* and states achieved by continuation of the deformation beyond this point states of reverse yield. In all cases, a loading or unloading process that has been interrupted can be resumed (loading continued in the same direction as before the interruption) without affecting the stress–strain path followed. Unloading from any state of forward yield (such as state A in Fig. 7.5a) begins with an elastic deformation and, if it proceeds beyond the reverse yield point (defined by the criterion $t_{11} - t_{22} = +Y$, cf. Eq. 7.42), continues as a process involving both elastic and plastic deformation, as discussed previously. Both loading and unloading of material in an elastic state (such as state B in Fig. 7.5a) begins as an elastic process and, if it proceeds beyond a yield point, continues as a process involving both elastic and plastic deformation. When material is in a state of reverse yield, as illustrated by state C on Fig. 7.5b, further unloading is simply a continuation of the first part of the unloading. Reloading from state C is an elastic process until it reaches the forward yield point, after which it proceeds as with the initial loading.

Although the foregoing discussion concerned deformations in the compressive quadrant of the stress–strain plane of Fig. 7.3, the same principles apply to analysis of deformation in the tensile quadrant, and similar results are obtained.

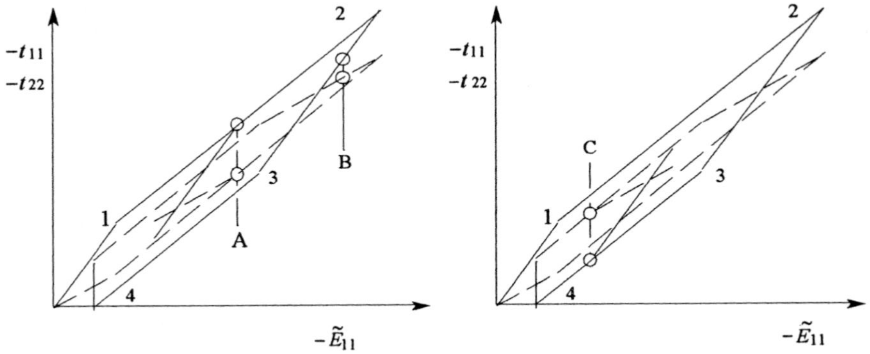

Figure 7.5. Stress–strain curves illustrating unloading and reloading processes. State A is one of forward yield, state B is in the elastic range, and state C is one of reverse yield.

7.2 Elastic–Plastic Response to Finite Uniaxial Deformation

The foregoing theory of ideal plasticity can be extended to allow finite deformations. In the general case, the theory becomes quite complicated, but a version restricted to uniaxial strains is not significantly more complicated than the small-deformation theory and is widely used in analyses of shock phenomena.

The kinematical description of finite elastic–plastic deformations follows the course set forth in Chap. 2, except that we must augment this description with a decomposition of the deformation into parts attributable to elastic and plastic responses. We shall view this decomposition as one in which elastic and plastic deformations occur sequentially. We shall also decompose the elastic part of the deformation into deviatoric and distortional parts, which leads to the sequence of configurations shown in Fig. 7.6. We suppose that the body is unstressed and undeformed when in its reference configuration, X, and is deformed plastically to another unstressed configuration, X^*. This is followed by an elastic deformation to the current configuration, x.

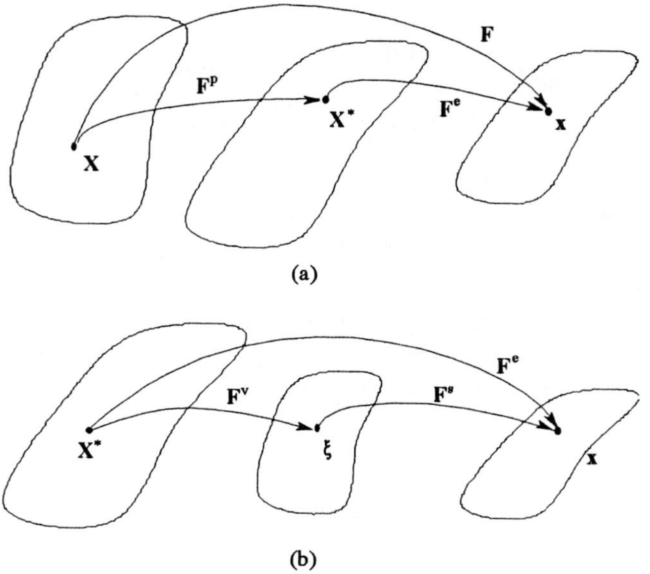

Figure 7.6. Configurations used to analyze finite elastic–plastic deformation of a body. (a) Decomposition of the deformation into elastic and plastic parts. (b) Decomposition of the elastic part of the deformation into dilatational and distortional parts.

The configuration X^* is a fictitious local state in which the body is plastically deformed but not stressed. The current configuration, x, is one in which the body has experienced both elastic and plastic deformation and is stressed. The motion taking the reference configuration into the current configuration is, as in Chap. 2, $x_i = x_i(\mathbf{X}, t)$. This is decomposed (locally) into a transformation $\mathbf{F}^{\mathrm{p}}$ taking the material from the reference configuration to the intermediate plastically deformed but unstressed configuration, X^*, followed by a transformation $\mathbf{F}^{\mathrm{e}}$ taking the material from the X^* configuration to the current configuration, x. The elastic deformation $\mathbf{F}^{\mathrm{e}}$ is decomposed as in Sect. 6.3.1 into a deformation $\mathbf{F}^{\mathrm{v}}$ in which the material has experienced dilatation but not shear, followed by a deformation $\mathbf{F}^{\mathrm{es}}$ representing the elastic shear. The deformation gradient can be calculated in the form

$$F_{iJ} = F^{\mathrm{e}}_{i\Gamma} F^{\mathrm{p}}_{\Gamma J}, \tag{7.54}$$

where

$$F^{\mathrm{e}}_{i\Gamma} = F^{\mathrm{es}}_{i\alpha} F^{\mathrm{v}}_{\alpha\Gamma}. \tag{7.55}$$

We see that, for finite elastic–plastic deformations interpreted as above, the decomposition is multiplicative rather than additive as it was for the small-deformation theory. The decomposition of Eq. 7.55 is introduced because the distortion is usually small and separating it from the dilatation allows us to take advantage of the simplifications that result when we retain only linear terms in the elastic shear strain.

As for small deformations, we require that the plastic contribution to the finite deformation be isochoric so that

$$\det \mathbf{F}^{\mathrm{p}} = 1\,, \tag{7.56}$$

and all of the dilatation is represented by $\mathbf{F}^{\mathrm{v}}$. Therefore, $\mathbf{F}^{\mathrm{v}} = (v/v_{\mathrm{R}})^{1/3}\,\mathbf{I}$.

A yield condition and a flow rule are required to complete the theory. The yield condition can be the same as for the small-deformation theory. The flow rule is more complicated, but is not needed for analysis of uniaxial deformations of materials exhibiting a rate-independent response. We shall consider the flow rule in Sect. 7.3.

Let us now consider the case of uniaxial deformation, for which

$$x = x(X,t)\,, \quad x_2 = X_2\,, \quad \text{and} \quad x_3 = X_3\,. \tag{7.57}$$

The only nonzero components of the deformation gradient are

$$F_{\mathrm{L}} = F_{11} = \partial x(X,t)/\partial X \quad \text{and} \quad F_{\mathrm{T}} = F_{22} = F_{33} = 1\,. \tag{7.58}$$

The reference, intermediate (plastically deformed), and current configurations are illustrated in Fig. 7.7. In the intermediate configuration the thickness of the body is decreased but the lateral dimensions are increased in the amount required to maintain the volume unchanged. In the current configuration, the body has the same lateral dimensions as in the reference configuration, reflecting its state of uniaxial deformation, but its thickness differs from that in the reference configuration because of the elastic volume change. As discussed in Sect. 2.2, this deformation also involves shear of the material. There is one other point: Since the only restriction on the intermediate configuration is that it have the symmetry described and be of the same volume as the reference configuration, it is arbitrary to within a rotation. In the present case we suppose that no rotation occurs, so we can write

$$\mathbf{F}^{\mathrm{p}} = \operatorname{diag}\left\| F_{\mathrm{L}}^{\mathrm{p}}, F_{\mathrm{T}}^{\mathrm{p}}, F_{\mathrm{T}}^{\mathrm{p}} \right\| \quad \text{and} \quad \mathbf{F}^{\mathrm{e}} = \operatorname{diag}\left\| F_{\mathrm{L}}^{\mathrm{e}}, F_{\mathrm{T}}^{\mathrm{e}}, F_{\mathrm{T}}^{\mathrm{e}} \right\|, \tag{7.59}$$

where $\operatorname{diag}\|\cdots\|$ designates a diagonal matrix having the indicated components in the 11, 22, and 33 positions, and where we must have

$$\det \mathbf{F}^{\mathrm{p}} = F_{\mathrm{L}}^{\mathrm{p}}(F_{\mathrm{T}}^{\mathrm{p}})^2 = 1\,, \quad \text{or} \quad F_{\mathrm{T}}^{\mathrm{p}} = (F_{\mathrm{L}}^{\mathrm{p}})^{-1/2}\,. \tag{7.60}$$

The deformation gradient, $\mathbf{F}$, takes the form of Eq. 7.58 if

$$F_{\mathrm{T}}^{\mathrm{e}} = (F_{\mathrm{L}}^{\mathrm{p}})^{1/2}\,. \tag{7.61}$$

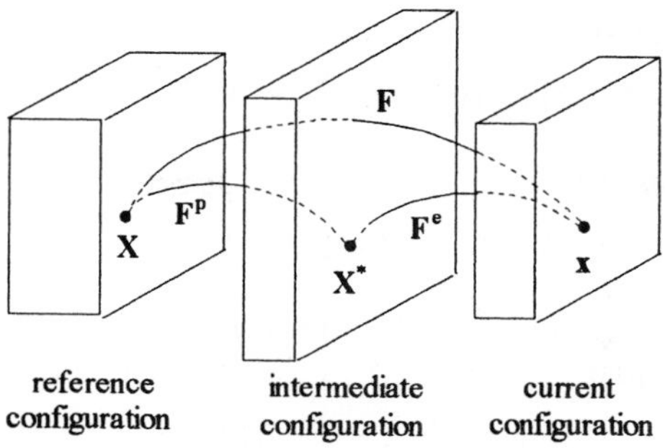

Figure 7.7. Configurations used to analyze finite elastic–plastic deformation of a body in states of uniaxial deformation.

Substituting Eqs. 7.60 and 7.61 into Eqs. 7.59 yields

$$\mathbf{F}^{p} = \text{diag}\left\| F_L^p,\ (F_L^p)^{-1/2},\ (F_L^p)^{-1/2} \right\|, \tag{7.62}$$

and

$$\mathbf{F}^{e} = \text{diag}\left\| F_L^e,\ (F_L^p)^{1/2},\ (F_L^p)^{1/2} \right\|. \tag{7.63}$$

We see that $\mathbf{F}^p$ and $\mathbf{F}^e$ can be expressed in terms of the two longitudinal components F_L^p and F_L^e.

Small Elastic Distortion Combined with Finite Dilatation. We now introduce the simplifications that result when the elastic shear strain is restricted to small values even though the compression may not be small. We note that this case differs from the uniaxial deformation considered in Chap. 6 because neither the elastic nor the plastic part of the elastic–plastic deformation is, itself, a uniaxial deformation. Nevertheless, we follow a procedure similar to that used previously, writing $\mathbf{F}^e$ in the form $\mathbf{F}^e = \mathbf{F}^{es}\,\mathbf{F}^{v}$ where

$$\mathbf{F}^{v} = (v/v_R)^{1/3}\,\mathbf{I}, \tag{7.64}$$

and

$$\mathbf{F}^{es} = \text{diag}\left\| 1+\widetilde{E}_L^{es},\ \ 1-\tfrac{1}{2}\widetilde{E}_L^{es},\ \ 1-\tfrac{1}{2}\widetilde{E}_L^{es} \right\|, \tag{7.65}$$

so that

$$\mathbf{F}^{e} = (v/v_R)^{1/3}\,\text{diag}\left\| 1+\widetilde{E}_L^{es},\ \ 1-\tfrac{1}{2}\widetilde{E}_L^{es},\ \ 1-\tfrac{1}{2}\widetilde{E}_L^{es} \right\| \tag{7.66}$$

to lowest order in the elastic shear strains. Also to lowest order, we have

$$\mathbf{E}^{es} = \text{diag}\left\| \widetilde{E}_L^{es},\ \ -\tfrac{1}{2}\widetilde{E}_L^{es},\ \ -\tfrac{1}{2}\widetilde{E}_L^{es} \right\|. \tag{7.67}$$

By Eq. 7.61 and 7.66 we have

$$\widetilde{E}_{\mathrm{L}}^{\mathrm{es}} = 2\,[1-(v/v_{\mathrm{R}})^{-1/3}(F_{\mathrm{L}}^{\mathrm{p}})^{1/2}], \tag{7.68}$$

and from Eq. 7.66 we see that

$$F_{\mathrm{L}}^{\mathrm{e}} = 3(v/v_{\mathrm{R}})^{1/3} - 2(F_{\mathrm{L}}^{\mathrm{p}})^{1/2}. \tag{7.69}$$

Knowledge of $\widetilde{E}_{\mathrm{L}}^{\mathrm{es}}$ and $F_{\mathrm{L}}^{\mathrm{p}}$ is sufficient for determination of all of the remaining measures of deformation.

Stress Relation. The stress is to be inferred from the thermoelastic part of the deformation gradient. The compression (all of which produces an elastic response) can be quite large, but the elastic distortion is limited to a rather small value. We considered this case in Chap. 6 and found the stress, as given by Eqs. 6.57 and 6.58, to be of the form

$$t_{ij} = -p(v,\eta)\,\delta_{ij} + 2\mu(v,\eta)\,\widetilde{E}_{ij}^{\mathrm{es}}, \tag{7.70}$$

where p is the thermodynamic pressure and $\widetilde{\mathbf{E}}^{\mathrm{es}}$ is the elastic part of the shear strain. The theory to be developed in this section is needed only for rather weak elastic–plastic waves because strong shocks can be analyzed using the compressible fluid theory of Chap. 5. For this reason we shall neglect thermal effects and consider only a purely mechanical theory. The functions p and μ appearing in Eq. 7.70 are then taken to be functions of v alone.

To proceed with the analysis it is necessary to consider separately the several branches of the stress–strain path shown in Fig. 7.8. Because we intend to apply the theory being developed to shock waves, we identify the yield point with the Hugoniot elastic limit.

Elastic Compression. For stresses below the Hugoniot elastic limit the compression and shear strain are both small and the stress relations

$$t_{11} = (\lambda_{\mathrm{R}} + 2\mu_{\mathrm{R}})\widetilde{E}_{11} \quad \text{and} \quad t_{22} = t_{33} = \lambda_{\mathrm{R}}\,\widetilde{E}_{11} \tag{7.71}$$

of linear elasticity theory are applicable. Since $|1-(v/v_{\mathrm{R}})| \ll 1$ in this range, we have $\widetilde{E}_{11} = -[1-(v/v_{\mathrm{R}})]$ and Eqs. 7.71 can be written

$$t_{11} = -(\lambda_{\mathrm{R}} + 2\mu_{\mathrm{R}})[1-(v/v_{\mathrm{R}})] \quad \text{and} \quad t_{22} = t_{33} = -\lambda_{\mathrm{R}}\,[1-(v/v_{\mathrm{R}})]. \tag{7.72}$$

The pressure and maximum shear stress are given by

$$\begin{aligned} p(v) &= -(\lambda_{\mathrm{R}} + \tfrac{2}{3}\mu_{\mathrm{R}})\widetilde{E}_{11} = (\lambda_{\mathrm{R}} + \tfrac{2}{3}\mu_{\mathrm{R}})[1-(v/v_{\mathrm{R}})] \\ \tau_{45^\circ} &= \tfrac{1}{2}(t_{11} - t_{22}) = \mu_{\mathrm{R}}\,\widetilde{E}_{11} = \tfrac{3}{2}\mu_{\mathrm{R}}\,\widetilde{E}_{\mathrm{L}}^{\mathrm{es}} = -\mu_{\mathrm{R}}\,[1-(v/v_{\mathrm{R}})]. \end{aligned} \tag{7.73}$$

These equations are valid in the range $0 \le -t_{11} \le t_{11}^{\mathrm{HEL}}$.

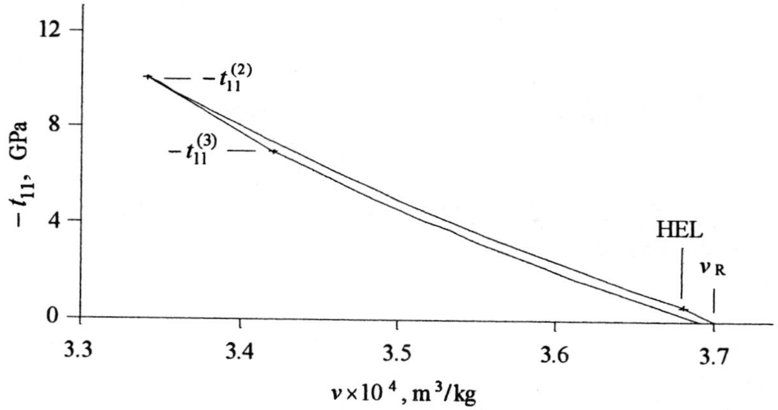

Figure 7.8. Stress–strain path for aluminum alloy 6061-T6 that is compressed by a plane shock of strength $-t_{11} = 10$ GPa and then relieved of this stress. The Hugoniot elastic limit is taken to be 0.6 GPa. The path from v_R to v^{HEL} is one of elastic compression, the path from $-t_{11}^{HEL}$ to $-t_{11}^{(2)}$ is one of plastic compression, the path from $-t_{11}^{(2)}$ to $-t_{11}^{(3)}$ is one of elastic decompression, and the remaining section from $-t_{11}^{(3)}$ to zero stress is one of plastic decompression.

In some cases it is convenient to use the nonlinear relation of Eq. 7.70 with $p(v)$ given by the equation of state that is to be adopted in the range of elastic–plastic compression and the shear modulus given by Eq. 6.62. In this case we have

$$-t_{11} = p(v) - \tfrac{4}{3}\mu(v)[1-(v/v_R)]. \tag{7.74}$$

Elastic–Plastic Compression. As in the small-deformation theory, the shear stress in compression is limited by the yield criterion, $\tau_{45°} = -Y/2$, where the value of Y, which is designated Y_0 in the initial state, may change (usually increase) as a result of metallurgical changes that occur in the course of a plastic-deformation process.

At the Hugoniot elastic limit, characterized by the stress

$$-t_{11} = t_{11}^{HEL} = [(\lambda/2\mu)+1]\,Y_0, \tag{7.75}$$

the specific volume is given by

$$v^{HEL} = v_R\left[1 - \frac{t_{11}^{HEL}}{\lambda_R + 2\mu_R}\right] = v_R\left[1 - \frac{Y_0}{2\mu_R}\right], \tag{7.76}$$

and the corresponding longitudinal strain is

$$E_{11}^{\mathrm{HEL}} = -Y_0/(2\mu_{\mathrm{R}}). \tag{7.77}$$

For stresses beyond the Hugoniot elastic limit, we use Eqs. 7.67 and 7.68 to write the stress relations 7.70 in the form

$$\begin{aligned} t_{11} &= -p(v) + 4\mu(v)\,[1 - (v/v_{\mathrm{R}})^{-1/3}(F_{\mathrm{L}}^{\mathrm{P}})^{1/2}] \\ t_{22} = t_{33} &= -p(v) - 2\mu(v)\,[1 - (v/v_{\mathrm{R}})^{-1/3}(F_{\mathrm{L}}^{\mathrm{P}})^{1/2}]. \end{aligned} \tag{7.78}$$

When the shear stress is expressed in terms of Y these equations give

$$(F_{\mathrm{L}}^{\mathrm{P}})^{1/2} = (v/v_{\mathrm{R}})^{1/3}[1 - \tfrac{1}{2}\widetilde{E}_{\mathrm{L}}^{\mathrm{es}}] = (v/v_{\mathrm{R}})^{1/3}\left[1 + \frac{Y}{6\mu(v)}\right], \tag{7.79}$$

and, as always,

$$t_{11} = -p(v) - \tfrac{2}{3}Y \quad \text{and} \quad t_{22} = t_{33} = -p(v) + \tfrac{1}{3}Y \tag{7.80}$$

for compression processes. These equations are the same as the corresponding equations for the small-deformation theory, except that the pressure varies nonlinearly with the compression rather than being related to it by a constant bulk modulus. The pressure is to be calculated from a suitable equation of state as discussed in Chap. 5. In practice, one usually uses the Hugoniot given by Eq. 3.12.

For non-hardening materials the yield strength is constant so the normal stress components differ from the pressure by a constant amount. However, it is often appropriate to incorporate deformation hardening into the constitutive description by allowing Y to depend upon the plastic strain, the plastic work done, or some other variable associated with the plastic deformation. One equation that has been used to represent deformation hardening is

$$Y = Y_0\left(1 + h\,|\gamma^{\mathrm{P}}|^{1/n}\right), \tag{7.81}$$

where γ^{P} is the plastic part of the shear angle, as defined in Fig. 2.2 and where $h \geq 0$ and $n > 1$ are material constants. Analysis of the figure (upon the premise that it represents the plastic part of the deformation) shows that γ^{P} and $F_{\mathrm{L}}^{\mathrm{P}}$ are related by the equation

$$\gamma^{\mathrm{P}} = 2\arctan\left[\frac{(F_{\mathrm{L}}^{\mathrm{P}})^{3/2} - 1}{(F_{\mathrm{L}}^{\mathrm{P}})^{3/2} + 1}\right]. \tag{7.82}$$

Equating the yield stress given by Eq. 7.79_2 to that given by Eq. 7.81 yields the equation

$$Y_0\left\{1 + h\left|2\arctan\left(\frac{(F_{\mathrm{L}}^{\mathrm{P}})^{3/2} - 1}{(F_{\mathrm{L}}^{\mathrm{P}})^{3/2} + 1}\right)\right|^{1/n}\right\} + 6\mu(v)\left[1 - (v/v_{\mathrm{R}})^{-1/3}(F_{\mathrm{L}}^{\mathrm{P}})^{1/2}\right] = 0, \tag{7.83}$$

relating the yield stress to the specific volume. This equation can be solved numerically for $F_L^p(v)$ when the coefficients Y_0, $\mu(v)$, h, and n are available. With this solution in hand, the yield stress $Y(v)$ follows immediately from Eq. 7.81. An example of the effect of changing the hardening exponent is shown in Fig. 7.9 and an example of a $t_{11} - v$ compression curve is illustrated in Fig. 7.10.

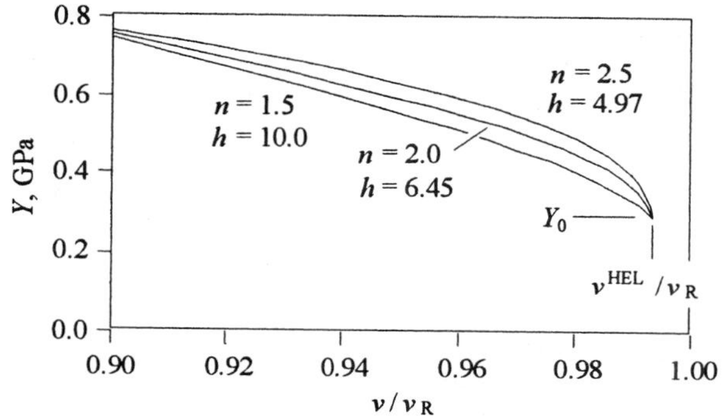

Figure 7.9. Curves showing the increase in Y for aluminum alloy 6061-T6 that follows from Eq. 7.81 for $Y_0 = 0.29$ GPa , $\mu(v) = \mu_R$, and several values of n. The parameter h was varied to achieve approximately the same terminal hardness in each case. In calculating these curves the values of F_L^p used are those related to v by Eq. 7.83.

If the material is compressed to the specific volume $v^{(2)}$ the deformation gradient, $F_L^{p(2)} = F_L^p(v^{(2)})$, is obtained from Eq. 7.83 and the yield stress, $Y^{(2)} = Y(v^{(2)})$, can then be obtained from Eq. 7.81. With these results the pressure, $p^{(2)} = p(v^{(2)})$ follows from the pressure equation of state and the longitudinal stress, $-t_{11}^{(2)}$, is given by Eq. 7.78$_1$. This state, which we shall call $s^{(2)}$, can serve as the starting point of an elastic decompression process.

Elastic Decompression. When the compressive stress producing a shock is removed, the immediate response is relief of the elastic deformation and, if the compressive stress was great enough, this is followed by a reverse plastic deformation. This, too, is the same as for the small-deformation theory except that the response is nonlinear.

The elastic decompression begins at a state characterized by the parameters $t_{11}^{(2)}$, $v^{(2)}$, and $F_L^p(v^{(2)})$ determined in the previous section as the limit of a plastic compression process. Because this part of the decompression process is elastic, F_L^p remains fixed so that $F_L^p = F_L^p(v^{(2)})$. The stress relations for the elastic decompression are those of Eqs. 7.78, so we have

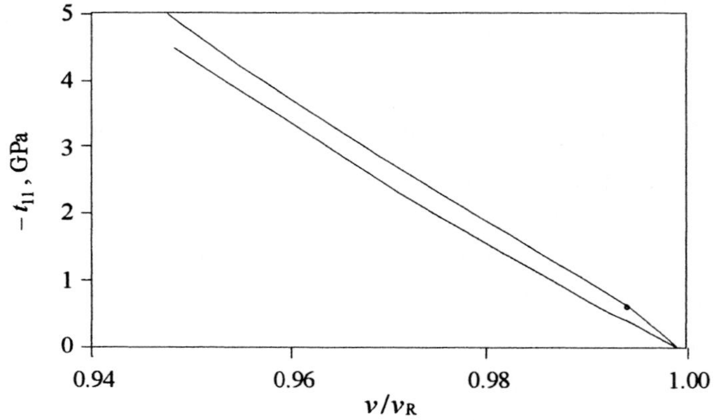

Figure 7.10. A Hugoniot curve corresponding to the $n = 2.5$ hardening curve shown on Fig. 7.9. The curve has a concave downward section immediately above the HEL. This means that, for the hardening equation and parameters used, plastic waves of moderate strength include a centered simple wave between the precursor shock and the plastic shock.

$$\begin{aligned} t_{11} &= -p(v) + 4\mu(v)\{1-(v/v_R)^{-1/3}\,[F_L^P(v^{(2)})]^{1/2}\} \\ t_{22} = t_{33} &= -p(v) - 2\mu(v)\{1-(v/v_R)^{-1/3}\,[F_L^P(v^{(2)})]^{1/2}\}. \end{aligned} \tag{7.84}$$

The elastic decompression process ends when $t_{11} = 0$ or when the shear stress reaches the reverse yield point, whichever occurs first. Let us suppose that the stress reaches the reverse yield point first. This will occur at the value of v, which we shall call $v^{(3)}$, for which

$$t_{11} - t_{22} = 6\mu(v)\{1-(v/v_R)^{-1/3}\,[F_L^P(v^{(2)})]^{1/2}\} = Y(v^{(2)}) . \tag{7.85}$$

Usually, one assumes that Y remains unchanged during the decompression process, and we shall do so here. However, there is no good reason to expect this behavior because the direction of the shear stress on the 45° planes is reversed when the decompression curve crosses the pressure curve. Many observations are inconsistent with the assumption that the forward and reverse yield stresses are the same. Often, reverse yielding occurs at a lower stress than forward yield, a response called a Bauschinger effect. Nevertheless, with the assumption that the forward and reverse yield stresses are of the same magnitude we can solve Eq. 7.85 for $v^{(3)}$ and, with this, we obtain $t_{11}^{(3)} \equiv t_{11}(v^{(3)})$ from Eqs. 7.84. Since the plastic strain maintains the value it had at the beginning of the elastic decompression, we have $F_L^P(v^{(3)}) = F_L^P(v^{(2)})$.

Elastic–Plastic Decompression. The stress in the plastic part of the decompression process begins at the state $s^{(3)}$ just determined and is given by the equations

$$t_{11} = -p(v) + \tfrac{2}{3}Y(v) \quad \text{and} \quad t_{22} = t_{33} = -p(v) - \tfrac{1}{3}Y(v).$$

In order to use these equations we must know the function $Y(v)$.

7.3 Finite Elastic–Viscoplastic Deformation

Both experimental observation and our understanding of the important physical mechanisms indicate that plastic deformation accumulates at a finite rate in response to imposition of shear stress. This rate effect is not captured by the theories presented in previous sections of this chapter, but is important to our understanding of shock processes because the characteristic times associated with the rate-dependent effects are comparable to those of a shock-wave experiment. They are responsible for some of the structural features of compression shocks that, in turn, influence evolutionary processes such as spall fracture.

There is no universally accepted theory of large elastic–plastic deformation; the subject is not in the settled state of the theory of elasticity. Although there are a number of carefully developed theories of elastic–viscoplastic response to finite deformations, it is difficult to validate or compare them because this requires both comprehensive data sets for a range of materials and an extensive computational program to simulate carefully executed and well-instrumented experiments. A variety of experimental results addressing specific physical responses have been reviewed by Asay and Chhabildas [6] and it is clear that some of the observed phenomena lie beyond the range of any theory of elastic–viscoplastic response. For example, experiments in which adiabatic shear bands form, flow, and then cool demonstrate changes of shear strength that depend upon time intervals between successive compression processes or between compression and decompression processes. These observations cannot be explained by any theory in which the stress depends only upon current values of strain and strain rate. Metallurgical examination of materials that have been recovered after shock-compression experiments shows that the deformation process is far more complicated than is described by the laminar motions contemplated in the discussions of this volume and the thermomechanical effect of these complications is not known [53,111].

The kinematical description of the deformation discussed in the previous section, including extensions that address rates of deformation [29], is widely accepted and has been adopted by most investigators developing continuum theories of finite elastic–viscoplastic response. Constitutive equations in use are more varied, partly as a result of the evolving state of the theories and partly as a result of the wide variety of responses that have been observed. In the case of shock physics, particularly, deformation accumulates at rates that exceed those usually studied in the laboratory and with which we are most experienced. We have noted that an important mechanism of plastic deformation is the motion of dislocations through the crystal lattice that forms the arrangement of the atoms in many materials. In this section we outline an evolutionary theory of plastic deformation. In previous parts of this volume, the presentation has been entirely in the framework of continuum mechanics, but much of what is known of shock physics is based upon considerations at the atomic scale, the scale of grains in polycrystalline materials, etc. In this section we shall make specific use of some elementary aspects of dislocation behavior, but at the scale of the continuum theory of dislocations (which is coarser than that of the individual dislocations). The theory developed will include such parameters as the density of dislocations and the average velocity with which they move. The equations obtained have been found to be of a form that yields predictions in good agreement with continuum-level experimental observations. However, this agreement is based upon the use of parameters inferred from continuum-level measurements and these parameters often seem inconsistent with related microscopic measurements. This is not unusual, as experience shows that continuum theories are often more broadly valid than their microscopic basis would suggest. Since it is well known that the actual mechanism of plastic deformation is far more complex than that of the dislocation-mechanical models, and since refinements of the details of these models have not led to improved agreement of theory and observation at the microscopic level, we use only the most basic aspects of the microscopic theory, with the objective of establishing forms of the continuum equations but not that of obtaining a theory that is valid at the microscopic level.

Kinematics. The kinematical description of finite elastic–plastic deformations that we shall use is that of Sect. 7.2. We shall need to augment the equations presented previously with some measures of deformation rate. The velocity gradient is given by Eq. 2.18,

$$l_{ij} = \dot{F}_{iI} \overset{-1}{F}_{Ij} , \tag{7.86}$$

and substitution of Eq. 7.54 into this equation yields the expression

$$l_{ij} = l_{ij}^{\mathrm{e}} + l_{ij}^{\mathrm{p}}, \tag{7.87}$$

where we have identified the parts of the velocity gradient associated with the elastic and the plastic contributions to the deformation as

$$l_{ij}^{\mathrm{e}} = \dot{F}_{i\Gamma}^{\mathrm{e}} \overset{-1}{F}{}_{\Gamma j}^{\mathrm{e}}, \quad \text{and} \quad l_{ij}^{\mathrm{p}} = F_{i\Gamma}^{\mathrm{e}} \Lambda_{\Gamma\Delta}^{\mathrm{p}} \overset{-1}{F}{}_{\Delta j}^{\mathrm{e}}, \tag{7.88}$$

respectively, with

$$\Lambda^{\mathrm{P}}_{\Gamma\Delta} = \dot{F}^{\mathrm{P}}_{\Gamma I} \overset{-1}{F}{}^{\mathrm{P}}_{I\Delta} . \tag{7.89}$$

The elastic and plastic parts of the stretching and spin are calculated as the symmetric and antisymmetric parts, respectively, of these tensors.

7.3.1 Constitutive Equations for Viscoplastic Flow

The development of this section is based upon the premise that all plastic flow is attributable to motion of edge dislocations. This allows us to use information available from studies of the motion of these defects to establish a specific form for the viscoplastic response function. The development presented is oversimplified from a microscopic point of view, but follows procedures that have proven successful for predicting observed responses in the conditions of interest [36,73,107]. Among the advantages of a theory based upon microscopic concepts rather than based solely upon continuum principles is that it is assembled from established equations that describe the separate phenomena contributing to the overall continuum response. It is easier to understand and, when necessary, modify each of the separate parts than to deal directly with the continuum response.

Dislocation Mechanics. Edge dislocation lines move through crystals on specific planes and in specific directions that are characteristic of the lattice arrangement of the crystal. A *slip system* in a crystal is defined by the plane on which the dislocation moves (i.e., on which slip occurs) and the direction of the relative displacement of the two parts of the crystal that are separated by the slip plane. A slip system is characterized by the normal vector to the slip plane and the *Burgers vector*, which lies in the direction in which a dislocation can move in this plane. The magnitude of the Burgers vector is the slip displacement that occurs when a single dislocation passes over the slip plane. Because the normal vector to the slip plane, **N**, and the Burgers vector, **B**, are associated with the crystallographic arrangement of the material, they are most naturally defined in the reference configuration. However, they can be equally well defined in the intermediate configuration, X^*, because the atoms of the plastically deformed material bear the same relationship to their (new) neighbors as they had (relative to their old ones) in the reference configuration.

Slip on a Single System. Figure 7.11 illustrates a simple shear produced by dislocation motion on the slip planes defined by the unit normal vector $\mathbf{N} = N(0,1,0)$ and with the Burgers vector $\mathbf{B} = B(1,0,0)$. The continuum description of the deformation is

$$X_1^* = X_1 + \gamma^{\mathrm{P}} X_2, \quad X_2^* = X_2, \quad X_3^* = X_3 , \tag{7.90}$$

from which we obtain the deformation gradient, its inverse and its material derivative

$$\mathbf{F}^{\mathrm{p}} = \begin{Vmatrix} 1 & \gamma^{\mathrm{p}} & 0 \\ 0 & 1 & 0 \\ 0 & 0 & 1 \end{Vmatrix}, \qquad \overset{-1}{\mathbf{F}}{}^{\mathrm{p}} = \begin{Vmatrix} 1 & -\gamma^{\mathrm{p}} & 0 \\ 0 & 1 & 0 \\ 0 & 0 & 1 \end{Vmatrix}, \qquad \dot{\mathbf{F}}^{\mathrm{p}} = \begin{Vmatrix} 0 & \dot{\gamma}^{\mathrm{p}} & 0 \\ 0 & 0 & 0 \\ 0 & 0 & 0 \end{Vmatrix}. \tag{7.91}$$

From Eq. 7.89 we also have

$$\mathbf{\Lambda}^{\mathrm{p}} = \dot{\mathbf{F}}^{\mathrm{p}} \overset{-1}{\mathbf{F}}{}^{\mathrm{p}} = \begin{Vmatrix} 0 & \dot{\gamma}^{\mathrm{p}} & 0 \\ 0 & 0 & 0 \\ 0 & 0 & 0 \end{Vmatrix}, \tag{7.92}$$

which can be written

$$\mathbf{\Lambda}^{\mathrm{p}} = \dot{\gamma}^{\mathrm{p}}\, \overline{\mathbf{B}} \otimes \overline{\mathbf{N}}, \tag{7.93}$$

where the overbar denotes normalization of the length of the vector and the outer product has the components $(\overline{\mathbf{B}} \otimes \overline{\mathbf{N}})_{\Gamma\Delta} = \overline{B}_{\Gamma}\, \overline{N}_{\Delta}$.

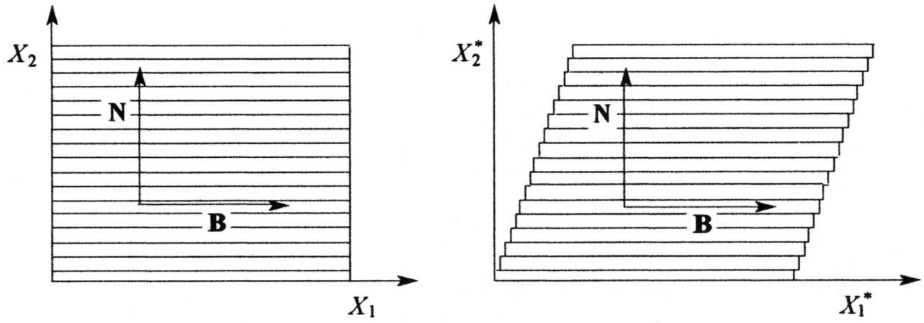

Figure 7.11. Dislocation model of simple shearing deformation from the reference configuration at the left to the plastically deformed configuration at the right. The horizontal lines represent slip planes.

Slip on Multiple Systems. The significance of the foregoing analysis is that the deformation rate tensor for a deformation produced by slip on several systems is obtained by adding the contribution for each of the individual systems (this procedure is not valid for the deformation gradient). As the illustration of Fig. 5.1 suggests, slip displacements that occur on several systems result in blocks of the material moving relative to one another but without stretching or rotation of the individual blocks.

A typical slip system is shown in Fig. 7.12; it is described by the normal vector to the slip plane

$$\mathbf{N} = N(\cos\varphi_2, \cos\varphi_1 \sin\varphi_2, \sin\varphi_1 \sin\varphi_2) \tag{7.94}$$

and the Burgers vector

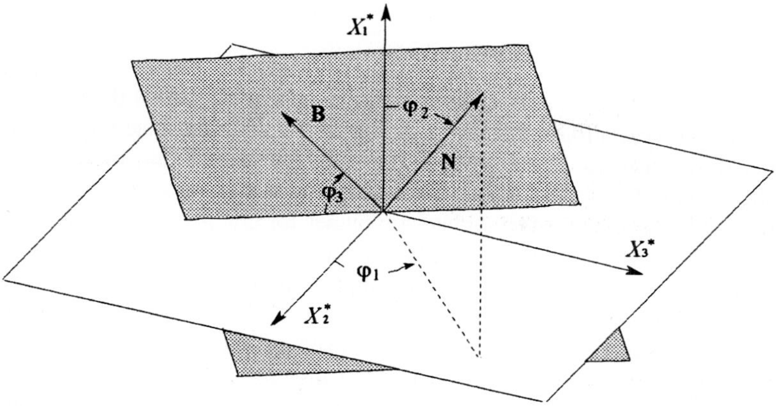

Figure 7.12. Slip system. The shaded area represents the slip plane, the vector **N** is its normal and the vector **B**, which lies in the plane, is the Burgers vector.

$$\mathbf{B} = B(\sin\varphi_2 \sin\varphi_3,\ \sin\varphi_1 \cos\varphi_3 - \cos\varphi_1 \cos\varphi_2 \sin\varphi_3 \\ - \cos\varphi_1 \cos\varphi_3 - \sin\varphi_1 \cos\varphi_2 \sin\varphi_3). \tag{7.95}$$

One can show that the normal vector is of length N, the Burgers vector is of length B, and **B** is perpendicular to **N**. To represent all possible slip planes we allow φ_1 to range over the angles $0 \le \varphi_1 < 2\pi$ and φ_2 to range over the angles $0 \le \varphi_2 \le \pi/2$. Each point of the surface of a unit hemisphere (for $X_1^* \ge 0$) centered on the origin of the X^* coordinates corresponds to a slip plane defined by the angles φ_1 and φ_2. For each of these planes there is a continuum of possible Burgers vectors corresponding to angles φ_3 in the range $0 \le \varphi_3 < \pi$. As shown here, the vectors are defined in the plastically deformed configuration but, as noted previously, are invariant to the plastic deformation that would have taken them to this configuration had they been defined in the reference configuration. Let us now consider the transformation of these vectors from the plastically deformed configuration to the current configuration. In a crystal the Burgers vector lies along a line of atoms and has a length related to the lattice spacing so it is transformed as a line element to its image, **b**, in the current configuration:

$$\mathbf{b} = \mathbf{F}^{\mathrm{e}} \mathbf{B}. \tag{7.96}$$

The normal vector to the slip plane represents a surface element of this plane and transforms to its image in the current configuration according to the equation [103, Eq. 20.8]:

$$\mathbf{n} = J \mathbf{N} \overset{-1}{\mathbf{F}^{\mathrm{e}}}. \tag{7.97}$$

In this case

$$J = \det \mathbf{F} = v / v_{\mathrm{R}} . \tag{7.98}$$

Slip in monocrystals occurs on known planes and directions that are characteristic of the crystallographic arrangement of the material. The slip planes are usually those in which the atoms are most closely packed and the direction of the slip is that of the most closely packed lattice vector. In any case, there is a known, modest, number of candidate slip planes for any given ductile crystalline material. In dealing with polycrystalline metals we shall assume that slip (interpreted in the continuum sense as an average response) can occur on any plane and in any direction in the plane. However, we shall introduce the approximation that it is sufficient to restrict attention to some finite number of slip systems, making the analysis the same as that for a monocrystal. In this case, a finite number of sets of angles, $(\varphi_1^{(k)}, \varphi_2^{(k)}, \varphi_3^{(k)})$ for $k = 1,\cdots,n$, define the candidate slip systems. The length of the dislocation lines per unit reference volume of material at $(\mathbf{X}^*, t)$ that falls on the kth slip system is designated $\mathcal{N}_{k\mathrm{T}}(\mathbf{X}^*, t)$ and the total length of dislocation line per unit volume, $\mathcal{N}_{\mathrm{T}}(\mathbf{X}^*, t)$, is the sum of the numbers $\mathcal{N}_{k\mathrm{T}}(\mathbf{X}^*, t)$ for each individual system. If the initial distribution of dislocations over the n slip systems is uniform, then

$$\mathcal{N}_{k\mathrm{T}}(\mathbf{X}^*, t) = \mathcal{N}_{\mathrm{T}}(\mathbf{X}^*, t) / n . \tag{7.99}$$

Experimental observation indicates that the motion of some of the dislocations on a given slip system may be prevented by obstacles in the lattice or they may become immobilized in the course of a deformation, leaving only a fraction, $f_{k\mathrm{M}}(\mathbf{X}^*, t)$, of the $\mathcal{N}_{k\mathrm{T}}(\mathbf{X}^*, t)$ dislocations that can contribute to deformation on the kth slip system. Therefore, the number of mobile dislocations on this system is

$$\mathcal{N}_{k\mathrm{M}}(\mathbf{X}^*, t) = f_{k\mathrm{M}}(\mathbf{X}^*, t)\, \mathcal{N}_{k\mathrm{T}}(\mathbf{X}^*, t) , \tag{7.100}$$

and the total number of mobile dislocations per unit reference state volume of the material, $\mathcal{N}_{\mathrm{M}}(\mathbf{X}^*, t)$, is the sum of $\mathcal{N}_{k\mathrm{M}}$ over the individual slip systems.

Since the velocity gradient for a deformation produced by slip on several systems is obtained by adding the contribution for each of the individual systems, we have

$$\mathbf{\Lambda}^{\mathrm{P}} = \sum_{k=1}^{n} \dot{\gamma}^{\mathrm{P}(k)}\, \overline{\mathbf{B}}^{(k)} \otimes \overline{\mathbf{N}}^{(k)} . \tag{7.101}$$

The connection between the stress and the plastic deformation rate is made through Orowan's equation,

$$\dot{\gamma}^{\mathrm{P}(k)} = B^{(k)}\, \mathcal{N}_{k\mathrm{M}}\, V_{X^*}^{(k)}(\tau) , \tag{7.102}$$

where $B^{(k)}$ is the length of the Burgers vector, $\mathcal{N}_{k\mathrm{M}}$ is the number of mobile dislocations on the kth slip system, $\tau^{(k)}$ is the shear traction on the slip plane in the direction of the Burgers vector, and $V_{X^*}(\tau^{(k)})$ is the average velocity with which these dislocations move in the direction $\mathbf{B}^{(k)}$. Substitution of Eq. 7.102 into Eq. 7.101 yields the result

$$\mathbf{\Lambda}^{\mathrm{p}} = \sum_{k=1}^{n} \mathcal{N}_{k\mathrm{M}}\, \mathbf{B}^{(k)} \otimes \overline{\mathbf{N}}^{(k)}\, V_{X^*}^{(k)}, \tag{7.103}$$

and the rate of plastic deformation is given by

$$\dot{\mathbf{F}}^{\mathrm{p}} = \mathbf{\Lambda}^{\mathrm{p}}\, \mathbf{F}^{\mathrm{p}} \ . \tag{7.104}$$

To calculate the dislocation velocity we need to know the component of the shear stress that falls on the slip plane and in the direction of the Burgers vector. Because the stress is defined in the current configuration we begin there. The first step is to transform the vectors $\mathbf{N}^{(k)}$ and $\mathbf{B}^{(k)}$ to the current configuration. This is done using Eqs. 7.96 and 7.97. In applying these equations we shall introduce approximations that follow from decomposing the elastic part of the deformation into a finite dilatation and a small distortion. When this decomposition is adopted we have

$$F_{i\Gamma}^{\mathrm{e}} = (v/v_{\mathrm{R}})^{1/3} F_{i\alpha}^{\mathrm{es}}\, \delta_{\alpha\Gamma}\,, \tag{7.105}$$

so

$$n_i^{(k)} = (v/v_{\mathrm{R}})^{2/3} N_\Gamma^{(k)}\, \delta_{\Gamma\alpha}\, \overset{-1}{F}{}_{\alpha i}^{\mathrm{es}} \tag{7.106}$$

and

$$b_i^{(k)} = (v/v_{\mathrm{R}})^{1/3} F_{i\alpha}^{\mathrm{es}}\, \delta_{\alpha\Gamma}\, B_\Gamma^{(k)}. \tag{7.107}$$

The magnitudes of these vectors are given by

$$n^{(k)} = (v/v_{\mathrm{R}})^{2/3} N^{(k)}\, [1 - (N^{(k)})^{-2} N_\Gamma^{(k)} N_\Delta^{(k)}\, \delta_{\Gamma\alpha}\, \delta_{\Delta\beta}\, \widetilde{E}_{\alpha\beta}^{\mathrm{es}} + \cdots] \tag{7.108}$$

and

$$b^{(k)} = (v/v_{\mathrm{R}})^{1/3} B^{(k)}\, [1 + (B^{(k)})^{-2} B_\Gamma^{(k)} B_\Delta^{(k)}\, \delta_{\Gamma\alpha}\, \delta_{\Delta\beta}\, \widetilde{E}_{\alpha\beta}^{\mathrm{es}} + \cdots] \tag{7.109}$$

when we neglect quadratic terms in $\widetilde{\mathbf{E}}^{\mathrm{es}}$. Unit vectors $\overline{\mathbf{n}}^{(k)}$ and $\overline{\mathbf{b}}^{(k)}$ are obtained by dividing the vector by its magnitude.

The traction vector on the surface element at $\mathbf{x}$ that has unit normal vector $\overline{\mathbf{n}}^{(k)}$ is given by

$$\mathbf{t}^{(k)} = \overline{\mathbf{n}}^{(k)}\, \mathbf{t}, \tag{7.110}$$

and the component of this traction in the direction $\mathbf{b}^{(k)}$ is

$$\boldsymbol{\tau}^{(k)} = (\mathbf{t}^{(k)} \cdot \overline{\mathbf{b}}^{(k)}) \overline{\mathbf{b}}^{(k)} . \tag{7.111}$$

The magnitude of the vector $\boldsymbol{\tau}^{(k)}$ is designated $\tau^{(k)}$: $\tau^{(k)} = \mathbf{t}^{(k)} \cdot \overline{\mathbf{b}}^{(k)}$.

Several expressions for the dislocation velocity in the current configuration, $V_x(\tau)$, have been used, including [36,60,73],

$$V_x^{(k)}(\tau^{(k)}) = \begin{cases} 0, & |\tau^{(k)}| \le \tau^{(k)*} \\ C_S \operatorname{sgn}(\tau^{(k)}) \dfrac{(s^{(k)})^m}{1+(s^{(k)})^m}, & |\tau^{(k)}| \ge \tau^{(k)*}, \end{cases} \tag{7.112}$$

where $s^{(k)} = (|\tau^{(k)}| - \tau^{(k)*})/\tau^{\dagger}$. In this equation the material parameters are $m > 0$, $C_S > 0$, the elastic shear wavespeed, $\tau^{\dagger} > 0$, a characteristic stress to be determined experimentally, and $\tau^{(k)*} > 0$, the *back stress* on the kth slip system. The back stress, which plays the role of a shear yield stress on the slip system and increases with increasing deformation, is given by

$$\tau^{(k)*} = \tau_0 \{ 1 + h \, |\gamma^{p(k)}|^{1/n} \} , \tag{7.113}$$

which can be placed in the form of an evolutionary equation for an internal state variable by differentiation.

The next task is transforming the dislocation velocity given by Eq. 7.112 from the current configuration to the plastically deformed configuration so that it can be substituted into Eq. 7.103. Since a dislocation moves over the same number of lattice sites in each configuration in a unit time, we have

$$V_{X^*}^{(k)} = (B^{(k)} / b^{(k)}) V_x^{(k)} , \tag{7.114}$$

with the ratio $B^{(k)} / b^{(k)}$ being determined from Eq. 7.109.

The factor $\mathcal{N}_{k\mathrm{M}}$ in Eq. 7.101 can be expressed in terms of $\mathcal{N}_{k\mathrm{T}}$ and $f_{k\mathrm{M}}$ by Eq. 7.100, with these factors being given by evolutionary equations for which we shall adopt the forms

$$\begin{aligned} \dot{\mathcal{N}}_{k\mathrm{T}} &= L_{\mathcal{N}} [\mathcal{N}_{\mathrm{T}}^{*} - \mathcal{N}_{k\mathrm{T}}] \mathcal{N}_{k\mathrm{T}} f_{k\mathrm{M}} \, | V_{X^*}^{(k)}(\tau^{(k)}) | \\ \dot{f}_{k\mathrm{M}} &= L_{\mathrm{M}} [f_{\mathrm{M}}^{*} - f_{k\mathrm{M}}] \mathcal{N}_{k\mathrm{T}} f_{k\mathrm{M}} V_{X^*}^{(k)}(\tau^{(k)}) \end{aligned} \tag{7.115}$$

suggested by Kelly and Gillis [64]. The quantities $\mathcal{N}_{\mathrm{T}}^{*} \ge \mathcal{N}_{k\mathrm{T}}$ and $f_{\mathrm{M}}^{*} \le f_{k\mathrm{M}}$ are saturation values beyond which the evolutionary processes cannot evolve and the coefficients $L_{\mathcal{N}}$ and L_{M} are positive material constants. According to these equations the total number of dislocations increases monotonically, although at a varying rate.

Among the reasons a dislocation-based theory of plasticity is attractive is the availability of equations such as Eqs. 7.102 –7.115, along with the insight into the deformation process that they provide.

7.3.2 Constitutive Equations for Thermoelastic Response

Our next task is incorporation of the foregoing description of viscoplastic deformation into a theory of thermomechanical response. An appropriate setting for this is the theory of thermodynamics with internal state variables developed by Coleman and Gurtin [24]. In adopting this theory, we are assuming that the thermomechanical response of the material depends only upon its current state, but that additional variables beyond the elastic strain and specific entropy are needed to characterize this state. These latter variables, called *internal state variables*, change as the deformation and specific entropy change, and are to be determined from first-order ordinary differential equations called evolutionary equations. The internal state variables identified in the previous section are $a_1 \equiv \mathcal{N}_{k\mathrm{T}}$, $a_2 \equiv f_{k\mathrm{M}}$, and $a_3 \equiv \tau^{(k)*}$. We call these variables structural variables because they are associated with defect structures in the lattice.

The specific internal energy function for this material depends upon the elastic strain, the specific entropy, and the internal state variables:

$$\varepsilon = \bar{\varepsilon}(E^{\mathrm{e}}_{\Gamma\Delta}, \eta, a)\,, \tag{7.116}$$

where a designates the list of internal state variables. Experimental observation indicates that the elastic response of metals is largely independent of prior plastic deformation, so we separate $\bar{\varepsilon}$ into a part that depends upon the strain but not the structural variables and a part that depends upon these variables but is independent of the strain, giving

$$\varepsilon = \hat{\varepsilon}_1\,(E^{\mathrm{e}}_{\Gamma\Delta}, \eta) + \hat{\varepsilon}_2\,(\eta, a)\,. \tag{7.117}$$

From this equation we obtain the stress equation of state

$$T_{\Gamma\Delta} = \rho_{\mathrm{R}} \frac{\partial \hat{\varepsilon}_1(\mathbf{E}^{\mathrm{e}}, \eta)}{\partial E^{\mathrm{e}}_{\Gamma\Delta}}\,, \tag{7.118}$$

and the temperature equation of state

$$\theta = \frac{\partial \hat{\varepsilon}_1(\mathbf{E}^{\mathrm{e}}, \eta)}{\partial \eta} + \frac{\partial \hat{\varepsilon}_2(\eta, a)}{\partial \eta}\,. \tag{7.119}$$

As we have seen in Chap. 6, the foregoing equations of state can be simplified if we assume the deformation from which $\mathbf{E}^{\mathrm{e}}$ is derived consists of a finite dilatation combined with a small distortion. In this case, the stress is given by Eqs. 6.57–6.62, which can be written

$$t_{ij} = -p(v, \eta)\delta_{ij} + 2\mu(v, \eta)\widetilde{E}^{s}_{ij}, \tag{7.120}$$

where the pressure can be derived from a suitable equation of state, as discussed in Chap. 5, and the shear modulus can be obtained from an equation such as 6.62.

Because the structural variables do not affect the elastic response the material that was initially isotropic remains so and $\hat{\varepsilon}_1(\mathbf{E}^e, \eta)$ is an isotropic function of $\mathbf{E}^e$. Normally, the structural variables will evolve differently on different slip systems, leading to anisotropic viscoplastic response.

7.3.3 Uniaxial Deformation

The kinematic description and stress equations for uniaxial deformation have been discussed in Sect. 7.2.

When isotropic materials are subject to uniaxial deformation along the 1 axes, the transverse stress components are the same, $t_{22} = t_{33}$, and the shear stress achieves its maximum absolute value, $|\tau_{45^\circ}|$, on planes lying at 45° to the x axis, with

$$\tau_{45^\circ} = \tfrac{1}{2}(t_{11} - t_{22}). \tag{7.121}$$

When we adopt Eq. 7.73 for the shear stress, Eq. 7.121 becomes

$$\tau_{45^\circ} = \tfrac{3}{2}\mu_R \widetilde{E}^{es}_{L}. \tag{7.122}$$

The dislocation velocity increases rapidly with an increase of the shear stress on the slip plane, so we shall assume that all of the slip occurs on these 45° planes and that its Burgers vector is in the direction of the vector of maximum shear traction, which we shall designate $\boldsymbol{\tau}_{45^\circ}$.

The unit normal vector characterizing these slip planes in the current configuration is of the form

$$\mathbf{n}(\varphi_1) = \tfrac{1}{\sqrt{2}}(1, \cos\varphi_1, \sin\varphi_1), \quad 0 \le \varphi_1 < 2\pi, \tag{7.123}$$

and the maximum shear traction vector on these planes is given by

$$\boldsymbol{\tau}_{45^\circ} = \tfrac{1}{\sqrt{2}}\tau_{45^\circ}(1, -\cos\varphi_1, -\sin\varphi_1), \quad 0 \le \varphi_1 < 2\pi. \tag{7.124}$$

Since we have postulated that slip occurs in the direction of the vector $\boldsymbol{\tau}_{45^\circ}$, the Burgers vector is

$$\mathbf{b}(\varphi_1) = \tfrac{1}{\sqrt{2}} b\,(1, -\cos\varphi_1, -\sin\varphi_1), \quad 0 \le \varphi_1 \le 2\pi. \tag{7.125}$$

The vectors $\overline{\mathbf{N}}$ and $\mathbf{B}$ related to $\mathbf{n}$ and $\mathbf{b}$ by Eqs. 7.96 and 7.97 are

$$\overline{\mathbf{N}} = \tfrac{1}{\sqrt{2}}\Big((1+\tfrac{3}{4}\widetilde{E}_{\mathrm{L}}^{\mathrm{es}}),\ (1-\tfrac{3}{4}\widetilde{E}_{\mathrm{L}}^{\mathrm{es}})\cos\varphi_1,\ (1-\tfrac{3}{4}\widetilde{E}_{\mathrm{L}}^{\mathrm{es}})\sin\varphi_1 \Big)$$
$$\mathbf{B} = \tfrac{1}{\sqrt{2}}B\Big((1-\tfrac{3}{4}\widetilde{E}_{\mathrm{L}}^{\mathrm{es}}),\ -(1+\tfrac{3}{4}\widetilde{E}_{\mathrm{L}}^{\mathrm{es}})\cos\varphi_1,\ -(1+\tfrac{3}{4}\widetilde{E}_{\mathrm{L}}^{\mathrm{es}})\sin\varphi_1 \Big), \qquad (7.126)$$

where we have used Eq. 7.66 and kept only linear terms in $\widetilde{E}_{\mathrm{L}}^{\mathrm{es}}$ and where

$$b = B(v/v_{\mathrm{R}})^{1/3}(1+\tfrac{1}{2}\widetilde{E}_{\mathrm{L}}^{\mathrm{es}}). \qquad (7.127)$$

Equation 7.101, from which we calculate the rate of plastic deformation, was developed for the case in which only a finite number of slip systems are considered. It can be applied to the present problem by selecting a finite number of values for φ_1, but it is easier to interpret φ_1 as a continuous variable, in which case the sum becomes the integral

$$\Lambda^{\mathrm{P}} = \int_0^{2\pi} \mathcal{N}_{\mathrm{M}}(\varphi)\mathbf{B}(\varphi)\otimes\overline{\mathbf{N}}(\varphi)\, V_{X^*}(\varphi)\,d\varphi. \qquad (7.128)$$

For the present case, $\tau^{(k)} = \tau_{45°}$, a value independent of k, and hence φ_1, so the dislocation velocity is also independent of φ_1. We shall also take $\mathcal{N}_{\mathrm{M}}$ to be independent of φ_1, so Eq. 7.128 can be written

$$\Lambda^{\mathrm{P}} = \mathcal{N}_{\mathrm{M}}\, V_{X^*}(\tau_{45°}) \int_0^{2\pi} \mathbf{B}(\varphi)\otimes\overline{\mathbf{N}}(\varphi)\,d\varphi, \qquad (7.129)$$

and from Eqs. 7.126 we find that

$$\Lambda^{\mathrm{P}} = \tfrac{1}{2}\pi B\mathcal{N}_{\mathrm{M}}\, V_{X^*}(\tau_{45°})\ \mathrm{diag}\,\|\,2,\ -1,\ -1\,\|. \qquad (7.130)$$

Substitution of this result into Eq. 7.104 yields the flow rule

$$\begin{aligned}\dot{F}_{\mathrm{L}}^{\mathrm{P}} &= \pi B\mathcal{N}_{\mathrm{M}}\, V_{X^*}(\tau_{45°})F_{\mathrm{L}}^{\mathrm{P}} \\ &= \pi B(v/v_{\mathrm{R}})^{-1/3}\mathcal{N}_{\mathrm{M}}\, V_x(\tau_{45°})(1+\tfrac{1}{4}\widetilde{E}_{\mathrm{L}}^{\mathrm{es}})F_{\mathrm{L}}^{\mathrm{P}},\end{aligned} \qquad (7.131)$$

giving the longitudinal component of the plastic deformation gradient.

Because only the $\varphi_2 = 45°$ planes play a role in the analysis, Eqs. 7.102 – 7.115 take the simplified forms

$$\begin{aligned}V_x(\tau_{45°}) &= \begin{cases} 0, & |\tau_{45°}| \le \tau^* \\ -C_{\mathrm{S}}\,\mathrm{sgn}(\tau_{45°})\dfrac{s^m}{1+s^m}, & |\tau_{45°}| > \tau^* \end{cases} \\ \tau^* &= \tau_0^*(1+h|\gamma^{\mathrm{P}}|^{1/n}) \\ \dot{\mathcal{N}}_{\mathrm{T}} &= -L_{\mathcal{N}}[\mathcal{N}_{\mathrm{T}}^* - \mathcal{N}_{\mathrm{T}}]\,\mathcal{N}_{\mathrm{T}}\, f_{\mathrm{M}}(v/v_{\mathrm{R}})^{-1/3}(1-\tfrac{1}{2}E_{\mathrm{L}}^{\mathrm{es}})V_x(\tau_{45°}) \\ \dot{f}_{\mathrm{M}} &= L_{\mathrm{M}}[f_{\mathrm{M}}^* - f_{\mathrm{M}}]\,\mathcal{N}_{\mathrm{T}}\, f_{\mathrm{M}}(v/v_{\mathrm{R}})^{-1/3}(1-\tfrac{1}{2}E_{\mathrm{L}}^{\mathrm{es}})V_x(\tau_{45°}),\end{aligned} \qquad (7.132)$$

where $s = (|\tau_{45^\circ}| - \tau^*)/\tau^\dagger$.

Using eq. 7.102, the evolutionary equations for $\mathcal{N}_T$ and f_M can be written

$$\begin{aligned} \dot{\mathcal{N}}_T &= (L_{\mathcal{N}}/B)\,[\mathcal{N}_T - \mathcal{N}_T^*]\,\dot{\gamma}^p \\ \dot{f}_M &= (L_M/B)[f_M^* - f_M]\,\dot{\gamma}^p\,. \end{aligned} \tag{7.133}$$

When quadratic and higher terms in $\widetilde{E}_L^{es}$ are neglected. These equations can be integrated to yield the expressions

$$\begin{aligned} \mathcal{N}_T &= \mathcal{N}_T^* + [\mathcal{N}_0 - \mathcal{N}_T^*]\exp[(L_{\mathcal{N}}/B)\,\gamma^p\,] \\ f_M &= f_M^* + [f_0 - f_M^*]\exp[(L_M/B)\,\gamma^p\,] \end{aligned} \tag{7.134}$$

when the initial conditions $\mathcal{N}_T = \mathcal{N}_0$ and $f_M = f_0$ are imposed at $\gamma^p = 0$. Determination of $\mathcal{N}_T$ and f_M is reduced to evaluation of Eqs. 7.134 with γ^p given by Eq. 7.82 and we have $\mathcal{N}_M = \mathcal{N}_T\, f_M$.

Since the Hugoniot curve represents the equilibrium response to shock compression, it does not depend upon the evolutionary equations and is the same as the one discussed in Sect. 2.2.

The richness of the theory lies in the changes in $\mathcal{N}_T$, f_M, and τ^* that occur as a result of the deformation, and in the way the dislocation velocity $V_x(\tau_{45^\circ})$ varies during the deformation process.

7.4 Exercises

7.4.1. Show that the maximum shear stress in a uniaxially strained body occurs on planes inclined at 45° to the strain axis.

7.4.2. Consider the extension of a slender rod in the approximation that the transverse stress components vanish (i.e., states of *uniaxial stress*). Calculate (using the small-strain theory) the longitudinal and transverse strain components and the elastic and plastic parts of the strain.

7.4.3. Derive Eq. 7.82.

7.4.4. Derive Eqs. 7.108 and 7.109.

7.4.5. Show that the flow rule of Eq. 7.21 is automatically satisfied for the uniaxial motions discussed in Sect. 7.1.1.

7.4.6. What is the slope of the $t_{11} - v$ elastic–plastic compression curve at the HEL for a metal that hardens according to Eq. 7.81? How does this slope compare with that of the elastic curve below the HEL? Since this analysis involves only the small strains that prevail near the HEL it is adequate to take $\mu(v) = \mu_R$.

CHAPTER 8

Weak Elastic Waves

8.1 Linear Theory of Elastic Waves

The linear theory of elasticity emerges when attention is restricted to the case in which deformations are small and stress is proportional to strain. The simplicity of this theory, combined with the applicability of many highly developed methods of linear analysis, makes it possible to solve complex multidimensional problems. It is not our purpose to address this entire body of work, however, but simply to consider the special case of linear plane waves of uniaxial strain. The theory of linear wave propagation is not usually regarded as a part of shock physics, but it does permit us to illustrate important aspects of wave propagation. The insight into the mechanics of wave propagation that follows from a study of this simple case is helpful as one seeks understanding of the more general problems.

When attention is restricted to uniaxial motions of isotropic materials, Eqs. 6.16 and 6.17 yield the stress relation

$$t_{11} = (\lambda_R + 2\mu_R)\frac{\partial u}{\partial x}, \tag{8.1}$$

which, according to Eq. 2.39, can be written

$$t_{11} = (\lambda_R + 2\mu_R)\frac{\partial U}{\partial X} \tag{8.2}$$

in the limit of small strains. The particle velocity, $\dot{x}$, is given by Eq. 2.36 which, by Eq. 2.40_2, can be written

$$\dot{x} = U_t\,. \tag{8.3}$$

Substitution of these results into the equation of conservation of momentum, Eq. 2.92_2 (with $f = 0$) produces the linear equation for governing the propagation of weak longitudinal elastic waves,

$$C_0^2\, U_{XX} = U_{tt}\ , \tag{8.4}$$

where

$$C_0^2 = (\lambda_R + 2\mu_R)/\rho_R \,. \tag{8.5}$$

For convenience, solutions of Eq. 8.4 are called *linear elastic waves* and C_0 is the wavespeed. In this case of small deformation, no distinction need be made between the Lagrangian and Eulerian coordinates, X and x.

Equation 8.4 has the same general character as the nonlinear equation that it approximates, but some important special features follow from its linearity. First, we have scaleability of solutions: if $f(X,t)$ is a solution, then so is $\alpha f(X,t)$ for any constant multiplier α. The second important property of the linear equation is superimposability of its solutions: If the functions $U_{(1)}(X,t)$ and $U_{(2)}(X,t)$ are solutions, then sum $U_{(1)} + U_{(2)}$ is also a solution. These properties indicate that linear waves of all amplitudes (and all parts of a given waveform) propagate at the same speed and that linear waves do not interact with one another. In particular, linear waves propagating in opposite directions can pass through one another and emerge unchanged in form.

It is easy to verify that the function

$$U(X,t) = U_{(R)}(X - C_0 t) + U_{(L)}(X + C_0 t) \tag{8.6}$$

is a solution of Eq. 8.4. The functions $U_{(R)}$ and $U_{(L)}$ are arbitrary, subject only to the restriction that they be twice differentiable. The function $U_{(R)}$ describes a wave propagating in the positive X direction and called a *right-propagating wave*. Similarly, the function $U_{(L)}$ describes a *left-propagating wave*.

Before considering these waves further, it is worthwhile to investigate shocks in this linear context. When the weak-shock limit of the jump conditions 2.110 and 2.113 is taken, it can be shown that the wavespeed and the material and spatial shockspeeds stand in the relation

$$C_0 = U_S = u_S - \dot{x}^- \,, \tag{8.7}$$

to within higher-order terms in the jumps.

With this result, the jump conditions for mass and momentum are found to be satisfied if

$$[\![U_t(X,t)]\!] + C_0 [\![U_X(X,t)]\!] = 0 \,. \tag{8.8}$$

The important point is that, in the weak-shock limit appropriate to the linear wave environment, the shock propagates at the same speed as the smooth wave. This means that a disturbance containing a shock propagates in the same way as a smooth solution, even though the derivatives appearing in the differential equation do not exist. Traveling-wave solutions of the form of Eq. 8.6 may involve shocks as well as smooth portions.

Let us now consider the function $U_{(R)}(X - C_0 t)$ corresponding to a right-propagating wave. It is a rather arbitrary curve—called a *waveform*—such as that shown in Fig. 8.1. The essential feature of this arbitrarily shaped wave is that, by virtue of the dependence of $U_{(R)}$ on $X - C_0 t$, it propagates to the right with no change in shape.

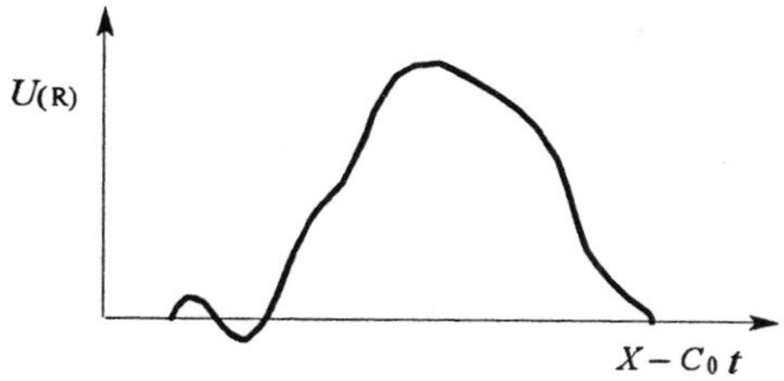

Figure 8.1. Waveform

The graph in Fig. 8.2 shows all parts of the disturbance being translated in the X direction by a distance equal to the wavespeed multiplied by the elapsed time interval. This is an example of the type of data that we would recover if we could take a sequence of stop-action photographs of the wave as it advanced through the material (bearing in mind that no distinction is made between X and x in the linear theory).

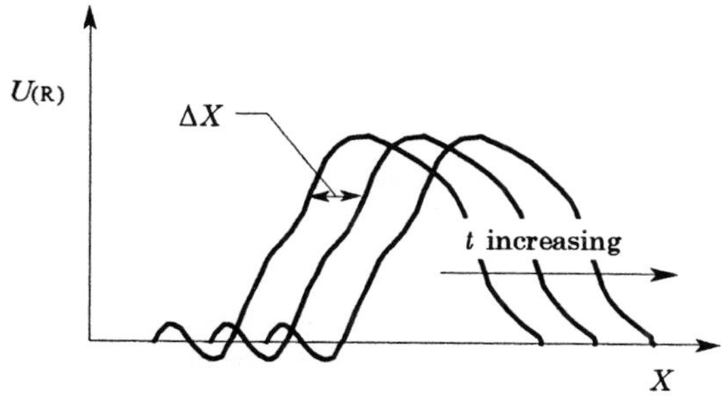

Figure 8.2. Propagating waveform

We can also plot $U_{(R)}$ as a function of t for several values of X, as shown in Fig. 8.3. Data of this sort are obtained by recording the information produced as the wave passes gauges fixed to various particles of the material.

The behavior of a wave propagating in the $-X$ direction (which we call a left-propagating wave) is entirely analogous to that of the right-propagating wave—only the direction of propagation is reversed.

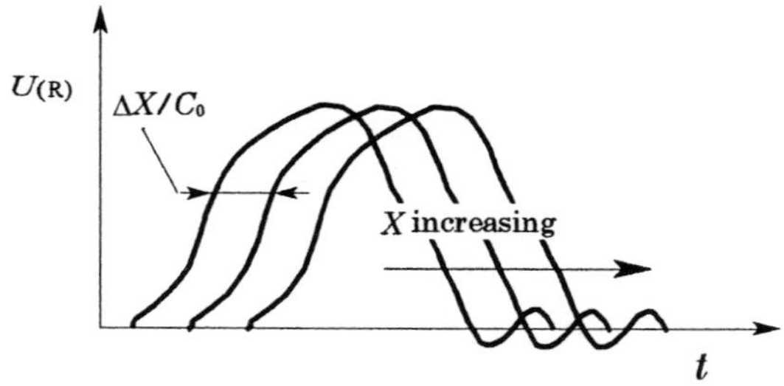

Figure 8.3. Temporal waveform. Note that, in this representation, the waveform is a reflection (about a vertical line) of the waveforms of Figs. 8.1 and 8.2.

It is useful to consider representation of wave phenomena in the X–t coordinate plane shown in Fig. 8.4. The value of the solution in a right-propagating wave is the same at all points along every line of the form $X - C_0 t = X_0$, and the analogous situation holds for left-propagating waves relative to lines of the form $X + C_0 t = X_1$. These lines, which are called *characteristic lines*, or simply *characteristics*, play a central role in discussions of all wave-propagation phenomena. Since each small part of the waveform propagates unchanged in form, it is useful to think of the wave-propagation process as one of transmission of a feature (often called a *wavelet*) of the right-propagating waveform along the characteristic line $X - C_0 t = X_0$, where the parameter X_0 is chosen to identify the specific characteristic line corresponding to the part of the waveform in question. Since a wavelet cannot propagate backward in time, the characteristics are oriented in the direction of increasing time. All points ahead of a point marking the present time are in the future, whereas those lying behind it are in the past. Events occurring in the past or at the present time can influence the future, but future events cannot, of course, change present or past events. Since wavelets can be transmitted in either direction from a point in the body, the mathematical standing of the coordinate axis is different from that of the time axis.

So far, we have discussed the propagation of an arbitrary disturbance, but have not addressed the relationship between a wave and the conditions producing it. In the context of the nondissipative theory under discussion, a wave will continue to propagate indefinitely. Waves can be introduced into a body by application of force or imposition of a velocity change at its surface or by energy deposition or through application of body forces by electrical, magnetic, or other means.

To begin our study of linear elastic waves, let us consider the simple case of waves in the half-space $X \geq 0$. The present time is designated $t = 0$. The future state of the body depends upon both the present state and stimuli to which the body is subjected in the future. The present (initial) state of the body is characterized by functions giving the strain (or equivalently, the stress) and the particle velocity at every point of the body:

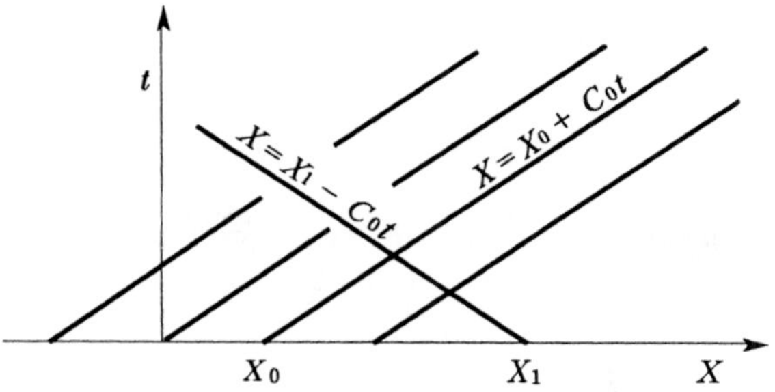

Figure 8.4. An X–t diagram. The lines $X = X_0 + C_0 t$ are trajectories of constant amplitude of a right-propagating waveform. Similarly, lines of the form $X = X_0 - C_0 t$ are trajectories of constant amplitude of a left-propagating waveform.

$$\begin{aligned} U_X(X,t)|_{t=0} &= S(X), \quad X \geq 0 \\ U_t(X,t)|_{t=0} &= C_0 V(X), \quad X \geq 0. \end{aligned} \tag{8.9}$$

These equations, giving the initial state of the body, are called *initial conditions* and the functions $S(X)$ and $C_0V(X)$ are *initial values*.

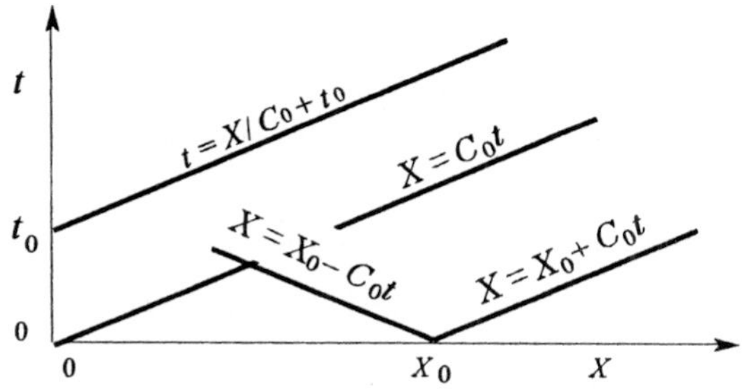

Figure 8.5. Typical characteristic curves.

Since two characteristics capable of transmitting information into the future emerge from each interior point $X > 0$ of the body, two conditions, such as are given by Eqs. 8.9, are required to completely determine the effect of the present state upon future states. The future states are also affected by influences to which the body is subjected at future times. Let us consider the case in which these influences are applied at the surface $X = 0$. Examination of the X–t plane of Fig. 8.5 shows that there is only one family of characteristics capable of transmitting boundary information into the future. Therefore, only one *boundary condition* can be specified. We might specify application of normal traction to the boundary, a condition that takes the form

$$t_{11}(0,t) = \rho_R C_0^2 U_X(X,t)|_{X=0} = \rho_R C_0^2 E(t), \quad t \geq 0. \tag{8.10}$$

Another possibility would be to specify the motion of the boundary in terms of the particle-velocity condition

$$\dot{x}(0, t) = U_t(X, t)\big|_{X=0} = C_0 H(t), \quad t \geq 0. \tag{8.11}$$

Another condition, not having to do with the wave-propagation process but simply specifying the reference position of the body, is

$$U(0, 0) = 0\,. \tag{8.12}$$

The characteristic line $X = C_0 t$ plays the very important role of separating the X–t plane into regions in which the waves are, and are not, affected by the boundary conditions. It is easy to see that information about events occurring on the boundary of the body at, or after, $t = 0$ cannot be transmitted to a point X in the interior before the time $t = X / C_0$. For this reason, the solution in the region $X > C_0 t$ is independent of the boundary condition, depending only upon the initial conditions.

8.1.1 Initial-value Problem

In the region $X > C_0 t$ of the X–t plane (let us call it "Region 1"), the solution of Eq. 8.4 depends only upon the initial conditions since insufficient time has elapsed to permit changes in conditions imposed on the boundary, $X = 0$, for $t > 0$, to produce an effect. In the following analysis we are assuming that the region is unbounded: $X \in [0, \infty)$. The case of slabs of finite thickness will be discussed in Sect. 8.3.

Let us consider a solution of the traveling-wave form

$$U_{(1)}(X, t) = U_{(1\mathrm{R})}(X - C_0 t) + U_{(1\mathrm{L})}(X + C_0 t), \quad X > C_0 t\,. \tag{8.13}$$

The functions $U_{(1\mathrm{R})}$ and $U_{(1\mathrm{L})}$, which are arbitrary at this point, are determined by requiring that $U_{(1)}$ satisfy the initial conditions 8.9. Differentiating, we have

$$\begin{aligned} \frac{\partial U_{(1)}(X, t)}{\partial X} &= U'_{(1\mathrm{R})}(X - C_0 t) + U'_{(1\mathrm{L})}(X + C_0 t) \\ \frac{\partial U_{(1)}(X, t)}{\partial t} &= -C_0 U'_{(1\mathrm{R})}(X - C_0 t) + C_0 U'_{(1\mathrm{L})}(X + C_0 t), \end{aligned} \tag{8.14}$$

where the prime denotes total differentiation with respect to the single argument (either $X + C_0 t$ or $X - C_0 t$) upon which the function depends. Substituting Eqs. 8.14 into Eqs. 8.9 leads to the results

$$\begin{aligned} U'_{(1\mathrm{R})}(X) &= \tfrac{1}{2}[S(X) - V(X)] \\ U'_{(1\mathrm{L})}(X) &= \tfrac{1}{2}[S(X) + V(X)], \end{aligned} \tag{8.15}$$

which can be integrated to yield

$$
\begin{aligned}
U_{(1R)}(X - C_0 t) &= U_{(1R)}(0) + \frac{1}{2}\int_0^{X - C_0 t} [S(\zeta) - V(\zeta)]\, d\zeta \\
U_{(1L)}(X + C_0 t) &= U_{(1L)}(0) + \frac{1}{2}\int_0^{X + C_0 t} [S(\zeta) + V(\zeta)]\, d\zeta.
\end{aligned}
\tag{8.16}
$$

Substituting these results into the general solution 8.13, applying the condition 8.12 to eliminate the constant term, and rewriting slightly, gives the final result

$$
\begin{aligned}
U_{(1)}(X, t) = &\int_0^{X} S(\zeta)\, d\zeta + \frac{1}{2}\int_X^{X + C_0 t} S(\zeta)\, d\zeta \\
&- \frac{1}{2}\int_{X - C_0 t}^{X} S(\zeta)\, d\zeta + \frac{1}{2}\int_{X - C_0 t}^{X + C_0 t} V(\zeta)\, d\zeta,
\end{aligned}
\tag{8.17}
$$

which shows that the displacement at the point (X, t) depends upon the initial conditions in the range $[X - C_0 t,\ X + C_0 t]$ and an offset displacement calculated by integrating the initial value of the strain from the boundary to the point X.

We are more often interested in the values of the particle velocity and strain (or stress) than the displacement, and these fields are easily seen to be given by the equations

$$
\begin{aligned}
\dot{x}(X, t) &= \frac{\partial U_1(X, t)}{\partial t} \\
&= \frac{1}{2} C_0 \left[S(X + C_0 t) + V(X + C_0 t) - S(X - C_0 t) + V(X - C_0 t) \right]
\end{aligned}
\tag{8.18}
$$

and

$$
\begin{aligned}
\widetilde{E}_{11}(X, t) &= \frac{\partial U_1(X, t)}{\partial X} \\
&= \frac{1}{2} \left[S(X + C_0 t) + V(X + C_0 t) + S(X - C_0 t) - V(X - C_0 t) \right].
\end{aligned}
\tag{8.19}
$$

Let us consider the X–t plane of Fig. 8.6. From the foregoing equations we see that the solution at the point of intersection (X^*, t^*) of the two characteristics shown can be expressed in the form

$$
\dot{x}(X^*, t^*) = \tfrac{1}{2} C_0 \left[S(X_1) + V(X_1) - S(X_0) + V(X_0) \right] \tag{8.20}
$$

and

$$\widetilde{E}_{1}(X^{*},t^{*})=\tfrac{1}{2}\left[S(X_{1})+V(X_{1})+S(X_{0})-V(X_{0})\right]. \tag{8.21}$$

This shows that the values of particle velocity and strain at any point in the region $X > C_0 t$ depend only upon the values of the initial conditions on the two characteristics that pass through the point in question.

Let us proceed to obtain the solution in the remaining part, $0 < X < C_0 t$, of the X–t plane.

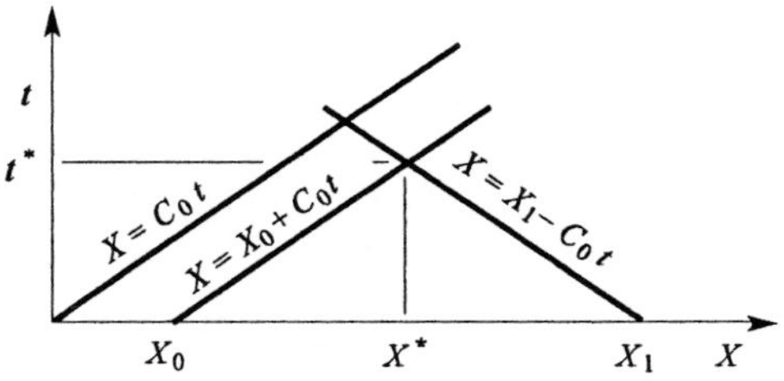

Figure 8.6. An X–t diagram to be used in conjunction with Eqs. 8.20 and 8.21 to see how the solution at (X^*, t^*) depends upon the initial conditions at X_0 and X_1.

8.1.2 Boundary-value Problem

The solution in the region $0 \le X < C_0 t$ of the X–t plane (let us call it "Region 2") can be expressed in the same traveling-wave form as before,

$$U_{(2)}(X,t)=U_{(2R)}(X-C_0 t)+U_{(2L)}(X+C_0 t),\quad X<C_0 t, \tag{8.22}$$

but with the functions $U_{(2R)}$ and $U_{(2L)}$ being determined by a boundary condition imposed at $X = 0$ and a condition that the solutions in the two regions match on the characteristic $X = C_0 t$ along which the regions are joined. For the boundary condition at $X = 0$, let us suppose that the normal traction is prescribed, as indicated in Eq. 8.10:

$$U_X(X,t)\big|_{X=0}=E(t),\quad t\ge 0. \tag{8.23}$$

Recall that the initial condition $U_X(X,t)\big|_{t=0} = S(X)$ has also been specified. At $X = 0$ and $t = 0$ these conditions may or may not agree. If they agree, i.e. if $S(0) = E(0)$, we have a smooth solution and the derivatives of U with respect to both X and t are continuous along $X = C_0 t$. If $S(0) \ne E(0)$, a shock is introduced at the boundary and propagates along this characteristic. The matching requirement is that the jump condition 8.8 be satisfied along this line. For this example, let us suppose that $S(0) \ne E(0)$ so a shock is produced. Then, the matching condition is

$$
0 = \left.\frac{\partial U_{(2)}(X,t)}{\partial t}\right|_{t=X/C_0} + C_0 \left.\frac{\partial U_{(2)}(X,t)}{\partial X}\right|_{t=X/C_0}
$$

$$
- \left.\frac{\partial U_{(1)}(X,t)}{\partial t}\right|_{t=X/C_0} - C_0 \left.\frac{\partial U_{(1)}(X,t)}{\partial X}\right|_{t=X/C_0}, \tag{8.24}
$$

which can be written in the form

$$
\left.\frac{\partial U_{(2)}(X,t)}{\partial t}\right|_{t=X/C_0} + C_0 \left.\frac{\partial U_{(2)}(X,t)}{\partial X}\right|_{t=X/C_0} = C_0 \left[S(2X)+V(2X)\right] \tag{8.25}
$$

by using Eqs. 8.18 and 8.19.

Imposing the stress boundary condition on the solution 8.22 yields the condition

$$
U'_{(2\mathrm{R})}(-C_0 t) + U'_{(2\mathrm{L})}(C_0 t) = E(t), \tag{8.26}
$$

and the matching condition of Eq. 8.8 requires that

$$
U'_{(2\mathrm{L})}(\zeta) = \tfrac{1}{2}[S(\zeta)+V(\zeta)] \tag{8.27}
$$

for $\zeta > 0$. Integration yields the displacement

$$
U_{(2\mathrm{L})}(X + C_0 t) = \frac{1}{2}\int_0^{X+C_0 t} [S(\zeta)+V(\zeta)]\, d\zeta + U_{(2\mathrm{L})}(0). \tag{8.28}
$$

With this result, Eq. 8.26 takes the form

$$
U'_{(2\mathrm{R})}(-C_0 t) = E(t) - \tfrac{1}{2}[S(C_0 t) + V(C_0 t)] \tag{8.29}
$$

and integration, followed by some simplification, gives

$$
U_{(2\mathrm{R})}(X - C_0 t) = +\frac{1}{2}\int_0^{-(X-C_0 t)} [S(\zeta)+V(\zeta)]\, d\zeta
$$

$$
- C_0 \int_0^{-(X-C_0 t)/C_0} E(\zeta)\, d\zeta + U_{(2\mathrm{R})}(0). \tag{8.30}
$$

Combining Eqs. 8.28 and 8.30 leads to the final result for the displacement in the region $0 \le X < C_0 t$:

$$U_{(2)}(X,t)=\frac{1}{2}\int_{-(X-C_0t)}^{X+C_0t}[S(\zeta)+V(\zeta)]\,d\zeta+\int_0^{-(X-C_0t)}[S(\zeta)+V(\zeta)]\,d\zeta$$
$$-C_0\int_0^{t-(X/C_0)}E(\zeta)\,d\zeta, \tag{8.31}$$

because, as before, the contribution $U_{(2R)}(0)+U_{(2L)}(0)$ vanishes by virtue of Eq. 8.12. The particle velocity and strain are given by the equations

$$\dot{x}(X,t)=-C_0\,E\left(t-\frac{X}{C_0}\right)$$
$$+\frac{C_0}{2}\left[S(X+C_0t)+V(X+C_0t)+S\big(-(X-C_0t)\big)+V\big(-(X-C_0t)\big)\right] \tag{8.32}$$

and

$$\widetilde{E}_{11}(X,t)=E\left(t-\frac{X}{C_0}\right)$$
$$+\frac{1}{2}\left[S(X+C_0t)+V(X+C_0t)-S\big(-(X-C_0t)\big)-V\big(-(X-C_0t)\big)\right]. \tag{8.33}$$

As for the solution in region 1, we can express the particle velocity and strain at a point X^*, t^* in terms of boundary and initial conditions at points where characteristics intersect the axes:

$$\dot{x}(X^*,t^*)=-C_0E(t_0)+\frac{C_0}{2}\left[S(X_1)+V(X_1)+S(X_0)+V(X_0)\right]$$
$$\widetilde{E}_{11}(X^*,t^*)=E(t_0)+\frac{1}{2}\left[S(X_1)+V(X_1)-S(X_0)-V(X_0)\right],$$

where the points X_0, X_1, and t_0 are defined in Fig. 8.7.

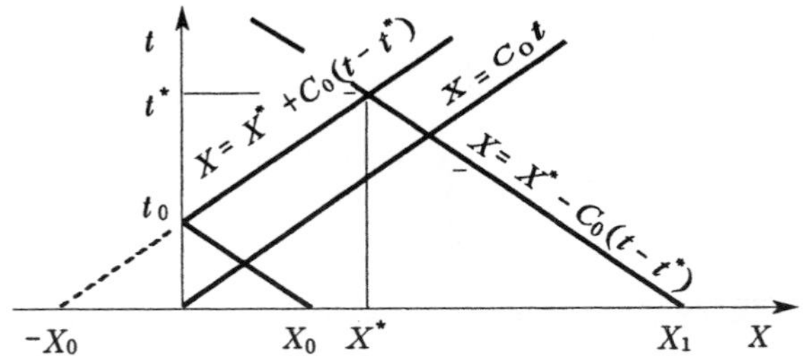

Figure 8.7. An X–t diagram to be used in conjunction with Eqs. 8.32 and 8.33 to see how the particle velocity and strain at (X^*,t^*) depends upon the initial conditions at X_0 and X_1 and the boundary condition at t_0.

8.1.3 Wave Propagation into an Undisturbed Body

Let us specialize the foregoing solution to the case of a body that is initially undeformed and at rest. The initial conditions for this case are

$$S(X)=0, \quad V(X)=0. \tag{8.34}$$

The solution in the region ahead of waves emanating from the boundary is

$$U_{(1)}(X,t)=0. \tag{8.35}$$

In this case, this region is called a *rest zone*. The solution in the region behind the leading characteristic of the wave follows immediately from Eq. 8.31 and the complete solution is

$$U(X,t)=\begin{cases} -C_0\displaystyle\int_0^{t-(X/C_0)} E(\zeta)d\zeta, & 0\le X\le C_0 t \\ 0, & X>C_0 t. \end{cases} \tag{8.36}$$

If $E(t)=0$ for all $t\ge 0$ the material is not subjected to any stimulus and, of course, remains quiescent. The boundary condition

$$E(t)=\begin{cases} 0, & t<0 \\ E_0, & t\ge 0, \end{cases} \tag{8.37}$$

with $E_0=\text{const.}$ corresponds to sudden application of a sustained traction. In this case the solution 8.36 is a shock and takes the form

$$U(X,t)=\begin{cases} 0, & t<0, X>0 \\ E_0(X-C_0 t), & 0\le X<C_0 t \\ 0, & X>C_0 t. \end{cases} \tag{8.38}$$

Examination of this solution shows that, as might be expected, the application of traction to the boundary causes it to move in the X direction (the direction of the force) by an amount $U(0,t)=-C_0 E_0 t$, i.e. at a rate $\dot{x}=-C_0 E_0$. The gradient $\partial U/\partial X=E_0$ in the region $0\le X<C_0 t$ is constant, corresponding to a constant stress in this region. The boundary between the rest zone and the disturbed zone is a shock of amplitude $[\![\partial U/\partial X]\!]=E_0$ (or, equivalently, $[\![\partial U/\partial t]\!]=-C_0 E_0$) propagating into the undisturbed material at the speed C_0.

If the boundary loading history is of the form

$$E(t)=\begin{cases} 0, & t<0 \\ E_0, & 0<t\le\tau \\ 0, & t>\tau, \end{cases} \tag{8.39}$$

the displacement field is found to be

$$U(X,t)=\begin{cases}0, & X>C_0t\\ E_0(X-C_0t), & C_0t>X>C_0(t-\tau)\\ 0, & C_0(t-\tau)>X>0.\end{cases} \tag{8.40}$$

This solution is simply a square pulse of strain or particle velocity that has duration τ and propagates unchanged in form at the speed C_0.

If the boundary load is applied for a finite time τ, as in the previous case, but is a smooth pulse so that $E(0)=0$ and $E(t)=0$ for $t\geq\tau$, the general solution 8.36 takes the form

$$U(X,t)=\begin{cases}0, \quad X\geq C_0t\\ -C_0\displaystyle\int_0^{t-(X/C_0)}E(\vartheta)d\vartheta, \quad C_0t>X>C_0(t-\tau)\\ -C_0\displaystyle\int_0^{\tau}E(\vartheta)d\vartheta, \quad 0\leq X\leq C_0(t-\tau).\end{cases} \tag{8.41}$$

Note that the material in the region $0\leq X\leq C_0(t-\tau)$ behind the pulse is displaced, but is at rest and unstrained.

8.1.4 Domains of Dependence and Influence

Because waves propagate at a finite speed, a disturbance introduced into a material body produces an effect at distant points only at later times. For one-dimensional problems such as that of uniaxial motion the issue is best studied in the X–t plane. Let us consider a point X^*, t^* in the initial response region

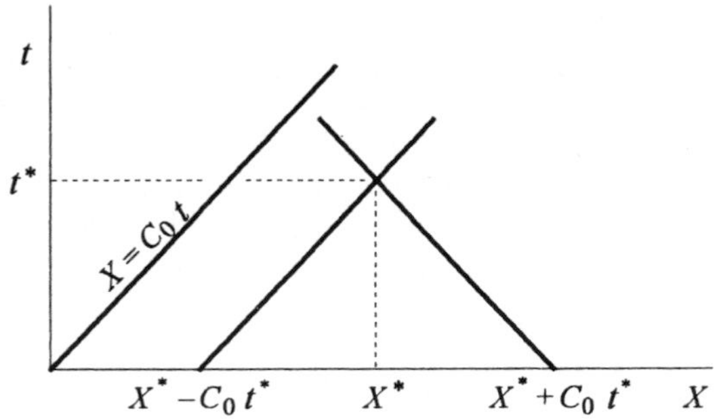

Figure 8.8. An X–t diagram showing domains of dependence and influence for the point X^*,t^*.

Figure 8.8 shows lines $X - X^* = \pm C_0(t - t^*)$. Examination of Eq. 8.17 shows that the displacement at X^*, t^* depends upon the initial conditions in the interval $[X^* - C_0 t^*, X^* + C_0 t^*]$. This interval is called the domain of dependence because the solution of Eq. 8.4 depends upon initial data in this interval but is independent of initial data outside this range. It is important to note that the derivatives of this solution, which give the strain and particle velocity, do not depend upon the entirety of the data in the domain of dependence, but only upon the data at the endpoints, as shown by Eqs. 8.18 and 8.19.

8.2 Characteristic Coordinates

The characteristic lines along which waveform information is transmitted play a fundamental role in the theory of partial differential equations. Since the linear case is so explicit and the solutions so transparent, the basic importance of the characteristics may be overlooked. The theory of nonlinear waves is less explicit and a full understanding of the properties of the characteristics is needed when these waves are studied.

As before, our investigation concerns the linear wave equation 8.4. A common and convenient procedure applied in studying differential equations of order higher than one (the wave equation is of order two because second derivatives occur) is to write them as systems of first-order equations in which the first derivatives of the unknown functions appear as dependent variables. In the case of the wave equation, we have

$$\begin{aligned} S(X,t) &= U_X(X,t) \\ V(X,t) &= \frac{1}{C_0} U_t(X,t). \end{aligned} \tag{8.42}$$

For the second derivatives to exist, we must have

$$U_{Xt}(X,t) = U_{tX}(X,t),$$

so

$$C_0 V_X(X,t) = S_t(X,t).$$

Substituting the relations 8.42 into the wave equation 8.4, we obtain

$$V_t(X,t) = -C_0 S_X(X,t)$$

so the first-order form of the wave equation is the system

$$\begin{aligned} S_t(X,t) - C_0 V_X(X,t) &= 0 \\ V_t(X,t) - C_0 S_X(X,t) &= 0. \end{aligned} \tag{8.43}$$

It is apparent that these equations are coupled, i.e. each equation involves both of the dependent variables, S and V. A great simplification would be achieved if the equations could be written in an uncoupled form, and this is exactly the form resulting if the coordinate lines $X =$ constant and $t =$ constant are replaced by coordinate lines made of the characteristics. Specifically, we choose coordinates ξ and η defined by the equations

$$\begin{aligned}\xi &= X + C_0 t \\ \eta &= X - C_0 t.\end{aligned} \tag{8.44}$$

From the chain rule for differentiation, we have

$$\begin{aligned}\frac{\partial S(X,t)}{\partial t} &= \frac{\partial S(\xi,\eta)}{\partial \xi}\frac{\partial \xi}{\partial t} + \frac{\partial S(\xi,\eta)}{\partial \eta}\frac{\partial \eta}{\partial t} = C_0\left(\frac{\partial S}{\partial \xi} - \frac{\partial S}{\partial \eta}\right) \\ \frac{\partial S(X,t)}{\partial X} &= \frac{\partial S(\xi,\eta)}{\partial \xi}\frac{\partial \xi}{\partial X} + \frac{\partial S(\xi,\eta)}{\partial \eta}\frac{\partial \eta}{\partial X} = \frac{\partial S}{\partial \xi} + \frac{\partial S}{\partial \eta},\end{aligned} \tag{8.45}$$

and analogous relations for the derivatives of $V(X,t)$. Substituting these expressions into Eqs. 8.43 and simplifying the result gives

$$\begin{aligned}S_\xi(\xi,\eta) - V_\xi(\xi,\eta) &= 0 \\ S_\eta(\xi,\eta) - V_\eta(\xi,\eta) &= 0,\end{aligned}$$

which we can write as

$$\begin{aligned}\frac{\partial}{\partial \xi}(S - V) &= 0 \\ \frac{\partial}{\partial \eta}(S + V) &= 0.\end{aligned} \tag{8.46}$$

These equations can be integrated immediately, with the result

$$\begin{aligned}S(\xi,\eta) - V(\xi,\eta) &= \Phi_R(\eta) \\ S(\xi,\eta) + V(\xi,\eta) &= \Phi_L(\xi).\end{aligned} \tag{8.47}$$

Equations 8.46 are said to be the *characteristic form* of the wave equation, and the functions $\Phi_L(\xi)$ and $\Phi_R(\eta)$ are the *Riemann invariants* of the system. Each of these expressions is called an invariant because its value is constant on the line $\xi =$ constant or $\eta =$ constant. Since these lines are characteristics for right- and left-propagating waves, respectively, the Riemann invariants are just the quantities that are transmitted by these waves. The traveling-wave form for solutions of the wave equation considered previously can be recovered from Eqs. 8.47 by solving for S and V:

$$\begin{aligned}S(\xi,\eta) &= \tfrac{1}{2}\left[\Phi_L(\xi) + \Phi_R(\eta)\right] \\ V(\xi,\eta) &= \tfrac{1}{2}\left[\Phi_L(\xi) - \Phi_R(\eta)\right].\end{aligned}$$

8.3 Plate of Finite Thickness

The preceding analysis concerned waves in a half-space, with specific examples being given for cases in which waves originated at the boundary and propagated into the interior of the material. The important issue of wave reflection does not arise in this case but does come to the fore when a wave encounters a boundary or an interface at which materials with differing properties are in contact.

8.3.1 Unrestrained Boundary

Let us consider the case in which the half-space is replaced by a plate of thickness X_1 that is undeformed and at rest. The solutions of Section 8.1.3 remain valid for this new domain for times $0 \le t < X_1/C_0$ prior to the time at which the wave first encounters the boundary at the point $X = X_1$.

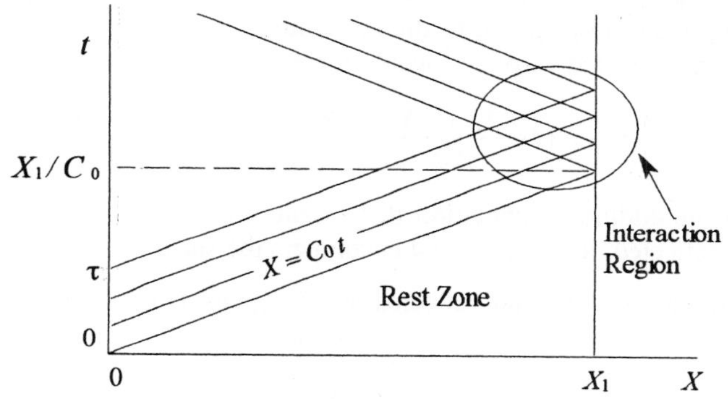

Figure 8.9. Lagrangian X–t diagram of a pulse of finite duration interacting with a boundary.

The X–t diagram of Fig. 8.9 illustrates the wave field for the case in which a traction

$$t_{11} = \begin{cases} 0, & t < 0 \\ \rho_0 C_0^2 E(t), & 0 \le t \le \tau \\ 0, & t > \tau, \end{cases} \tag{8.48}$$

with $E(0) = E(\tau) = 0$ (thus precluding introduction of a shock), is applied to the boundary at $X = 0$. Focusing upon the incident wave region, we have

$$U_{(1)}(X, t) = -C_0 \int_0^{t-(X/C_0)} E(\zeta)\, d\zeta, \quad C_0(t - \tau) \le X \le C_0 t. \tag{8.49}$$

As before, the strain and particle velocity associated with this disturbance are given by

$$\frac{\partial U_{(I)}}{\partial X} = E\left(t - \frac{X}{C_0}\right)$$
$$\frac{\partial U_{(I)}}{\partial t} = -C_0\, E\left(t - \frac{X}{C_0}\right). \tag{8.50}$$

Because the boundary at $X = X_1$ is free of traction, a second wave, called a reflected wave, must arise to cancel the stress transmitted to the boundary by the incident wave. When the incident wave encounters the boundary at $t = X_1/C_0$ the reflection process begins. Clearly, the reflected wave is a left-propagating wave having the property that it exactly offsets the stress that the incident wave would produce at X_1 if the material extended beyond the boundary at that point. If we imagine the material to extend beyond X_1 to $X = 2X_1$, then we see that a wave of the same temporal shape as the incident wave, but carrying stress of the opposite sign and propagating to the left from $X = 2X_1$ would effect the cancellation. The reflected wave U_L would therefore be such that

$$\frac{\partial U_{(L)}(X,t)}{\partial X} = -E\left(t - \frac{2X_1 - X}{C_0}\right), \quad \frac{2X_1 - X}{C_0} \le t \le \frac{2X_1 - X}{C_0} + \tau\,, \tag{8.51}$$

subject to the additional condition that t cannot exceed $2X_1/C_0$, the time at which the reflected disturbance first encounters the surface at $X = 0$.

Combining incident and reflected waves gives the displacement gradient field

$$\frac{\partial U(X,t)}{\partial X} = E\left(t - \frac{X}{C_0}\right) - E\left(t - \frac{2X_1 - X}{C_0}\right) \tag{8.52}$$

in the region where both are defined. This *interaction region* is encircled in Fig. 8.9. Clearly, the gradient U_X, hence the stress, vanishes at $X = X_1$, thus satisfying the condition that the boundary be free of stress. The stress field is obtained by substituting Eq. 8.52 into Eq. 8.1. If we are careful to define $E(t)$ to vanish everywhere outside the interval $(0, \tau)$, then the inequalities limiting the domain of the foregoing equations are unnecessary and Eq. 8.52 is valid everywhere, as are the remaining equations in this section.

The displacement field associated with the reflected wave is obtained by integrating Eq. 8.51. We obtain

$$U_{(L)}(X,t) = -C_0 \int_0^{t-[(2X_1 - X)/C_0]} E(\zeta)\,d\zeta, \quad \text{for } 2X_1 - C_0 t \le X \le 2X_1 - C_0(t - \tau).$$

The strain and particle velocity associated with this reflected wave are given by

$$\frac{\partial U_{(L)}}{\partial X} = -E\left(t - \frac{2X_1 - X}{C_0}\right)$$

$$\frac{\partial U_{(L)}}{\partial t} = -C_0\, E\left(t - \frac{2X_1 - X}{C_0}\right).$$

The particle velocity in the interaction region is

$$\dot{x} = \frac{\partial U_{(1)}}{\partial t} + \frac{\partial U_{(L)}}{\partial t} = -C_0\left[E\left(t - \frac{X}{C_0}\right) + E\left(t - \frac{2X_1 - X}{C_0}\right)\right] \quad (8.53)$$

and the value at the interface, $X = X_1$, is

$$\dot{x}(X, t) = -2C_0\, E\left(t - \frac{X_1}{C_0}\right), \quad \frac{X_1}{C_0} \le t \le \frac{X_1}{C_0} + \tau.$$

Note that the wave reflection accelerates the material particles on a stress-free surface to twice the velocity of the particles in the incident wave. It is worth re-emphasizing the point that the reflected wave transmits stress of opposite sign to that of the incident wave: A tensile wave is reflected from a stress-free boundary as a compressive wave, and vice versa.

Reflection of a wave incident on an immovable interface can be analyzed as was done for the stress-free surface, except that the analysis is carried out in terms of the particle-velocity waveform rather than the stress or strain waveform.

Graphical Solution. The foregoing discussion and analysis suggests a graphical method of analyzing the wave interaction at a stress-free surface. Equation 8.52, representing the superposition of a left- and a right-propagating wave, with the left-propagating wave being an image of the right-propagating wave that has been reflected right-to-left, changed in sign, and translated to the right of the reflecting boundary by an amount that matches the distance of the left-propagating wave from this boundary. These waves propagate in their respective directions at a speed C_0, and the complete solution of the problem is their sum. This suggests the graphical procedure illustrated in Fig. 8.10 for solving the reflection problem. The graphical procedure involves marching the incident and reflected waveforms toward, and through, the boundary from each side. To obtain the solution, the waveforms are advanced toward the boundary in equal steps, beginning at the same distance. In the region in which they overlap, the interaction region, the solution is simply the sum of the two waveforms and can be obtained by adding the amplitudes of the incident and reflected contributions at each value of X. Once the left-propagating wave moves past the right-propagating wave, the reflection is complete and the left-propagating wave comprises the entire solution, which is valid until this wave first encounters the boundary at $X = 0$. The subsequent reflection at this boundary can, of course, be analyzed by a procedure similar to that just discussed.

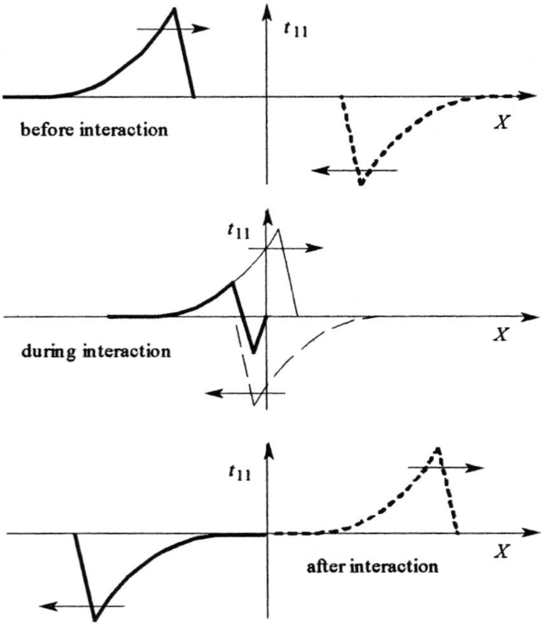

Figure 8.10. Snapshot sequence of a stress pulse interacting with an unrestrained surface. The figure illustrates the graphical method of solving problems of this class. The portions of the figure to the left of the vertical line designating the interface correspond to the real situation, whereas the portions to the right of the interface correspond to the imaginary waves used to generate a solution satisfying the condition that the normal stress vanish on the boundary.

The graphical method can be applied to analysis of reflection of a wave incident on an immovable boundary as was done for the unrestrained surface, except that the figures involve the particle-velocity waveform rather than the strain waveform.

8.4 Wave Interaction at a Material Interface

When an elastic wave encounters an interface at which materials having different properties are joined, an interaction occurs that produces both transmitted and reflected waves.

Consider the case shown in Fig. 8.11. It shows two material slabs of differing density and/or longitudinal elastic modulus that are joined at $X = X_1$ and that are unstressed and at rest in the region ahead of a wave incident on the interface.

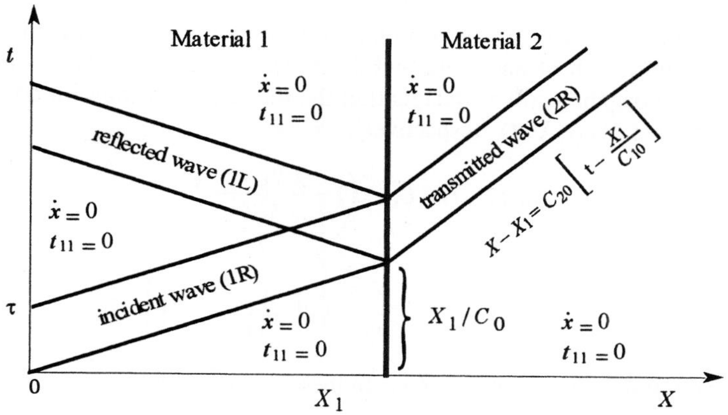

Figure 8.11. Reflection at a material interface.

A boundary condition to be imposed at the interface is that the two materials remain in contact,

$$\dot{x}(X_1^+,\ t) = \dot{x}(X_1^-,\ t)\,. \tag{8.54}$$

Since the interface has no mass, the stress must also be continuous at the interface. The stress is given by the equation

$$t_{11} = \rho_R\, C_0^2 U_X(X, t) \tag{8.55}$$

for the material on each side of the interface, but with the values $\rho_0 = \rho_{1R}$ and $C_0 = C_{10}$ for material 1 and $\rho_0 = \rho_{2R}$ and $C_0 = C_{20}$ for material 2. With this, the stress continuity condition becomes

$$\rho_{1R}\, C_{10}^2\, U_X\,(X_{1+}, t) = \rho_{2R}\, C_{20}^2\, U_X(X_{1-}, t)\,. \tag{8.56}$$

The incident wave is given by Eq. 8.41,

$$U_{(1R)}(X, t) = -\,C_{10} \int_0^{t-(X/C_{10})} E(\vartheta)\,d\vartheta\,, \tag{8.57}$$

which is valid in the incident-wave region shown on Fig. 8.11. Differentiation provides the strain and particle-velocity waveforms

$$\frac{\partial U_{(1R)}}{\partial X} = E\left(t - \frac{X}{C_{10}}\right)$$
$$\frac{\partial U_{(1R)}}{\partial t} = - C_{10}\, E\left(t - \frac{X}{C_{10}}\right). \tag{8.58}$$

The transmitted wave is a right-propagating wave excited by the interfacial boundary conditions 8.54 and 8.56, and the reflected wave will be a similarly excited left-propagating wave. Designating the reflected and transmitted waves $U_{(R2)}(X, t)$ and $U_{(L1)}(X, t)$, respectively, we have

$$U_{(1L)}(X, t) = \overline{U}_{(1L)}\left(t - \frac{2X_1 - X}{C_{10}}\right)$$

and

$$U_{(2R)}(X, t) = \overline{U}_{(2R)}\left[t - \frac{X}{C_{20}} - X_1\left(\frac{1}{C_{10}} - \frac{1}{C_{20}}\right)\right]$$

in the wave regions shown on Fig. 8.11. In these expressions

$$\overline{U}_{(L1)}(0) = -\int_0^{\tau} E(\xi)\, d\xi\,, \quad \overline{U}_{(R2)}(0) = 0\,,$$

and the constant terms have been chosen so that the argument values are positive.

Substituting these results into the condition for velocity continuity, Eq. 8.54, yields

$$\frac{\partial U_{(1R)}(X_1, t)}{\partial t} + \frac{\partial U_{(1L)}(X_1, t)}{\partial t} = \frac{\partial U_{(2R)}(X_1, t)}{\partial t}\,,$$

or

$$-C_{10}\, E(\xi) + \overline{U}'_{(1L)}(\xi) = \overline{U}'_{(2R)}(\xi)\,, \tag{8.59}$$

for $C_{10} X_1 \le t \le C_{10} X_1 + \tau$, and where $\xi = t - (X_1 / C_{10})$. Similarly, the condition for continuity of stress at the interface becomes

$$\rho_{1R} C_{10}^2 \left(\frac{\partial U_{(1R)}(X_1, t)}{\partial X} + \frac{\partial U_{(1R)}(X_1, t)}{\partial X}\right) = \rho_{2R}\, C_{20}^2\, \frac{\partial U_{(2R)}(X_1, t)}{\partial X}\,,$$

where $C_{10} X_1 \le t \le C_{10} X_1 + \tau$, or

$$\rho_{1R}\, C_{10}^2 \left(E(\xi) + \frac{1}{C_{10}} \overline{U}'_{(1L)}(\xi)\right) = -\rho_{2R}\, C_{20}\, \overline{U}'_{(2R)}(\xi)\,. \tag{8.60}$$

Simultaneous solution of Eqs. 8.59 and 8.60 yields

$$\overline{U}'_{(1L)}(\xi) = -C_{10}\,\frac{Z_1 - Z_2}{Z_1 + Z_2}\,E(\xi)$$
$$\overline{U}'_{(2R)}(\xi) = -\frac{2C_{10}Z_1}{Z_1 + Z_2}\,E(\xi)\,, \tag{8.61}$$

where we have introduced the *mechanical impedances* (or simply *impedances*) $Z_1 = \rho_{1R}\,C_{10}$ and $Z_2 = \rho_{2R}\,C_{20}$.

With this, the stresses become

$$t_{11(1R)}(X,t) = Z_1\,C_{10}\;E\left(t - \frac{X}{C_1}\right)$$
$$t_{11(2R)}(X,t) = C_{10}\,\frac{2Z_1Z_2}{Z_1 + Z_2}\,E\left[t - \frac{X}{C_{20}} + X_1\left(\frac{C_{10} - C_{20}}{C_{10}\,C_{20}}\right)\right] \tag{8.62}$$
$$t_{11(1L)}(X,t) = -C_{10}\,\frac{Z_1(Z_1 - Z_2)}{Z_1 + Z_2}\,E\left(t - \frac{2X_1 - X}{C_{10}}\right)$$

in the wave regions. Similar relations can be obtained for the particle velocities:

$$\dot{x}_{(1R)}(X,t) = -C_{10}\;E\left(t - \frac{X}{C_{10}}\right)$$
$$\dot{x}_{(2R)}(X,t) = -C_{10}\,\frac{2Z_1}{Z_1 + Z_2}\,E\left[t - \frac{X}{C_{20}} + X_1\left(\frac{C_{10} - C_{20}}{C_{10}\,C_{20}}\right)\right] \tag{8.63}$$
$$\dot{x}_{(1L)}(X,t) = -C_{10}\,\frac{Z_1 - Z_2}{Z_1 + Z_2}\,E\left(t - \frac{2X_1 - X}{C_{10}}\right)$$

in the wave regions. To obtain the relative strengths of the various waves it is useful to evaluate these functions at the material interface $X = X_1$. We obtain

$$t_{11(1R)}(X_1,t) = C_{10}\,Z_1\;E\left(t - \frac{X_1}{C_{10}}\right)$$
$$t_{11(2R)}(X_1,t) = C_{10}\,\frac{2Z_1Z_2}{Z_1 + Z_2}\,E\left(t - \frac{X_1}{C_{10}}\right) \tag{8.64}$$
$$t_{11(1L)}(X_1,t) = -C_{10}\,\frac{Z_1(Z_1 - Z_2)}{Z_1 + Z_2}\,E\left(t - \frac{X_1}{C_{10}}\right)$$

and

$$\dot{x}_{(1R)}(X_1, t) = -C_{10}\ E\left(t - \frac{X_1}{C_{10}}\right)$$

$$\dot{x}_{(2R)}(X_1, t) = -C_{10}\ \frac{2Z_1}{Z_1 + Z_2}\ E\left(t - \frac{X_1}{C_{10}}\right) \tag{8.65}$$

$$\dot{x}_{(1L)}(X_1, t) = -C_{10}\ \frac{Z_1 - Z_2}{Z_1 + Z_2}\ E\left(t - \frac{X_1}{C_{10}}\right).$$

From these relations we see that the strengths of the transmitted and reflected waves are related to the incident wave by the equations

$$t_{11(2R)}(X_1, t) = \frac{2Z_2}{Z_1 + Z_2}\ t_{11(1R)}(X_1, t)$$

$$t_{11(1L)}(X_1, t) = -\frac{Z_1 - Z_2}{Z_1 + Z_2}\ t_{11(1R)}(X_1, t), \tag{8.66}$$

and

$$\dot{x}_{(2R)}(X_1, t) = \frac{2Z_1}{Z_1 + Z_2}\ \dot{x}_{(1R)}(X_1, t)$$

$$\dot{x}_{(1L)}(X_1, t) = \frac{Z_1 - Z_2}{Z_1 + Z_2}\ \dot{x}_{(1R)}(X_1, t), \tag{8.67}$$

where t falls in the interval $C_{10}\ X_1 \le t \le C_{10}\ X_1 + \tau$, and where the impedance ratios are called transmission and reflection coefficients, respectively.

Note that when $Z_1 = Z_2$ no wave interaction occurs: The transmitted wave is the same as the incident wave and there is no reflected wave. (However, the speed of the transmitted wave may differ from that of the incident wave unless the density and wavespeed are *both* the same for the two materials.) The case of an unrestrained boundary lies at one limit of the range of possible materials, that in which the incident wave encounters a void. At the other extreme, the incident wave propagates through an elastic material toward a rigid, immovable body.

In the limit that material 2 is a void its impedance is zero, $Z_2 = 0$. The foregoing results show that the reflected wave is of the same amplitude as the incident wave, but of opposite sign:

$$t_{11(1L)}(X_1, t) = -\ t_{11(1R)}(X_1, t), \quad \dot{x}_{(1L)}(X_1, t) = -\ \dot{x}_{(1R)}(X_1, t).$$

Of course, the transmitted wave is meaningless in the absence of material in which it can propagate. Since the interfacial stress is the sum of the incident and reflected stresses, we see that it vanishes, as expected at an unrestrained surface. Similarly, the particle velocity of the interface during the reflection is twice that of the incident wave.

In the limit that material 2 is a rigid, immovable body, its impedance is infinite and the reflected wave is of the same amplitude as the incident wave:

$$t_{11(1L)}(X_1, t) = t_{11(1R)}(X_1, t), \quad \dot{x}_{(1L)}(X_1, t) = -\dot{x}_{(1R)}(X_1, t).$$

The interfacial stress during reflection is twice that of the incident wave and the particle velocity is zero.

8.5 Exercises

8.5.1. Derive Eq. 8.4.

8.5.2. Make the calculation required to obtain Eq. 8.17 from the results preceding it in the text.

8.5.3. Prepare an analysis analogous to that of Section 8.3.1 for an immovable boundary, i.e., one at which $\dot{x} = 0$ for all t. Note that the peak stress at this boundary during the interaction is twice that of the incident wave.

8.5.4. Devise a graphical method of solving the problem of wave reflection at an immovable boundary and work out an example.

8.5.5. Describe, using the graphical method, the response of a stack consisting of numerous plates all of the same material and each of thickness L to impact by a single plate of thickness L. What if the impacting plate is of thickness $2L$ or $3L$?

8.5.6. The presentation in the text of the graphical method of solving linear plane-wave problems was based upon Eq. 8.52 for the displacement gradient and, thus, focused upon the stress field. One can develop a similar graphical method based upon Eq. 8.53, which deals with the particle-velocity field. Discuss, using this graphical method, the response of a stack of thin plates to an incident triangular compression pulse of width corresponding to several plate thicknesses.

8.5.7. Explain how the graphical method can be used to solve the problem of a half-space in which an initial stress distribution is present in a region near the boundary. This situation can arise when a very brief pulse of energy is deposited by a laser or x-ray source. When the energy is absorbed in accordance with Lambert's law of absorption the specific internal energy distribution decays exponentially with distance into the material.

CHAPTER 9

Finite-amplitude Elastic Waves

In Chap. 3 we discussed propagation of plane longitudinal shocks, a specific aspect of nonlinear wave propagation. In Chap. 8, we discussed more general waveforms, but in the approximation of small (infinitesimal) deformation that reduced the problem to one of solving a linear equation. Now we consider smooth uniaxial motions in a more general case in which the equation to be solved is not linear. We shall not be able to obtain results as complete and detailed as in the foregoing cases, but much can be learned.

The partial differential equations describing nonlinear longitudinal elastic waves are of a type called quasilinear hyperbolic partial differential equations (systems) of second order. They have been widely studied, with extensive results having been reported (see, for example, [28,108]). Much of the analysis in these references is in the context of gas dynamics, but the application to plane, longitudinal waves in elastic solids is immediate.

Once the general principles are understood and something is known of the phenomena we can expect to encounter, we can feel comfortable solving specific problems by numerical means. One-dimensional problems of the sort that we shall discuss in this chapter are amenable to numerical solution by an ordinary personal computer.

9.1 Nonlinear Wave Equation

Lagrangian Representation. When a spatially uniform, time-dependent load is applied to the surface of an isotropic elastic body occupying the halfspace $X \geq 0$, a plane longitudinal wave is introduced into the material. This motion is one of uniaxial strain, as discussed in Sect. 2.2 and described by the equations

$$x = X + U(X,t), \quad x_2 = X_2, \quad x_3 = X_3. \tag{9.1}$$

From this we obtain

$$U_X = F_{11} - 1 = (\rho_R/\rho) - 1 \equiv G, \quad \dot{x} = U_t, \tag{9.2}$$

where the subscripts X and t indicate partial differentiation with respect to the variable shown, holding the other constant. The variable G is introduced as a convenient simplification of notation. Some useful quantities associated with this motion have been listed in the section cited above.

The governing equations for this case have been given in the Lagrangian form as Eqs. 2.92. Let us begin our considerations with the equation for conservation of energy, which is

$$\rho_R \dot{\varepsilon} - t_{11} \dot{x}_X = 0 , \tag{9.3}$$

when $Q = 0$ and $r = 0$, as we shall assume throughout this chapter.* From the equation of state, which we can write in the form $\varepsilon = \hat{\varepsilon}(G, \eta)$, we have

$$\dot{\varepsilon} = \frac{\partial \hat{\varepsilon}}{\partial G}\dot{G} + \frac{\partial \hat{\varepsilon}}{\partial \eta}\dot{\eta} = \frac{1}{\rho_R} t_{11} \dot{G} + \theta \dot{\eta} , \tag{9.4}$$

so Eq. 9.3 can be written

$$t_{11} \dot{G} + \rho_R \theta \dot{\eta} - t_{11} \dot{x}_X = 0 , \tag{9.5}$$

but this can be reduced to

$$\dot{\eta} = 0 \tag{9.6}$$

by using Eq. 2.92_1. This means that η is a function of X alone. If, as we shall assume, the material is initially in a uniform state, then η has a constant value, the stress relation is independent of X, and corresponds to the isentrope through this state,

$$t_{11} = t_{11}^{(\eta)}(G) . \tag{9.7}$$

The equation for conservation of energy plays no further role in the analysis, and the problem is set in terms of the remaining two equations, $2.92_{1,2}$, which we can write

$$\begin{aligned} \dot{x}_t - C_L^2(G)\, G_X &= 0 \\ G_t - \dot{x}_X &= 0 , \end{aligned} \tag{9.8}$$

where

$$C_L^2 = \frac{1}{\rho}\frac{dt_{11}^{(\eta)}}{dE_{11}} = \frac{1}{\rho_R}\frac{dt_{11}^{(\eta)}}{dU_X} = \frac{1}{\rho_R}\frac{dt_{11}^{(\eta)}}{dG} . \tag{9.9}$$

When the material is a fluid, it is usual to write these equations in one of the forms

---

*Although we shall not discuss the case $r \neq 0$, it does arise in a variety of practical problems involving heating by electrical current pulses, pulsed lasers, x-ray sources, particle beam accelerators, and other devices that can cause rapid deposition of energy into the interior of material bodies. The restriction $Q = 0$ implies, formally, that the material is a nonconductor of heat and, practically, that the effects of heat conduction are assumed to be negligible.

$$C_{\mathrm{L}}^2 = -\frac{1}{\rho_{\mathrm{R}}^2}\frac{dp^{(\eta)}}{dv} = \left(\frac{\rho}{\rho_{\mathrm{R}}}\right)^2 \frac{dp^{(\eta)}}{d\rho}. \tag{9.10}$$

For thermodynamically stable materials the derivatives appearing in Eqs. 9.9 and 9.10 are positive [18, p. 135], so C_{L} is a real number (taken to be positive) that we shall be able to identify as the *Lagrangian soundspeed.*

Equations 9.8 can be written in either of the equivalent and useful forms

$$\begin{aligned} C_{\mathrm{L}}\,(G_t + C_{\mathrm{L}}\,G_X) - (\dot{x}_t + C_{\mathrm{L}}\,\dot{x}_X) &= 0 \\ C_{\mathrm{L}}\,(G_t - C_{\mathrm{L}}\,G_X) + (\dot{x}_t - C_{\mathrm{L}}\,\dot{x}_X) &= 0, \end{aligned} \tag{9.11}$$

or

$$C_{\mathrm{L}}^2\,U_{XX} = U_{tt}. \tag{9.12}$$

When experiments are conducted to study solids, sensors are usually fixed to the material and thus record what happens to the particle at which the sensor is located. It is for this reason, along with the fact that the equations are simpler, that problems of solid mechanics are most often analyzed in the *Lagrangian* frame.

Eulerian Representation. The Eulerian form of the equations governing nonlinear plane-wave propagation have been given as Eqs. 2.87, which we write

$$\begin{aligned} \rho_t + (\rho\,\dot{x})_x &= 0 \\ (\rho\,\dot{x})_t + (\rho\,\dot{x}^2)_x - (t_{11})_x &= 0. \end{aligned} \tag{9.13}$$

The energy-conservation equation has been omitted from consideration for the reasons discussed above. Equations 9.13 can be written in the simplified form

$$\begin{aligned} \rho_t + \rho_x\,\dot{x} + \rho\,\dot{x}_x &= 0 \\ \rho\,\dot{x}_t + \rho\,\dot{x}\,\dot{x}_x + c_{\mathrm{L}}^2\,\rho_x &= 0, \end{aligned} \tag{9.14}$$

where we have related the stress to the uniaxial deformation (represented without loss of generality by the density) by the isentrope

$$t_{11} = t_{11}^{(\eta)}(\rho). \tag{9.15}$$

Using this equation, we have

$$(-t_{11})_x = c_{\mathrm{L}}^2(\rho)\,\rho_x, \tag{9.16}$$

where

$$c_{\mathrm{L}}^2(\rho) = -\frac{dt_{11}^{(\eta)}(\rho)}{d\rho}. \tag{9.17}$$

We take $c_{\mathrm{L}} > 0$ and will be able to identify $c_{\mathrm{L}}(\rho)$ as the *Eulerian soundspeed.*

The Eulerian soundspeed measures the rate of progress of a wavelet along the x axis relative to the motion of the material moving at the velocity $\dot{x}$, and is related to the *Lagrangian* soundspeed C_{L} by the equation

$$c_{\mathrm{L}} = (1+G)C_{\mathrm{L}} = (\rho_{\mathrm{R}}/\rho)C_{\mathrm{L}}, \tag{9.18}$$

reflecting the different distance between corresponding points in the reference and current configurations that is to be covered in the same time interval.

When Eq. 9.14_1 is multiplied by c_{L} and the result both added to and subtracted from Eq. 9.14_2 we arrive at the equivalent system

$$\begin{aligned} c_{\mathrm{L}}\left[\rho_t + (c_{\mathrm{L}} + \dot{x})\rho_x\right] + \rho\left[\dot{x}_t + (c_{\mathrm{L}} + \dot{x})\dot{x}_x\right] = 0 \\ c_{\mathrm{L}}\left[\rho_t - (c_{\mathrm{L}} - \dot{x})\rho_x\right] - \rho\left[\dot{x}_t - (c_{\mathrm{L}} - \dot{x})\dot{x}_x\right] = 0. \end{aligned} \tag{9.19}$$

The second-order representation of Eqs. 9.14 does not have the simple form of its *Lagrangian* counterpart, Eq. 9.12, but the linear approximation to this equation does take the simple form

$$u_{tt} - c_{\mathrm{L}}^2 u_{xx} = 0, \tag{9.20}$$

where $c_{\mathrm{L}} = c_{\mathrm{L}}(\rho_{\mathrm{R}})$. This is the equation discussed in Chap. 8.

When experiments are conducted upon fluids, for example in a wind tunnel, sensors are usually fixed in space and thus record what happens at the sensor location, x, as the particles move past. It is for this reason, among others, that the theory presented in fluid dynamics books is usually set in an Eulerian framework.

9.1.1 Qualitative Discussion of Elastic Wave Propagation

Equation 9.12 has the same form as the corresponding equation of the linear theory, Eq. 8.4, except that the wavespeed is no longer a constant. The nonlinear equation is similar to the linear equation in that solutions take the form of left- and right-propagating waves that move through the material at a finite speed. Characteristic curves define trajectories along which wavelets are transmitted in the X–t or x–t planes. The boundary conditions that can be imposed are the same as for the linear equation. Riemann invariants can be determined and used as in the linear theory.

It is here that the similarity of the two theories ends. Solutions of the nonlinear equation do not enjoy the scaling and superposition properties that facilitated solution of the linear equation. Functions of the form $f(X \pm C_{\mathrm{L}} t)$ do not satisfy the nonlinear equation, left- and right-propagating waves do not pass through one another without interaction, and reflection from boundaries changes the amplitude and character of the wave in ways that are more complicated than was the case for the linear equation. The fact that $f(X \pm C_L t)$ is not a solution of the nonlinear equation suggests that the disturbance changes form as it propagates, and this will prove to be the case. Smooth waves can spread or can become steeper to form shocks and shocks can propagate intact or can spread to form smooth waves. The behavior of solutions of the nonlinear equation can usefully be discussed by comparison to those of the linear equation, but the range of behavior of nonlinear waves is richer and more interesting. Unfortunately, analysis of these waves is sufficiently complicated that computers using finite-difference or finite-element programs are required for solution of most practical problems. However, application of this technology is facilitated if one is familiar with the results obtained by analysis of some important simple problems.

9.1.2 Characteristic Curves

Lagrangian Analysis. In our study of the linear wave equation, we discovered that wavelets are transmitted along lines $X \pm C_0 t = \text{const.}$, and we called these curves characteristic lines for the equation. Since the coordinate lines $X = \text{const.}$ and $t = \text{const.}$ could be transformed to lines of the form $X \pm C_0 t = \text{const.}$, the latter could be used as a coordinate system replacing the former. A similar situation arises in connection with the nonlinear equation, except that the characteristic curves for the nonlinear equation are not straight lines, reflecting the fact that the wavespeed is not constant. Indeed, the slope of the characteristics at a point depends upon the displacement gradient at that point and, since the displacement gradient is not known until the problem is solved, the characteristics must be determined as part of the solution.

The analysis of this subsection is carried out using the Lagrangian form of the theory as represented by Eqs. 9.8. Let the characteristics be curves $X^* = \text{const.}$ and $t^* = \text{const.}$ that are related to X and t by equations of the form

$$X^* = X^*(X, t) \quad \text{and} \quad t^* = t^*(X, t). \tag{9.21}$$

It is apparent that these characteristics must be such that each point (X, t) is uniquely related to a point (X^*, t^*) if we are to be able to proceed with the analysis. This means that Eqs. 9.21 must be invertible to yield expressions of the form

$$X = X(X^*, t^*) \quad \text{and} \quad t = t(X^*, t^*). \tag{9.22}$$

We shall see that this condition is met in regions where the motion is smooth. The condition for invertibility is that the Jacobian of either transformation does not vanish. For example, we can require that

$$\mathcal{J} \equiv X_{X^*} t_{t^*} - X_{t^*} t_{X^*} \neq 0. \tag{9.23}$$

We expect the characteristic curves to have slopes associated with the wavespeed, as in the linear case. In the nonlinear case the wavespeed varies, so we are restricted to use of the local analog of Eqs. 8.44, which we can write

$$\frac{dX}{dt} = C_L(G) \quad \text{and} \quad \frac{dX}{dt} = -C_L(G)\,, \tag{9.24}$$

respectively. When these equations are expressed in terms of the characteristic coordinates, we have

$$\begin{aligned} X_{X^*} - C_L(G)\,t_{X^*} &= 0 \quad \text{on} \quad t^* = \text{const.} \\ X_{t^*} + C_L(G)\,t_{t^*} &= 0 \quad \text{on} \quad X^* = \text{const.} \end{aligned} \tag{9.25}$$

Note that when $C_L(G) = C_R = \text{const}$ we recover Eqs. 8.44 by integration of Eqs. 9.25.

Differentiation of $\dot{x}(X^*, t^*)$ and $G(X^*, t^*)$ with respect to X and t, using the chain rule, gives

$$\begin{aligned} \dot{x}_t &= \dot{x}_{X^*} X_t^* + \dot{x}_{t^*} t_t^*, \quad & \dot{x}_X &= \dot{x}_{X^*} X_X^* + \dot{x}_{t^*} t_X^* \\ G_t &= G_{X^*} X_t^* + G_{t^*} t_t^*, \quad & G_X &= G_{X^*} X_X^* + G_{t^*} t_X^*. \end{aligned} \tag{9.26}$$

From Eqs. 9.21 we have

$$\begin{aligned} 1 &= X_X^* X_{X^*} + X_t^* t_{X^*}, \quad & 0 &= X_X^* X_{t^*} + X_t^* t_{t^*} \\ 1 &= t_X^* X_{t^*} + t_t^* t_{t^*}, \quad & 0 &= t_X^* X_{X^*} + t_t^* t_{X^*}, \end{aligned} \tag{9.27}$$

so

$$\begin{aligned} X_X^* &= t_{t^*}/\mathcal{J}, \quad & X_t^* &= -X_{t^*}/\mathcal{J} \\ t_X^* &= -t_{X^*}/\mathcal{J}, \quad & t_t^* &= X_{X^*}/\mathcal{J}, \end{aligned} \tag{9.28}$$

and Eqs. 9.8 can be written in the form

$$\begin{aligned} (\dot{x}_{X^*} - C_L\,G_{X^*})\,t_{t^*} + (\dot{x}_{t^*} + C_L\,G_{t^*})\,t_{X^*} &= 0 \\ (\dot{x}_{X^*} - C_L\,G_{X^*})\,t_{t^*} - (\dot{x}_{t^*} + C_L\,G_{t^*})\,t_{X^*} &= 0 \end{aligned} \tag{9.29}$$

by use of Eqs. 9.28 and 9.25. Adding and subtracting these equations, and adjoining the results to Eqs. 9.25, gives the final expression of the problem in terms of the equations

$$\begin{aligned} \dot{x}_{X^*} - C_L\,G_{X^*} &= 0 \\ \dot{x}_{t^*} + C_L\,G_{t^*} &= 0 \end{aligned} \tag{9.30}$$

for the field variables $\dot{x}(X^*, t^*)$ and $G(X^*, t^*)$ and Eqs. 9.25 for the characteristic curves $X = X(X^*, t^*)$, and $t = t(X^*, t^*)$.

The equations of motion 9.30 can be integrated immediately to give

$$
\begin{aligned}
\Phi^{+}(t^{*}) &= \dot{x} - \int_{0}^{G} C_{\mathrm{L}}(G')\,dG' \\
\Phi^{-}(X^{*}) &= \dot{x} + \int_{0}^{G} C_{\mathrm{L}}(G')\,dG'.
\end{aligned} \qquad (9.31)
$$

These functions, called *Riemann invariants*, are determined by suitable boundary and initial conditions. They are called invariants because they are constant along curves of constant t^{*} and X^{*}, respectively. This means, for example, that a value determined for the function

$$\dot{x} - \int_{0}^{G} C_{\mathrm{L}}(G')\,dG'$$

at any point on a curve of constant t^{*} remains the same at all points on this curve.

Adding and subtracting Eqs. 9.31 yields the equations

$$
\begin{aligned}
\dot{x} &= \tfrac{1}{2}\left[\Phi^{-}(X^{*}) + \Phi^{+}(t^{*})\right] \\
\int_{0}^{G} C_{\mathrm{L}}(G')\,dG' &= \tfrac{1}{2}\left[\Phi^{-}(X^{*}) - \Phi^{+}(t^{*})\right]
\end{aligned} \qquad (9.32)
$$

for $\dot{x}$ and G as functions of X^{*} and t^{*}.

The general utility of these solutions is limited because the relationship of the characteristic coordinates to X and t has yet to be determined, and this determination requires solution of partial differential equations. However, the complexity of the characteristic form 9.30 of the field equations relative to Eq. 9.12 or Eq. 9.8 is reduced when we restrict our attention to a particularly important class of solutions called *simple waves*, which we shall discuss in Sect. 9.2.

Eulerian Analysis. One can obtain results essentially the same as those of the foregoing section by analysis of Eqs. 9.19, the Eulerian form of the wave equation.

Let us transform Eqs. 9.19 to characteristic coordinates, x^{*} and t^{*}. On the characteristic curves we have

$$
\begin{aligned}
x_{x^{*}} - (c_{\mathrm{L}} + \dot{x})\,t_{x^{*}} &= 0 \\
x_{t^{*}} + (c_{\mathrm{L}} - \dot{x})\,t_{t^{*}} &= 0,
\end{aligned} \qquad (9.33)
$$

or

$$\frac{dx}{dt} = c_L(\rho) + \dot{x} \quad \text{on} \quad t^* = \text{const.}$$
$$\frac{dx}{dt} = -c_L(\rho) + \dot{x} \quad \text{on} \quad x^* = \text{const.}, \tag{9.34}$$

where we have chosen to express the deformation in terms of the material density, ρ. The fields can be expressed in terms of characteristic coordinates by equations of the form

$$\rho = \rho(x^*, t^*)$$
$$\dot{x} = \dot{x}(x^*, t^*). \tag{9.35}$$

We proceed with the transformation exactly as in the Lagrangian case, arriving at the result

$$\rho_t = (-\rho_{x^*} x_{t^*} + \rho_{t^*} x_{x^*})/j$$
$$\rho_x = (-\rho_{x^*} t_{t^*} + \rho_{t^*} t_{x^*})/j$$
$$\dot{x}_t = (-\dot{x}_{x^*} x_{t^*} + \dot{x}_{t^*} x_{x^*})/j$$
$$\dot{x}_x = (-\dot{x}_{x^*} t_{t^*} + \dot{x}_{t^*} t_{x^*})/j, \tag{9.36}$$

where $j = x_{t^*} t_{x^*} - x_{x^*} t_{t^*} \neq 0$ for smooth motions. When these results are substituted into Eqs. 9.19 and cancellation is performed, we arrive at the field equations in terms of characteristic coordinates,

$$\dot{x}_{x^*} + \frac{c_L(\rho)}{\rho} \rho_{x^*} = 0$$
$$\dot{x}_{t^*} - \frac{c_L(\rho)}{\rho} \rho_{t^*} = 0. \tag{9.37}$$

These equations can be written in the form

$$\frac{d}{dx^*}\left(\dot{x} + \int_{\rho_R}^{\rho} \frac{c_L(\rho')}{\rho'} d\rho'\right) = 0$$
$$\frac{d}{dt^*}\left(\dot{x} - \int_{\rho_R}^{\rho} \frac{c_L(\rho')}{\rho'} d\rho'\right) = 0, \tag{9.38}$$

and integration gives the Riemann invariants

$$\dot{x} + \int_{\rho_R}^{\rho} \frac{c_L(\rho')}{\rho'} d\rho' = \Phi^+(t^*)$$
$$\dot{x} - \int_{\rho_R}^{\rho} \frac{c_L(\rho')}{\rho'} d\rho' = \Phi^-(x^*). \tag{9.39}$$

Adding and subtracting these equations gives the expressions

$$\dot{x} = \frac{1}{2}\left[\Phi^+(t^*) + \Phi^-(x^*)\right]$$
$$\int_{\rho_R}^{\rho} \frac{c_L(\rho')}{\rho'} d\rho' = \frac{1}{2}\left[\Phi^+(t^*) - \Phi^-(x^*)\right] \tag{9.40}$$

for the field variables in characteristic coordinates.

9.2 Simple Waves

When a continuously varying stress or particle-velocity history is imposed on the boundary of a body in a state of uniform deformation and motion, a simple wave is introduced into the material. A *simple wave* is a solution of the field equations in a region of the X–t or x–t plane adjacent to another region in which the variables $\dot{x}$ and either G or ρ are constant (it is easy to see that constants form a trivial solution of the field equations). The mathematical justification of the term "simple" is that the field quantities in a simple wave are functions of only one independent variable. The simple wave provides a convenient setting in which to explore some general aspects of nonlinear wave propagation. Simple waves are also of interest because they arise in such practical situations as shock reflection from an unrestrained boundary. A disturbance introduced into an elastic body can propagate as a smooth wave or can become steeper to form a shock. Which of these disturbances is produced depends upon whether the boundary loading produces an expansion or contraction of the material and upon the slope of the function $C_L(G)$ or $c_L(\rho)$.

Let us consider a wave propagating in the $+X$ or $+x$ direction into a region of uniform motion characterized by $\dot{x} = \dot{x}^- =$ const. and $G = G^- =$ const. The wave is produced by imposing the velocity history

$$\dot{x}(0, t) = \begin{cases} \dot{x}^-, & t \le t^- \\ f(t), & t^- \le t \le t^+ \\ \dot{x}^+, & t \ge t^+ \end{cases} \tag{9.41}$$

on the boundary $X = 0$, with the conditions $f(t^-) = \dot{x}^-$ and $f(t^+) = \dot{x}^+$ imposed to ensure continuity.

9.2.1 Lagrangian Analysis

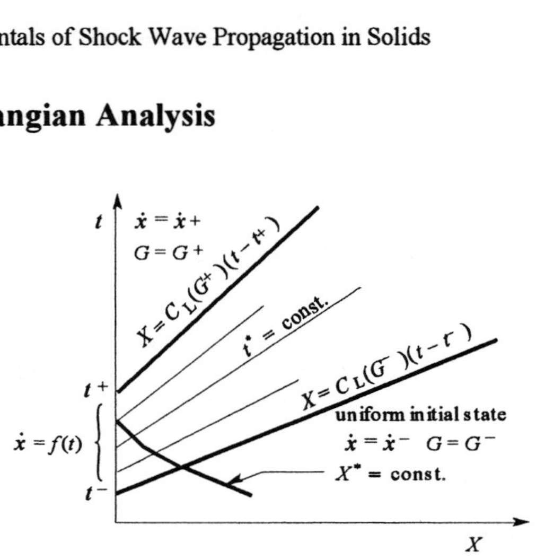

Figure 9.1. The X–t diagram for a right-traveling simple wave.

The X–t diagram for the simple wave problem is shown in Fig. 9.1. The characteristics in the region ahead of the wave can be written

$$X - C_L(G^-)t = \Psi^+(t^*), \quad X + C_L(G^-)t = \Psi^-(X^*). \tag{9.42}$$

Let us choose t^* to be the time at which the $t^* = \text{const.}$ characteristic intersects the boundary $X = 0$. Then $\Psi^+(t^*) = -C_L(G^-)t^*$. Similarly, we shall choose X^* to be the coordinate value, X, at which the $X^* = \text{const.}$ characteristic intersects the line $t = 0$. Then $\Psi^-(X^*) = X^*$ and, in the region ahead of the disturbance, the characteristics of Eqs. 9.42 take the form

$$X = C_L(G^-)(t - t^*), \quad X + C_L(G^-)t = X^*, \tag{9.43}$$

where t^* and X^* are the characteristic coordinates.

The disturbance introduced into the body at the time t^- propagates into the region of uniform initial state at the velocity $C_L(G^-)$ and the leading characteristic is given by Eq. 9.43_1 with $t^* = t^-$.

Now let us consider the X^* characteristics for $0 \le X^* < \infty$. One characteristic of this family passes through each point of the positive quadrant of the X–t plane and, on each of these characteristics, Eq. 9.31_2 takes the form

$$\Phi^-(X^*) = \dot{x}^- + \int_0^{G^-} C_L(G')\,dG', \tag{9.44}$$

a value that is actually *independent* of X^* because $\dot{x}^-$ and G^- each have the same constant value throughout the region ahead of the wave. This value is maintained on each X^* characteristic as it enters and passes through the disturbance propagating from the boundary into the material, so Eqs. 9.32 can be written

$$\dot{x} = \frac{1}{2}\left[\Phi^{+}(t^{*}) + \dot{x}^{-} + \int_{0}^{G^{-}} C_{L}(G')\,dG'\right]$$
$$\int_{0}^{G} C_{L}(G')\,dG' = \frac{1}{2}\left[-\Phi^{+}(t^{*}) - \dot{x}^{-} + \int_{0}^{G^{-}} C_{L}(G')\,dG'\right], \tag{9.45}$$

and we see that the solution in the region occupied by the wave depends only upon the single characteristic coordinate, t^*. The foregoing development shows that any disturbance propagating into a region in a uniform state is a simple wave.

The Riemann invariant on each of the X^* characteristics has the value

$$\dot{x}^{-} + \int_{0}^{G^{-}} C_{L}(G')\,dG'\,. \tag{9.46}$$

Therefore, we have

$$\dot{x} + \int_{0}^{G} C_{L}(G')\,dG' = \dot{x}^{-} + \int_{0}^{G^{-}} C_{L}(G')\,dG' \tag{9.47}$$

in the simple wave. We know that $\dot{x}(0,t^{*}) = f(t^{*})$ on the boundary at the time $t = t^{*}$, so we can determine the value of $G(0,t^{*}) \equiv G^{*}$ from Eq. 9.47:

$$\int_{0}^{G^{*}} C_{L}(G')\,dG' = \dot{x}^{-} - f(t^{*}) + \int_{0}^{G^{-}} C_{L}(G')\,dG'. \tag{9.48}$$

We evaluate the Riemann invariant $\Phi^{+}(t^{*})$ for characteristics in the simple wave from the known conditions at the time at which the characteristic intersects the boundary. From Eq. 9.31_1 we have

$$\dot{x} - \int_{0}^{G} C_{L}(G')\,dG' = f(t^{*}) - \int_{0}^{G^{*}} C_{L}(G')\,dG' \tag{9.49}$$

at all points along the t^* characteristic, in particular at the point where an X^* and a t^* characteristic intersect, so adding Eq. 9.47 to Eq. 9.49 and using Eq. 9.48 gives

$$\dot{x} = f(t^{*}) \tag{9.50}$$

at all points along the t^* characteristic. Substituting this result into Eq. 9.49 gives

$$\int_0^G C_L(G')\,dG' = \int_0^{G^*} C_L(G')\,dG', \tag{9.51}$$

and we see that the value of G along this characteristic is just its value on the boundary:

$$G = G(0, t^*) \equiv G^*, \tag{9.52}$$

with G^* being given as the solution of Eq. 9.48.

Since G is constant along each of the t^* characteristics in the wave region, they are straight lines:

$$X = C_L(G^*(t^*))\,(t - t^*)\,. \tag{9.53}$$

We now know both $\dot{x}$ and G at all points in the wave, and see that they are functions of only the single variable t^*. The leading characteristic (corresponding to the first arrival of the wave at a point) is $X = C_L(G^-)(t - t^-)$ and the trailing characteristic (corresponding to passage of the disturbance) is $X = C_L(G^+)(t - t^+)$. Consideration of the region $0 \le X \le C_L(G^+)\,(t - t^+)$ behind the trailing characteristic shows that $\dot{x} = \dot{x}^+$ and $G = G(0, t^+) \equiv G^+$ throughout this region.

To plot waveforms as a functions of X for a given value of t, one selects several values of t^*, determines the associated values of the variable G and of $\dot{x}$, t_{11}, or any other variable defining the waveform of interest. Then, a set of values of X determined from Eq. 9.53 forms the basis for plotting the waveform. An analogous procedure can be adopted to plot waveforms as functions of t for various values of X.

For some purposes it is necessary to determine the X–t trajectories of the X^* characteristics. In the region ahead of the wave these characteristics are the straight lines $X = -C_L(G^-)t + X^*$. They enter the wave region at the point $X = X_0$, $t = t_0$, given by

$$X_0 = \tfrac{1}{2}\left[X^* - C_L(G^-)t^-\right]$$
$$t_0 = \frac{X^* + C_L(G^-)t^-}{2C_L(G^-)}\,. \tag{9.54}$$

When an X^* characteristic enters the wave region, the value of G changes with position along the characteristic and the characteristic becomes curved, with its shape depending upon the function $G(X, t)$ in the wave region. The X^* characteristic is given as the solution of Eq. 9.25_2, which we write in the form

$$X_{t^*} + C_L(G^*(t^*))\, t_{t^*}. \tag{9.55}$$

Differentiation of Eq. 9.53 with respect to t^* gives

$$\frac{dX}{dt^*} = \frac{dC_L(t^*)}{dt^*}(t - t^*) + C_L(t^*)\left(\frac{dt}{dt^*} - 1\right), \tag{9.56}$$

and substitution of Eq. 9.55 yields

$$\frac{dt}{dt^*} + \frac{1}{2C_L(t^*)}\frac{dC_L(t^*)}{dt^*}t = \frac{1}{2}\left(1 + \frac{1}{C_L(t^*)}\frac{dC_L(t^*)}{dt^*}\right). \tag{9.57}$$

The wavespeed, C_L, depends upon t^* through the relation $G(t^*)$ so we can write

$$\frac{dC_L}{dt^*} = \frac{dC_L}{dG}\frac{dG}{dt^*},$$

but, by differentiation of Eq. 9.48, we find that

$$\frac{dG(t^*)}{dt^*} = -\frac{1}{C_L(G(t^*))}\frac{df(t^*)}{dt^*}.$$

With this, Eq. 9.57 takes the form

$$\frac{dt}{dt^*} + \varphi(t^*)\, t = \psi(t^*), \tag{9.58}$$

where

$$\begin{aligned} \varphi(t^*) &= -\frac{1}{2C_L^2(t^*)}\frac{dC_L(G)}{dG}\frac{df(t^*)}{dt^*} \\ \psi(t^*) &= \frac{1}{2}\left(1 - \frac{1}{C_L^2(t^*)}\frac{dC_L(G)}{dG}\frac{df(t^*)}{dt^*}t^*\right). \end{aligned} \tag{9.59}$$

Equation 9.58 is a linear first-order ordinary differential equation having the familiar solution

$$t(t^*) = \exp\left[-\int_{t^-}^{t^*}\varphi(t')dt'\right]\left\{t_0(X^*) + \int_{t^-}^{t^*}\psi(t')\exp\left[\int_{t^-}^{t'}\varphi(t'')dt''\right]dt'\right\}, \tag{9.60}$$

where we have chosen the constant of integration $t_0(X^*)$ to be the time at which the characteristic intercepts the leading edge of the wave. Equations 9.53 and 9.60 express the X–t trajectory of the X^* characteristic in terms of the parameter t^*.

Throughout the foregoing development we have assumed that the wave spreads as it propagates, as indicated in Fig. 9.1. The condition for this is simply that the wavespeed $C_L(G)$ decrease as each succeeding part of the waveform passes a given point, i.e. that

$$C_L(G^{**}) < C_L(G^{*}) \tag{9.61}$$

for all t^* and t^{**} such that $0 \le t^* < t^{**} \le t^+$. When this condition is not met, a shock will form, as will be discussed in Sect. 9.5.1. Since $C_L(G)$ is calculated from the stress relation $t_{11}(G)$ according to Eq. 9.9, the inequality 9.61 imposes some restrictions upon this function and upon the values of G taken in the wave.

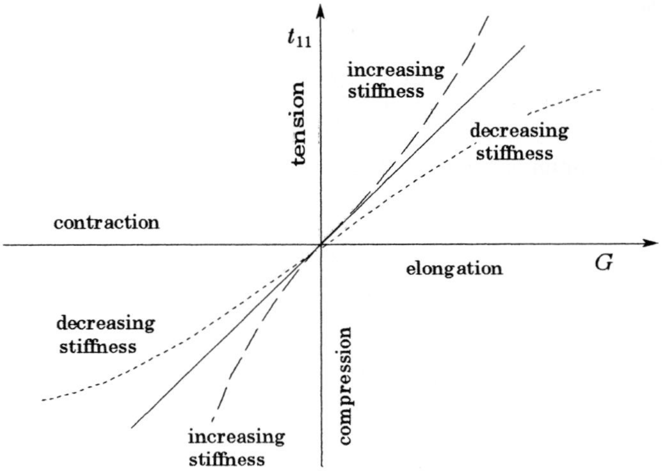

Figure 9.2. Stress–deformation curves of types that produce both shocks and smooth waves. The various cases that can arise are listed in Table 9.1.

Several stress–deformation relations are shown in Fig. 9.2. In view of Eq. 9.9_2, we see that the soundspeed for the material in the state of deformation G, corresponds to the shape of the curve at that deformation level. The curves that bend away from the abscissa yield soundspeeds that increase with increasing deformation, whereas curves that bend toward this axis yield soundspeeds that decrease with increasing deformation. Several cases arise in connection with a given wave-propagation problem. The stress response function can have any of the shapes shown in the figure, and the transition between points on the response curve can be in either direction. This leads to the possibilities outlined in Table 9.1.

Table 9.1. Waveforms associated with given stress–deformation curve shape for waves producing either contraction or expansion

Curve shape	Stress range	Wave type	
		Contraction wave	Expansion wave
Increasing stiffness	Tension quadrant	Smooth	Shock
	Compression quadrant	Shock	Smooth
Decreasing stiffness	Compression quadrant	Smooth	Shock
	Tension quadrant	Shock	Smooth

Example: Third-order Elasticity. Because of the implicit form of the solution for the simple wave, it is useful to study an example to see how wave profiles are actually determined. Let us consider the case of a solid described by the stress relation 6.21₂, which we write in terms of G (recall that smooth elastic waves are isentropic),

$$t_{11} = C_{11}\,G + \tfrac{1}{2}(3C_{11} + C_{111})\,G^2 + \ldots = C_{11}\,(1 + CG + \ldots)G\,, \tag{9.62}$$

where

$$C = \frac{3C_{11} + C_{111}}{2C_{11}}\,,$$

a material constant measuring the nonlinearity of the response, is negative for normal materials.

The soundspeed associated with this relation is

$$C_{\mathrm{L}} = C_{\mathrm{R}}(1 + CG + \ldots)\,, \tag{9.63}$$

where

$$C_{\mathrm{R}} = (C_{11}/\rho_{\mathrm{R}})^{1/2} \tag{9.64}$$

is the soundspeed of the linear theory.

We continue our consideration of the wave introduced into material that is in a uniform state of deformation and motion characterized by $G = G^-$ and $\dot{x} = \dot{x}^-$, by imposition of the velocity history of Eq. 9.41 on the boundary at $X = 0$.

The first step in calculating the simple waveform is determination of the displacement gradient $G(0, t^*) \equiv G^*$ on the boundary at a time $t = t^*$ in the interval $0 \le t \le t^+$. This quantity is obtained from Eq. 9.48 which, for the case at hand, becomes

$$\left(1+\frac{1}{2}CG^*+\ldots\right)G^* = \frac{\dot{x}^- - f(t^*)}{C_R} + \left(1+\frac{1}{2}CG^-+\ldots\right)G^-, \quad t^- \le t^* \le t^+, \tag{9.65}$$

or,

$$G^* = \left(\frac{\dot{x}^- - f(t^*)}{C_R}\right) + G^- - \frac{1}{2}C\left(\frac{\dot{x}^- - f(t^*)}{C_R}\right)^2 - C\left(\frac{\dot{x}^- - f(t^*)}{C_R}\right)G^- + \ldots, \tag{9.66}$$

where we have taken the expansion to second order in the small quantities G^- and $[\dot{x}^- - f(t^*)]/C_R$. From Eq. 9.50 we have

$$\dot{x}^* = f(t^*), \quad t^- \le t^* \le t^+. \tag{9.67}$$

The t^* characteristics are given by

$$X = C_L(G^*)(t - t^*), \quad t^- \le t^* \le t^+. \tag{9.68}$$

As we have seen, G and $\dot{x}$ are constant along each of these characteristics, maintaining the values they have on the boundary, G^* and $\dot{x}^*$, respectively. Substituting Eq. 9.66 into Eq. 9.63 gives

$$C_L(G^*) = C_R\left[1 + C\left(\frac{\dot{x}^- - f(t^*)}{C_R} + G^-\right) + \ldots\right], \tag{9.69}$$

so the advancing characteristic intersecting the boundary at $t = t^*$ is given by

$$X = C_R\left[1 + C\left(\frac{\dot{x}^- - f(t^*)}{C_R} + G^-\right) + \ldots\right](t - t^*). \tag{9.70}$$

The leading characteristic, corresponding to $t^* = t^-$ is, since $f(t^-) = \dot{x}^-$,

$$X = C_R[1 + C\,G^- + \ldots](t - t^-), \tag{9.71}$$

and the trailing characteristic is

$$X = C_R\left[1 + C\left(\frac{\dot{x}^- - \dot{x}^+}{C_R} + G^-\right) + \ldots\right](t - t^+). \tag{9.72}$$

As we noted previously, the wave must spread as it propagates if it is to be a simple wave as presented. Accordingly, we must have

$$-C\,f(t^*) \ge -C\,f(t^{**}) \tag{9.73}$$

for all t^* and t^{**} such that $0 \le t^* < t^{**} \le t^+$. As is apparent, this imposes conditions on both the material response, as represented by C, and the imposed disturbance, as represented by $f(t)$. For normal materials, $C \le 0$ and we must have $f(t^*) \ge f(t^{**})$, whereas, for the unusual materials for which $C > 0$, we must have $f(t^*) \le f(t^{**})$ if the disturbance is to propagate as a spreading wave.

As an example, we present an X–t diagram and a sequence of waveforms for the case of a boundary loading given by

$$\dot{x}(0,t) = \begin{cases} 0, & t \le 0 \\ \dot{x}^+(t/t^+), & 0 \le t \le t^+ \\ \dot{x}^+, & t \ge t^+; \end{cases} \tag{9.74}$$

where t^+ and $\dot{x}^+$ are given constants. Evaluating Eq. 9.66 for this case gives

$$G^* = G^- - \frac{\dot{x}^+}{C_R}\frac{t^*}{t^+} + C\left(\frac{\dot{x}^+}{C_R}\frac{t^*}{t^+}\right)G^- - \frac{C}{2}\left(\frac{\dot{x}^+}{C_R}\frac{t^*}{t^+}\right)^2 + \cdots, \tag{9.75}$$

and, from Eq. 9.67,

$$\dot{x}^* = \dot{x}^+ \frac{t^*}{t^+}. \tag{9.76}$$

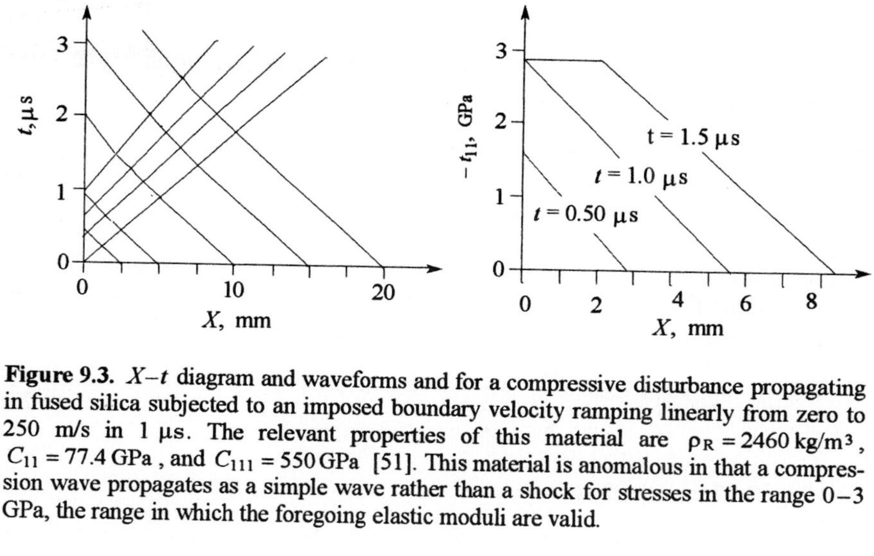

Figure 9.3. X–t diagram and waveforms and for a compressive disturbance propagating in fused silica subjected to an imposed boundary velocity ramping linearly from zero to 250 m/s in 1 μs. The relevant properties of this material are $\rho_R = 2460\ \text{kg/m}^3$, $C_{11} = 77.4\ \text{GPa}$, and $C_{111} = 550\ \text{GPa}$ [51]. This material is anomalous in that a compression wave propagates as a simple wave rather than a shock for stresses in the range 0–3 GPa, the range in which the foregoing elastic moduli are valid.

9.2.2 Eulerian Analysis

We now consider the simple wave problem posed at the beginning of Sect. 9.2, but using the Eulerian coordinate, x, in place of the Lagrangian coordinate, X. Let us suppose that the boundary is subject to an imposed velocity history, as given by Eq. 9.41, with $f(t) > 0$. This boundary condition produces a smooth compression wave. In a normal material, the Eulerian wavespeed, c_{L} increases with increasing compression, leading to an x–t diagram such as that shown in Fig. 9.4. The converging characteristics will eventually intersect, leading to shock formation. We shall address this issue in Sect. 9.5.1.

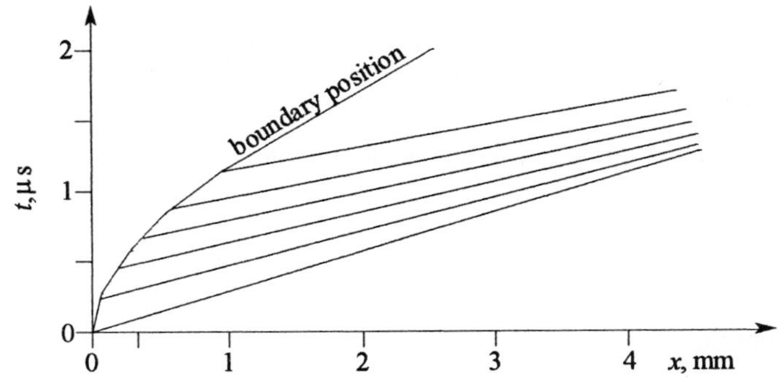

Figure 9.4. The x–t diagram for a right-traveling simple wave. The drawing illustrates a convergence of the characteristic rays that is associated with steepening of the waveform.

Using Eq. 9.34, we see that the characteristics in the region ahead of the wave take the form

$$\begin{aligned} x &= [\, c_{\mathrm{L}}(\rho^-) + \dot{x}^- \,]\, t + \psi^+(t^*) \quad \text{on} \quad t^* = \text{const.} \\ x &= [\, -c_{\mathrm{L}}(\rho^-) + \dot{x}^- \,]\, t + \psi^-(x^*) \quad \text{on} \quad x^* = \text{const.} \end{aligned} \tag{9.77}$$

Let us parameterize these characteristics so that t^* is the time at which the characteristic $t^* = \text{const.}$ intersects the boundary and x^* is the value of x at which the characteristic $x^* = \text{const.}$ intersects the line $t = 0$. Then

$$\begin{aligned} \psi^+(t^*) &= x_{\mathrm{B}}(t^*) - [c_{\mathrm{L}}(\rho^-) + \dot{x}^-]\, t^* \\ \psi^+(x^*) &= x^*, \end{aligned} \tag{9.78}$$

and Eqs. 9.77 take the form

$$\begin{aligned} x &= [\, c_{\mathrm{L}}(\rho^-) + \dot{x}^- \,](t - t^*) + \dot{x}_{\mathrm{B}}(t^*) \quad \text{on} \quad t^* = \text{const.} \\ x &= [\, -c_{\mathrm{L}}(\rho^-) + \dot{x}^- \,](t + \dot{x}^*) \quad \text{on} \quad x^* = \text{const.} \end{aligned} \tag{9.79}$$

Evaluating the Riemann invariants (Eqs. 9.39) in the region ahead of the wave gives

$$\dot{x}^{-} + \int_{\rho_R}^{\rho^{-}} \frac{c_L(\rho')}{\rho'} d\rho' = \Phi^{+}(t^{*})$$
$$\dot{x}^{-} - \int_{\rho_R}^{\rho^{-}} \frac{c_L(\rho')}{\rho'} d\rho' = \Phi^{-}(x^{*})\,. \tag{9.80}$$

Since the Riemann invariant maintains its value when the associated $x^* = \text{const.}$ characteristic is extended into the region of the x–t plane occupied by the wave, we have

$$\dot{x} - \int_{\rho_R}^{\rho} \frac{c_L(\rho')}{\rho'} d\rho' = \dot{x}^{-} - \int_{\rho_R}^{\rho^{-}} \frac{c_L(\rho')}{\rho'} d\rho' \tag{9.81}$$

in the simple wave.

Since $\dot{x} = f(t)$ on the boundary for $0 \le t \le t^{+}$, we have the equation

$$x_B(t) = \int_0^t f(t')\,dt' \quad \text{for} \quad 0 \le t \le t^{+} \tag{9.82}$$

for the position of the boundary at early times. On this boundary, Eq. 9.81 becomes

$$\int_{\rho_R}^{\rho} \frac{c_L(\rho')}{\rho'} d\rho' = f(t) - \dot{x}^{-} + \int_{\rho_R}^{\rho^{-}} \frac{c_L(\rho')}{\rho'} d\rho'\,, \tag{9.83}$$

a result giving values of $\rho(t)$ on the boundary. Since we are identifying the time at which the characteristics for right-propagating wavelets intersect the boundary as t^*, it is useful to write Eq. 9.82 in terms of this parameter:

$$\int_{\rho_R}^{\rho^{*}} \frac{c_L(\rho')}{\rho'} d\rho' = f(t^{*}) - \dot{x}^{-} + \int_{\rho_R}^{\rho^{-}} \frac{c_L(\rho')}{\rho'} d\rho'\,, \tag{9.84}$$

where $\rho^{*} = \rho(t^{*})$.

We evaluate the Riemann invariant $\Phi^{+}(t^{*})$ for characteristics in the simple wave from known conditions at the intersection of the characteristic with the boundary. From Eq. 9.39$_1$ we have

$$\Phi^{+}(t^{*}) = \dot{x}^{*} + \int_{\rho_R}^{\rho^{*}} \frac{c_L(\rho')}{\rho'} d\rho'\,,$$

so this equation can be written

$$\dot{x} + \int_{\rho_R}^{\rho} \frac{c_L(\rho')}{\rho'} d\rho' = f(t^*) + \int_{\rho_R}^{\rho^*} \frac{c_L(\rho')}{\rho'} d\rho', \tag{9.85}$$

where $\dot{x}^* = \dot{x}(t^*) = f(t^*)$. From Eqs. 9.81, 9.85, and 9.84 we find that

$$\dot{x} = f(t^*), \tag{9.86}$$

i.e., the value of $\dot{x}$ along the entire characteristic t^* is constant and equal to the value imposed on the boundary at the time $t = t^*$. Substituting this result into Eq. 9.85 gives

$$\int_{\rho_R}^{\rho} \frac{c_L(\rho')}{\rho'} d\rho' = \int_{\rho_R}^{\rho^*} \frac{c_L(\rho')}{\rho'} d\rho',$$

showing that $\rho = \rho^*$, i.e., the density is also constant along the characteristic and equal to the value attained on the boundary, as calculated from Eq. 9.84.

Now, let us return to Eq. 9.34$_1$ for the t^* characteristic. Since $x = x^*$ and $\rho = \rho^*$ along the entire characteristic the equation becomes

$$\frac{dx}{dt} = c_L(\rho^*) + \dot{x}^*,$$

and integration gives

$$x = [c_L(\rho^*) + \dot{x}^*](t - t^*) + x_B(t^*) \tag{9.87}$$

as the equation defining the t^* characteristic.

Although calculation of the trajectories of the x^* characteristics is not required to obtain the simple wave solution, they will prove useful in discussing shock formation. The x^* characteristics, called *cross characteristics* because their trajectory crosses the simple wave, are determined by solving Eq. 9.34$_2$. Since x^* and c are most naturally expressed as functions of t^*, it is convenient to adopt the representation

$$x = x(t^*) \tag{9.88}$$

for these characteristics. The derivative dx/dt can be written

$$\frac{dx}{dt} = \frac{\dfrac{dx}{dt^*}}{\dfrac{dt^*}{dt}}, \tag{9.89}$$

so we have

$$\frac{dx}{dt^*} = \frac{dx}{dt}\frac{dt}{dt^*}, \tag{9.90}$$

and substitution of this result into Eq. 9.34$_2$ gives

$$\frac{dx}{dt^*} = [\,\dot{x}(t^*) - c(t^*)\,]\frac{dt}{dt^*}. \tag{9.91}$$

Substituting the derivative of Eq. 9.87 with respect to t^* into Eq. 9.91 gives

$$\frac{dx}{dt^*} + \Phi(t^*)t = \Psi(t^*), \tag{9.92}$$

with

$$\Phi(t^*) = \frac{1}{2c(t^*)}\left[\frac{d\dot{x}(t^*)}{dt^*} + \frac{dc(t^*)}{dt^*}\right]$$
$$\Psi(t^*) = \frac{1}{2} + \Phi(t^*)t^*. \tag{9.93}$$

Equation 9.92 is a linear, first-order, ordinary differential equation that has the solution

$$t(t^*) = \exp\left(-\int_0^{t^*}\Phi(t)dt\right)\left\{\frac{x^*}{2C_B} + \int_0^{t^*}\left[\Psi(t')\exp\left(\int_0^{t'}\Phi(t)\,dt\right)\right]dt'\right\}, \tag{9.94}$$

where we have imposed the initial condition

$$t(0) = \frac{x^*}{2C_B}, \tag{9.95}$$

with C_B being the soundspeed of the material ahead of the wave. All of the functions required to evaluate $\Phi(t^*)$ and $\Psi(t^*)$ can be obtained from the isentrope, the associated isentropic soundspeed, and the simple wave solution. Numerical evaluation of Eq. 9.94 then leads to the cross characteristics.

A similar analysis leads to trajectories of the material particle positions. In this case, the differential equation defining the trajectory is

$$\frac{dx}{dt} = \dot{x}, \tag{9.96}$$

and the analysis leads to the trajectory

$$t(t^*) = \exp\left(-2\int_0^{t^*}\Phi(t')dt'\right)\left\{\frac{x(0)}{C_B} + \int_0^{t^*}\left[[2\vartheta\Phi(t'')+1]\exp\left(2\int_0^{t''}\Phi(t')dt'\right)\right]dt''\right\} \tag{9.97}$$

through the simple wave, where $x(0)$ is the particle position at $t = 0$.

Example. Let us consider the evolution of a smooth compression wave propagating in a normal material characterized by the Hugoniot $U_S = C_B + S\dot{x}$. and a Grüneisen parameter of the form $\gamma = \gamma_R(\rho_R/\rho)$. Shear stresses are neglected. We suppose that the material is undeformed, unstressed, and at rest in its initial state. We take

$$\dot{x}_b(t) = \begin{cases} 0, & t<0 \\ \dot{x}^+ t/t^*, & 0 \le t \le t^+ \\ \dot{x}^+, & t > t^+ \end{cases}$$

for the boundary condition producing the wave. The x–t diagram for the problem is shown in the left panel of Fig. 9.5.

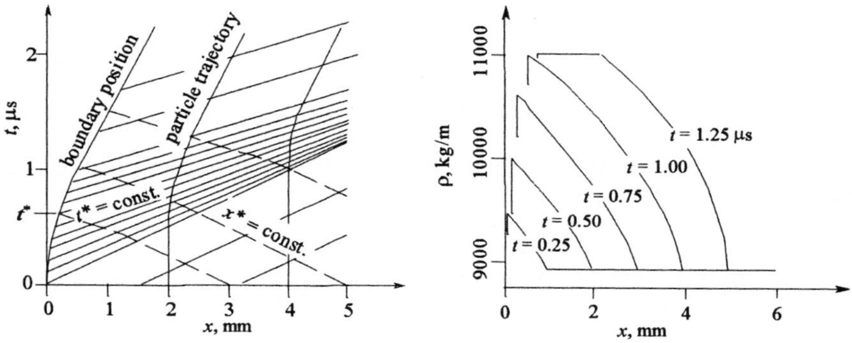

Figure 9.5. The x–t diagram and density waveforms for a right-propagating simple wave in copper. The boundary velocity is assumed to increase linearly from zero to 1000 m/s over the time interval from zero to 1 µs and to remain constant at 1000 m/s thereafter.

Since the disturbance is a smooth wave, the compression process is isentropic, with the isentrope being given by the equation

$$p^{(\eta)}(\Delta) = \rho_R C_B^2 \exp(\gamma_R \Delta) \int_0^{\Delta} \frac{1+(S-\gamma_R)\Delta'}{(1-S\Delta')^3} \exp(-\gamma_R \Delta')\, d\Delta', \tag{9.98}$$

where $\Delta = 1-(\rho_R/\rho)$. The soundspeed derived from Eq. 9.98 is given by

$$c_L^2(\Delta) = \frac{\rho_R}{\rho^2}\left[\gamma_R\, p^{(\eta)}(\Delta) + \rho_R C_B^2 \frac{1+(S-\gamma_R)\Delta}{(1-S\Delta)^3}\right]. \tag{9.99}$$

The density at the boundary is obtained from Eq. 9.83 which, for the case at hand, is

$$\int_{\rho_R}^{\rho(t^*)} \frac{c_L(\rho')}{\rho'} d\rho' = f(t^*). \tag{9.100}$$

The integration required to solve this equation must be performed numerically, but once this is done, $\rho(t^*)$ is easily obtained. The particle velocity on the t^* characteristic is $\dot{x} = f(t^*)$, as given by Eq. 9.86. With these quantities in hand the characteristics, as given by Eq. 9.87, can be plotted. A sequence of density waveforms representing this solution is shown in the right panel of Fig. 9.5.

An important point to note from this figure is that the waveform becomes steeper as the wave propagates into the material. This is characteristic of compression waves propagating in normal materials, and leads eventually to formation of a shock, as will be discussed in Sect. 9.5.1. As a preliminary to this, we calculate the time and place at which the waveform develops a vertical tangent.

The particle-velocity waveform at a given time, t, (regarded as a fixed parameter) can be represented by an equation of the form

$$\dot{x} = \dot{x}(t^*(x;t)),$$

so its slope can be written

$$\frac{\partial \dot{x}(x,t)}{\partial x} = \frac{\dfrac{d\dot{x}(t^*)}{dt^*}}{\dfrac{dx(t,t^*)}{dt^*}}. \tag{9.101}$$

The denominator in this expression is obtained by differentiating Eq. 9.87, and is

$$\frac{\partial x(t,t^*)}{\partial t^*} = \left[\frac{dc_L(t^*)}{dt^*} + \frac{d\dot{x}(t^*)}{dt^*}\right](t-t^*) - c_L(t^*) - \dot{x}(t^*) + \frac{dx_B(t^*)}{dt^*}. \tag{9.102}$$

The last two terms cancel in accord with the solution in the simple wave region, and substituting this result into Eq. 9.101 gives the equation

$$\frac{\partial \dot{x}(x,t)}{\partial x} = \frac{\dfrac{d\dot{x}(t^*)}{dt^*}}{\left[\dfrac{dc_L(t^*)}{dt^*} + \dfrac{d\dot{x}(t^*)}{dt^*}\right](t-t^*) - c_L(t^*)} \tag{9.103}$$

for the slope of the waveform. The vertical tangent occurs at the time when the denominator vanishes,

$$t = t^* + \frac{c_L(t^*)}{\dfrac{dc_L(t^*)}{dt^*} + \dfrac{d\dot{x}(t^*)}{dt^*}}. \tag{9.104}$$

The corresponding value of x, as obtained from Eq. 9.92, is

$$x = x_B(t^*) + \frac{c_L(t^*)[c_L(t^*) + \dot{x}(t^*)]}{\dfrac{dc_L(t^*)}{dt^*} + \dfrac{d\dot{x}(t^*)}{dt^*}}. \tag{9.105}$$

For the example problem that we have been discussing, this point is on the characteristic $t^* = 0$, and falls at $t = 1.315\,\mu s$ and $x = 5.181\,mm$.

9.3 The Centered Simple Wave

In the preceding section we considered the simple wave produced by application of a continuous stress or particle-velocity history to the boundary. We now want to consider the case of sudden load application. When a stress or particle-velocity step is applied to the surface of the halfspace a shock, a centered simple wave, or a wave combining these two types of disturbances, is introduced into the body. Which of these disturbances is produced depends, as with simple waves generally, upon whether the boundary loading causes an expansion or a contraction of the material and upon the shape of the function $C_L(G)$.

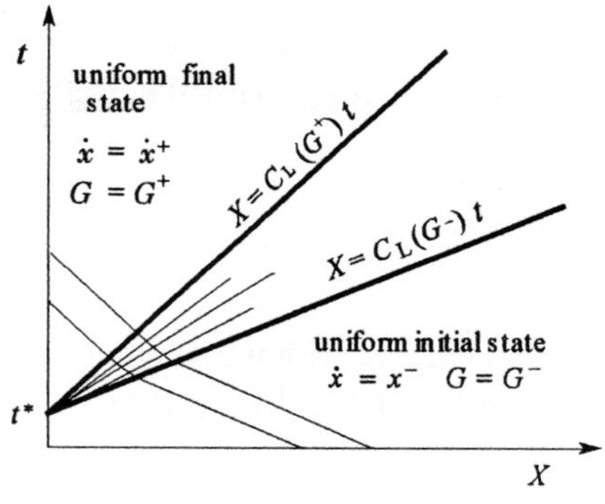

Figure 9.6. The X–t diagram for a centered simple wave.

Lagrangian Analysis. The X–t diagram of Fig. 9.6 shows a wave that is spreading as it propagates, i.e. the leading edge of the waveform propagates faster than the trailing portion. If the opposite were true the disturbance would be a shock.

The X–t diagram of Fig. 9.6 corresponds to initial conditions

$$\dot{x}(X,t)=\dot{x}^{-},\quad G(X,0)=G^{-}\quad \text{for}\ \begin{cases}0\le t\le t^{*}\\ X\ge 0,\end{cases} \tag{9.106}$$

describing a material in a state of uniform deformation and motion. The boundary condition under consideration is

$$\dot{x}(0,t)=\begin{cases}\dot{x}^{-}, & t<t^{*}\\ \dot{x}^{+}, & t\ge t^{*}.\end{cases} \tag{9.107}$$

As in the case of the linear problem of Chap. 8 and the nonlinear problem of the previous section, the wavefront advances into the material at a constant speed $C_{\mathrm{L}}(G^{-})$. The material ahead of the wave, i.e., material in the region $X>C_{\mathrm{L}}(G^{-})(t-t^{*})$, remains in its initial state. The material behind the disturbance will also be in a uniform state corresponding to the velocity imposed on the boundary and some, as yet undetermined, deformation G^{+}. Since we are seeking a smooth, spreading waveform, let us consider the case in which $C_{\mathrm{L}}(G\ )\ge C_{\mathrm{L}}(G)\ge C_{\mathrm{L}}(G^{+})$ for all G in the interval between G^{-} and G^{+}. The trailing edge of the wave advances into the material at the rate $C_{\mathrm{L}}(G^{+})$, so the wave itself occupies the wedge $C_{\mathrm{L}}(G^{+})(t-t^{*})\le X\le C_{\mathrm{L}}(G^{-})(t-t^{*})$ shown on the diagram. In the case of the simple wave studied in the previous section, the leading and trailing characteristics of the wave intersected the boundary at the times of initiation and termination of the imposed load variation, with the characteristics within the wave intersecting at intermediate points. The discontinuous loading corresponds to the limit in which the initial and terminal characteristics, as well as those in between, intersect the boundary at the time of application of the load step. Following the lead of the previous work, we seek a solution of Eqs. 9.8 that is constant along the advancing characteristics, which can only be rays through the point $X=0$, $t=t^{*}$. A solution of this form will depend upon the single variable

$$Z=\frac{X}{t-t^{*}}.$$

In this case, the wave equation, in the form 9.8, becomes

$$\begin{aligned} Z\frac{d\dot{x}(Z)}{dZ}+C_{\mathrm{L}}^{2}(G)\frac{dG(Z)}{dZ}&=0\\ Z\frac{dG(Z)}{dZ}+\frac{d\dot{x}(Z)}{dZ}&=0.\end{aligned} \tag{9.108}$$

Elimination of $d\dot{x}/dZ$ from the first equation of this pair gives

$$[Z^{2}-C_{\mathrm{L}}^{2}(G)]\frac{dG(Z)}{dZ}=0,$$

which can have a nontrivial solution $G(Z)$ only if the coefficient vanishes: $Z^2 - C_L^2 = 0$. Since we are seeking a solution in the form of a right-propagating wave, we choose the positive square root, thus obtaining an implicit solution

$$C_L(G) = +Z \tag{9.109}$$

for $G(Z)$. Substitution of Eq. 9.109 into Eq. 9.108₁ gives

$$\frac{d\dot{x}(Z)}{dZ} + Z\frac{dG(Z)}{dZ} = 0,$$

or

$$\frac{d\dot{x}}{dG} = -Z, \tag{9.110}$$

from which we obtain

$$\dot{x} = -\int_{G^-}^{G} C_L(G')\,dG' + \dot{x}^-, \tag{9.111}$$

where the constant of integration has been chosen to ensure satisfaction of the condition $\dot{x} = \dot{x}^-$ and $G = G^-$ that matches this solution continuously to the solution in the region ahead of the wave. Denoting by G^+ the value of G in the region behind the wave, we find that it is given implicitly by

$$\dot{x}^+ = -\int_{G^-}^{G^+} C_L(G')\,dG' + \dot{x}^-, \tag{9.112}$$

where $\dot{x}^+$ is the imposed boundary velocity. This completes the solution of the problem, since Eq. 9.109 can be inverted to give G as a function of $X/(t-t^*)$. Then, for any particle and time X and t in the wave, the integral in Eq. 9.111 can be calculated, thus yielding $\dot{x}$. In the next section, we shall consider the possibilities that arise if Eq. 9.109 cannot be inverted to yield $G[X/(t-t^*)]$.

In the wave region the X^* characteristic must satisfy Eq. 9.25₂. Since the solution in the wave region is given by Eq. 9.109, $C_L(G) = X/(t-t^*)$, we seek the solution of

$$\frac{dX}{dt} = -\frac{X}{t-t^*} \tag{9.113}$$

that passes through the point (X_0, t_0) given by Eq. 9.54. The solution to this equation is

$$X = \frac{t_0 - t^*}{t - t^*} X_0, \tag{9.114}$$

and we see that the X^* characteristics are hyperbolas. If we write X_0 and t_0 in terms of X^*, t^*, and G^- this result takes the form

$$X = \frac{\left[X^* - C_L(G^-)t^*\right]^2}{4C_L(G^-)(t-t^*)}. \tag{9.115}$$

This characteristic leaves the wave region at the point (X_1, t_1) given by

$$\begin{aligned} t_1 &= t^* + \frac{X^* - C_L(G^-)t^*}{2\left[C_L(G^-)\,C_L(G^+)\right]^{1/2}} \\ X_1 &= \frac{1}{2}\left[\frac{C_L(G^+)}{C_L(G^-)}\right]^{1/2}\left[X^* - C_L(G^-)t^*\right]. \end{aligned} \tag{9.116}$$

Example: Third-order Elastic Response. To make the foregoing analysis more explicit, let us consider the case when the soundspeed is given by Eq. 9.63. The displacement gradient is obtained as a function of X and t by substitution of this relation for the soundspeed into Eq. 9.109 and solving for G:

$$G = \frac{1}{CC_R}\left[\frac{X}{t-t^*} - C_R\right]. \tag{9.117}$$

Substitution of Eq. 9.63 into Eq. 9.111 and integration gives

$$\dot{x} = \dot{x}^- - C_R\left[1 + \tfrac{1}{2}C(G+G^-)\right](G-G^-), \tag{9.118}$$

or,

$$\dot{x} = \dot{x}^- + \frac{1}{2CC_R}\left[C_R^2 - \left(\frac{X}{t-t^*}\right)^2\right] + C_R\left(1 + \frac{1}{2}CG^-\right)G^-. \tag{9.119}$$

The value of G in the region behind the wave is determined by substituting the boundary condition $\dot{x}(0, t^+) = \dot{x}^+$ into Eq. 9.118 and solving for $G = G^+$. The result is

$$G^+ = \left[1 + C\left(\frac{\dot{x}^+ - x^-}{C_R}\right)\right]G^- - \left[1 + \frac{C}{2}\left(\frac{\dot{x}^+ - x^-}{C_R}\right)\right]\left(\frac{\dot{x}^+ - x^-}{C_R}\right). \tag{9.120}$$

With this, the equation for the leading characteristic is

$$X = C_R(1 + CG^-)(t - t^+), \tag{9.121}$$

and for the trailing characteristic we have

$$X = C_R(1 + CG^+)(t - t^+).$$

The X–t diagram and waveforms for a specific case are plotted in Fig. 9.7.

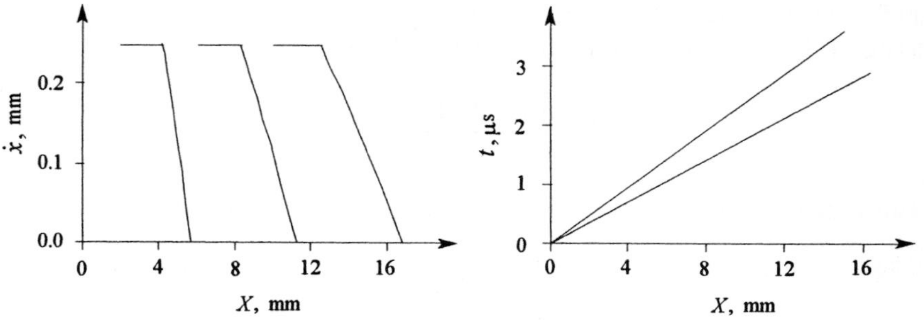

Figure 9.7. When a slab of fused silica that is at rest and undeformed is impacted in such a way that a shock producing a state characterized by the particle velocity $\dot{x}^+ = 250$ m/s is introduced into the material, it spreads into a centered simple wave of the sort discussed. This figure shows the X–t diagram and several waveforms for the problem.

Eulerian Analysis. This same centered simple wave problem can be analyzed using Eqs. 9.19, the Eulerian form of the wave equation. As with the Lagrangian analysis, we seek a solution that depends upon the single variable

$$z = \frac{x}{t}. \tag{9.122}$$

In this case, Eqs. 9.19 become

$$\begin{aligned} &\left(\frac{c_L(\rho)}{\rho}\rho_z + \dot{x}_z\right)(c_L + \dot{x} - z) = 0 \\ &\left(\frac{c_L(\rho)}{\rho}\rho_z - \dot{x}_z\right)(c_L - \dot{x} + z) = 0. \end{aligned} \tag{9.123}$$

On the advancing characteristics $z = c_L + \dot{x}$ so we must have

$$\dot{x}_z - \frac{c_L(\rho)}{\rho}\rho_z = 0, \tag{9.124}$$

or,

$$\dot{x} - \int_{\rho^-}^{\rho} \frac{c_L(\rho')}{\rho'} d\rho' = \dot{x}^- \tag{9.125}$$

on $z = \text{const}$.

Example: Ideal Gas. Since the Eulerian analysis is usually applied to gas flows, let us consider the case where the centered simple wave is one of decompression of an ideal gas. Let the initial state be one in which the density is ρ^-, the particle velocity is $\dot{x}^-$, and the pressure is p^-. For an ideal gas, we have

$$p = p^- \left(\frac{\rho}{\rho^-}\right)^\Gamma, \tag{9.126}$$

leading, via Eq. 9.17, to the Eulerian soundspeed

$$c_L = \sqrt{\frac{\Gamma p}{\rho}}. \tag{9.127}$$

The soundspeed in the material into which the wave is advancing is given by

$$c_L^- = \sqrt{\frac{\Gamma p^-}{\rho^-}}, \tag{9.128}$$

so Eq. 9.125 becomes

$$\dot{x} = \dot{x}^- - \frac{2c_L^-}{\Gamma - 1}\left[1 - \left(\frac{\rho}{\rho^-}\right)^{(\Gamma-1)/2}\right]. \tag{9.129}$$

When Eq. 9.126 is substituted into this relation we obtain an expression for $\dot{x}$ in terms of the pressure in the wave:

$$\dot{x} = \dot{x}^- - \frac{2c_L^-}{\Gamma - 1}\left[1 - \left(\frac{p}{p^-}\right)^{(\Gamma-1)/(2\Gamma)}\right]. \tag{9.130}$$

Substituting the boundary pressure for p gives an expression for the particle velocity on the trailing characteristic of the wave

$$\dot{x}_B = \dot{x}^- - \frac{2c_L^-}{\Gamma - 1}\left[1 - \left(\frac{p_B}{p^-}\right)^{(\Gamma-1)/(2\Gamma)}\right]. \tag{9.131}$$

On the advancing characteristics $z = c_L + \dot{x}$, or

$$z = c_L^-\left(\frac{\rho}{\rho^-}\right)^{(\Gamma-1)/2} + \dot{x} = c_L^-\left(\frac{p}{p^-}\right)^{(\Gamma-1)/2\Gamma} + \dot{x}. \tag{9.132}$$

Substituting this into Eq. 9.129 gives

$$\dot{x} = \frac{2}{\Gamma+1}(z - c_L^-) + \frac{\Gamma-1}{\Gamma+1}\dot{x}^-, \tag{9.133}$$

an expression for the particle velocity in the wave as a function of $z = x/t$. Substitution of this result into Eq. 9.130 gives the pressure relation

$$\frac{p}{p^-} = \left(\frac{\Gamma-1}{\Gamma+1}\frac{z-\dot{x}^-}{c_L^-} + \frac{2}{\Gamma+1}\right)^{2\Gamma/(\Gamma-1)}, \tag{9.134}$$

and, using Eq. 9.126, we obtain the density relation

$$\frac{\rho}{\rho^-} = \left(\frac{\Gamma - 1}{\Gamma + 1} \frac{z - \dot{x}^-}{c_L^-} + \frac{2}{\Gamma + 1} \right)^{2/(\Gamma - 1)} . \tag{9.135}$$

To complete the solution we need to determine the range of z over which it is valid. This is done by substituting the boundary condition $p = p_B$ into Eq. 9.134 and solving for the value of z defining the trailing edge of the wave. We obtain

$$z_B = \dot{x}^- - \frac{c_L^-}{\Gamma - 1} \left[2 - (\Gamma + 1) \left(\frac{p_B}{p^-} \right)^{(\Gamma - 1)/(2\Gamma)} \right] . \tag{9.136}$$

The solution occupies the range

$$\dot{x}^- - \frac{c_L^-}{\Gamma - 1} \left[2 - (\Gamma - 1) \left(\frac{p_B}{p^-} \right)^{(\Gamma - 1)/(2\Gamma)} \right] \le \frac{x}{t} \le c_L^- + \dot{x}^- . \tag{9.137}$$

The region behind the wave is one in which $p = p_B$ and $\dot{x} = \dot{x}_B$. The boundary of the material is at $x_B = \dot{x}_B t$ for $t > 0$.

Eulerian Analysis of Waves in a Shock Tube. A shock tube is a device in which two columns of quiescent gas are held at different pressure by a separating diaphragm. The initial pressure distribution is as shown in Fig. 9.8. Wave motion is initiated when the diaphragm is suddenly shattered, placing the two gas columns in contact. The wave motion is produced as the high-pressure gas begins to expand toward the low-pressure gas. The expansion of the high-pressure gas takes place in a right-propagating centered simple wave and compression of the low-pressure gas is effected by a left-propagating shock; the x–t diagram for this process is shown in Fig. 9.9

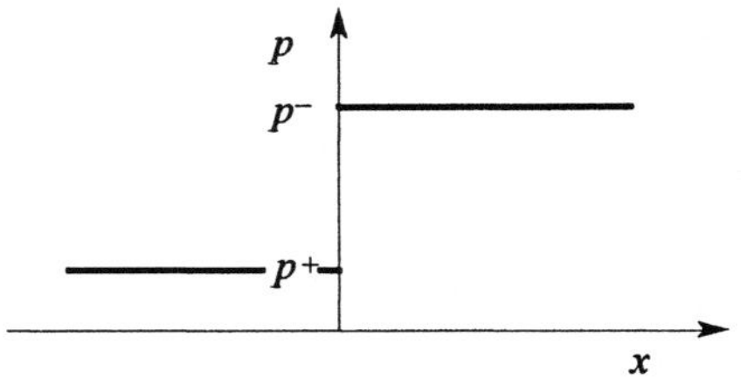

Figure 9.8. Pressure distribution in a shock tube at the instant, $t = 0$, of diaphragm rupture.

The pressure and particle velocity have the (as yet undetermined) constant, uniform values p_B and $\dot{x}_B$ in the region between the waves. The surface at which the two gas columns are in contact moves at the particle velocity $\dot{x}_B$. Although the pressure and particle velocity are continuous at the contact surface, the mass density (and the temperature, specific entropy, and specific internal

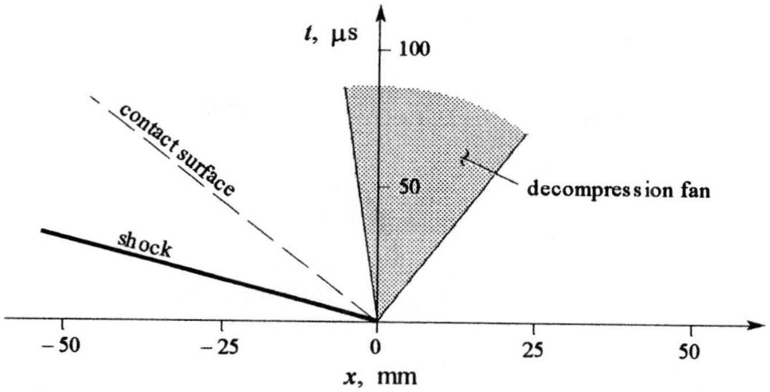

Figure 9.9. Eulerian space–time diagram of wave propagation in a shock tube. This drawing is made for the case in which air at a pressure of 100 atm and a temperature of 293 K expands into a column of air at a pressure of 1 atm and the same temperature.

energy) is discontinuous at this surface. The state of the material in the region between the shock and the trailing characteristic of the simple wave is established by matching these two solutions. The jumps at the shock must satisfy the conditions $2.110_{1,2}$, which we can write

$$\begin{aligned}(\rho_B^+ - \rho^+)\, u_S &= \rho_B^+ \dot{x}_B \\ \rho_B^+ \dot{x}_B u_S &= \rho_B^+ \dot{x}_B^2 + p_B - p^+ .\end{aligned} \tag{9.138}$$

In addition, the state of the shock-compressed gas must lie on the Hugoniot, which is given by Eq. 5.59. When this equation is evaluated at the state in question, it provides the relation

$$p_B = p^+ \frac{(\Gamma+1)\rho_B^+ - (\Gamma-1)\rho^+}{(\Gamma+1)\rho^+ - (\Gamma-1)\rho_B^+} \tag{9.139}$$

between the pressure in the region and the mass density in the part of the region between the shock and the contact interface.

In order for the pressure and particle velocity to be continuous at the contact interface the values of these variables obtained from analysis of the shock, as given by Eq. 9.142_1, must agree with the values obtained from the simple wave analysis (Eq. 9.130 with $\dot{x}^- = 0$). The pressure and particle velocity at the trailing edge of the centered simple wave satisfy the equation

$$\dot{x}_B = -\frac{2c_L^-}{\Gamma-1}\left[1-\left(\frac{p_B}{p^-}\right)^{(\Gamma-1)/(2\Gamma)}\right]. \tag{9.140}$$

The four equations 9.138–9.140 relate the unknown quantities ρ_B^+, p_B, $\dot{x}_B$, and u_S.

Let us begin solving these equations by expressing ρ_B^+ and u_S in terms of ρ_B^+ and p_B. Equations 9.138 can be written in the form

$$u_S = \frac{\rho_B^+ \dot{x}_B}{\rho_B^+ - \rho^+}$$
$$\dot{x}_B^2 = \frac{p^+}{\rho^+}\left(\frac{p_B}{p^+} - 1\right)\left(1 - \frac{\rho^+}{\rho_B^+}\right), \tag{9.141}$$

and Eq. 9.139 can be used to eliminate ρ_B^+ from these expressions. The results are

$$\dot{x}_B^2 = \frac{2p^+}{\rho^+} \frac{\left(\frac{p_B}{p^+} - 1\right)^2}{(\Gamma+1)\frac{p_B}{p^+} + (\Gamma - 1)}$$
$$u_S^2 = \frac{p^+}{2\rho^+}\left[(\Gamma+1)\frac{p_B}{p^+} + (\Gamma-1)\right]. \tag{9.142}$$

Equation 9.142_1 is the $p - \dot{x}$ Hugoniot for the gas, centered on the state ahead of the shock. Equation 9.140 represents the $p - \dot{x}$ path of the isentropic decompression that occurs in the simple wave. Simultaneous solution of these equations, which is best accomplished numerically, completes the analysis. A plot of a Hugoniot and isentropic paths starting at three different initial states is given in Fig. 9.10.

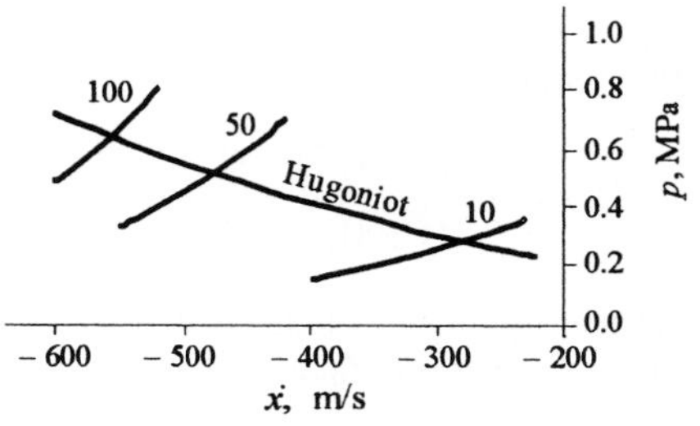

Figure 9.10. Hugoniot and isentropes through pressures of 10, 50, and 100 atm. on the Hugoniot for a polytropic gas for which $\Gamma = 1.4$ (air).

Plots of pressure and particle velocity distributions at 10 and 50 μs are shown in Fig. 9.11 for the case in which the high and low pressures are 1 and 100 atm, respectively.

Expansion to Zero Pressure. A special case of the shock-tube problem arises when the high-pressure gas expands into a void. In this case there is no shock and $p_B = 0$. The leading characteristic of the expansion is $x = c_L^- t$, and the trailing characteristic is that for which $p = 0$. From Eq. 9.130 we get

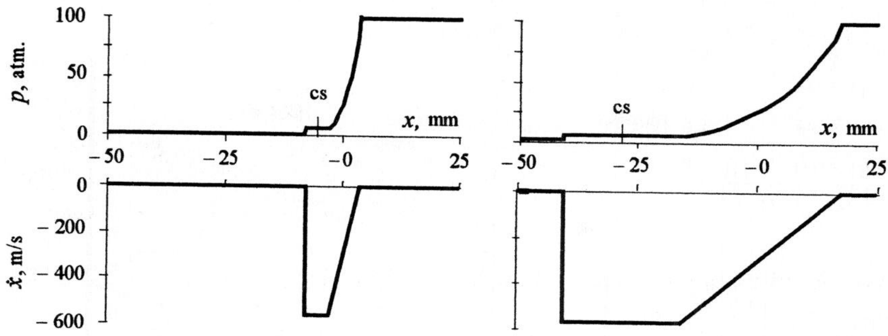

Figure 9.11. Pressure and particle-velocity distributions at 10 and 50 μs for the case in which the high and low pressures are 1 and 100 atm, respectively. The contact surface is located at the positions marked CS.

$$\dot{x}_B = -\frac{2c_L^-}{\Gamma - 1} \tag{9.143}$$

for this case, a value called the *escape velocity*. Since

$$c_L = c_L^- \left(\frac{p}{p^-}\right)^{(\Gamma-1)/(2\Gamma)}, \tag{9.144}$$

and $z = c_L + \dot{x}$ we have

$$\begin{aligned} \dot{x} &= -\frac{2c_L^-}{\Gamma+1}\left(1 - \frac{z}{c_L^-}\right) \\ p &= p^-\left(\frac{2}{\Gamma+1} + \frac{\Gamma-1}{\Gamma+1}\frac{z}{c_L^-}\right)^{2\Gamma/(\Gamma-1)}, \end{aligned} \tag{9.145}$$

results that are valid in the range

$$-\frac{2c_L^-}{\Gamma+1} \le z \le c_L^- . \tag{9.146}$$

9.3.1 Shock Reflection from an Unrestrained Boundary

In Sect. 3.7.2 we analyzed the interaction that occurs when a shock encounters an unrestrained (stress free) boundary. We saw that the reflected disturbance produces a reduction in pressure. Assuming that the incident shock is stable, this reflected disturbance cannot propagate as a stable shock. It is, instead, a centered simple wave. The *X–t* diagram of Fig. 9.12 illustrates this situation.

We shall assume that the material is unstressed, undeformed, and at rest ahead of the incident shock:

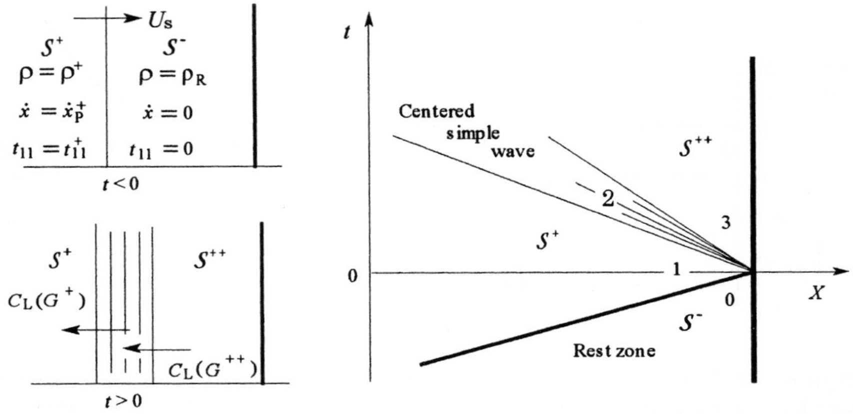

Figure 9.12. The X–t diagram for reflection of a shock as a centered simple wave. The upper left part of the figure shows the state before the shock encounters the boundary, and the lower left part of the figure shows the state after the reflection. The right part of the figure shows the part of the X–t diagram near the interaction.

$$G^- = 0, \quad \dot{x}^- = 0, \quad t_{11}^- = 0. \tag{9.147}$$

We also assume that the state of the material behind the shock, the velocity of the shock, and the mechanical properties of the material are known:

$$U_X = U_X^+ = G^+, \quad \dot{x} = \dot{x}^+, \quad t_{11} = t_{11}^+. \tag{9.148}$$

These variables are related to one another by the jump conditions 2.113 and the stress response function for the material.

Since the material boundary is unrestrained, the stress component t_{11} must vanish in region 3 behind the reflected wave. The stress response function can then be used to determine the value $G = G^{++}$ in this region. To do this properly, it is necessary to calculate the entropy jump across the shock before solving for G^{++}. This can be done, but we shall take advantage of the fact that the effect is small and adopt a purely mechanical theory.

The solution in the simple wave region is obtained as in the previous section (taking account of the reversed direction of propagation), and is

$$C_L(G) = -X/t$$

$$\dot{x} = +\int_{G^+}^{G} |C_L(G')|\, dG' + \dot{x}^+. \tag{9.149}$$

Having determined the value of G in the region behind the reflected wave from the stress response function, we now obtain the particle velocity in this region as

$$\dot{x}^{++} = + \int_{G^+}^{G^{++}} |C_L(G')| \, dG' + \dot{x}^+. \tag{9.150}$$

The leading characteristic of the reflected wave is

$$X = -C_L(G^+)\, t \tag{9.151}$$

and the trailing characteristic of this wave is

$$X = -C_L(G^{++})\, t, \tag{9.152}$$

completing the solution. Since we have neglected the entropy jump across the incident shock in obtaining these results, the stress-response function used in analyzing the simple wave is the same as the Hugoniot used in analyzing the shock.

Example. Let us consider the foregoing problem for the case of a material governed by the Hugoniot $U_S = C_B + S\dot{x}$. Let us also adopt a purely mechanical theory, i.e., neglect the entropy jump at the shock so that the stress response function is the same as the p–G Hugoniot,

$$p = -\frac{\rho_R C_B^2 G}{(1+SG)^2}. \tag{9.153}$$

The shock is propagating into material that is undeformed and at rest, and we suppose it to have a given strength $\dot{x}^+ > 0$. Then

$$G^+ = -\frac{\dot{x}^+}{C_B + S\dot{x}^+}. \tag{9.154}$$

Substituting Eq. 9.153 into Eq. 9.9$_3$ we find

$$C_L(G) = -C_B \frac{(1-SG)^{1/2}}{(1+SG)^{3/2}}, \tag{9.155}$$

where we have chosen the negative square root since the reflected wave is propagating in the $-X$ direction.

The receding characteristics $t^* = \text{const.}$ in the centered simple decompression wave are given by

$$X = C_L(G)t = -C_B \frac{(1-SG)^{1/2}}{(1+SG)^{3/2}} t, \quad G^+ \le G \le G^{++}, \tag{9.156}$$

so the leading and trailing characteristics are

$$X = -C_{\mathrm{B}} \frac{(1-SG^{+})^{1/2}}{(1+SG^{+})^{3/2}} t \tag{9.157}$$

and

$$X = -C_{\mathrm{B}} t, \tag{9.158}$$

respectively, where this last result follows from the fact that $G^{++} = 0$ in the mechanical theory.

The value of G at a point (X^*, t^*) in the wave is obtained by solving Eq. 9.156, which we can write in the form

$$-\frac{X^*}{C_{\mathrm{B}} t^*} = \frac{(1-SG)^{1/2}}{(1+SG)^{3/2}}. \tag{9.159}$$

When this equation is squared it becomes cubic in $G(X/C_{\mathrm{B}} t)$ and can be shown to have one real root given by

$$G = \frac{1}{S}\left\{-1 + \left[R^2 + R^2\left(1 + \frac{R^2}{27}\right)\right]^{1/3} + \left[R^2 - R^2\left(1 + \frac{R^2}{27}\right)\right]^{1/3}\right\}, \tag{9.160}$$

where $R = -C_{\mathrm{B}} t^*/X^*$. This equation gives G at any point (X^*, t^*) of the wave. Stress waveforms are obtained by substituting G into Eq. 9.153. To complete the solution, we must find $\dot{x}(X^*, t^*)$. Substitution of Eq. 9.155 into Eq. 9.149 gives

$$\dot{x} - \dot{x}^+ = C_{\mathrm{B}} \int_{G^+}^{G} \frac{(1-SG')^{1/2}}{(1+SG')^{3/2}} dG', \tag{9.161}$$

and evaluation of the integral leads to the result

$$\begin{aligned} \frac{\dot{x}}{C_{\mathrm{B}}} = \frac{\dot{x}^+}{C_{\mathrm{B}}} &- \frac{2}{S}\left\{\sqrt{\frac{1-SG}{1+SG}} - \tan^{-1}\sqrt{\frac{1-SG}{1+SG}}\right\} \\ &+ \frac{2}{S}\left\{\sqrt{\frac{1-SG^+}{1+SG^+}} - \tan^{-1}\sqrt{\frac{1-SG^+}{1+SG^+}}\right\}, \end{aligned} \tag{9.162}$$

giving $\dot{x}$ as a function of X and t, through the dependence (Eq. 9.160) of G upon these variables. Some results of this analysis are plotted in Fig. 9.13.

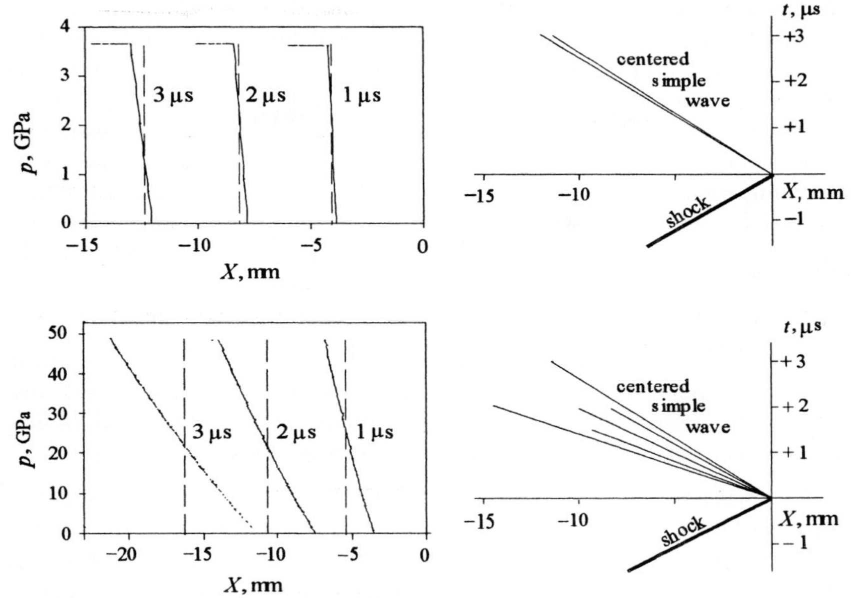

Figure 9.13. The centered simple decompression wave in copper ($\rho_R = 8930\,\mathrm{kg/m^3}$, $C_B = 3940\,\mathrm{m/s}$, $S = 1.489$ [77]) that results from shock reflection at an unrestrained surface. The top part of the figure shows results for a shock of strength $\dot{x}^+ = 100\,\mathrm{m/s}$ and the bottom part shows similar results for a shock of strength $\dot{x}^+ = 1000\,\mathrm{m/s}$. The broken vertical lines on the waveforms show the location of the decompression shock calculated as in Chap. 3.

9.3.2 Combined Centered Simple Waves and Shocks

In various of the foregoing sections of this book we have shown that stable compression shocks propagate in cases where the stress relation is curved like that shown in Fig. 9.14a, and simple waves propagate under similar conditions if the stress relation is curved as in Fig. 9.14b. The key point is that, in the first case (corresponding to a shock), the line connecting the initial and final states lies above the curve, whereas it lies below the curve in the second case (corresponding to a smooth wave). One could consider more complicated cases in which the stress relation is inflected so that the line connecting initial and final states lies partly above and partly below the curve. When this happens, the waveform will be a shock combined with a centered simple wave, as indicated in parts (c) and (d) of the figure. In the case in which the line lies above the first portion of the curve, the solution consists of a shock corresponding to a transiltion from the initial state to the point at which a line through the initial point is tangent to the stress relation, followed by a centered simple wave from this state to the final state. In the other case, that depicted in part (d) of the figure, the shock follows the simple wave, and provides a transition from the tangent point of a line through the final state to the curve. As suggested in the X–t diagrams, the shocks and centered waves are directly adjacent to one another; there is no intervening region. The solution is obtained by fitting the shock and simple wave together so that the stress and particle velocity are continuous. Since the slope of the curve and the Rayleigh line are the same at the tangent point, the edge of the simple wave adjacent to the shock will propagate at the same speed as the shock.

One can see that, in a somewhat more extreme version of the case shown in part (d) of the figure a line originating at the initial state might pass to the final state without intersecting the curve, even though the latter was inflected. In this, the transition from the initial to the final state takes the form of a stable shock.

Waveforms like those depicted in parts (c) and (d) of the figure are often observed, but the explanation *does not* usually turn upon inflections in the equilibrium stress response curve, as do the cases discussed here. Rather, the behavior is attributed to the occurrence of phase transformations, the onset of plastic deformation, an increase of yield strength as plastic deformation accumulates, or viscoelastic effects.

9.4 Comparison of Transitions Through Simple Waves and Weak Shocks

As discussed in Chap. 3, the change in field variables associated with passage of a simple wave is often calculated using the shock-jump conditions, i.e. ignoring the shock instability that gives rise to the simple wave. Since the simple wave is an isentropic process, whereas the shock is a dissipative process, some error must be incurred in this method of analysis, and it is useful to estimate its magnitude. For an elastic material, the entropy jump across the shock and the error in calculating stress and temperature have been shown to be of third order in $[\![v]\!]^3$ or $[\![G]\!]^3$. When the error in estimating the particle velocity change is evaluated by this method we find that

$$\dot{x}^+ - \dot{x}^-\Big|_{\text{simple wave}} = [\![\dot{x}]\!]_{\text{shock}} + \frac{1}{24}\frac{[C_L'(G^-)]^2}{C_L(G^-)}(G^+ - G^-)^3 + \ldots. \quad (9\ 163)$$

Normally, it is in the case of a decompression process, i.e. one for which $G^+ - G^- > 0$, that a simple wave is treated as a shock. When this is so, we see from the foregoing equation that the change in particle velocity inferred from the jump condition underestimates the true value by the term of third order in $G^+ - G^-$ that is given.

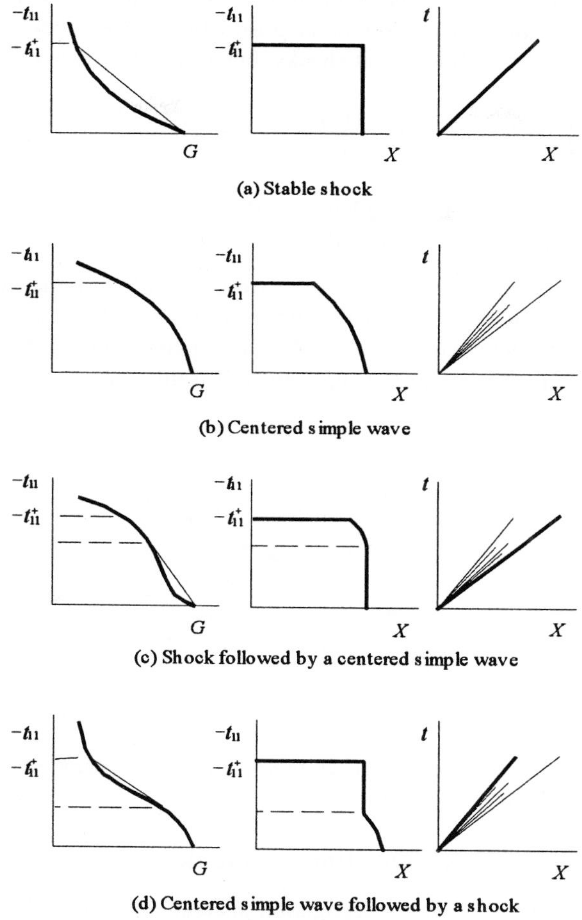

Figure 9.14. Stress–deformation curves with associated waveforms and X–t diagrams.

Figure 9.15 shows an X–t diagram of a shock of constant strength propagating into a region of uniform state. The characteristics drawn on the diagram reflect the fact that the wavespeed in the material ahead of the shock is less than the shockspeed and the wavespeed in the material behind the shock exceeds the shockspeed. We know that the quantity

$$\dot{x} + \int_0^G C_L(G')\,dG' = \varphi(X^*) \tag{9.164}$$

is constant along the receding characteristics in smooth fields. We do not expect it to remain invariant as a characteristic passes through a shock, but we can show that the jump is small for weak shocks. In particular, we have

$$\left[\!\left[\dot{x} + \int_0^G C_L(G')\,dG' \right]\!\right] = -\frac{1}{24}\frac{[C_L'(G^-)]^2}{C_L(G^-)}(G^+ - G^-)^3 + \ldots, \qquad (9.165)$$

so we see again that the difference between the result for a smooth wave and a shock is of third order in shock strength.

In summary, we see that the change in stress, specific volume, temperature, entropy, particle velocity, and the Riemann invariant at a shock and at a simple wave of the same strength $G^+ - G^-$ differ by a quantity of third order in this strength. Some numerical examples were given in Table 3.2.

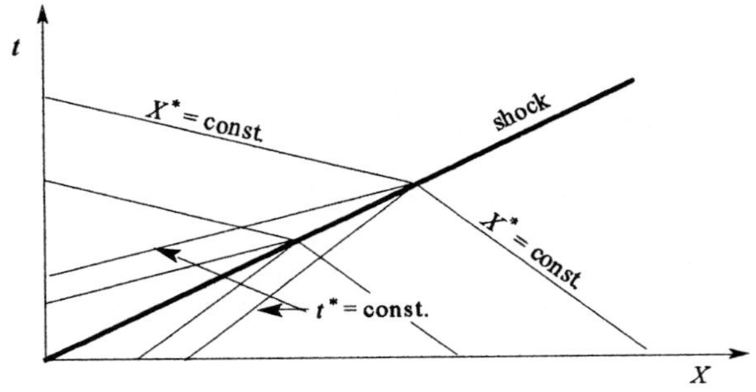

Figure 9.15. The X–t diagram of a shock propagating into a region of uniform state, showing characteristic curves.

9.5 Formation and Attenuation of Shocks

Many important nonlinear wave-propagation phenomena involve flows in which the entropy varies with position. Particular cases of interest involve unsteady shocks—those that are changing strength as they propagate. Simple analytical methods fail in these cases, but it is possible to develop some understanding of the issue if attention is restricted to disturbances of only moderate strength. In this case, the isentrope can be approximated by a function that is quadratic in Δ. To this degree of approximation, the Hugoniot is the same as the isentrope, so the entropy jump at the shock is neglected. This leaves us with an isentropic flow. Two cases involving shocks of varying strength are considered in the following subsections.

9.5.1 Shock Formation

Examination of the analysis of the simple compression wave discussed in Sect. 9.2.2 shows that the characteristics are converging and must eventually intersect. The extension of the diagram to later times that is given in Fig. 9.16a shows these intersections. The waveforms shown in Fig. 9.5b have been calculated for times before the characteristics intersect, but the same procedure can be applied to calculation of waveforms for times within, and beyond, the region of intersection. Several such waveforms have been plotted in Fig. 9.16b. One sees that the solution is triple-valued for points beyond that at which the first intersection occurs (given by Eqs. 9.104 and 9.105). The low-density branch corresponds to the state ahead of the wave, and the high-density branch corresponds to the final state achieved behind the wave. At intermediate densities, the waveform bulges forward, giving rise to a multivalued solution.

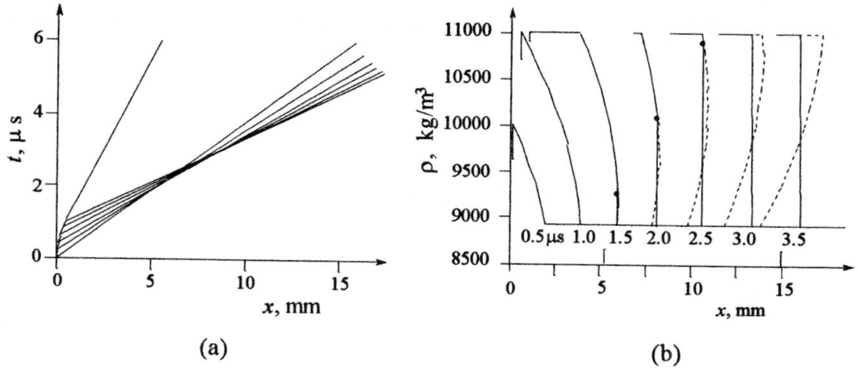

Figure 9.16. (a) The x–t diagram of Fig. 9.4, extended to later times at which the characteristics intersect. Not shown are the characteristics in the region ahead of and behind the wave. In the region beyond the point where the characteristics begin to intersect, these characteristics overlay those shown, leading to a triple-valued solution. (b) The waveforms of Fig. 9.5b, extended to later times at which the solution to the wave equation becomes multivalued. The dotted lines designate the inadmissible triple-valued section of the waveforms. The shocks shown have been embedded in the field in such a way that the multivalued portion of the solution is eliminated and mass is conserved. The vertical marks at the left indicate the boundary position for each waveform.

It is clear that this result is inadmissible on physical grounds: One cannot accept a solution that yields three different values for a field variable at the same time and place. The resolution of this problem is to embed a shock into the waveform so that the solution is single valued except for the usual indeterminacy at the shock itself. The shock is inserted into the flow in such a way that mass is conserved. The equation for mass conservation is

$$\int_{x_b(t)}^{x_S(t)} \rho(x,t)\,dx = \rho_R\, x_S(t)\,, \tag{9.166}$$

where $x_b(t)$ is the position of the boundary and $x_S(t)$ is the position of the shock. Once a characteristic analysis has been used to determine the density field at several times, this equation is easily solved for shock position at each of these times. The shock amplitude is obtained by connecting the upper and lower branches of the triple-valued solution. It is essential to this analysis that the entropy jump at the shock is neglected, thus restricting the validity of the result to shocks of only moderate strength.

9.5.2 Shock Attenuation

Shock attenuation (often called *hydrodynamic attenuation*) occurs when a shock is overtaken by a smooth decompression wave. Consider the case in which an unstressed projectile plate of thickness L and having velocity $\dot{x}_P$ impacts an unstressed halfspace that is at rest and made of the same normal material as the projectile plate. The disturbance produced is illustrated in the X–t diagram of Fig. 9.17.

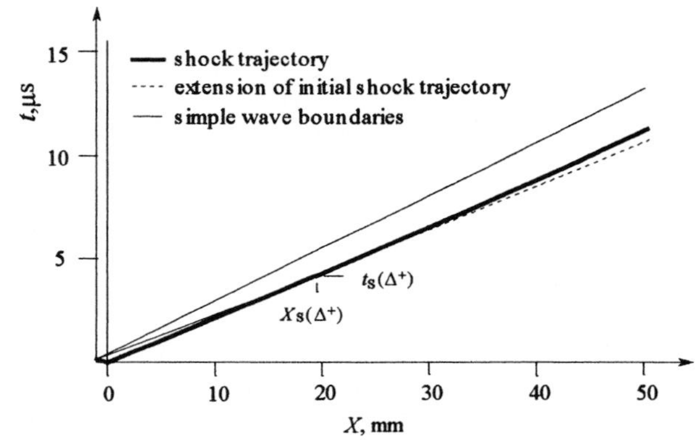

Figure 9.17. An X–t diagram illustrating the overtaking of a shock by a centered simple decompression wave. The diagram is drawn for the case in which a copper projectile plate 1-mm thick and moving at 1000 m/s impacts a stationary copper target.

At impact, shock waves form at the interface and propagate forward into the target and backward into the projectile plate. The state behind these shocks is characterized by the parameters

$$\dot{x}^+ = \dot{x}_P/2, \quad p^+ = \rho_R U_S \dot{x}^+, \quad \text{and} \quad \Delta^+ = \dot{x}^+/U_S. \tag{9.167}$$

When the receding shock encounters the unrestrained back surface of the projectile plate, a centered simple decompression wave forms and propagates forward. Since the soundspeed in the compressed material behind the advancing shock exceeds the shock velocity, the smooth wave will eventually overtake the shock. In this wave the advancing characteristics are given by

$$X = C(\Delta)(t - t^*) - L, \tag{9.168}$$

and the particle velocity is related to the compression by the equation

$$\dot{x} = \dot{x}^{+} - \int_{\Delta}^{\Delta^{+}} C(\Delta')\, d\Delta' . \tag{9.169}$$

Because the overtaking wave is one of decompression, we expect that the interaction will cause a decrease in shock strength and velocity. Because of the varying shock strength, the flow behind the shock will not be isentropic. This complication is avoided in the weak-shock approximation that we shall adopt. To determine the shock trajectory and the waveform, we need to calculate the soundspeed from the second-order expression for the isentrope. If we base this upon the Hugoniot $U_S = C_B + S\dot{x}$, we obtain

$$p^{(\eta)}(\Delta) = \rho_R\, C_B^2\, \Delta\, (1 + 2S\Delta)\,, \tag{9.170}$$

so the Lagrangian soundspeed is given by

$$C(\Delta) = C_B\, (1 + 2S\Delta)\,, \tag{9.171}$$

and

$$\int_{\Delta}^{\Delta^{+}} C(\Delta')\, d\Delta' = C_B(\Delta^{+} - \Delta)\left[1 + S(\Delta^{+} - \Delta)\right]. \tag{9.172}$$

With this, Eq. 9.169 becomes

$$\dot{x} = \dot{x}^{+} - C_B\left\{\Delta^{+} - \Delta^{-} + S\left[(\Delta^{+})^2 - \Delta^2\right]\right\}. \tag{9.173}$$

The shock trajectory (following the point at which the interaction begins) can be expressed parametrically by giving t and X as a functions of Δ in the form $t = t_S(\Delta)$ and $X = X_S(\Delta)$. Substitution of these equations into Eq. 9.168 gives

$$X_S(\Delta) = C(\Delta)\left[t_S(\Delta) - t^{*}\right] - L\,. \tag{9.174}$$

Through differentiation, we obtain

$$\frac{dX_S(\Delta)}{d\Delta} = \frac{dC(\Delta)}{d\Delta}\left[t_S(\Delta) - t^{*}\right] + C(\Delta)\frac{dt_S(\Delta)}{d\Delta}\,. \tag{9.175}$$

Along the shock trajectory

$$dX = U_S(\Delta)\, dt\,, \tag{9.176}$$

so Eq. 9.175 becomes the linear, first-order ordinary differential equation,

$$\frac{dt_S(\Delta)}{d\Delta} + \varphi(\Delta)\, t_S(\Delta) = \varphi(\Delta)\, t^{*}\,, \tag{9.177}$$

where we have written

$$\varphi(\Delta) \equiv \frac{dC(\Delta)}{d\Delta} \frac{1}{C(\Delta) - U_S(\Delta)}. \tag{9.178}$$

Combining expansions of the various quantities appearing in Eq. 9.178 shows that

$$\varphi(\Delta) = \frac{2}{\Delta}\left(1 - \frac{1}{2} S\Delta\right) \tag{9.179}$$

to within the accuracy of the analysis.

The solution of Eq. 9.177 is

$$t_S(\Delta) = t^* + \left(\frac{\Delta^+}{\Delta}\right)^2 [t_S(\Delta^+) - t^*] \exp[-2S(\Delta^+ - \Delta)], \tag{9.180}$$

where

$$t_S(\Delta^+) = \frac{C(\Delta^+)\, t^* + L}{C(\Delta^+) - U_S^+} \tag{9.181}$$

is the time at which the leading characteristic of the decompression wave intersects the shock. The associated value of X on the shock trajectory is given by Eq. 9.174.

From the known speed of the initial shock and the soundspeed behind this shock, the X–t diagram can be plotted up to the time at which the decompression wave begins to interact with the shock. The diagram is completed when Eqs. 9.180 and 9.174 are used to plot the trajectory of the attenuating shock.

The amplitude of the attenuating shock at a point on its trajectory can be characterized by the value of Δ at the point. If one selects several values of Δ in the range $0 \le \Delta \le \Delta^+$, the associated points on the trajectory can be plotted using Eqs. 9.180 and 9.174. The value of Δ associated with each of these points can be used to calculate the pressure (using Eq. 9.170) and the particle velocity (using Eq. 9.173). The waveform behind the shock is given by the simple wave solution. Several pressure waveforms associated with the problem described in Fig. 9.17 are plotted in Fig. 9.18.

9.6 Collision of Two Centered Simple Decompression Waves

Let us consider the waves and wave interactions that arise when two slabs of like material collide. Suppose the collision involves an unstressed projectile of thickness L_P moving at velocity $\dot{x}_P$ colliding with a target of thickness $L_T > L_P$ that is unstressed and at rest. The waves produced are illustrated in the X–t plot in Fig. 9.19.

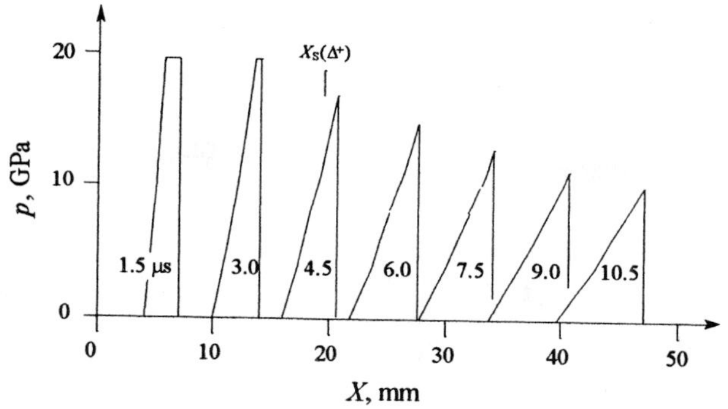

Figure 9.18. A sequence of pressure–distance waveforms illustrating shock attenuation. The time corresponding to each waveform is given by the associated number on the drawing. The distance at which the leading characteristic of the decompression wave catches up to the shock and the attenuation process begins is marked $X_S(\Delta^+)$.

We shall analyze this event for a material for which the Hugoniot takes the form

$$U_S = C_B + S\dot{x} \quad \text{or equivalently,} \quad p^{(H)}(\Delta) = \frac{\rho_R C_B^2 \Delta}{(1 - S\Delta)^2}. \tag{9.182}$$

We shall also need the isentrope through the shock-compressed state for which $p = p^+$ and $\Delta = \Delta^+$. It is given by

$$p^{(\eta)}(\Delta) = \chi_c(\Delta)\left\{ p^+ + \int_{\Delta^+}^{\Delta} \frac{\kappa_c(\Delta')}{\chi_c(\Delta')} d\Delta' \right\}, \tag{9.183}$$

where

$$\chi_c(\Delta) = \exp[\gamma_R(\Delta - \Delta^+)] \tag{9.184}$$

and

$$\kappa_c(\Delta) = \left(1 - \frac{\gamma_R}{2}\Delta\right)\frac{dp^{(H)}(\Delta)}{d\Delta} - \frac{\gamma_R}{2} p^{(H)}(\Delta) = \rho_R C_B^2 \frac{1 + (S - \gamma_R)\Delta}{(1 - S\Delta)^3}, \tag{9.185}$$

with the second of these equations arising through introduction of Eq. 9.182 for the Hugoniot.

$$0 = p^+ + \int_{\Delta^+}^{\Delta^{(2)}} \frac{\kappa_c(\Delta')}{\chi_c(\Delta')} d\Delta' \tag{9.186}$$

for a range of values of $\Delta^{(2)}$ until a solution is found. For the example of Fig. 9.19, we obtain $\Delta^{(2)} = -9.134 \times 10^{-4}$. This expansion is attributable to the irreversible heating associated with the shock compression.

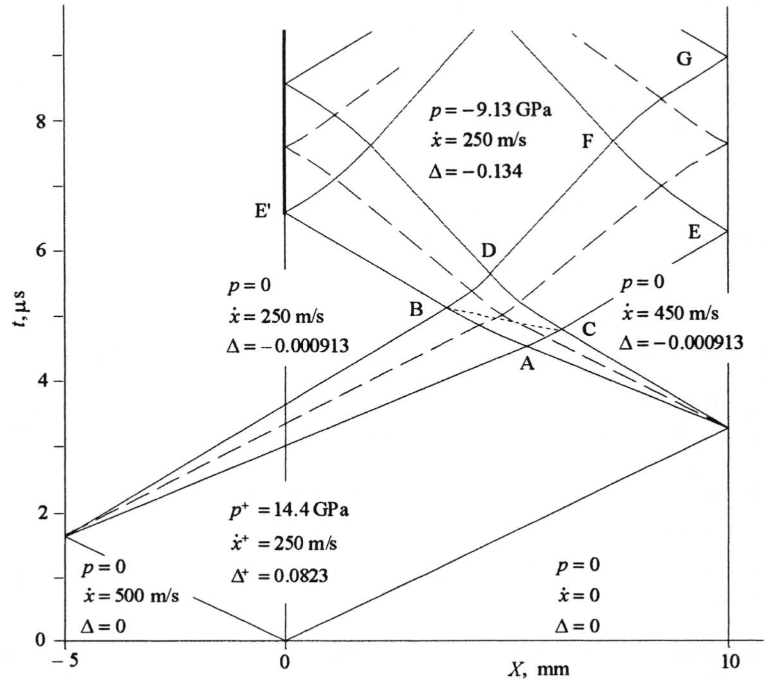

Figure 9.19. Waves resulting from the collision of an impactor of thickness L_P against a target of thickness $L_T > L_P$. The drawing is made for a uranium impactor of thickness 5 mm impacting a uranium target of thickness 10 mm at a velocity of 500 m/s. Uranium parameters are $\rho_R = 18950\ \mathrm{kg/m^3}$, $C_B = 2487$ m/s, and $S = 2.200$. The impact produces compression of about 8% and pressure $p^+ = 14.4$ GPa. The several wavespeeds are $U_S = 3037$ m/s, $C_L(\Delta^+) = 3611$ m/s, and $C_L(\Delta^{(2)}) = 2479$ m/s. Uranium was chosen for this example because, with $S = 2.200$, it is highly nonlinear so the effects of the interaction are emphasized. The broken lines are typical characteristic curves. The dotted line passing through the interaction region between points A and B separates the part of the region BCD that is in tension from the part ABC that is in compression. The bold line lying along the impact interface above point E′ indicates the separation of the interface that occurs when the tensile region expands to meet the interface.

We shall need to determine the compression, $\Delta^{(2)}$, at the point on the decompression isentrope for which $p = 0$. This is done by numerical evaluation of the integral.

The Lagrangian soundspeed for material in states on this isentrope is given by

$$C^{(\eta)}(\Delta) = \left(\frac{1}{\rho_R} \frac{dp^{(\eta)}(\Delta)}{d\Delta} \right)^{1/2}, \tag{9.187}$$

and substitution of Eq. 9.183 into this equation yields the result

$$C^{(\eta)}(\Delta) = \left[\frac{1}{\rho_R} [\gamma_R p^{(\eta)}(\Delta) + \kappa_c(\Delta)] \right]^{1/2} . \quad (9.188)$$

Shocks. Because of the symmetric impact, the shocks introduced into the projectile and target produce the state characterized by $\dot{x}^+ = \dot{x}_P/2$. The jump conditions for the shock propagating into the target are

$$\begin{aligned} \dot{x}^+ &= U_S\, \Delta^+ \\ p^+ &= \rho_R\, U_S\, \dot{x}^+ , \end{aligned} \quad (9.189)$$

or,

$$p^+ = \frac{\rho_R}{\Delta^+} (\dot{x}^+)^2 = \frac{\rho_R}{4\,\Delta^+} (\dot{x}_P)^2 . \quad (9.190)$$

Centered Simple Decompression Waves. Let us begin our analysis by considering the right-traveling wave formed at the unrestrained back face of the impactor plate when the shock encounters this surface. In accord with the boundary condition, the material is unstressed in the region behind the wave, so the compression, $\Delta^{(2)}$, in this region, is the value on the isentrope when the pressure is zero.

The characteristic lines along which the fields are constant in the right-propagating wave are represented by the equation

$$X + L_P = C_L(\Delta^*) \left(t - \frac{L_P}{U_S} \right) , \quad (9.191)$$

where we have identified Δ^* as the characteristic coordinate. The quantity Δ^* has the range $\Delta^+ \geq \Delta^* \geq \Delta^{(2)}$. The leading and trailing characteristics of this wave are obtained by substituting Δ^+ and $\Delta^{(2)}$, respectively, into Eq. 9.191.

Within the wave the fields satisfy the equation

$$\dot{x}(\Delta^*) + \int_{\Delta^*}^{\Delta^+} C_L(\Delta')\, d\Delta' = \dot{x}^+ , \quad (9.192)$$

and, using Eq. 9.191, we can determine $\dot{x}(X,t)$ and $\Delta(X,t)$ within the domain occupied by the simple wave.

To determine the fields in the interaction region we need to know the Riemann invariants on the ray characteristics for each of the decompression waves.

For the right-propagating wave the Riemann invariant, Eq. 9.31_1, is

$$\Phi^+(t^*) = \dot{x} + \int_{\Delta^{(2)}}^{\Delta} C_L(\Delta')d\Delta' . \tag{9.193}$$

Evaluating this quantity using Eq. 9.192 yields the value

$$\Phi^+(t^*) = \dot{x}^+ - \int_{\Delta^*}^{\Delta^+} C_L(\Delta')\,d\Delta' + \int_{\Delta^{(2)}}^{\Delta^*} C_L(\Delta')\,d\Delta' \tag{9.194}$$

for the invariant within the wave. Since this value is preserved on the continuation of the characteristic into and through the interaction region, we can equate the right members of the foregoing two equations to obtain the result

$$\dot{x} + \int_{\Delta^{(2)}}^{\Delta} C_L(\Delta')\,d\Delta' = \dot{x}^+ - \int_{\Delta^*}^{\Delta^+} C_L(\Delta')\,d\Delta' + \int_{\Delta^{(2)}}^{\Delta^*} C_L(\Delta')\,d\Delta' , \tag{9.195}$$

even though the characteristic curve is no longer represented by Eq. 9.191.

The left-traveling wave originating at the downstream face of the target plate is analyzed in the same way, with the result that, in the region behind this wave,

$$p = 0, \quad \Delta = \Delta^{(2)}, \quad \text{and} \quad \dot{x} = \dot{x}^+ + \int_{\Delta^{(2)}}^{\Delta^+} C_L(\Delta)\,d\Delta . \tag{9.196}$$

The characteristic line on which the wave fields are constant is

$$X - L_T = -C_L(\Delta^{**})\left(t - \frac{L_T}{U_S}\right) \tag{9.197}$$

and we identify Δ^{**} (for $\Delta^+ \geq \Delta^{**} \geq \Delta^{(2)}$) as the characteristic coordinate. The leading and trailing characteristics are obtained by substituting Δ^+ and $\Delta^{(2)}$, respectively, into Eq. 9.197.

The particle velocity within the wave region, $\Delta^+ \geq \Delta^{**} \geq \Delta^{(2)}$, is given by

$$\dot{x}(\Delta^{**}) = \dot{x}^+ + \int_{\Delta^{**}}^{\Delta^+} C_L(\Delta')\,d\Delta' . \tag{9.198}$$

The Riemann invariant on the ray characteristics of the left-propagating wave is

$$\Phi^+(t^{**}) = \dot{x} - \int_{\Delta^{(2)}}^{\Delta} C_L(\Delta')d\Delta' . \tag{9.199}$$

Evaluating this function, using Eq. 9.198, yields the value

$$\Phi^+(t^{**}) = \dot{x}^+ + \int_{\Delta^{**}}^{\Delta^+} C_L(\Delta')d\Delta' - \int_{\Delta(2)}^{\Delta^{**}} C_L(\Delta')d\Delta' . \tag{9.200}$$

When this value is substituted into Eq. 9.199 we obtain the result

$$\dot{x} - \int_{\Delta(2)}^{\Delta} C_L(\Delta')d\Delta' = \dot{x}^+ + \int_{\Delta^{**}}^{\Delta^+} C_L(\Delta')d\Delta' - \int_{\Delta(2)}^{\Delta^{**}} C_L(\Delta')d\Delta' , \tag{9.201}$$

which holds on the entire Δ^{**} characteristic.

Interaction Region. Now let us consider the region in which the two decompression waves interact. The characteristics in this region are continuations of the characteristic rays defining each of the two intersecting waves. Since the known values of the Riemann invariants are preserved on the extensions of these characteristic curves into, and through, the interaction region, we can determine the fields in this region as functions of the characteristic coordinates. This is done by sequentially adding and subtracting Eqs. 9.195 and 9.201 to obtain

$$\begin{aligned} \dot{x}(\Delta^*, \Delta^{**}) = \dot{x}^+ + \frac{1}{2}\Bigg[& \int_{\Delta^{**}}^{\Delta^+} C_L(\Delta')\,d\Delta' - \int_{\Delta(2)}^{\Delta^{**}} C_L(\Delta')\,d\Delta' \\ & - \int_{\Delta^*}^{\Delta^+} C_L(\Delta')\,d\Delta' + \int_{\Delta(2)}^{\Delta^*} C_L(\Delta')\,d\Delta' \Bigg] \\ \int_{\Delta(2)}^{\Delta} C_L(\Delta')\,d\Delta' = \frac{1}{2}\Bigg[& \int_{\Delta(2)}^{\Delta^*} C_L(\Delta')\,d\Delta' + \int_{\Delta(2)}^{\Delta^{**}} C_L(\Delta')\,d\Delta' \\ & - \int_{\Delta^*}^{\Delta^+} C_L(\Delta')\,d\Delta' - \int_{\Delta^{**}}^{\Delta^+} C_L(\Delta')\,d\Delta' \Bigg]. \end{aligned} \tag{9.202}$$

The second of these equations gives the solution $\Delta(\Delta^*, \Delta^{**})$ in implicit form and the first equation gives $\dot{x}(\Delta^*, \Delta^{**})$ explicitly, once $\Delta(\Delta^*, \Delta^{**})$ has been determined.

To express this solution in X, t coordinates we must calculate the X, t trajectories of the characteristic curves. The region in which the two decompression waves interact is bounded by the cross characteristics that are extensions of the leading and trailing rays of the centered simple decompression waves. The first point of interaction of the decompression waves occurs at the point of intersection of the leading characteristics, which we designate point A. The X, t coordinates of this point are

$$X_A = \frac{L_T - L_P}{2}\left[1 + \frac{C_L(\Delta^+)}{U_S}\right]$$
$$t_A = \frac{L_T + L_P}{2}\left[\frac{1}{C_L(\Delta^+)} + \frac{1}{U_S}\right]. \tag{9.203}$$

The characteristic through this point that crosses the right-propagating wave is given by Eq. 9.115, which takes the form

$$X = -L_P + (X_A + L_P)\frac{t_A - \dfrac{L_P}{U_S}}{t - \dfrac{L_P}{U_S}} \tag{9.204}$$

in the present case. Similarly, the characteristic through this point that crosses the left-propagating wave is given by

$$X = L_T - (L_T - X_A)\frac{t_A - \dfrac{L_T}{U_S}}{t - \dfrac{L_T}{U_S}}. \tag{9.205}$$

Point B, corresponding to $\Delta^* = 0$ and $\Delta^{**} = \Delta^+$, has the X, t coordinates

$$X_B = -L_P + \left[C_L(\Delta^{(2)})(X_A + L_P)\left(t_A - \frac{L_P}{U_S}\right)\right]^{1/2}$$
$$t_B = \frac{L_P}{U_S} + \frac{X_B + L_P}{C_L(\Delta^{(2)})}, \tag{9.206}$$

and Point C, corresponding to $\Delta^* = \Delta^+$ and $\Delta^{**} = 0$, has the X, t coordinates

$$X_C = L_T - \left[C_R\,(L_T - X_A)\left(t_A - \frac{L_T}{U_S}\right)\right]^{1/2}$$
$$t_C = \frac{L_T}{U_S} + \frac{L_T - X_C}{C_R}. \tag{9.207}$$

The X, t coordinates of point D, corresponding to $\Delta^* = 0$ and $\Delta^{**} = 0$, are not yet known, but the particle velocity, deformation gradient, and stress can be calculated from the known values of Δ^* and Δ^{**}. One finds that $\dot{x}_D = \dot{x}_P/2$ and that the compression and pressure at this point are negative, i.e., the material is in tension. For the example described in Fig. 9.19 one finds $\Delta_D = -13.4\%$ from which we calculate that the tension at this point is 9.13 GPa, and the wavespeed for this deformation gradient is $C_L = 1359$ m/s.

To obtain the fields in the interaction region as functions of X and t, it is necessary to calculate the X, t trajectories of the characteristic curves $\Delta^* = \text{const.}$ and $\Delta^{**} = \text{const.}$ These trajectories are solutions of Eqs. 9.30 which, in the present case, take the form

$$\begin{aligned}
&\frac{\partial X(\Delta^*, \Delta^{**})}{\partial \Delta^*} + C_L(\Delta^*, \Delta^{**}) \frac{\partial t(\Delta^*, \Delta^{**})}{\partial \Delta^*} = 0 \quad \text{on} \quad \Delta^{**} = \text{const.} \\
&\frac{\partial X(\Delta^*, \Delta^{**})}{\partial \Delta^{**}} - C_L(\Delta^*, \Delta^{**}) \frac{\partial t(\Delta^*, \Delta^{**})}{\partial \Delta^{**}} = 0 \quad \text{on} \quad \Delta^* = \text{const.},
\end{aligned} \tag{9.208}$$

where

$$C_L = C_R (1 + 2C\Delta)^{1/2}, \tag{9.209}$$

with Δ being the solution of Eq. 9.202.

Equations 9.208 can be solved for $X(\Delta^*, \Delta^{**})$ and $t(\Delta^*, \Delta^{**})$ by a simple finite-difference procedure, yielding the X, t trajectories of the characteristic curves (within the interaction zone), and thereby the coordinates of point D and the values of the field variables as functions of X and t. The analysis was conducted on a 10×10 point grid and checked for convergence by refining the analysis to a 19×19 grid. Since the intersection of each of the calculated characteristic curves with the interaction region boundary is obtained from this solution, and the wavespeed at each intersection point is also known, the characteristics can be extended through the waves emanating from the interaction region. The stress histories at the X coordinate through points A, C, and D are shown on Fig. 9.20. The discontinuities of these histories fall at the points where the lines of constant X pass from one wave region to another (see Fig. 9.19).

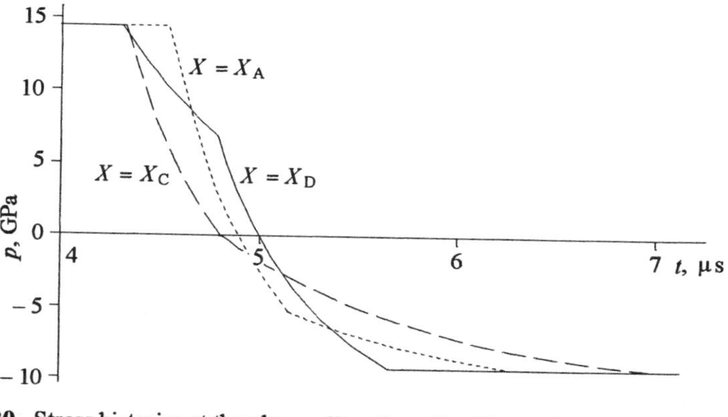

Figure 9.20. Stress histories at the planes $X = X_A$, $X = X_C$, and $X = X_D$ for the wave interaction shown in Fig. 9.19.

Transmitted Waves. The waves emanating from the interaction region are simple waves because they are bounded by regions of the X–t plane in which the field values are constants. The ray characteristics defining these waves are straight-line extensions of the characteristics intersecting the boundaries of the interaction zone. Because the values of Δ and $\dot{x}$ are constants on each of the rays, they take the values on the curves BD and CD at the point of their intersection. These values have already been calculated as part of the solution in the interaction region. At the time corresponding to the point E′, the left-propagating wave begins to interact with the target plate. Since the stress in this wave is tensile, and since the interface is free to separate, the reflection occurs as at an unrestrained surface. A similar reflection occurs at the unrestrained back surface of the target plate. These reflections are entirely similar and we shall concentrate on the one that occurs at the back surface of the target plate.

Reflection at the Unrestrained Back Surface of the Target. The right-propagating simple wave that emerges from the interaction region is reflected from the unrestrained back surface of the target plate. This reflection process is analyzed in essentially the same way as the fields in the interaction region were analyzed. The Riemann invariant on each of the ray characteristics of the right-propagating simple wave emanating from the interaction region maintain the value established previously. The boundary condition on the unrestrained surface is one of zero pressure, so the compression on this surface is $\Delta = \Delta^{(2)}$. The particle velocity is determined by substituting this compression into the Riemann invariant for the right-propagating simple wave. With the compression and particle velocity on this surface having been established, the Riemann invariant defining the wave reflected from the surface can be evaluated and the fields in the reflection region determined as functions of the characteristic coordinates. The X, t trajectories of these characteristics are then determined by numerical solution of Eqs. 9.208 in the same way as for the interaction region. The result is illustrated in Fig. 9.19. The velocity history of the unrestrained surface that is obtained from this solution is plotted in Fig. 9.21.

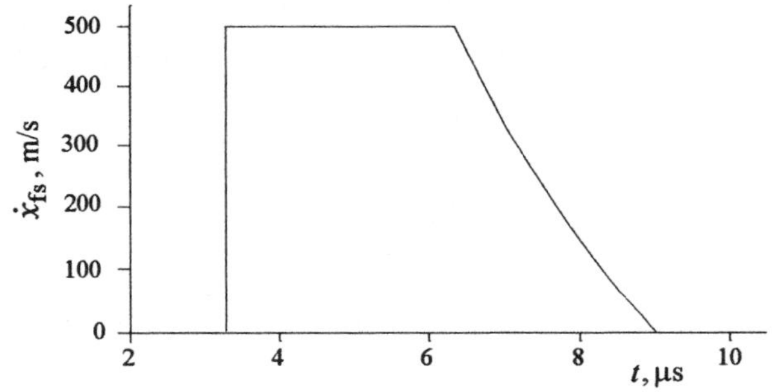

Figure 9.21. Back surface velocity history for the problem illustrated in Fig. 9.19.

Spall Fracture. In the example problem discussed in this section, the wave interaction has been shown to lead to states of tension in the interior of the target plate. If the tension exceeds a critical value for the material, called the *spall strength*, fracture will occur. This results in a layer of the material separating from the target plate and the solution presented will cease to be valid. We shall address this issue in Chap. 12.

9.7 Exercises

9.7.1. Referring to the discussion of waveforms propagated in materials described by third-order elastic coefficients, what can be said regarding the case $3C_{11} + C_{111} = 0$? For a commentary on this matter, see Thurston [97].

9.7.2. Solve the shock-reflection problem discussed in Sect. 9.3.1 for the case in which the shock encounters a material of lower impedance than that in which the incident wave is propagating. Let the two materials be governed by Hugoniots of the form of Eq. 3.10.

9.7.3. The speed with which the leading edge of a compressed slab expands into void space when the constraining pressure is suddenly released is called the *escape velocity*. What is the escape velocity of an isentropically compressed material for which the expansion isentrope is given by Eq. 9.153?

9.7.4. Derive Eq. 9.163. Hint: expand Eqs. 9.47 and 2.116_2 in powers of $G^+ - G^-$ and compare the results.

9.7.5. Derive Eq. 9.165.

9.7.6. Discuss the implications of the results of Fig. 9.13 for use of a shock as an approximation to a centered simple wave.

9.7.7. Derive an expression for the X^* characteristics crossing a simple wave.

9.7.8. Use the definitions of the Lagrangian and Eulerian wavespeeds (the relation of the wavespeed to the derivative of pressure with respect to a measure of uniaxial deformation) to prove Eq. 9.18.

9.7.9. Show that the Riemann invariants given by Eqs. 9.31 are the same as those given by 9.38.

9.7.10. Derive Eq. 9.20.

CHAPTER 10

Elastic–Plastic and Elastic–Viscoplastic Waves

10.1 Weak Elastic–Plastic Waves

We consider a homogeneous and isotropic elastic–plastic plate subject to a small uniaxial deformation. Because of the assumption that the strain is small, no distinction need be made between Lagrangian and Eulerian coordinates and, since the only coordinate appearing in the following analysis is X_1, we shall use the abbreviation $X_1 = X$.

The requirements for balance of mass and momentum at a shock are represented by the jump equations 2.113 1,2, which we write in the form

$$[\![-\widetilde{E}_{11}]\!] U_S = [\![\dot{x}]\!], \qquad \rho_R U_S [\![\dot{x}]\!] = [\![-t_{11}]\!]. \tag{10.1}$$

In writing the first of these equations we have omitted the small term involving $\widetilde{E}_{11}^2$. The principles of balance of energy and production of entropy must also be satisfied at a shock, but we are restricting attention to weak shocks so thermal variables can be neglected, leaving us with a purely mechanical theory.

When applied to a shock propagating into material in a known state $S^- = \{t_{11}^-, \widetilde{E}_{11}^-, \dot{x}^-\}$, the two jump equations involve the unknown quantities $S^+ = \{t_{11}^+, \widetilde{E}_{11}^+, \dot{x}^+\}$ and U_S. One of the quantities comprising S^+ is normally specified as a measure of the stimulus producing the shock.* The relevant material properties are characterized by Hugoniot curves appropriate to the theory of ideal elastic–plastic response. For the analysis in this chapter it is adequate to restrict attention to the $t_{11} - \widetilde{E}_{11}$, $t_{22} - \widetilde{E}_{11}$, and $t_{11} - \dot{x}$ Hugoniots. Neglect of thermal effects means that the response of a material to shock-induced deformation is the same as its response to the same deformation effected quasi-statically, i.e., the $t_{11} - \widetilde{E}_{11}$ and $t_{22} - \widetilde{E}_{11}$ Hugoniots are the same as the static stress–strain curves of Fig 7.3.

---

* In the case of ideal elastic–plastic materials in which the elastic response is linear, the shock velocity is excluded from a role in describing the material state because relations between the shock velocity and the other variables are not invertible.

Bilinear Hugoniots such as are shown in Fig. 7.4 correspond to shocks that are unstable (for both compression and decompression) when their amplitude is such as to include the slope discontinuity in the path. These shocks separate into distinct marginally stable elastic and plastic shocks. All responses to instantaneous load application or release take the form of propagating shocks.

Several quantitative examples are presented in the remaining sections of this chapter. To permit comparison of cases, they are all calculated for a steel. Specifically, this material is characterized by the elastic moduli $\lambda_R = 123.1$ GPa and $\mu_R = 79.29$ GPa (corresponding to a Young's modulus of 30×10^6 psi and a shear modulus of 11.5×10^6 psi), a density $\rho_R = 7870$ kg/m^3, and a yield stress $Y = 1.00$ GPa (145,000 psi). These parameters lead to the elastic wavespeeds $C_0 = 5983$ m/s and $C_B = 4728$ m/s. These material constants were used in all example calculations where a result is said to be for "steel".

10.1.1 Compression Shocks

Let us begin by analysis of the waves produced by sudden application of a uniform compressive load to the face $X = 0$ of a plate.

Examination of the Hugoniot shown in Fig. 10.1a shows that two amplitude ranges must be considered. Waves propagating into undeformed material and having compressive amplitude less than the Hugoniot elastic limit involve only elastic response. A shock of greater amplitude is unstable, causing it to separate into two shocks. The leading shock, called the *elastic precursor*, is of amplitude $t_{11} = -t_{11}^{\mathrm{HEL}}$ and propagates at the longitudinal elastic wavespeed

$$C_0 = \left(\frac{\lambda_R + 2\mu_R}{\rho_R}\right)^{1/2}. \tag{10.2}$$

The second shock, called the *plastic shock* or *plastic wave*, takes the material to the state imposed on the boundary and propagates at the bulk wavespeed

$$C_B = \left(\frac{\lambda_R + (2/3)\mu_R}{\rho_R}\right)^{1/2} = \left(\frac{B_R}{\rho_R}\right)^{1/2}. \tag{10.3}$$

As is obvious from these equations, $C_0 > C_B$. This waveform is shown in Fig. 10.1b.

The elastic and plastic shocks, and their interactions with each other, material interfaces, etc., can be analyzed using the jump equations 10.1. The shock velocities to be used in these equations are $U_S = \pm C_0$ for the elastic shock, and $U_S = \pm C_B$ for the plastic shock. Applying these equations to shocks in the elastic range that are propagating into undeformed material at rest gives

$$\begin{aligned} t_{11} &= \rho_R C_0^2 \widetilde{E}_{11}, \quad 0 \le -\widetilde{E}_{11} \le \widetilde{E}_{11}^{\mathrm{HEL}} \\ \dot{x} &= -t_{11}/(\rho_R C_0), \quad 0 \le -t_{11} \le t_{11}^{\mathrm{HEL}}. \end{aligned} \tag{10.4}$$

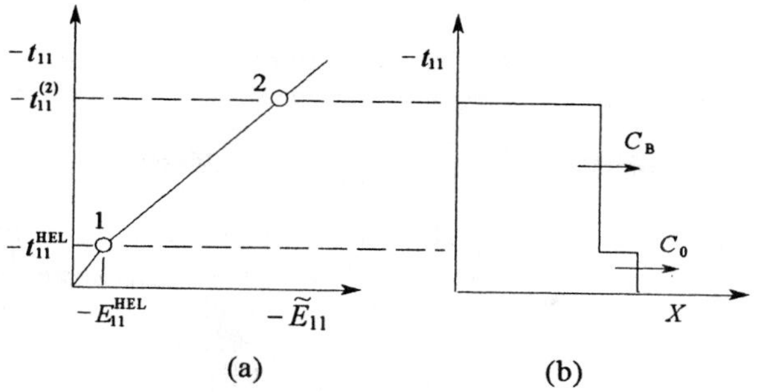

Figure 10.1. (a) Uniaxial-strain Hugoniot. The slope of the elastic segment of the Hugoniot is $\rho_R C_0^2$ and that of the plastic segment is $\rho_R C_B^2$. (b) elastic–plastic waveform of stress amplitude $t_{11}^{(2)}$ propagating into undeformed material.

Since the amplitude of the elastic precursor is $-t_{11}^{HEL}$, the particle velocity and strain of the material behind it are given by

$$\dot{x}^{(1)} = \frac{1}{\rho_R C_0} t_{11}^{HEL} = C_0 \tilde{E}_{11}^{HEL} \equiv \dot{x}^{HEL} \tag{10.5}$$

$$\tilde{E}_{11}^{(1)} = -\frac{1}{C_0}\dot{x}^{HEL} = -\frac{1}{\rho_R C_0^2} t_{11}^{HEL} \equiv -\tilde{E}_{11}^{HEL}, \tag{10.6}$$

and the transverse stress component is

$$t_{22}^{(1)} = -t_{22}^{HEL} = -\frac{3C_B^2 - C_0^2}{2C_0^2} t_{11}^{HEL}. \tag{10.7}$$

The yield stress measures $\tilde{E}_{11}^{HEL}$, $\dot{x}^{HEL}$, t_{11}^{HEL}, and t_{22}^{HEL} can be expressed in other forms, as listed in Table 10.1.

Suppose the state behind the plastic wave is characterized by the constant field values $\mathcal{S}^{(2)} = \{t_{11}^{(2)}, \tilde{E}_{11}^{(2)}, \dot{x}^{(2)}\}$. Equations 10.1, applied to the plastic wave, yield the two equations

$$\begin{aligned} C_B(-\tilde{E}_{11}^{(2)} - \tilde{E}_{11}^{HEL}) &= \dot{x}^{(2)} - \dot{x}^{HEL} \\ \rho_R C_B(\dot{x}^{(2)} - \dot{x}^{HEL}) &= -t_{11}^{(2)} - t_{11}^{HEL}, \end{aligned} \tag{10.8}$$

relating the variables defining the state $\mathcal{S}^{(2)}$. If $\dot{x}^{(2)}$ is given as a boundary condition it is convenient to write these equations in the form

$$\begin{aligned} \tilde{E}_{11}^{(2)} &= -\frac{1}{C_B}\left\{\dot{x}^{(2)} - \left(\frac{C_0 - C_B}{C_0}\right)\dot{x}^{HEL}\right\}, \quad \text{for} \quad \dot{x}^{(2)} \geq \dot{x}^{HEL}. \\ t_{11}^{(2)} &= -\rho C_B\, x^{(2)} - \rho(C_0 - C_B)\,\dot{x}^{HEL} \end{aligned} \tag{10.9}$$

Table 10.1. Expressions for the Hugoniot elastic limit

	t_{11}^{HEL}	t_{22}^{HEL}	$\widetilde{E}_{11}^{\mathrm{HEL}}$	$\dot{x}^{\mathrm{HEL}}$
λ, μ	$\dfrac{\lambda_R + 2\mu_R}{2\mu_R} Y$	$\dfrac{\lambda}{2\mu_R} Y$	$\dfrac{1}{2\mu_R} Y$	$\left[\dfrac{\lambda_R + 2\mu_R}{4\rho_R \mu_R^2}\right]^{1/2} Y$
B, μ	$\left[\dfrac{B_R}{2\mu_R} + \dfrac{2}{3}\right] Y$	$\left[\dfrac{B_R}{2\mu_R} - \dfrac{1}{3}\right] Y$	$\dfrac{1}{2\mu_R} Y$	$\left[\dfrac{B_R + \frac{4}{3}\mu_R}{4\rho_R \mu_R^2}\right]^{1/2} Y$
C_0, C_B	$\dfrac{2C_0^2 Y}{3(C_0^2 - C_B^2)}$	$\dfrac{3C_B^2 - C_0^2}{3(C_0^2 - C_B^2)} Y$	$\dfrac{2Y}{3\rho_R (C_0^2 - C_B^2)}$	$\dfrac{2C_0 Y}{3\rho_R (C_0^2 - C_B^2)}$
t_{11}^{HEL}	t_{11}^{HEL}	$\dfrac{3C_B^2 - C_0^2}{2C_0^2} t_{11}^{\mathrm{HEL}}$	$\dfrac{1}{\rho_R C_0^2} t_{11}^{\mathrm{HEL}}$	$\dfrac{1}{\rho_R C_0} t_{11}^{\mathrm{HEL}}$
t_{22}^{HEL}	$\dfrac{2C_0^2}{3C_B^2 - C_0^2} t_{22}^{\mathrm{HEL}}$	t_{22}^{HEL}	$\dfrac{2}{\rho_R (3C_B^2 - C_0^2)} t_{22}^{\mathrm{HEL}}$	$\dfrac{2C_0 t_{22}^{\mathrm{HEL}}}{\rho_R (3C_B^2 - C_0^2)}$
$\widetilde{E}_{11}^{\mathrm{HEL}}$	$\rho_R C_0^2 \widetilde{E}_{11}^{\mathrm{HEL}}$	$\dfrac{\rho_R}{2}(3C_B^2 - C_0^2) \widetilde{E}_{11}^{\mathrm{HEL}}$	$\widetilde{E}_{11}^{\mathrm{HEL}}$	$C_0 \widetilde{E}_{11}^{\mathrm{HEL}}$
v_1^{HEL}	$\rho_R C_0 \dot{x}^{\mathrm{HEL}}$	$\dfrac{\rho_R (3C_B^2 - C_0^2)}{2C_0} \dot{x}^{\mathrm{HEL}}$	$\dfrac{\dot{x}^{\mathrm{HEL}}}{C_0}$	$\dot{x}^{\mathrm{HEL}}$

Similarly, if $t_{11}^{(2)}$ is given, we have

$$\left.\begin{aligned} \widetilde{E}_{11}^{(2)} &= \frac{1}{\rho_R C_B^2}\left\{ t_{11}^{(2)} + \left(\frac{C_0^2 - C_B^2}{C_0^2}\right) t_{11}^{\mathrm{HEL}} \right\} \\ \dot{x}^{(2)} &= \frac{1}{\rho_R C_B}\left\{ -t_{11}^{(2)} - \left(\frac{C_0 - C_B}{C_0}\right) t_{11}^{\mathrm{HEL}} \right\} \end{aligned}\right\}, \quad \text{for} \quad -t_{11}^{(2)} \ge t_{11}^{\mathrm{HEL}}. \tag{10.10}$$

We can use these equations to plot the $t_{11}-\dot{x}$ Hugoniot corresponding to the $t_{11}-\widetilde{E}_{11}$ Hugoniot of Fig. 7.4. In the elastic range, Eq. 10.4$_2$ gives

$$-t_{11} = \rho_R C_0 \dot{x}, \quad \text{for } \dot{x} \le \dot{x}^{\mathrm{HEL}} \text{ or } -t_{11} \le t_{11}^{\mathrm{HEL}}, \tag{10.11}$$

and in the plastic range Eq. 10.9$_2$ can be written

$$-t_{11} = \rho_R C_B \dot{x} + \left(\frac{C_0 - C_B}{C_0}\right) t_{11}^{\mathrm{HEL}}, \quad \text{for } \dot{x} \ge \dot{x}^{\mathrm{HEL}} \text{ or } -t_{11} \ge t_{11}^{\mathrm{HEL}}. \tag{10.12}$$

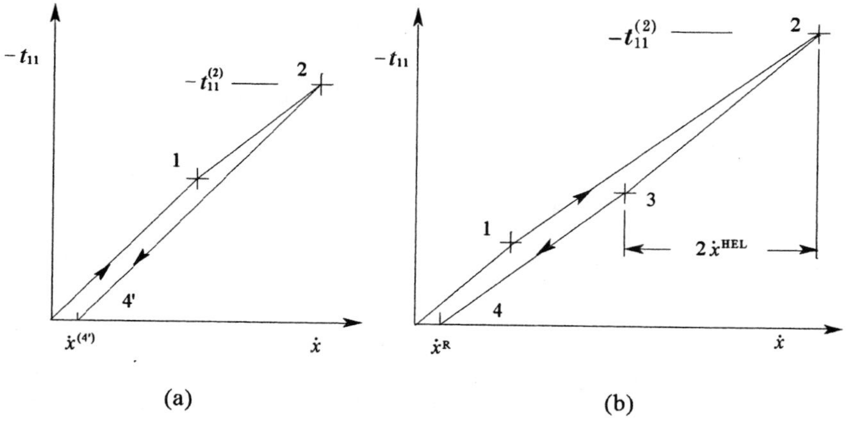

Figure 10.2. Stress–particle-velocity Hugoniot curves corresponding to the elastic–plastic stress relations of Fig. 7.4. Part (a) of the figure is drawn for the case $-t_{11}^{(2)} \le 2t_{11}^{\mathrm{HEL}}$ and part (b) of the figure is drawn for $-t_{11}^{(2)} > 2t_{11}^{\mathrm{HEL}}$.

The result is shown as the compression branch on each of the diagrams in Fig. 10.2.

Finally, using Eq. 7.53, we can write the transverse stress components in either of the equivalent forms

$$t_{22}^{(2)} = t_{11}^{(2)} + \frac{3(C_0^2 - C_B^2)}{2C_0^2}\, t_{11}^{\mathrm{HEL}} = t_{11}^{(2)} + Y\,, \tag{10.13}$$

or

$$t_{22}^{(2)} = -\rho_R C_B \dot{x}^{(2)} + \frac{\rho_R (C_0 + 3C_B)(C_0 - C_B)}{2C_0}\, \dot{x}^{\mathrm{HEL}}\,. \tag{10.14}$$

10.1.2 Impact of Thick Plates

Waves generated by planar impact of one elastic–plastic plate on another are analyzed as discussed in Chap. 3. The state at the impact interface is defined by the intersection of the t_{11}–$\dot{x}$ Hugoniots for the plate materials. These Hugoniot curves are to be centered at the appropriate initial conditions and oriented for wave propagation away from the interface, as shown in Fig. 10.3a. The X–t diagram for the shocks produced by a typical impact is given in Fig. 10.3b, which shows the two shocks produced in each plate due to the instability of the shocks initially produced by the impact. Finally, a plot of the resulting waveform is given in Fig. 10.4.

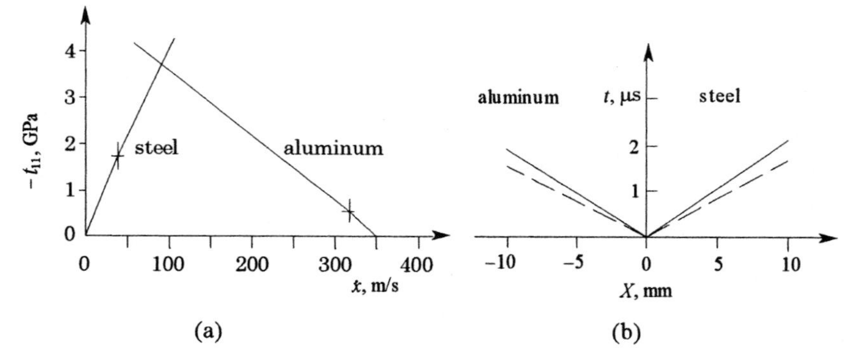

Figure 10.3. (a) Stress–particle-velocity Hugoniots for a 350 m/s impact of an aluminum alloy 6061-T6 plate against a steel plate and (b) The X–t diagram for the shocks produced by this impact.

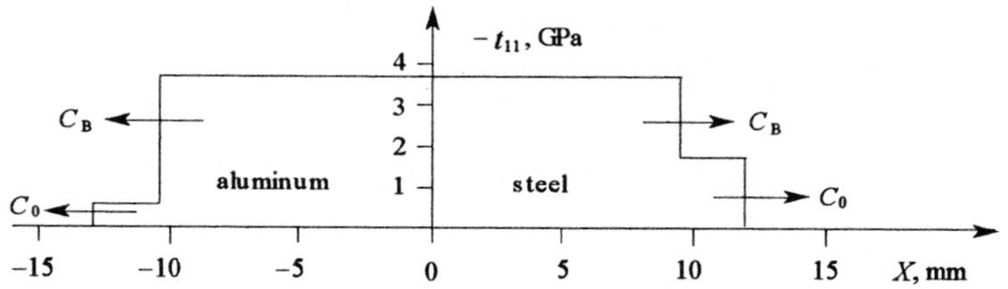

Figure 10.4. The waveform produced by impact of an aluminum alloy 6061-T6 plate on a steel plate at 350 m/s. the plot is made for the time $t = 2\ \mu s$.

10.1.3 Decompression Shocks

Let us now consider the case in which the applied stress producing the waveform of Fig. 10.1b is suddenly removed from the boundary of the slab after a time interval t^*. In obtaining a solution to this problem, three amplitude ranges must be considered: i) $-t_{11}^{(2)} \le t_{11}^{\mathrm{HEL}}$, ii) $t_{11}^{\mathrm{HEL}} < -t_{11}^{(2)} \le 2\,t_{11}^{\mathrm{HEL}}$, iii) $2\,t_{11}^{\mathrm{HEL}} < -t_{11}^{(2)}$. The first range involves only linear elastic response. The propagating disturbance takes the form of a non-attenuating flat-topped pulse advancing at the velocity C_0. We shall not discuss this case further. In the second range decompression to $t_{11} = 0$ occurs by means of a single, elastic decompression shock. In the third range the decompression shock is unstable, separating into elastic and plastic shocks. Discontinuous loading and unloading problems can all be solved using the shock jump equations.

Elastic Decompression. We begin our consideration of this low-stress case of the attenuation problem by determining the state behind a decompression shock propagating into compressed material in the state discussed in Sect. 10.1.1 The case is defined by the requirement that the peak stress in the compressed state (called state 2) is restricted to the range $t_{11}^{\text{HEL}} < -t_{11}^{(2)} \leq 2t_{11}^{\text{HEL}}$. Examination of the Hugoniot curves for this case, as given in Fig. 10.5a,b, shows that decompression to zero longitudinal stress can be accomplished by a single elastic shock taking the material to a state 4′ that is a function of state 2. Application of the jump equations 10.1 to this shock gives the particle velocity, strain and transverse stress in the decompressed state (state 4′) as

$$v_1^{(4')} = -\frac{C_0 - C_B}{\rho_R C_0 C_B}\left[t_{11}^{(2)} + t_{11}^{\text{HEL}}\right]$$

$$\widetilde{E}_{11}^{(4')} = \frac{C_0^2 - C_B^2}{\rho_R C_0^2 C_B^2}\left[t_{11}^{(2)} + t_{11}^{\text{HEL}}\right] \tag{10.15}$$

$$t_{22}^{(4')} = \frac{3(C_0^2 - C_B^2)}{2\rho_R C_B^2}\left[t_{11}^{(2)} + t_{11}^{\text{HEL}}\right].$$

Examination of the X–t diagram of Fig. 10.5c indicates that the elastic decompression wave will overtake the plastic part of the compression wave at

$$t_a = \frac{C_0}{C_0 - C_B}t^*, \quad x_a = \frac{C_0 C_B}{C_0 - C_B}t^*, \tag{10.16}$$

where t^* is the duration of load application. The waveform produced in this case is shown in the first column of Fig. 10.7.

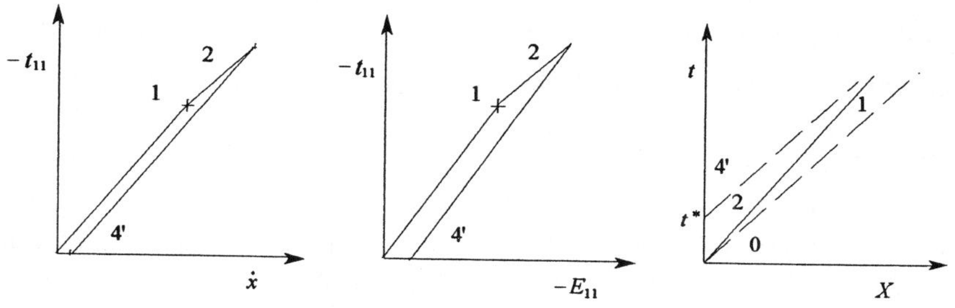

Figure 10.5. Hugoniot curves and an X–t diagram for the case in which the peak stress satisfies $t^{\text{HEL}} < -t_{11}^{(2)} < 2t^{\text{HEL}}$. In this and all following X–t diagrams the solid lines designate plastic waves and the broken lines designate elastic waves.

Elastic–Plastic Decompression. Stress–strain paths for unloading from a state $t_{11}^{(2)}$–$\widetilde{E}_{11}^{(2)}$ have been given in Fig. 7.4. The associated t_{11}–$\dot{x}$ Hugoniot of Eqs. 10.11 and 10.12 is plotted in Fig. 10.2b, along with an unloading Hugoniot for right-propagating waves, which can be found using the jump equations as above, or by graphical means. The residual strain, particle velocity, and transverse component of stress (corresponding to state 4 on the figures) are given by

$$\widetilde{E}_{11}^{(4)} = -\widetilde{E}_{11}^{\mathrm{R}} = -\frac{2Y}{3B_{\mathrm{R}}}$$

$$\dot{x}^{(4)} = \dot{x}^{\mathrm{R}} = \frac{C_0 - C_{\mathrm{B}}}{\rho_{\mathrm{R}}\, C_0\, C_{\mathrm{B}}}\, t_{11}^{\mathrm{HEL}} \qquad (10.17)$$

$$t_{22}^{(4)} = -\, t_{22}^{\mathrm{R}} = -\,Y.$$

Several other expressions for these quantities are

$$\widetilde{E}_{11}^{\mathrm{R}} = \left(\frac{C_0^2 - C_{\mathrm{B}}^2}{\rho_{\mathrm{R}}\, C_0^2\, C_{\mathrm{B}}^2}\right) t_{11}^{\mathrm{HEL}} = \left(\frac{C_0^2 - C_{\mathrm{B}}^2}{C_0\, C_{\mathrm{B}}^2}\right) \dot{x}^{\mathrm{HEL}} = \left(\frac{C_0^2 - C_{\mathrm{B}}^2}{C_{\mathrm{B}}^2}\right) \widetilde{E}_{11}^{\mathrm{HEL}}$$

$$(10.18)$$

$$\dot{x}^{\mathrm{R}} = \left(\frac{C_0 - C_{\mathrm{B}}}{\rho_{\mathrm{R}}\, C_0\, C_{\mathrm{B}}}\right) t_{11}^{\mathrm{HEL}} = \left(\frac{C_0 - C_{\mathrm{B}}}{C_{\mathrm{B}}}\right) \dot{x}^{\mathrm{HEL}} = \frac{C_0}{C_{\mathrm{B}}}\,(C_0 - C_{\mathrm{B}})\, \widetilde{E}_{11}^{\mathrm{HEL}}.$$

Note that, in contrast to the previous case of elastic decompression in which the decompressed state varied with varying peak stress, the state 4 obtained upon release of material compressed into the plastic region can be completely described in terms of material properties and is independent of peak stress.

Let us suppose that a stress $-t_{11}^{(2)} > 2t_{11}^{\mathrm{HEL}}$ is applied at the boundary at the time $t = 0$ and removed at the time $t = t^*$. The $t_{11} - \dot{x}$ Hugoniot and the X–t diagram for this problem are given in Fig. 10.6.

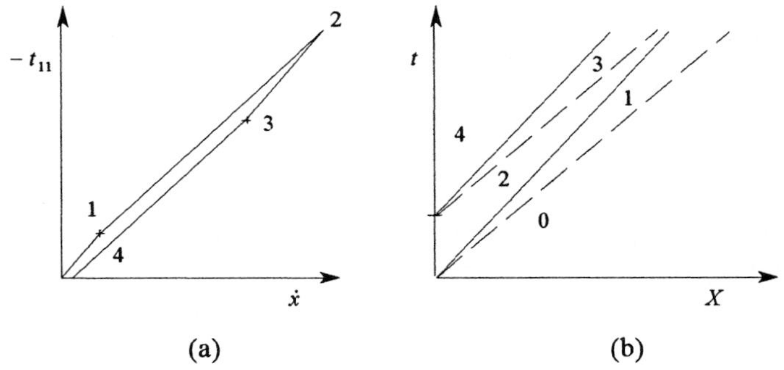

Figure 10.6. Stress–particle-velocity Hugoniot and X–t diagram for a compression-decompression process. Hugoniot states and the regions of X–t space in which they prevail are matched by the numbers.

Two examples of the waveform as it evolves in time are shown in Figs. 10.7b,c. It is apparent from the X–t diagram that the first step of the unloading wave will overtake the second step of the loading wave at the time and place given by Eqs. 10.16.

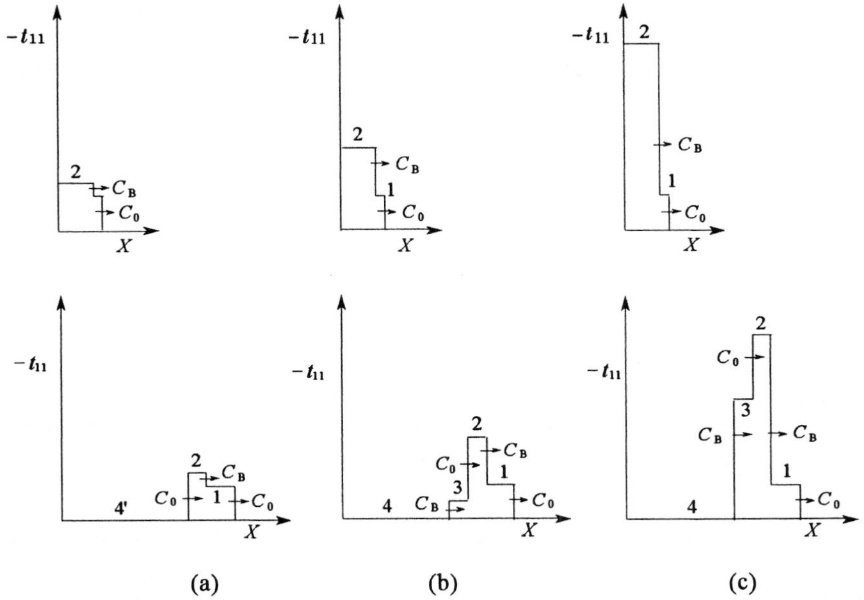

Figure 10.7. Decompression wave overtaking a compression wave. The waves are of amplitude (a) $t_{11}^{\mathrm{HEL}} < -t_{11}^{(2)} < 2t_{11}^{\mathrm{HEL}}$ (–2.5 GPa), (b) $2t_{11}^{\mathrm{HEL}} < -t_{11}^{(2)} < 3t_{11}^{\mathrm{HEL}}$ (–4.5 GPa), and (c) $-t_{11}^{(2)} > 3t_{11}^{\mathrm{HEL}}$(– 10 GPa) propagating in steel. Waveforms in the top row correspond to a time of $t^* = 1\,\mu\mathrm{s}$ and waveforms in the lower row are plotted for $t = 3\ \mu\mathrm{s}$.

The field values in region 1 of Fig. 10.6b are given by Eqs. 10.5 and 10.6, and in region 2, we have fields given by Eqs. 10.9 or 10.10. In region 3 the fields are determined by application of Eqs. 10.1 to a right-propagating elastic wave ($U_S = +C_0$) taking material from the state $t_{11}^{(2)} - \dot{x}^{(2)}$ to a state in which the stress is $t_{11}^{(2)} + 2t_{11}^{\mathrm{HEL}}$. The result is

$$t_{11}^{(3)} = t_{11}^{(2)} + 2t_{11}^{\mathrm{HEL}}$$

$$\dot{x}^{(3)} = \dot{x}^{(2)} - 2\dot{x}^{\mathrm{HEL}} = -\frac{1}{\rho_R C_B}\left[t_{11}^{(2)} + \left(\frac{C_0 + C_B}{C_0}\right)t_{11}^{\mathrm{HEL}}\right] \tag{10.19}$$

$$\tilde{E}_{11}^{(3)} = \tilde{E}_{11}^{(2)} + 2\tilde{E}_{11}^{\mathrm{HEL}} = \frac{1}{\rho_R C_B^2}\left[t_{11}^{(2)} + \left(\frac{C_0^2 + C_B^2}{C_0^2}\right)t_{11}^{\mathrm{HEL}}\right].$$

In writing these results, we have expressed the strength of the loading disturbance in terms of its stress amplitude $t_{11}^{(2)}$, and have expressed all of the measures of the HEL in terms of t_{11}^{HEL}. One could, of course, express the same results in terms of other measures of the applied load and material yield point.

It is important to note that the yield condition, which is always given by Eq. 7.39, changes from the expression $t_{11} - t_{22} = -Y$ used for the compression processes to the expression $t_{11} - t_{22} = +Y$ for the decompression processes. Using this yield condition and the previous results, we obtain the transverse stress in state 3 as

$$t_{22}^{(3)} = t_{22}^{(2)} + 2t_{22}^{\mathrm{HEL}} = t_{11}^{(3)} - Y = t_{11}^{(2)} + \frac{C_0^2 + 3C_{\mathrm{B}}^2}{2C_0^2} t_{11}^{\mathrm{HEL}}. \tag{10.20}$$

10.1.4 Reflection from an Immovable Boundary

Analysis of the reflection of an elastic–plastic compression wave from an immovable boundary proceeds in the same way as calculation of other shock interactions. It is only necessary to think of the boundary as an interface between the elastic–plastic material and an incompressible body. The $t_{11}-\dot{x}$ Hugoniot of the latter material is a vertical line through $\dot{x} = 0$: no amount of stress can impart a velocity to this material.

For times prior to the encounter of the elastic precursor with the boundary at $X = X_{\mathrm{L}}$, the solution is as given in Sect. 10.1.1 The precursor reflects from the boundary at the time $t = X_{\mathrm{L}}/C_0$. The new state behind the reflected wave (call it state 3) lies at the intersection of the Hugoniot for a left-propagating shock centered on state 1 and the line $\dot{x} = 0$ in the $t_{11}-\dot{x}$ plane. Since the transition from state 1 (a forward-yield state) to state 3 is compressive, the reflected shock is a plastic transition. Application of the jump equations to this shock gives

$$\begin{aligned} t_{11}^{(3)} &= -\frac{C_0 + C_{\mathrm{B}}}{C_0} t_{11}^{\mathrm{HEL}} \\ \tilde{E}_{11}^{(3)} &= -\frac{C_0 + C_{\mathrm{B}}}{\rho_{\mathrm{R}} C_0^2 C_{\mathrm{B}}} t_{11}^{\mathrm{HEL}}, \end{aligned} \tag{10.21}$$

since $\dot{x}^{(3)} = 0$.

The next wave interaction is the encounter of this reflected plastic shock with the incident plastic shock. The state produced by this interaction (let us call it state 4) involves further compression so both the transition from state 2 to state 4 and the transition from state 3 to state 4 are plastic shocks, as shown in Fig. 10.8. Application of the jump equations to each of these shocks gives

$$t_{11}^{(4)} = t_{11}^{(2)} - \frac{C_B}{C_0} t_{11}^{\mathrm{HEL}}$$

$$\dot{x}^{(4)} = \frac{1}{\rho_R C_B}\left[-t_{11}^{(2)} - t_{11}^{\mathrm{HEL}}\right] \tag{10.22}$$

$$\widetilde{E}_{11}^{(4)} = \frac{1}{\rho_R C_B^2}\left[t_{11}^{(2)} + \frac{C_0^2 - C_0 C_B - C_B^2}{C_0} t_{11}^{\mathrm{HEL}}\right].$$

The final interaction that we shall consider is that of the main plastic compression shock with the boundary. The result of this reflection is a further compression, with formation of a left-propagating shock transition from state 4 to a new state (which we shall call state 5). Because this transition is a further compression, it is a plastic shock. The analysis of state 5 is routine and gives the results:

$$\dot{x}^{(5)} = 0$$

$$t_{11}^{(5)} = 2t_{11}^{(2)} + \frac{C_0 - C_B}{C_0} t_{11}^{\mathrm{HEL}} \tag{10.23}$$

$$\widetilde{E}_{11}^{(5)} = \frac{1}{\rho_R C_B^2}\left[-2t_{11}^{(2)} + \frac{C_B (C_0 - C_B)}{C_0^2} t_{11}^{\mathrm{HEL}}\right].$$

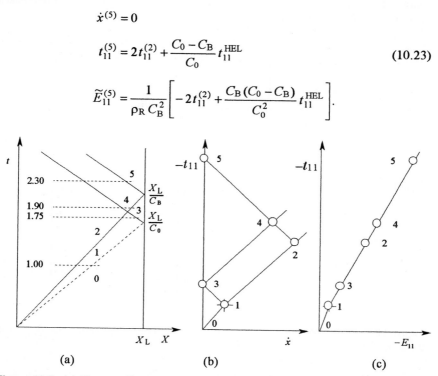

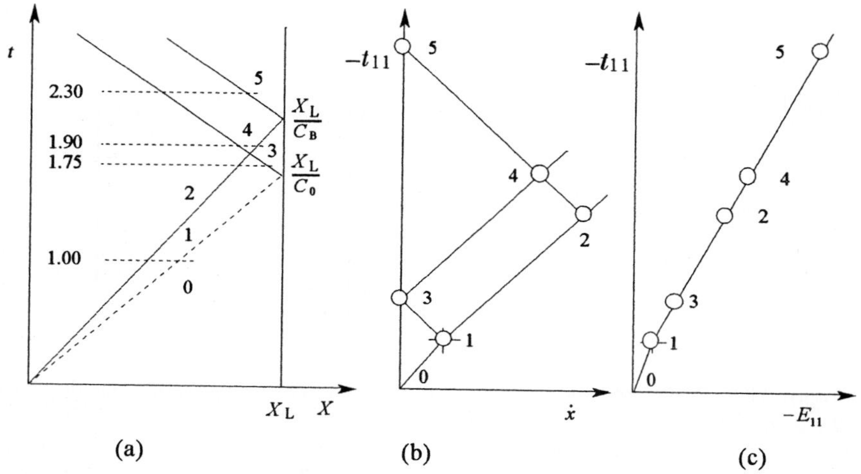

Figure 10.8. (a) The x–t diagram, (b) t_{11}–$\dot{x}$ Hugoniot plot, and (c) t_{11}–$\widetilde{E}_{11}$ Hugoniot plot for shock reflection from an immovable boundary. The drawings are made for an incident wave of $t_{11} = -6.25$ GPa amplitude propagating in steel, and the propagation distance to the reflecting boundary is 10 mm. The numbers marked along the ordinate in (a) are times in μs for which waveforms are plotted in Fig. 10.9.

A sequence of waveforms is plotted for a specific example problem in Fig. 10.9. Further interactions can be expected when the left-propagating waves reach the boundary at $X = 0$ but we have not specified the nature of this boundary and shall not address the issue here.

In Chap. 8 we found that an elastic wave described by the linear theory is reflected from an immovable boundary with doubled amplitude. In Chap. 9 we found that, when the nonlinear theory is used, the amplitude of a reflected elastic wave is more than twice that of the incident wave (for normal materials). In the present case the elastic precursor is reflected as a plastic shock rather than as an elastic shock and its amplitude, $-t_{11}^{(3)}$, is less than twice that of the incident precursor (for steel it is 1.8 times that of the precursor). The longitudinal compressive stress at the boundary after the reflection process is completed, $-t_{11}^{(5)}$, is less than twice the amplitude, $-t_{11}^{(2)}$, of the incident wave. In the case of steel, the error in assuming the stress doubles upon reflection is about 10 % when the amplitude of the incident wave is near the Hugoniot elastic limit, and decreases with increasing strength of the wave.

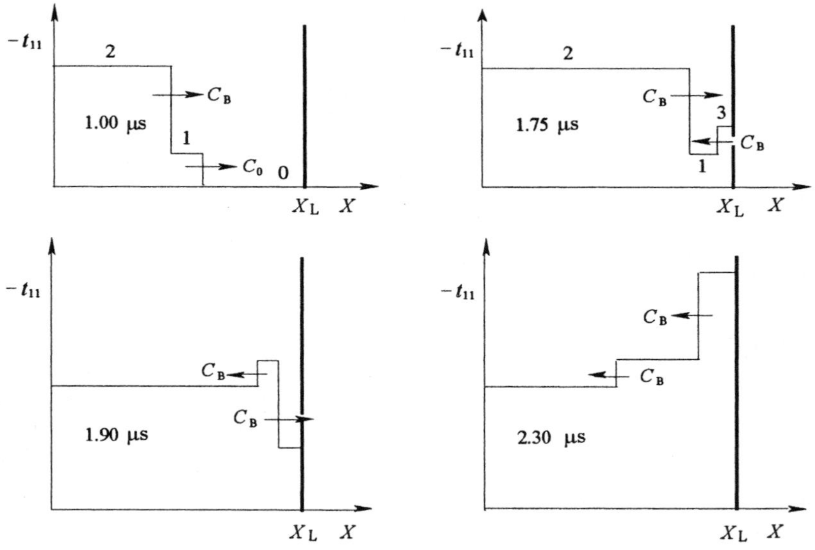

Figure 10.9. Waveforms for the example of Fig. 10.8. The heavy vertical line designates the immovable boundary.

10.1.5 Reflection From an Unrestrained Boundary

Analysis of the reflection of an elastic–plastic compression wave from an unrestrained (also called "stress-free") boundary proceeds in the same way as calculation of other shock interactions. It is only necessary to think of the boundary as an interface between the elastic–plastic material and a solid that does not resist compression. The $t_{11}-\dot{x}$ Hugoniot of the latter material is a horizontal line through $t_{11}=0$: no stress can be produced in this material.

For times prior to the encounter of the elastic precursor with the boundary at $X=X_L$, the solution is as given in Sect. 10.1.1 In contrast to the preceding analysis of reflection from an immovable boundary, it is not possible to present a solution for reflection from an unrestrained boundary that is valid for an incident elastic–plastic wave of any strength. The wave interactions arising during the reflection process change qualitatively as the strength of the incident wave under consideration passes from one range of similar responses to another. We shall discuss the response in four stress ranges. The range of low stresses, $t_{11} \le t_{11}^{\mathrm{HEL}}$, is that of elastic response discussed in Chap. 8. The response in two intermediate ranges, which we shall call the low– and high–intermediate stress ranges is also discussed. Finally, we shall comment upon response in the range $t_{11} > 2t_{11}^{\mathrm{HEL}}$ of high stress.

10.1.5.1 Reflection in the Low–intermediate Stress Range

The low-intermediate stress range is defined by the inequalities

$$t_{11}^{\mathrm{HEL}} < -t_{11}^{(2)} \le \frac{C_0+3C_B}{C_0+C_B}\,t_{11}^{\mathrm{HEL}}\,. \tag{10.24}$$

As the analysis proceeds we shall see how the upper limit to the range arises. The X–t diagram for this case is given in Fig. 10.10 and the $t_{11}-\dot{x}$ and $t_{11}-\widetilde{E}_{11}$ Hugoniot diagrams describing the interactions are given in Fig. 10.11. Although the complete diagrams are given in these figures, it is important to realize that they must be drawn in steps as the interactions are analyzed.

The first wave interaction occurs when the precursor reflects from the boundary at the time X_L/C_0. The state behind the reflected wave (call it state 3) lies at the intersection of the Hugoniot for a left-propagating shock centered on state 1 and the line $t_{11}=0$ in the $t_{11}-\dot{x}$ plane. Since the transition from state 1 (a forward-yield state) to state 3 is a decompression, the reflected shock is an elastic transition to zero longitudinal stress. Application of the jump equations to this case gives

$$t_{11}^{(3)}=0, \quad \dot{x}^{(3)}=2\dot{x}^{\mathrm{HEL}}, \quad \widetilde{E}_{11}^{(3)}=0\,. \tag{10.25}$$

The next wave interaction is the encounter of this reflected elastic shock with the incident plastic shock. This interaction produces a state that we have designated state 4. This state is represented on the diagrams of Fig. 10.11. It is defined by the intersection point of the Hugoniot for left-propagating waves centered on state 2 and right-propagating waves centered on state 3. As discussed previously, state 4 lies at this intersection because this is the only state satisfying the jump equations, i.e., lying on the Hugoniots in question, and leading to the required continuity of the fields t_{11} and $\dot{x}$ in the region. It is important to recognize, however, that there is no requirement that $\widetilde{E}_{11}$, $\widetilde{E}_{22}$, or t_{22} be continuous in this region. In the present case these latter fields are not continuous, exhibiting a non-propagating discontinuity at the material surface where the interaction occurs. This discontinuity is called a *contact discontinuity* and the surface a *contact surface* or *contact interface*. Let us analyze the interaction by application of the jump equations 10.1. Application of these equations to the right-propagating wave centered on state 3 gives

$$C_0\left(-\widetilde{E}_{11}^{(4+)}+\widetilde{E}_{11}^{(3)}\right)=\dot{x}^{(4)}-\dot{x}^{(3)}$$
$$\rho_R C_0\left(\dot{x}^{(4)}-\dot{x}^{(3)}\right)=-t_{11}^{(4)}+t_{11}^{(3)}, \tag{10.26}$$

and a similar application to the left-propagating wave centered on state 2 gives

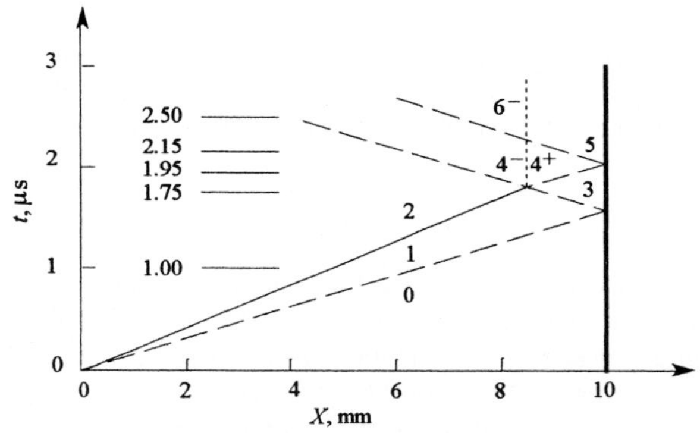

Figure 10.10. The X–t diagram for reflection of a shock of strength in the low–intermediate stress range from an unrestrained boundary.

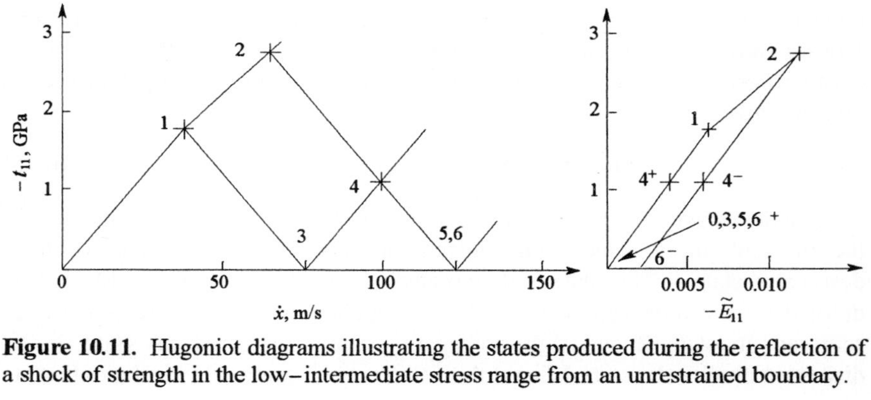

Figure 10.11. Hugoniot diagrams illustrating the states produced during the reflection of a shock of strength in the low–intermediate stress range from an unrestrained boundary.

$$-C_0\left(-\widetilde{E}_{11}^{(4-)}+\widetilde{E}_{11}^{(2)}\right)=\dot{x}^{(4)}-\dot{x}^{(2)}$$
$$-\rho_R C_0\left(\dot{x}^{(4)}-\dot{x}^{(2)}\right)=-t_{11}^{(4)}+t_{11}^{(2)}. \tag{10.27}$$

In writing these equations we have allowed for the possibility that $\widetilde{E}_{11}$ is discontinuous and have recognized from the Hugoniots that the two shocks produced by the interaction will both propagate at the elastic wavespeed (but in opposite directions).

The values of the field variables in state 2 are known, being given by Eqs. 10.10, and the values in state 3 are given by Eqs. 10.25. When these values are substituted into Eqs. 10.26 and 10.27 and the latter solved we obtain the result

$$t_{11}^{(4)}=\frac{C_0+C_B}{2C_B}\left(t_{11}^{(2)}+t_{11}^{\mathrm{HEL}}\right)$$
$$\dot{x}^{(4)}=\frac{1}{2\rho_R C_0 C_B}\left(-(C_0+C_B)t_{11}^{(2)}+(3C_B-C_0)t_{11}^{\mathrm{HEL}}\right)$$
$$\widetilde{E}_{11}^{(4+)}=\frac{C_0+C_B}{2\rho_R C_0^2 C_B}\left(t_{11}^{(2)}+t_{11}^{\mathrm{HEL}}\right) \tag{10.28}$$
$$\widetilde{E}_{11}^{(4-)}=\frac{2C_0^2+C_0C_B-C_B^2}{2\rho_R C_0^2 C_B^2}\left(t_{11}^{(2)}+t_{11}^{\mathrm{HEL}}\right).$$

It is important to note that state 4 lies below the elastic limit on the Hugoniot through state 3. A different and more complicated solution arises when yielding occurs in the transition from state 3 to state 4, and it is the need to separate these two cases that establishes the upper limit on the low-intermediate stress range.

Region 5 arises as a result of reflection of the right-propagating transition from region 3 to region 4$^+$ from the unrestrained surface. Because $t_{11}^{(5)}=0$ on this surface, examination of the Hugoniot curves of Fig. 10.11, shows that $\tilde{E}_{11}^{(5)}=0$ as well, and that the transition propagates at the elastic wavespeed. Accordingly, application of the jump equations to this shock gives

$$\dot{x}^{(5)}=\frac{-1}{\rho_R C_0 C_B}\left((C_0+C_B)t_{11}^{(2)}+(C_0-C_B)t_{11}^{\mathrm{HEL}}\right). \tag{10.29}$$

When this decompression encounters the contact surface formed at X_a another interaction producing a region, which we have designated region 6, occurs. Analysis of this region gives

$$t_{11}^{(6)}=0,\quad \dot{x}^{(6)}=\dot{x}^{(5)},\quad \widetilde{E}_{11}^{(6+)}=0$$
$$E_{11}^{(6-)}=\frac{C_0^2-C_B^2}{\rho_R C_0^2 C_B^2}\left(t_{11}^{(2)}-t_{11}^{\mathrm{HEL}}\right), \tag{10.30}$$

and we see that region 6^+ is really just a continuation of region 5; no discontinuity is formed. Since there are no waves in the domain that could produce further interactions, analysis of the reflection is now complete. Plots of the waveform at various times are given in Fig. 10.12.

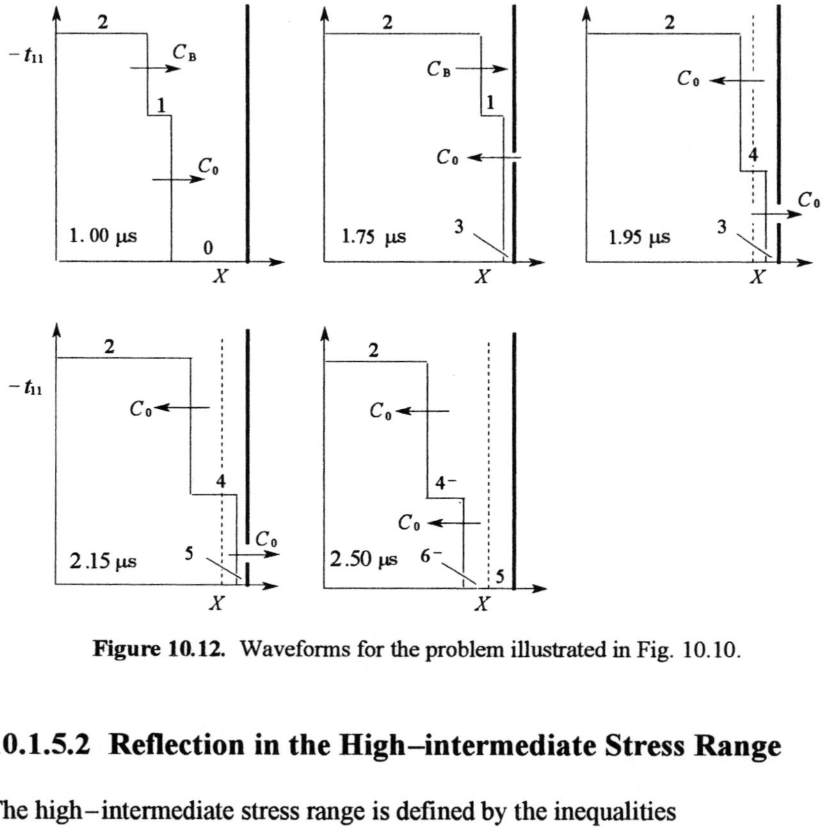

Figure 10.12. Waveforms for the problem illustrated in Fig. 10.10.

10.1.5.2 Reflection in the High–intermediate Stress Range

The high–intermediate stress range is defined by the inequalities

$$\frac{C_0 + 3C_B}{C_0 + C_B} t_{11}^{\mathrm{HEL}} < -t_{11}^{(2)} \le 2t_{11}^{\mathrm{HEL}} . \tag{10.31}$$

This is a very narrow range in which the Hugoniot segment from state 3 to state 4 of the previous example crosses a yield point, thus causing a splitting of the associated transition into an elastic and a plastic shock. The upper limit of the range is set by the condition that the decompression from state 9 (see Fig. 10.13) to a state of zero stress does not involve reverse yielding.

We shall illustrate this case by an X–t diagram, the usual Hugoniot diagrams, and some illustrations of the waveform. The essential difference between this case and that of the low–intermediate stress range is the wave splitting mentioned and the creation of a second contact surface associated with an interaction of the plastic part of the split wave.

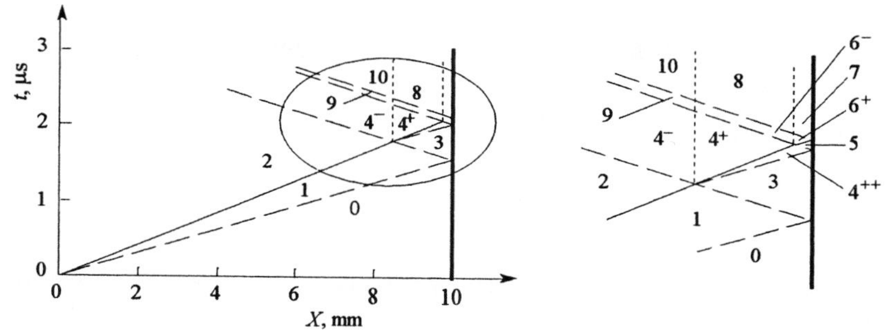

Figure 10.13. The X–t diagram for reflection of a compression wave in the high–intermediate stress range from an unrestrained surface of a 10-mm thick slab. The region in the ellipse is shown enlarged at the right. This figure is scaled for steel.

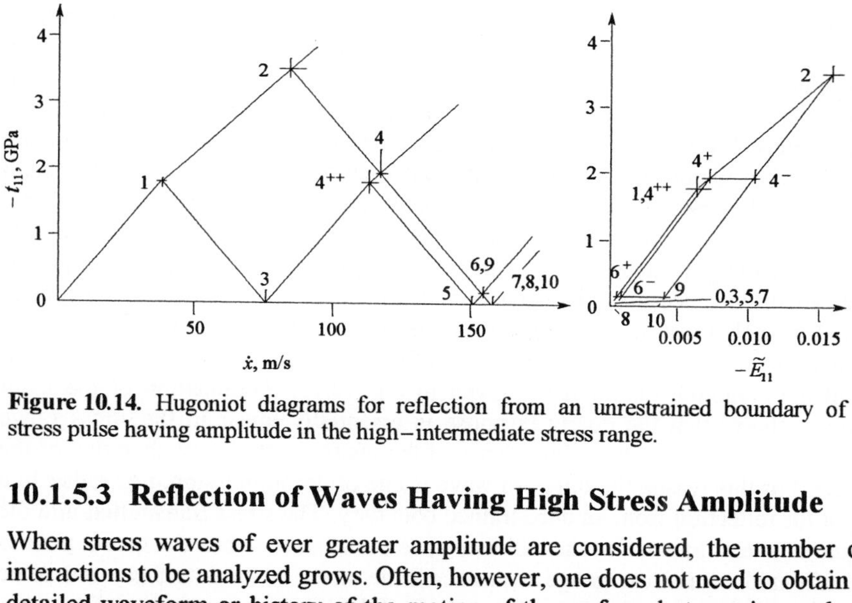

Figure 10.14. Hugoniot diagrams for reflection from an unrestrained boundary of a stress pulse having amplitude in the high–intermediate stress range.

10.1.5.3 Reflection of Waves Having High Stress Amplitude

When stress waves of ever greater amplitude are considered, the number of interactions to be analyzed grows. Often, however, one does not need to obtain a detailed waveform or history of the motion of the surface, but requires only a determination of the final velocity achieved. This can be done quite easily; the free-surface velocity approaches, through a complicated process, the same value that would be realized if decompression simply took place on the Hugoniot for left-propagating waves centered on state 2. The result is easily seen to be

$$\dot{x}^{(\mathrm{fs})} = 2\,\dot{x}^{(2)} - \frac{C_0 - C_{\mathrm{B}}}{C_{\mathrm{B}}}\,\dot{x}^{\mathrm{HEL}}\,. \tag{10.32}$$

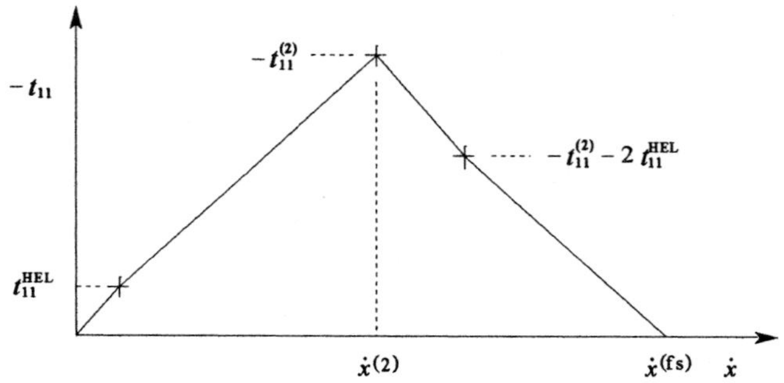

Figure 10.15 Simplified $t_{11} - \dot{x}$ Hugoniot for reflection of a compression wave of high amplitude.

10.1.6 Interaction with a Material Interface

We have been considering reflection of a compressive elastic–plastic wave from a boundary that is either completely fixed or completely free to move. Other cases of interest arise when the wave reflects from an interface with another material. In these cases both reflected and transmitted waves are produced by the interaction. Analysis of these problems proceeds much as has already been discussed. The interfacial state lies at the intersection of the appropriate t_{11}–$\dot{x}$ Hugoniots for the materials in question. Because numerous parameters are involved, it is not practical to classify and analyze all possible cases. We shall consider the special case in which a wave passing through an elastic–plastic slab encounters an interface with an X-cut slab of crystalline quartz ($\rho_R = 2649\ \text{kg/m}^3$ and $C_0 = 5749\ \text{m/s}$ for the quartz). This problem is of practical importance because the quartz crystal, which is piezoelectric, can be configured in such a way that it produces an electrical signal that is a measure of the stress at the interface. The X–t diagram for this problem is given in Fig. 10.16 and the Hugoniot diagrams are given in Fig. 10.17. Examination of the latter figure shows that the quartz response is essentially linear in the stress range plotted, and that the shock impedance of the quartz is much less than that of the steel. For this reason the reflected wave is one of decompression, i.e., rather like that for reflection from an unrestrained boundary. The stress transmitted into the quartz is less than that in the incident wave in the steel. When the quartz plate is configured as a stress gauge it measures the stress at the interface, and some calculation is required to infer the strength of the incident wave.

10.1.7 Impact of Plates of Finite Thickness

When the impacting plates just considered are of finite thickness, further wave interactions occur. In many cases, these interactions result in production of tensile stress in the interior of the thicker plate. The analysis of elastic–plastic waves generated by the impact of plates of finite thickness provides the oppor-

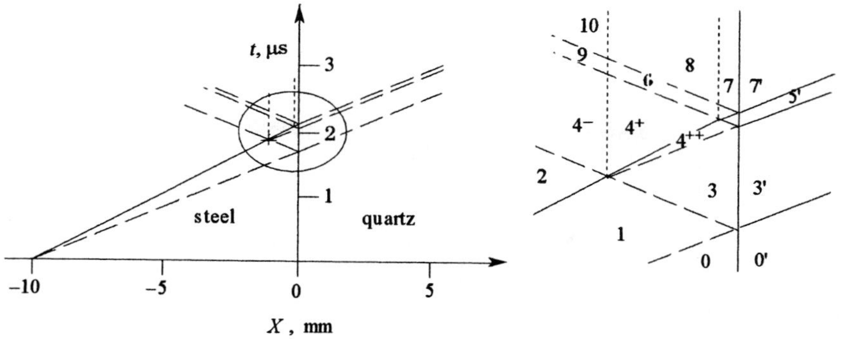

Figure 10.16. The X–t diagram for transmitted and reflected waves produced when an elastic–plastic disturbance encounters an interface with an elastic material. The drawing at the right is an enlarged view of the region in the circle at the left. The scale is for steel and X-cut quartz.

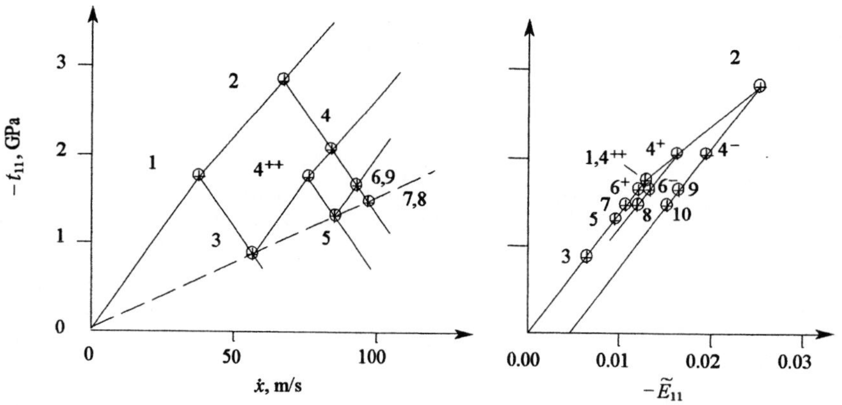

Figure 10.17. Hugoniot curves showing the encounter of an elastic–plastic wave with an interface with a lower impedance elastic material. The broken line in the stress–particle-velocity diagram is the quartz Hugoniot; all of the other lines are for the steel.

tunity to discuss tensile waves and is of considerable practical importance because of the possibility that fracture will occur.

In this section we present the calculation of a specific example that illustrates the phenomena that can be expected to occur. We consider the impact of a 5-mm thick steel plate on a 10-mm thick plate of the same material at a velocity of 230 m/s. The calculation proceeds as outlined in the preceding sections, so it will be sufficient to summarize the results. The X–t, t_{11}–$\dot{x}$, and t_{11}–$\widetilde{E}_{11}$ diagrams for the problem are presented in Figs. 10.19, 10.20, and 10.21, respectively.

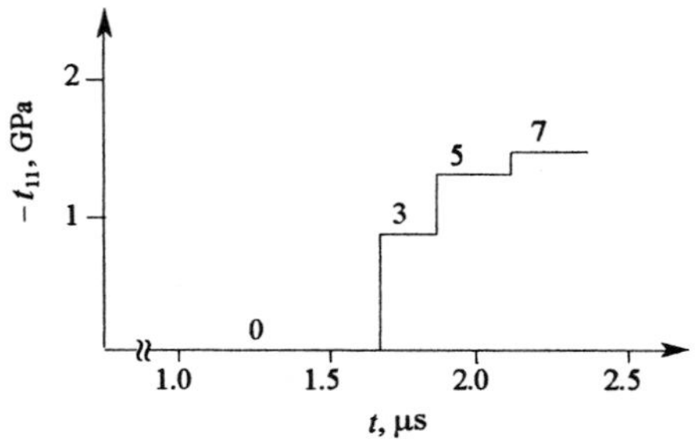

Figure 10.18. The interfacial stress history resulting from the interaction. This corresponds to the record that would be obtained if the quartz plate were configured as a stress gauge. Numbers on the curve correspond to states on the Hugoniot diagrams.

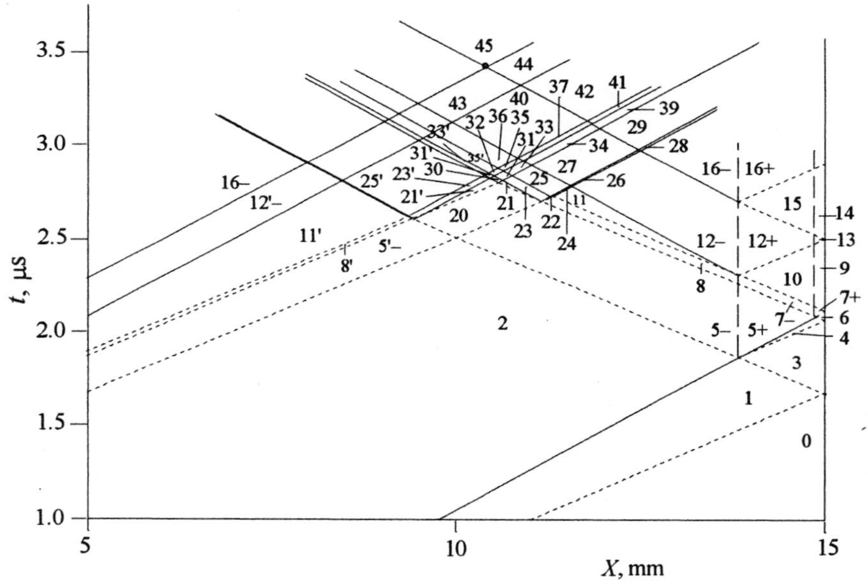

Figure 10.19. Space–time diagram showing the wave interactions that result from the impact of a 5-mm thick steel plate on a 10-mm thick plate at the velocity 230 m/s. The interactions that occur at the left and right stress-free boundaries are reflected images of each other, so only one need be analyzed. Calculation of the interaction at the boundary is terminated when the stress decreases to a value that is zero, to within the accuracy of the calculation.

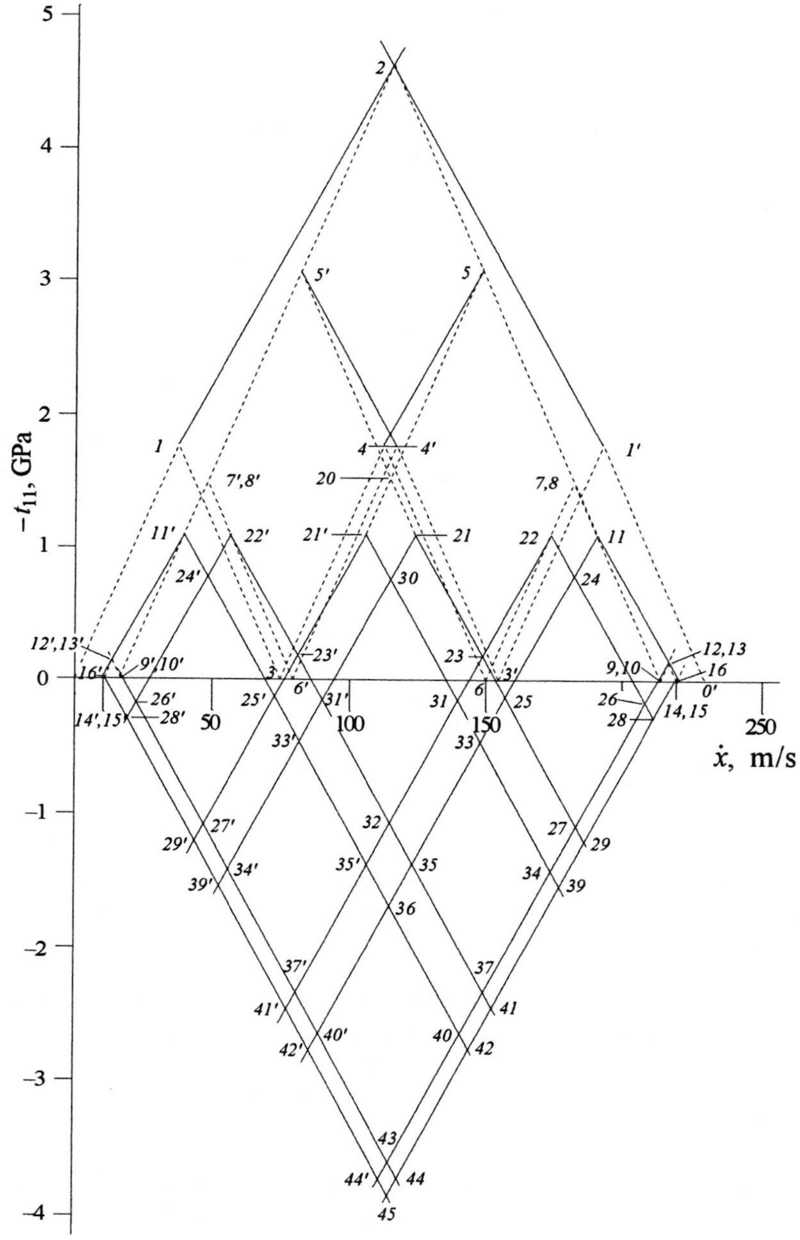

Figure 10.20. Stress–particle-velocity diagram of the wave interactions occurring in the problem described above. Numbers on the curves correspond to X–t regions designated in Fig. 10.19.

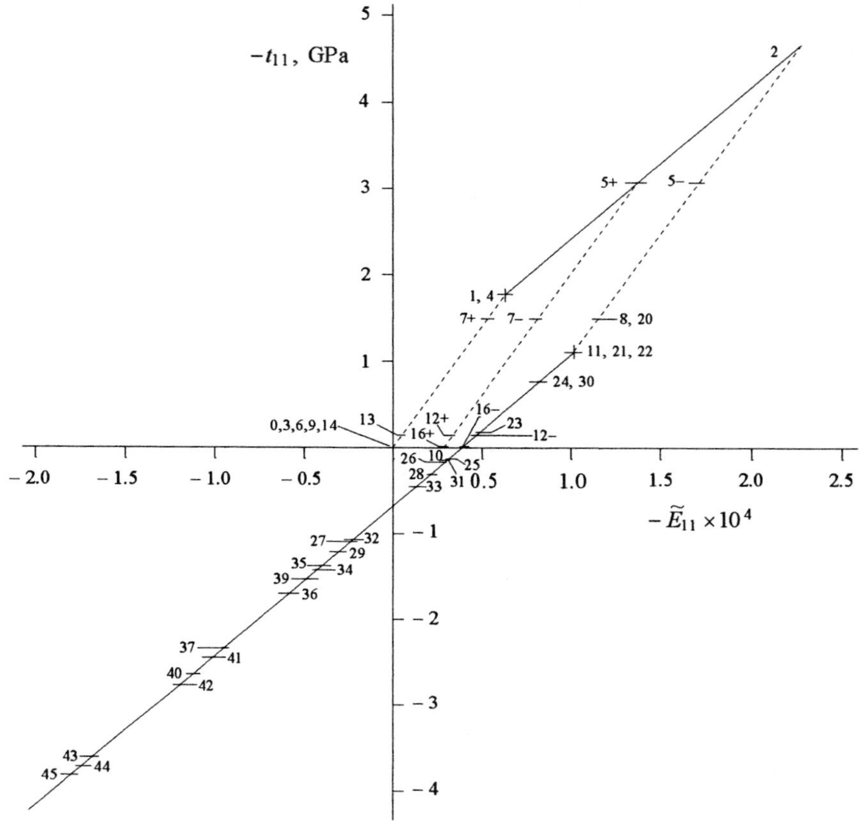

Figure 10.21. Stress–strain diagram of the wave interactions occurring in the problem described above. Numbers on the curves correspond to X–t regions designated in Fig. 10.19.

The particular point of this problem is that the material of the target plate goes into tension as a result of the interaction of the decompression waves arising when the compression waves are reflected from the unrestrained boundaries of the two plates. As shown in Fig. 10.22, the tensile stress develops incrementally as the interaction evolves. This is in contrast to the case discussed in Chap. 3, where the tension was produced instantaneously by collision of decompression shocks and the case discussed in Chap. 9, in which the transition from compression to tension proceeded smoothly in the interaction zone of two smooth decompression waves. The analysis given here is based upon the assumption that the maximum tension that arises in the target plate is insufficient to cause fracture, but fracture often results from wave interactions such as those discussed, and this possibility is addressed in Chap. 12.

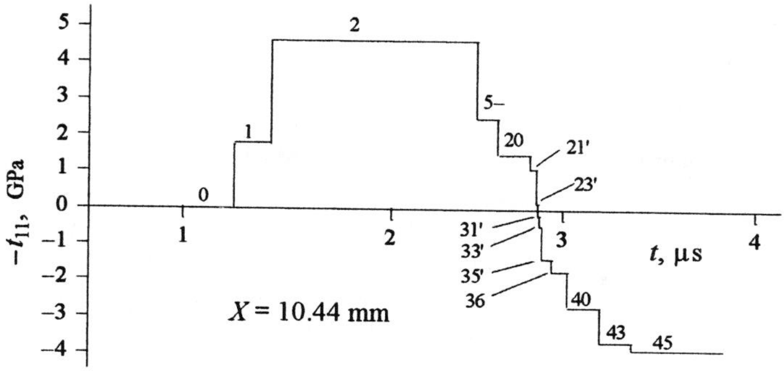

Figure 10.22. Stress history at the plane $X = 10.44$ mm (the plane at which the maximum tensile stress first appears. Numbers on the curves correspond to X–t regions designated in Fig. 10.19.

10.1.8 Pulse Attenuation

Attenuation of an elastic–plastic pulse such as we have discussed in Sect. 10.3 occurs in discrete steps when elastic decompression shocks overtake plastic shocks. Solution of the overtake problem is complicated by the need for sequential analysis of numerous shock interactions, with analysis of a given case being concluded only when it has been carried to a time after which no further interactions are possible. Elastic–plastic pulses do not attenuate to zero strength, but only to the amplitude of the elastic precursor.

The details of the shock interactions change as $t_{11}^{(2)}$ occupies various ranges that depend on t_{11}^{HEL} and the wavespeeds. The lowest-stress range, $0 < -t_{11}^{(2)} \le t_{11}^{\mathrm{HEL}}$, is one of elastic response. Since no attenuation occurs in this range we shall ignore it. In the next-higher range, $t_{11}^{\mathrm{HEL}} < -t_{11}^{(2)} \le 2t_{11}^{\mathrm{HEL}}$, complete decompression is achieved by a single elastic shock and a simple, direct analysis applies throughout the range. When the peak compressive stress exceeds $2\,t_{11}^{\mathrm{HEL}}$ the decompression process involves both elastic and plastic transitions and presents a sequence of cases distinguished by qualitative changes in the shock interactions that occur. The cases are defined in terms of a range of the peak stress, $-t_{11}^{(2)}$, throughout which the interactions are qualitatively the same so that a single analytical solution is valid. Pulses in the stress range

$$\frac{C_0 + 3C_{\mathrm{B}}}{2C_0}\, t_{11}^{\mathrm{HEL}} < -t_{11}^{(2)} \le \frac{C_0 + 5C_{\mathrm{B}}}{C_0 + C_{\mathrm{B}}}\, t_{11}^{\mathrm{HEL}} \qquad (10.33)$$

(−4.91 to −3.55 GPa in steel) are analyzed in the following paragraphs. Other cases are discussed in [30]. The analysis is accompanied by quantitative calculation of field values, waveforms, etc. for a pulse of 1 μs duration introduced into the steel described earlier. The X–t diagram for this problem is given in Fig. 10.23, and the solution in regions 1–4 is as given in Sects. 10.2 and 10.3.

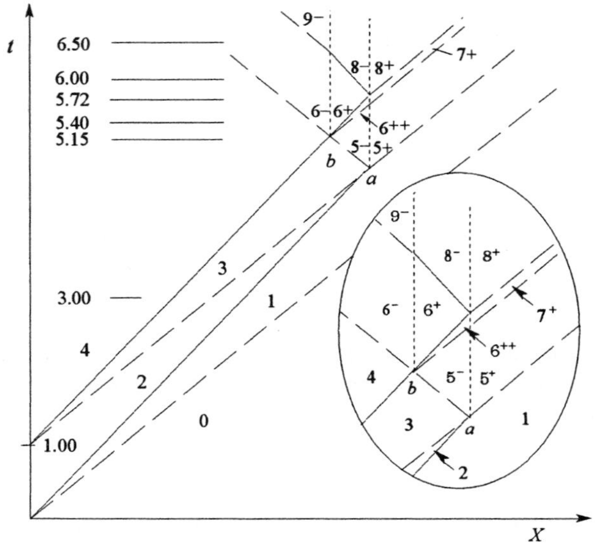

Figure 10.23. Space–time diagram for pulse propagation in stress range D. The scale is for steel, the solid lines designate plastic waves and the broken lines designate elastic waves.

The attenuation process begins when the elastic part of the decompression wave overtakes the plastic part of the compression wave (see Figs. 10.7b and 10.23), an interaction producing a right-propagating shock moving into the material in state 1 and a left-propagating shock moving into the material in state 3. The analysis of this interaction proceeds exactly as described in Sect. 10.6.2, and leads to the same solution. The waveform at times slightly before and after the formation of region 5 is plotted in Fig. 10.24.

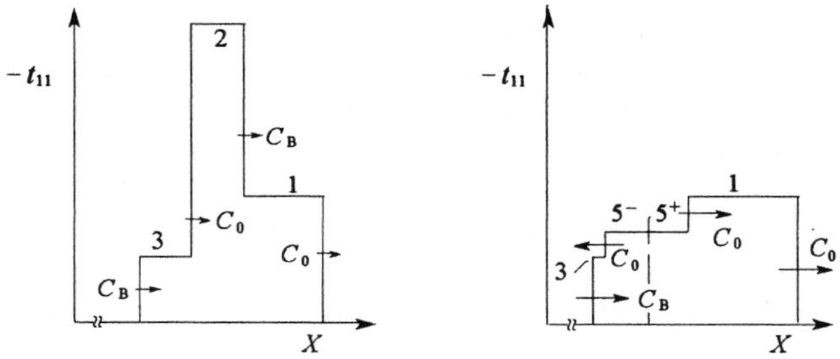

Figure 10.24. Waveform at times ($t = 3.00$ and $5.15\,\mu s$) slightly before and after the first wave interaction for a pulse of amplitude in stress range D. The example refers to a stress pulse of 1 μs duration and amplitude −4.5 GPa propagating in steel.

Note that, if $-t_{11}^{(2)}$ were somewhat larger, $-t_{11}^{(5)}$ would exceed t_{11}^{HEL} and the Hugoniot connecting state 1 to state 5 would encounter the plastic rather than the elastic segment of the compression Hugoniot. The need to separate response ranges at this point determines the upper limit on peak compressive stress defining the response range of the inequality 10.33.

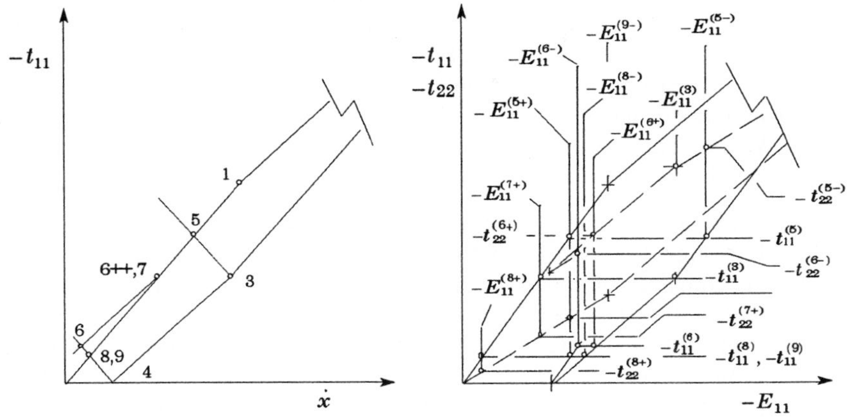

(a) (b)

Figure 10.25. (a) Stress–particle-velocity and (b) stress–strain Hugoniots for attenuation of a compression pulse. The example is drawn for a stress pulse of amplitude −4.5 GPa in steel with only the lower portions of the Hugoniots being shown.

The next interaction (see Fig. 10.23) is the collision of the left-propagating shock taking the material from state 3 to state 5^- with the right-propagating shock taking material from state 3 to state 4. To analyze this interaction we need the $t_{11}-v$ Hugoniots for right-propagating shocks centered on state 5^- and for left-propagating shocks centered on state 4. Determination of these Hugoniots requires examination of the t_{11}–$\widetilde{E}_{11}$ and $t_{22}-\widetilde{E}_{11}$ Hugoniots in order to establish whether the transition process at issue is elastic or plastic. State 4 is at the reverse yield point, so the Hugoniot begins with an elastic segment that extends to the forward yield point. Further loading produces plastic deformation. Since state 5^- is not at yield (see Fig. 10.24b), the Hugoniot centered on it begins with an elastic segment taking the material to the reverse yield point and continues with a plastic segment. Let us call the state at the intersection of these Hugoniots state 6 (which we shall assume may contain a contact surface separating substates 6^- and 6^+). Examination of Fig. 10.25 indicates that the intersection of these Hugoniots is such that the segment of the Hugoniot connecting state 5^- to state 6^+ passes over the reverse yield point. This means that the decompression shock will be unstable, separating into an elastic decompression from state 5^- to the yield state (which we call state 6^{++}) followed by a plastic decompression from state 6^{++} to state 6^{+}. The transition from state 4 to state 6^{-} is an elastic compression.

Applying the jump equations 10.1 to the right-propagating elastic shock transition from state 5^{-} to state 6^{++} yields the values for the field variables in the latter region. The analysis of the interaction is now completed by applying the jump equations to the right-propagating plastic shock taking the material from state 6^{++} to the state 6^{+} and the left-propagating elastic shock taking the material from state 4 to state 6^{-}. The result is recorded in [30, App. B] and plotted in Fig. 10.25. It is easy to verify that state 6^{++} lies on the yield surface and that the stresses and strains are the same as in state 3, although the particle velocity is lower. We find that state 6^{+} lies below the HEL and is on the yield surface. State 6^{-} involves even lower stresses and lies inside the yield surface.

Continuation of the solution requires analysis of several more interactions, as indicated in Fig. 10.23. Some of these interactions are of shocks with other shocks and some are of shocks with contact interfaces. In the following paragraphs we briefly sketch the results of the analysis of the remaining regions shown on Fig. 10.23. The state points associated with each of these regions are plotted on the Hugoniots of Fig. 10.25. The field values in each of the regions shown on the figure are given in [30, App. B] and illustrated in the stress–strain and stress–particle-velocity planes and as stress profiles in Fig. 10.26.

State 7 is formed when the right-propagating transition from state 5^{-} to state 6^{++} encounters the contact surface separating regions 5^{-} and 5^{+}. We expect formation of a left-propagating transition from state 6^{++} to state 7^{-} and a right-propagating transition from state 5^{+} to state 7^{+}. Since state 5^{+} is in the elastic range, both compression and decompression from this state begin with an elastic deformation. State 6^{++} is at the reverse yield point so compression is elastic and decompression is plastic. The $t_{11} - v$ Hugoniots for a right-propagating shock centered on state 5^{+} and a left-propagating shock centered on state 6^{++} intersect at the point corresponding to state 6^{++}, so we have the solution

$$t_{11}^{(7)} = t_{11}^{(6++)}, \quad v^{(7)} = v^{(6^{++})}. \tag{10.34}$$

The left-propagating shock is of vanishing strength since there is no jump in either t_{11} or v. From the jump condition 10.1_1 we have

$$\widetilde{E}_{11}^{(7-)} = \widetilde{E}_{11}^{(6++)}, \tag{10.35}$$

and we find from Eq. 10.12 that

$$t_{22}^{(7-)} = t_{22}^{(6++)}. \tag{10.36}$$

In effect, there is no distinct region 7^{-}; region 6^{++} is simply extended to the contact surface. Application of the jump condition 10.1_1 to the transition from region 5^+ to region 7^+ yields $\widetilde{E}_{11}^{(7+)}$ as given in [30, App. B]. Waveforms at various times after formation of region 6 are shown in Fig. 10.26.

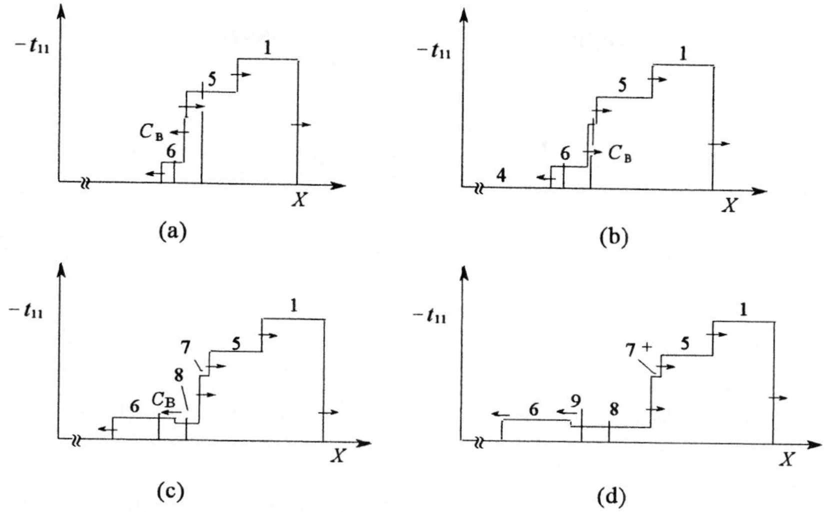

Figure 10.26. This is a continuation of the sequence of waveforms begun in Fig. 10.7b and 10.24, but with an enlargement of both horizontal and vertical scale. Waveforms are for the times (a) $t = 5.40$, (b) $t = 5.72$, (c) $t = 6.00$, and (d) $t = 6.50\,\mu s$. Arrows designate the direction of wave propagation. Elastic shocks carry no notation, but plastic waves are marked with their wavespeed, C_B. Numerals relate portions of the waveform to the corresponding region of the X–t diagram.

State 8 is formed as a result of the encounter of the plastic decompression wave with the contact surface through region 5. We expect formation of a left-propagating transition from state 6^+ to state 8 and a right-propagating transition from state 7^+ to state 8. State 6^+ is at reverse yield so compression is an elastic process and decompression is a plastic process. State 7^+ is elastic: a small range of elastic compression and a larger range of elastic decompression is available before an elastic–plastic transition point is reached and deformation continues as a plastic process. The intersection of the two Hugoniots of interest falls on the main compression branch of the $t_{11}-v$ and $t_{11}-\widetilde{E}_{11}$ paths at a point well below state 6^{++}. Application of the jump equations to these shocks leads to the solution given in [30, App. B].

State 9 is analyzed in the same way as the regions discussed previously. Like the interaction producing region 7, this interaction produces only one nontrivial shock, a left-propagating elastic shock. At the completion of this interaction, all shocks in the field are elastic shocks; they propagate at the same speed, C_0, and all are propagating away from the contact interfaces, so there can be no further interactions.

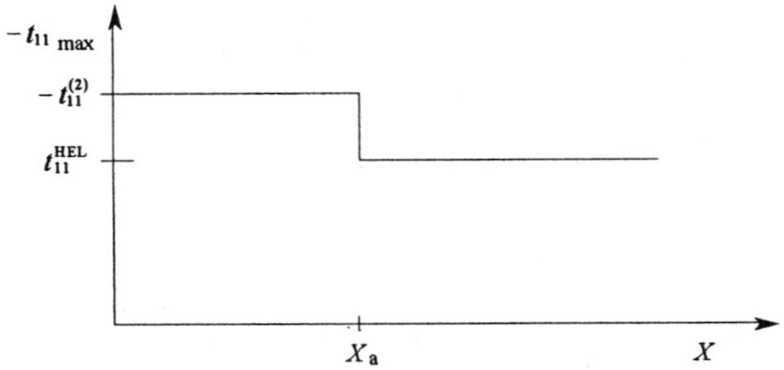

Figure 10.27. Attenuation curve for a pulse in stress range D. The point x_a is defined in Fig. 10.23 (steel).

The end result of a pulse-attenuation calculation is often presented in the form of an *attenuation curve* giving the maximum stress reached at each point x, without regard to the time or duration of its application. Examination of Fig. 10.23 shows that, for any point to the left of the interaction point X_a, the stress is successively 0, $-t_{11}^{(1)}$, $-t_{11}^{(2)}$, and 0. Beyond the interaction point, the succession of stress values is 0, $-t_{11}^{(1)}$, and $-t_{11}^{(5)}$. Since $0 < -t_{11}^{(5)} < -t_{11}^{(1)} < -t_{11}^{(2)}$, the attenuation curve is as given in Fig. 10.27. Note that the pulse amplitude never falls below $-t_{11}^{(1)} = t_{11}^{HEL}$ since no attenuating disturbance can overtake the leading elastic wave.

It is worth pointing out that the attenuation process, as represented by the peak-stress criterion used in drawing the attenuation curve, is completed with formation of region 5. The several subsequent interactions that must be analyzed to complete the description of the entire waveform cannot cause reduction of the peak compressive stress below t_{11}^{HEL}.

Attenuation of Higher-amplitude Pulses. The method outlined in the previous section suffices for analysis of attenuation of a pulse of any amplitude. Unfortunately, because of the multiplicity of stress ranges and the large number of interactions that must be considered, the analyses become increasingly awkward as attention is turned to pulses of ever higher amplitude. Several such pulses have been analyzed in [30].

10.1.9 Numerical Solution of Weak Elastic–Plastic Wave-propagation Problems

Computer software has long been available to carry out calculations of the sort presented in the foregoing sections of this chapter. These results are exact within the context of the theory. The more usual approach to these analyses, however, is to apply software implementing finite-difference methods. This software is widely used and capable of solving wave-propagation problems in the context of much more comprehensive theories of material response than the theory of ideal plasticity at small strain. A shortcoming of this approach is that it is Eqs. 2.87 or Eqs. 2.92 that are solved. These equations are applicable only to smooth waves and cannot be used for the analysis of shocks such as occur in the solutions presented in the preceding sections of this chapter. To circumvent this difficulty an artificial viscosity term is included in the material description. This results in the shocks being replaced by smooth, but steep, waves. Usually, the results obtained by this method are satisfactory, but the many small steps that comprise the wave are blended together to produce a smooth waveform. It is unclear what the effect might be of wave interactions and contact interfaces that are closer together than the resolution of the finite-difference analysis. These issues are further obscured because shocks in real materials are not the mathematical discontinuities of the theory, and measurements (particularly those made with low temporal resolution) may further spread the recorded waveform. When experimental observations are compared to numerical simulations, it is difficult to separate effects of the many small wave interactions that occur from dispersion resulting from material behavior or artificial viscosity in the numerical simulation.

10.2. Finite-amplitude Elastic–Plastic Waves

Finite-amplitude elastic–plastic waves propagating in non-hardening materials are analyzed in much the same way as the weak elastic–plastic waves, but the Hugoniot of Fig. 7.8 and the associated decompression path are used. Because the compression–decompression loop is so slender, a schematic illustration showing a broader loop is given in Fig. 10.28. The analysis of compression waves is conducted using the nonlinear Hugoniot. Decompression waves separate into two parts, an elastic wave taking the material from the shock-compressed state to the state in which the shear stress reaches the reverse yield point, followed by a plastic wave in which the compressive stress is further reduced. The elastic decompression wave is of higher amplitude and propagates faster than the elastic compression wave because of the nonlinear elastic response. Because the decompression path is concave upward, the plastic part of the process will occur in the form of a simple wave, although it is often analyzed as a shock, as was done in Chap. 3.

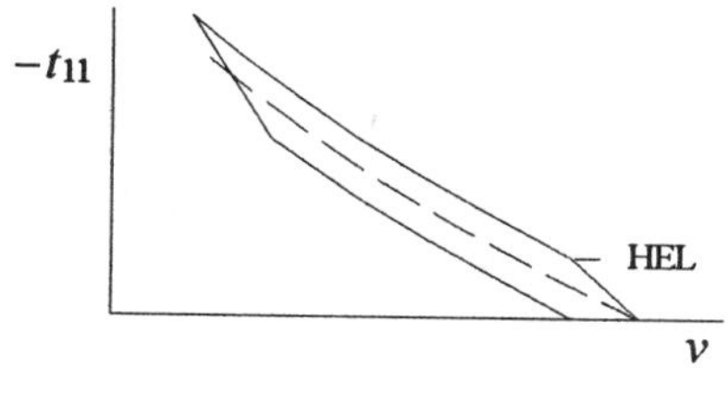

Figure 10.28. Schematic illustration of an elastic–plastic Hugoniot and decompression isentrope for a non-hardening material.

Effect of Deformation Hardening on Waveforms. A schematic illustration of a compression curve for a deformation hardening material is presented in Fig. 10.29. As discussed in Chap. 7, hardening usually leads to a curve that is concave downward in an interval above the HEL. Processes that produce monotonic compression can be analyzed in the same way as was done in Chap. 9 for elastic waves. Transitions from the HEL to a state such as B on the concave upward portion of the curve (but below its intersection with an extension of the elastic curve below the HEL) comprise three or four parts. The leading part of the wave is a transition from the unstressed state to the HEL.* This will be followed by a plateau if an abrupt decrease in wavespeed occurs at the HEL and a centered simple wave transition from the plateau to the point A, defined as the point of tangency of a line from point B to the curve. Finally, a shock transition will take the material from state A to the boundary state B. If the hardening curve is tangent to the elastic line below the HEL at their point of intersection, the plateau vanishes and the waveform rises smoothly from the HEL. This wave is illustrated in Fig. 10.30.

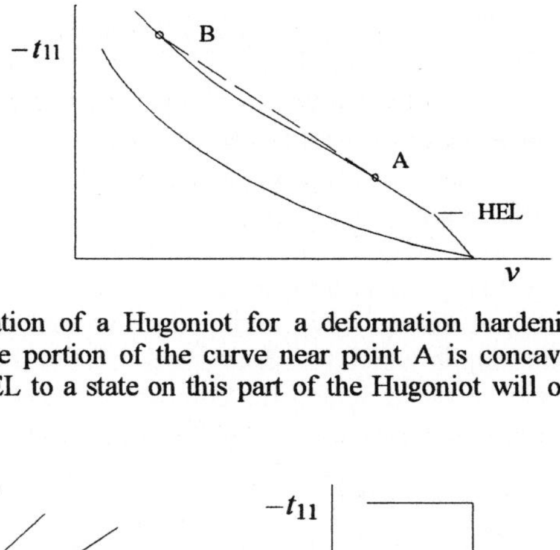

Figure 10.29. Illustration of a Hugoniot for a deformation hardening elastic–plastic material. Note that the portion of the curve near point A is concave downward so a transition from the HEL to a state on this part of the Hugoniot will occur in a centered simple wave.

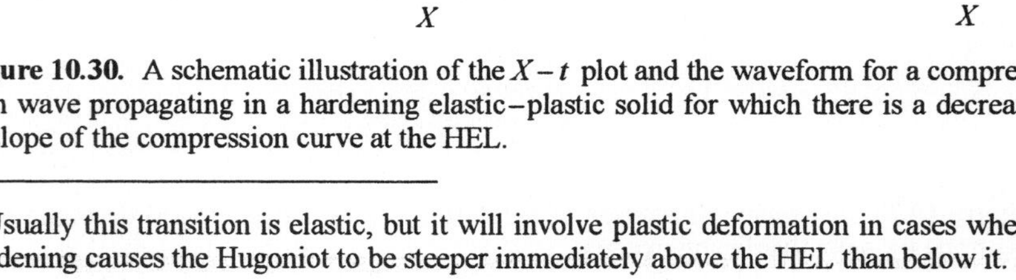

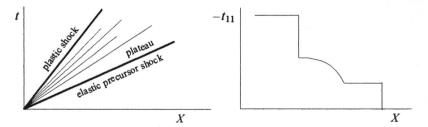

Figure 10.30. A schematic illustration of the $X-t$ plot and the waveform for a compression wave propagating in a hardening elastic–plastic solid for which there is a decrease in slope of the compression curve at the HEL.

* Usually this transition is elastic, but it will involve plastic deformation in cases where hardening causes the Hugoniot to be steeper immediately above the HEL than below it.

10.3 Finite-amplitude Elastic–Viscoplastic Waves

Solution of wave-propagation and -interaction problems of the sort discussed earlier in this chapter can only be achieved numerically for elastic–viscoplastic materials. Nevertheless, there are two problems that can be discussed and that illustrate important aspects of the behavior of elastic–viscoplastic solids.

10.3.1 Analysis of the Precursor Shock

We have seen that the shock produced in a plate-impact experiment involving an ideal elastic–plastic solid immediately splits into a precursor shock producing an elastic transition to the HEL state and a second shock producing a plastic transition to the state imposed by the boundary condition. The response of an elastic–viscoplastic material is somewhat different because the instantaneous and equilibrium responses differ. At the moment of impact the material experiences an elastic shock transition to the state imposed by the boundary condition. The shear stress in the material behind this shock initiates dislocation motion producing a gradual relaxation of this shear stress and a decrease of the amplitude of the precursor. The precursor shock propagates at the speed appropriate to its amplitude, decreasing as the shock is attenuated.

We begin by deriving an equation (called a *shock-change equation*) for the rate of change of amplitude of a shock as it advances through the material*.

Shock-change Equation. Analysis of the changing amplitude of the precursor shock is based on an equation giving the rate of change of the stress as perceived by an observer moving with the shock. This problem can be analyzed using either the Eulerian or the Lagrangian coordinate, but we shall adopt the latter. The time derivative along a shock trajectory of a field $\varphi(X,t)$ (written $D\varphi(X,t)/Dt$, where X and t are related by the equation defining the shock trajectory) is given by the equation

$$\frac{D\varphi(X,t)}{Dt} = \frac{\partial\varphi}{\partial t} + U_S \frac{\partial\varphi}{\partial X},$$

where U_S is the Lagrangian velocity of the shock. When applied to the stress and particle velocity immediately behind the shock, this equation becomes

---

* Shock-change equations are useful in contexts other than the present one of elastic precursor decay. Any case in which a shock initiates an evolutionary process that affects the shock itself is a candidate for application of a shock-change equation. Examples include initiation or extinguishing of a chemical reaction in an explosive [84], kinetics of a shock-induced phase transformation, and attenuation of shocks in viscoelastic and viscoplastic materials [85].

$$\frac{Dt_{11}}{Dt} = \frac{\partial t_{11}(X,t)}{\partial t} + U_S \frac{\partial t_{11}(X,t)}{\partial X}$$
$$\frac{D\dot{x}}{Dt} = \frac{\partial \dot{x}(X,t)}{\partial t} + U_S \frac{\partial \dot{x}(X,t)}{\partial X}. \tag{10.37}$$

In addition, we have the conservation equations

$$\frac{\partial \dot{x}}{\partial X} - \rho_R \frac{\partial v}{\partial t} = 0$$
$$\frac{\partial t_{11}}{\partial X} - \rho_R \frac{\partial \dot{x}}{\partial t} = 0 \tag{10.38}$$

in the region of smooth flow behind the shock and the jump equations

$$\dot{x} = \rho_R U_S (v_R - v)$$
$$t_{11} = -\rho_R U_S \dot{x} \tag{10.39}$$

for a shock propagating into material that is at its reference specific volume, is unstressed, and is at rest. The field variables in the jump equations take the values immediately behind the shock.

For our analysis we use the stress relation of Eq. 7.78$_1$,

$$t_{11} = -p(v) + 4\,\mu(v)\left[1 - (v/v_R)^{-1/3}(F_L^P)^{1/2}\right], \tag{10.40}$$

and we shall need the material derivative of this stress component,

$$\frac{\partial t_{11}(X,t)}{\partial t} = \rho_R^2 C_L^2 \frac{\partial v(X,t)}{\partial t} - \kappa \frac{\partial F_L^P(X,t)}{\partial t}, \tag{10.41}$$

where C_L^2, the square of the frozen soundspeed (the value for F_L^P held fixed), is given by

$$\rho_R^2 C_L^2 = -p'(v) + \tfrac{4}{3}\rho_R\,\mu(v)(v/v_R)^{-4/3}(F_L^P)^{1/2}$$
$$+ 4\mu'(v)[1 - (v/v_R)^{-1/3}(F_L^P)^{1/2}], \tag{10.42}$$

and where

$$\kappa = 2\mu(v)(v/v_R)^{-1/3}(F_L^P)^{-1/2}. \tag{10.43}$$

Using Eq. 10.41 to eliminate $\partial v/\partial t$ from Eq. 10.38$_1$, we obtain

$$\frac{Dt_{11}}{Dt} - U_S \frac{\partial t_{11}}{\partial X} - \rho_R C_L^2 \frac{\partial \dot{x}}{\partial X} = -\kappa \dot{F}_L^P, \tag{10.44}$$

and using this equation to eliminate $\partial t_{11}/\partial t$ from Eq. 10.37 gives

$$\frac{D\dot{x}}{Dt} - U_S \frac{\partial \dot{x}}{\partial X} - \frac{1}{\rho_R} \frac{\partial t_{11}}{\partial X} = 0 . \qquad (10.45)$$

When Eqs. 10.44 and 10.45 are combined to eliminate $\partial \dot{x} / \partial X$ we obtain the result

$$\frac{Dt_{11}}{Dt} - \frac{\rho_R C_L^2}{U_S} \frac{D\dot{x}}{Dt} + \frac{1}{U_S}(C_L^2 - U_S^2)\frac{\partial t_{11}}{\partial X} = -\kappa \dot{F}_L^P . \qquad (10.46)$$

We now introduce the $\dot{x} - t_{11}$ Hugoniot and note that differentiation along this Hugoniot gives

$$\frac{D\dot{x}}{Dt} = \frac{d\dot{x}^{(H)}(t_{11})}{dt_{11}} \frac{Dt_{11}}{Dt} , \qquad (10.47)$$

which can be used to eliminate $D\dot{x} / Dt$ from Eq. 10.46 so that we have

$$\frac{Dt_{11}}{Dt} = - \frac{(C_L^2 - U_S^2)\dfrac{\partial t_{11}}{\partial X} + \kappa U_S \dot{F}_L^P}{U_S - \rho_R C_L^2 \dfrac{d\dot{x}^{(H)}(t_{11})}{dt_{11}}} , \qquad (10.48)$$

which is the *shock-change equation* sought. It shows that the stress rate at the precursor shock depends upon the stress gradient and the plastic-deformation rate behind the shock in addition to the material properties. In the next section we shall pursue this matter in more detail, but we first address some approximations that have proven useful in applications.

Isentropic Elastic Response. We have seen that a Hugoniot differs little from the isentrope through the same initial state when the compression is small, as is usually the case behind an elastic-precursor shock. Accordingly, we may replace the Hugoniot appearing in Eq. 10.48 with the isentrope. Therefore, both the Hugoniot and the isentrope are given by Eq. 10.40, with $F_L^P = 1$ since there is no plastic deformation in the precursor shock. The soundspeed immediately behind the precursor is given by

$$C_L^2 = \frac{1}{\rho_R^2} \frac{dt_{11}(v)}{dv} , \qquad (10.49)$$

where the function $t_{11}(v)$ is the isentrope in this case, which is taken to be the same as the Hugoniot.

Elimination of U_S from the jump equations 10.39 yields the result

$$[\dot{x}^{(H)}(v)]^2 = (v - v_R) t_{11}^{(H)}(v) . \qquad (10.50)$$

Differentiation of this equation and use of Eq. 10.49 yields the result

$$\frac{d\dot{x}^{(\mathrm{H})}(v)}{dv}=\frac{1}{2}\left[\frac{t_{11}^{(\mathrm{H})}(v)}{\dot{x}^{(\mathrm{H})}(v)}+\rho_{\mathrm{R}}\,C_{\mathrm{L}}^{2}\frac{(v-v_{\mathrm{R}})}{\dot{x}^{(\mathrm{H})}(v)}\right], \tag{10.51}$$

which can be expressed in the form

$$\frac{d\dot{x}^{(\mathrm{H})}(v)}{dv}=\frac{\rho_{\mathrm{R}}}{2U_{\mathrm{S}}}(C_{\mathrm{L}}^{2}-U_{\mathrm{S}}^{2}) \tag{10.52}$$

by use of the jump equations. When substituted into the soundspeed equation

$$\rho_{\mathrm{R}}^{2}\,C_{\mathrm{L}}^{2}=\frac{dt_{11}^{(\mathrm{H})}(v)}{dv}=\frac{dt_{11}^{(\mathrm{H})}(\dot{x})}{d\dot{x}}\frac{d\dot{x}^{(\mathrm{H})}(v)}{dv} \tag{10.53}$$

that follows from identifying the isentrope with the Hugoniot, this gives

$$\frac{dt_{11}^{(\mathrm{H})}(\dot{x})}{d\dot{x}}=\frac{2\rho_{\mathrm{R}}\,C_{\mathrm{L}}^{2}\,U_{\mathrm{S}}}{C_{\mathrm{L}}^{2}-U_{\mathrm{S}}^{2}}, \tag{10.54}$$

so, Eq. 10.48 becomes

$$\frac{Dt_{11}}{Dt}=-\frac{(C_{\mathrm{L}}^{2}-U_{\mathrm{S}}^{2})\dfrac{\partial t_{11}}{\partial X}+\kappa\,U_{\mathrm{S}}\,\dot{F}_{\mathrm{L}}^{\mathrm{P}}}{U_{\mathrm{S}}\left[\dfrac{3}{2}-\dfrac{1}{2}\dfrac{C_{\mathrm{L}}^{2}}{U_{\mathrm{S}}^{2}}\right]} \tag{10.55}$$

in the isentropic approximation.

Linear Elastic Response. When the elastic response is taken to be linear the shockspeed and the soundspeed are equal and Eq. 10.55 takes the much simpler form

$$\frac{Dt_{11}}{Dt}=-\kappa\,\dot{F}_{\mathrm{L}}^{\mathrm{P}}=-2\mu_{\mathrm{R}}\,\dot{F}_{\mathrm{L}}^{\mathrm{P}}\,. \tag{10.56}$$

Constitutive equations for $\dot{F}_{\mathrm{L}}^{\mathrm{P}}$ depend on the stress, so we need to develop linear approximations to the stress relations of Eqs. 7.79, which are

$$\begin{aligned} t_{11}&=-p(v)-4\mu(v)\left[1-(v/v_{\mathrm{R}})^{-1/3}\right]\\ t_{22}&=-p(v)+2\mu(v)\left[1-(v/v_{\mathrm{R}})^{-1/3}\right], \end{aligned} \tag{10.57}$$

when $F_{\mathrm{L}}^{\mathrm{P}}=1$ as is the case immediately behind the precursor shock. In the linear approximation, the pressure term in these equations is

$$p(v)=(\lambda_{\mathrm{R}}+\tfrac{2}{3}\mu_{\mathrm{R}})[1-(v/v_{\mathrm{R}})]\,, \tag{10.58}$$

where λ_R and μ_R are the constant Lamé elastic coefficients. To first order in $1-(v/v_R)$ we have

$$1-(v/v_R)^{-1/3} = -\tfrac{1}{3}[1-(v/v_R)], \tag{10.59}$$

so Eqs. 10.57 become

$$\begin{aligned} t_{11} &= -(\lambda_R + 2\mu_R)\left[1-(v/v_R)\right] \\ t_{22} &= -\lambda_R\left[1-(v/v_R)\right]. \end{aligned} \tag{10.60}$$

We shall need the maximum shear stress associated with this stress field, which is

$$\tau_{45°} = \tfrac{1}{2}(t_{11} - t_{22}) = \frac{\mu_R}{\lambda_R + 2\mu_R} t_{11}. \tag{10.61}$$

With this definition $\tau_{45°}$ is positive in tension and negative in compression.

Precursor Shock Attenuation. Let us begin by examining Eq. 10.55. We note that C_L and U_S are both positive, with C_L being moderately greater than U_S, so both the denominator of Eq. 10.55 and the coefficient of the stress-gradient term are positive. The coefficient κ is also positive, so the term $\kappa U_S \dot{F}_L^P$ has the same sign as $\dot{F}_L^P$, which is negative for a compression shock. The stress-gradient term accounts for the effect on the precursor amplitude of wavelets overtaking the shock from the rear and the term involving $\dot{F}_L^P$ accounts for the effect of viscoplastic flow on the amplitude. When the material is being compressed the effect of the dissipation term is always to make Dt_{11}/Dt positive, i.e., to make t_{11} become less negative. There are two cases to consider when evaluating the effect of the stress gradient term, as shown in Fig. 10.31. Part (a) of the figure shows a waveform with positive stress gradient immediately behind the shock. In this case the stress-gradient term tends to make Dt_{11}/Dt negative, and thereby make the stress become more negative. The opposite trend prevails for the waveform in part (b) of the figure.

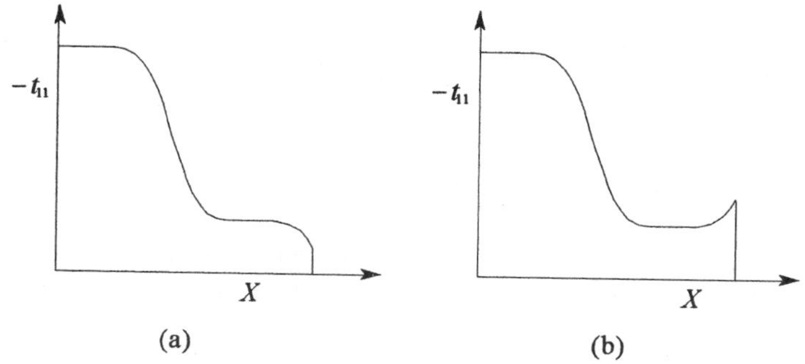

Figure 10.31. Elastic–viscoplastic waveforms showing gradients of different sign behind the precursor shock. (a) $\partial t_{11}/\partial X > 0$ and (b) $\partial t_{11}/\partial X < 0$.

The complexity of the elastic–viscoplastic constitutive equations makes it necessary to resort to numerical simulation to obtain solutions that would exhibit waveforms of the sort illustrated in Fig. 10.31. Finite-difference and finite-element methods are usually used to solve problems of this class, but it may be necessary to solve the shock-change equation in conjunction with this method of analysis if the precursor behavior is to be simulated satisfactorily. However, when the precursor attenuation problem is posed in terms of the simplified shock-change equation 10.56 it is often possible to obtain the attenuation curve analytically.

The plastic part of the deformation gradient is related to the stress by Eq. 7.131, which we write

$$\dot{F}_{\mathrm{L}}^{\mathrm{p}} = \pi B (v / v_{\mathrm{R}})^{-1/3} \mathcal{N}_0 \, f_0 \, V_x(\tau_{45^\circ})(1 + \tfrac{1}{4} E_{\mathrm{L}}^{\mathrm{es}}) F_{\mathrm{L}}^{\mathrm{p}} , \qquad (10.62)$$

in view of the fact that $\mathcal{N}_{\mathrm{M}} = \mathcal{N}_0$, $f_{\mathrm{M}} = f_0$, and $F_{\mathrm{L}}^{\mathrm{p}} = 1$ at the precursor shock. Among the several equations for $V(\tau)$ that have been proposed we shall adopt one of the simpler ones,

$$V(\tau_{45^\circ}) = - V_0 \frac{\tau_{45^\circ} - \tau_0}{\tau_0} = - V_0 \frac{t_{11} - t_{11}^0}{t_{11}^0} , \qquad (10.63)$$

for $t_{11} \le t_{11}^0 \le 0$. When Eq. 10.63 is substituted into Eq. 10.62 and that result substituted into Eq. 10.56 we obtain the equation

$$\frac{D t_{11}}{D t} = - \frac{\mu_{\mathrm{R}}}{t_0} \frac{t_{11} - t_{11}^0}{t_{11}^0} , \qquad (10.64)$$

with t_0 being given by

$$1 / t_0 = 2 \pi B (v / v_{\mathrm{R}})^{-1/3} \mathcal{N}_0 \, f_0 \, V_0 \, (1 + \tfrac{1}{4} \widetilde{E}_{\mathrm{L}}^{\mathrm{es}}) . \qquad (10.65)$$

From Eq. 10.60 we find that

$$\frac{v}{v_{\mathrm{R}}} = 1 + \frac{t_{11}}{\lambda_{\mathrm{R}} + 2 \mu_{\mathrm{R}}} . \qquad (10.66)$$

Since the second term of this equation is very small relative to the first at the stress level of the precursor we shall neglect it. Similarly, $| \widetilde{E}_{\mathrm{L}}^{\mathrm{es}} | \ll 1$ and we shall neglect it as well, leaving us with

$$1 / t_0 = 2 \pi B \mathcal{N}_0 \, f_0 \, V_0 . \qquad (10.67)$$

When t_0 is given by Eq. 10.67, the solution of Eq. 10.64 is

$$t_{11} = t_{11}^0 + (t_{11}^{\mathrm{B}} - t_{11}^0) \exp \left(\frac{\mu_{\mathrm{R}}}{t_{11}^0} \frac{t}{t_0} \right) , \qquad (10.68)$$

where t_{11}^{B} is the stress imposed on the boundary at $t=0$. Since $t_{11}^{0}<0$ this is a decaying exponential function of time. Since the precursor is propagating at the constant longitudinal elastic wavespeed, the temporal attenuation curve of Eq. 10.68 is equivalent to the spatial attenuation curve

$$t_{11}=t_{11}^{0}+(t_{11}^{B}-t_{11}^{0})\exp\left(\frac{\mu_R}{t_{11}^{0}}\frac{X}{C_L t_0}\right). \tag{10.69}$$

This equation is compared to some data for a mild steel in Fig. 10.32. Examination of this figure shows that the predicted attenuation rate is less than that observed in the material layer nearest the surface. Three explanations for this have been offered. The first is simply that the linear elastic-response equation 10.56 neglects the attenuation due to the overtaking decompression wave associated with a precursor spike as in Fig. 10.31b. This issue has been studied by Herrmann [55] and Clifton [22]. A second possibility is that the dependence of dislocation velocity on the shear stress given by the equation used here, Eq. 10.63, is incorrect. Various theories of dislocation velocity have been offered and the effect of changing the velocity equation is easily investigated. Another potential explanation that has been advanced is that the dislocation density is higher near the surface of the material tested than in its interior. This increased density would result from damage introduced into the lattice during sample preparation and to slight, but unavoidable, surface roughness. As shown by Eq. 10.65, the increased dislocation density results in a decreased characteristic response time and, thus, a more rapid attenuation rate. A short segment of an attenuation curve calculated using a 40% higher dislocation density (a very small change relative to the usual variation of this parameter) is shown on the figure and can be seen to fit the data better in the region near the boundary. We see from this discussion how the interplay of continuum analysis with the motivating microscopic basis of the theory can lead to improved understanding of the physical process. The most precise and comprehensive investigations of precursor attenuation have involved measurements of waveforms propagating in carefully characterized lithium fluoride monocrystals. When the measurements are compared to theoretical predictions, it is found that the continuum analysis is in good agreement with the observations, but dislocation-mechanical equations and parameters employed to achieve this agreement are inconsistent with the best understanding of this aspect of the physics. A brief review of this work, with additional references, is given in [59, p. 226]. It is a common observation that continuum theories motivated by microscopic considerations prove more broadly valid than the underlying microscopic theory.

10.3.2 Steady Waves in Elastic–Viscoplastic Solids

In Sect. 2.5.3 we considered the possibility that a structured longitudinal wave could propagate at a constant velocity and without changing form, i.e., the field

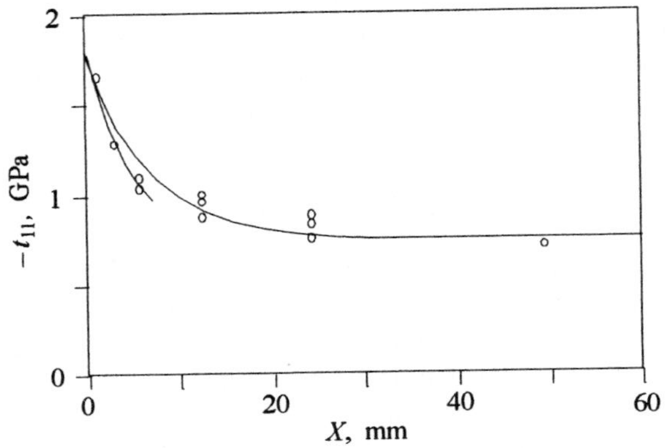

Figure 10.32. Spatial attenuation data for a mild steel [23] in comparison to a plot of Eq. 10.69.

variables could be expressed as functions of the single independent variable $Z = X - U_S t$, where U_S is a constant Lagrangian wavespeed. We know that shocks propagate in this way, so the point of this discussion is whether or not there are also smooth steady waves. In this section we shall show that smooth steady waves can propagate in some elastic–viscoplastic materials. In Sect. 10.2 we noted that the compression curve for hardening materials is usually concave downward in a stress interval immediately beyond the yield point and that this causes a centered simple wave to form between the elastic precursor and the main shock. When we consider an elastic–viscoplastic material the situation becomes more complicated because the motion that occurs immediately behind the precursor can no longer be a simple wave. The main shock will be structured as a result of the viscoplastic flow, and the interaction among the various parts of the waveform can only be determined by numerical solution of the governing partial differential equations. When work hardening is absent, we will find that steady waveforms exist and it is this problem that we shall address.

Our first step in solving this problem is determining the Hugoniot curve, the locus of endstates behind steady waves of various strengths. The next step is determination of the field variables in the wave as functions of the specific volume. Finally, we must determine the waveform itself, i.e., the fields as functions of the steady wave coordinate Z. We shall see that the Hugoniot is determined in the usual way and is independent of the waveform or the rate at which the deformation is produced. Determination of the field variables as functions of v appears, at first, to involve the rate of deformation, but this proves not to be the case. The waveform does depend on the rate equations but, since this is the last step of the analysis, it is easy to investigate the effect of changing the evolutionary equations without reworking the entire analysis.

The equations to be satisfied in a steady wave were shown to be

$$\begin{aligned}\frac{d}{dZ}(\dot{x}+\rho_{\mathrm{R}}\,U_{\mathrm{S}}\,v)&=0\\ \frac{d}{dZ}(t_{11}+\rho_{\mathrm{R}}\,U_{\mathrm{S}}\,\dot{x})&=0.\end{aligned}\tag{10.70}$$

Integration of these equations shows that the jumps between any two states in a steady wave satisfy the same equations as for a shock transition between these two states. In the case of a transition from a known Hugoniot elastic limit to any other state in the waveform, we have

$$\begin{aligned}\dot{x}+\rho_{\mathrm{R}}\,U_{\mathrm{S}}\,v&=\dot{x}^{\mathrm{HEL}}+\rho_{\mathrm{R}}\,U_{\mathrm{S}}\,v^{\mathrm{HEL}}\\ t_{11}+\rho_{\mathrm{R}}\,U_{\mathrm{S}}\,\dot{x}&=-\,t_{11}^{\mathrm{HEL}}+\rho_{\mathrm{R}}\,U_{\mathrm{S}}\,\dot{x}^{\mathrm{HEL}}.\end{aligned}\tag{10.71}$$

We expect that the steady waveform will extend over the entire range $-\infty<Z<\infty$. The state approached as $Z\rightarrow\infty$ will be the HEL state and that approached as $Z\rightarrow-\infty$ is to be specified as a boundary condition. When Eqs. 10.71 are evaluated at the endstate behind the wave, i.e., in the limit $Z\rightarrow-\infty$, we have

$$\begin{aligned}\dot{x}^{+}+\rho_{\mathrm{R}}\,U_{\mathrm{S}}\,v^{+}&=\dot{x}^{\mathrm{HEL}}+\rho_{\mathrm{R}}\,U_{\mathrm{S}}\,v^{\mathrm{HEL}}\\ t_{11}^{+}+\rho_{\mathrm{R}}\,U_{\mathrm{S}}\,\dot{x}^{+}&=-\,t_{11}^{\mathrm{HEL}}+\rho_{\mathrm{R}}\,U_{\mathrm{S}}\,\dot{x}^{\mathrm{HEL}}.\end{aligned}\tag{10.72}$$

These are two equations relating the four variables U_{S} $\dot{x}^{+}$, t_{11}^{+}, and v^{+}, one of which is specified as a boundary condition. To complete the solution it is necessary to select constitutive equations, and we shall use those of Sect. 7.3.

It is important to note that the Hugoniot depends upon the yield stress but not upon the evolutionary equations for the dislocation-mechanical parameters $\mathcal{N}_{\mathrm{M}}$ and f_{M} or on the plastic deformation rate $\dot{F}_{\mathrm{L}}^{\mathrm{P}}$. These quantities determine the waveform, but not the endstate.

Variation on the Rayleigh Line. Let us now turn to determination of the field variables as functions of v within the waveform, i.e., along the Rayleigh line. The evolutionary equation for $F_{\mathrm{L}}^{\mathrm{P}}$, Eq. 7.131, can be written

$$\dot{F}_{\mathrm{L}}^{\mathrm{P}}=P(v,F_{\mathrm{L}}^{\mathrm{P}})\,F_{\mathrm{L}}^{\mathrm{P}},\tag{10.73}$$

where

$$P(v,F_{\mathrm{L}}^{\mathrm{P}})=\tfrac{1}{2}\pi B(v/v_{\mathrm{R}})^{-1/3}\mathcal{N}_0\,f_0\,V_x(\tau_{45^\circ})\,[3-(v/v_R)^{-1/3}(F_{\mathrm{L}}^{\mathrm{P}})^{1/2}]F_{\mathrm{L}}^{\mathrm{P}}.\tag{10.74}$$

In writing this equation we have used Eq. 7.68 to eliminate $E_{\mathrm{L}}^{\mathrm{es}}$. We shall also need the material derivative of t_{11} that, as given by Eq. 10.41, is

$$\dot{t}_{11} = \rho_R^2 C_L^2 \dot{v} - \kappa \dot{F}_L^P, \tag{10.75}$$

where C_L^2 and κ are functions of v and F_L^P that are given by Eqs. 10.42 and 10.43, respectively.

In a steady wave the material derivative is given by

$$\left.\frac{\partial(\cdot)}{\partial t}\right|_X = -U_S \frac{d(\cdot)}{dZ},$$

so Eqs. 10.73 and 10.75 take the forms

$$\frac{dF_L^P}{dZ} = -\frac{1}{U_S} P F_L^P \tag{10.76}$$

and

$$\frac{dv}{dZ} = \frac{1}{\rho_R^2 C_L^2}\left[\frac{dt_{11}}{dZ} + \kappa \frac{dF_L^P}{dZ}\right], \tag{10.77}$$

respectively. Substitution of Eq. 10.76 into 10.77 and use of Eqs. 10.70 to eliminate dt_{11}/dZ yields the result

$$\frac{dv}{dZ} = -\frac{\kappa P F_L^P}{\rho_R^2 U_S (C_L^2 - U_S^2)}. \tag{10.78}$$

Substitution of Eq. 10.76 into this equation places it in the form

$$\frac{dv}{dZ} = \frac{\kappa}{\rho_R^2 (C_L^2 - U_S^2)} \frac{dF_L^P}{dZ}, \tag{10.79}$$

from which we obtain the differential equation

$$\frac{dF_L^P}{dv} = \frac{\rho_R^2}{\kappa}(C_L^2 - U_S^2) \tag{10.80}$$

for $F_L^P(v)$. Note that this equation involves only the variables F_L^P and v so its solution gives $F_L^P(v)$ without consideration of the rate of deformation or the dislocation-mechanical variables. We expect to have $dF_L^P/dv > 0$ in a compression wave so, according to Eq. 10.80, the steady wavespeed must fall in the range $0 < U_S < C_L$. When a value in this range is selected the equation can be integrated to yield the function $F_L^P(v)$.

Substitution of $C_L^2(v, F_L^P)$, as given by Eq. 10.42, and $\kappa(v, F_L^P)$, as given by Eq. 10.43, into Eq. 10.80 yields the equation

$$\frac{dF_{\rm L}^{\rm p}}{dv}+2f_1(v)F_{\rm L}^{\rm p}=2f_2(v)(F_{\rm L}^{\rm p})^{1/2}, \tag{10.81}$$

where

$$\begin{aligned} f_1(v)&=\frac{\mu'(v)}{\mu(v)}-\frac{1}{3v} \\ f_2(v)&=\frac{1}{4\mu(v)}(v/v_{\rm R})^{1/3}\,[-p'(v)+4\mu'(v)-\rho_{\rm R}^2U_{\rm S}^2]. \end{aligned} \tag{10.82}$$

This equation is a special case of a first-order ordinary differential equation called *Bernoulli's equation* that can be transformed to the linear equation

$$y'(v)+f_1(v)\,y=f_2(v) \tag{10.83}$$

by introducing the new dependent variable $y=(F_{\rm L}^{\rm p})^{1/2}$. The solution of this familiar equation is

$$(F_{\rm L}^{\rm p})^{1/2}=y=\varphi(v)\left[1+\int_{v^{\rm HEL}}^{v}\frac{f_2(v')}{\varphi(v')}dv'\right], \tag{10.84}$$

where

$$\varphi(v)=\exp\left[-\int_{v^{\rm HEL}}^{v}\varphi(v')dv'\right], \tag{10.85}$$

and where we have imposed the boundary condition $F_{\rm L}^{\rm p}(v^{\rm HEL})=1$. Evaluation of Eq. 10.85 yields the result

$$\varphi(v)=\frac{\mu(v^{\rm HEL})}{\mu(v)}\left(\frac{v}{v^{\rm HEL}}\right)^{1/3}, \tag{10.86}$$

and evaluation of the other integral gives

$$\begin{aligned} \int_{v^{\rm HEL}}^{v}\frac{f_2(v')}{\varphi(v')}dv'=-\frac{1}{4\mu(v^{\rm HEL})}\left(\frac{v^{\rm HEL}}{v_{\rm R}}\right)^{1/3}&[p(v)-p(v^{\rm HEL}) \\ &-4\mu(v)+4\mu(v^{\rm HEL})-\rho_{\rm R}^2U_{\rm R}^2(v^{\rm HEL}-v)], \end{aligned} \tag{10.87}$$

so the solution we seek is

$$\begin{aligned} (F_{\rm L}^{\rm p})^{1/2}=\frac{\mu(v^{\rm HEL})}{\mu(v)}\left(\frac{v}{v^{\rm HEL}}\right)^{1/3}-\frac{1}{4\mu(v)}\left(\frac{v}{v_{\rm R}}\right)^{1/3}&[p(v)-p(v^{\rm HEL}) \\ -4\mu(v)+4\mu(v^{\rm HEL})-\rho_{\rm R}^2U_{\rm S}^2(v^{\rm HEL}-v)]. \end{aligned} \tag{10.88}$$

With this solution in hand, Eqs. 7.79 give the stress components at points in the waveform as functions of v. The particle velocity $\dot{x}(v)$ can be obtained from the jump equations, which completes the determination of the mechanical field variables as functions of the specific volume at points on the Rayleigh line.

From Eqs. 7.134 we have equations for the dislocation density and mobile fraction as functions of v:

$$\begin{aligned} \mathcal{N}_T &= \mathcal{N}_T^* + [\mathcal{N}_0 - \mathcal{N}_T^*] \exp(L_{\mathcal{N}} \gamma^P / B) \\ f_M &= f_M^* + [f_0 - f_M^*] \exp(L_M \gamma^P / B), \end{aligned} \tag{10.89}$$

where γ^P is given by Eq. 7.82 and when the initial conditions $\mathcal{N}_T = \mathcal{N}_0$ and $f_M = f_0$ are imposed at $\gamma^P = 0$.

Example Calculation. The foregoing equations have been solved numerically for steady waves of various amplitudes propagating in aluminum alloy 6061-T6. The parameters used are given in Table 10.2 and some of the particulars of the several calculations are given in Table 10.3. Dislocations were allowed to multiply from an initial value of 10^{11} m^{-2} to a possible maximum of 10^{14} m^{-2}. The dislocation density increase with compression is illustrated in Fig. 10.33. Several of the $\dot{x}(t)$ waveforms are presented in Fig. 10.34 and the peak strain rate as defined by Swegle and Grady [94] is plotted in Fig. 10.35.

Table 10.2. Material parameters used for steady-wave analysis

$\rho_R = 2703\,\mathrm{kg/m^3}$	$\tau^\dagger = 10.5\,\mathrm{GPa}$
$C_B = 5190\,\mathrm{m/s}$	$B = 2.860 \times 10^{-10}\,\mathrm{m}$
$S = 1.338$	$\mathcal{N}_0 = 10^{12}\,\mathrm{m}^{-2}$
$\mu(v) = \mu_R = 27.6\,\mathrm{GPa}$	$\mathcal{N}_T^* = 10^{18}\,\mathrm{m}^{-2}$
$t_{11}^{HEL} = 0.5810\,\mathrm{GPa}$	$f_0 = 1$
$v^{HEL} = 3.6802 \times 10^{-4}\,\mathrm{m^3/kg}$	$f_M^* = 0.4$
$\dot{x}^{HEL} = 33.56\,\mathrm{m/s}$	$L_{\mathcal{N}}/B = 10$
$\tau_0^* = 0.1453\,\mathrm{GPa}$	$L_M/B = 100$

Table 10.3. Steady-wave parameters

U_S, m/s	$-t_{11}^+$, GPa	$\dot{x}^+$, m/s	F_L^{P+}	$\mathcal{N}_M^+$, m^{-2}	$\dot{\eta}_{max}$, µs^{-1}
5400	2.064	169	0.9874	1.38×10^{16}	0.103
5450	2.625	207	0.9829	1.70	0.288
5500	3.224	245	0.9782	2.05	0.845
5550	3.781	283	0.9741	2.36	2.017
5600	4.375	320	0.9698	2.71	4.178
5700	5.590	394	0.9614	3.40	13.584

The quantity η is the compression relative to the HEL, $\eta = 1 - (v/v^{HEL})$, so $\dot{\eta} = -\dot{v}/v^{HEL}$. To calculate $\dot{\eta}_{max}$, which Swegle and Grady call the *maximum strain rate*, we note that $\dot{v} = -U_S(dv/dZ)$ where the derivative is given by Eq. 10.79. One simply selects the maximum value of this derivative from the calculation and uses the foregoing equations to calculate $\dot{\eta}_{max}$. Values of $\dot{\eta}_{max}$ determined by direct examination of the waveform in either graphical or numerical form exhibit considerable scatter, and this uncertainty must also arise in interpreting experimental results.

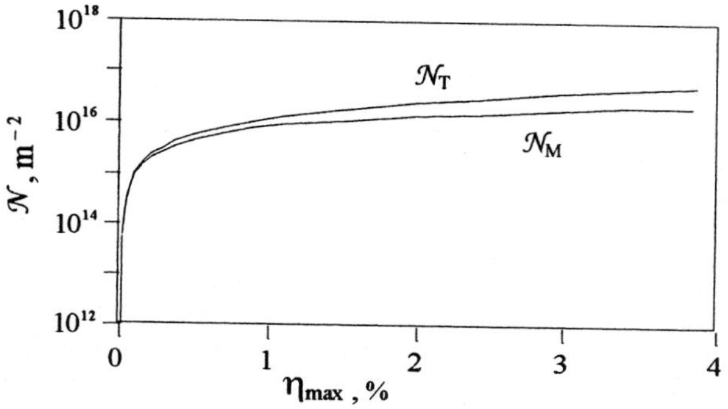

Figure 10.33. Dislocation multiplication and immobilization as the material is deformed during passage of a steady wave propagating at 5550 m/s. The compression is measured relative to v^{HEL}, $\eta = 1 - (v/v^{HEL})$.

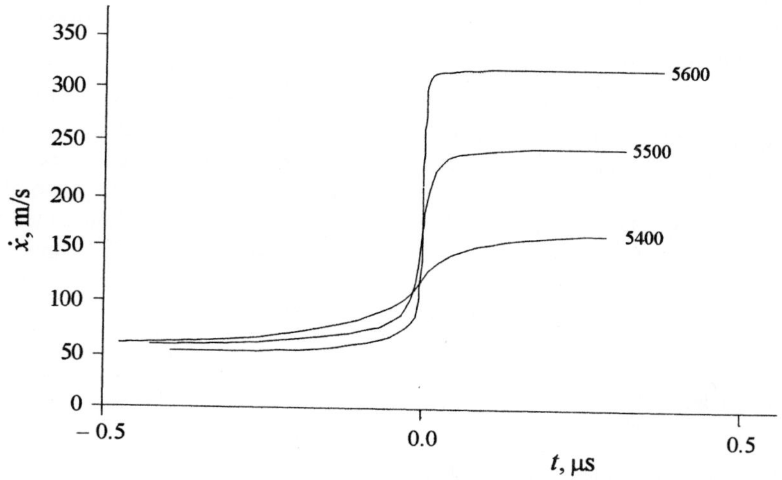

Figure 10.34. Typical steady waveforms calculated using the theory and parameters discussed in this section. The numbers below the curves identify them by the value of U_S in m/s.

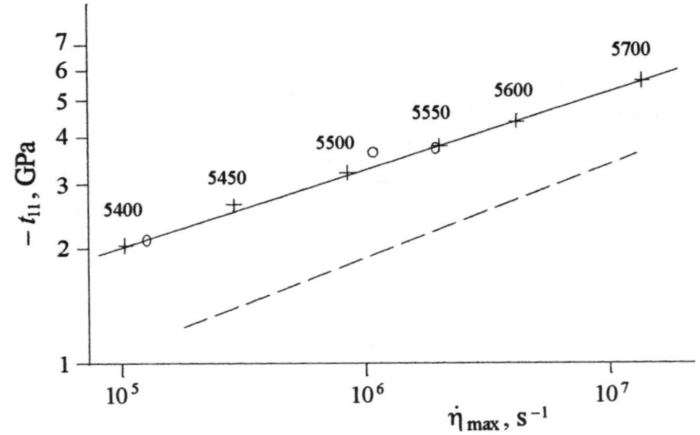

Figure 10.35. Peak shock stress plotted as a function of strain rate as defined by Swegle and Grady [94]. The data plotted by circles are from the foregoing paper but, as noted by these investigators, the waveform yielding point at the lowest stress was probably not steady and the point is, therefore, likely to be set at too high a strain rate. The solid line through the plotted points is simply an indication of the trend of the calculated results and the broken line indicates the slope corresponding to the fourth-power law discovered by these authors. One can see that the calculated line is slightly less steep than the fourth-power law.

10.4 Exercises

10.4.1. Using the theory of weak elastic–plastic waves, derive equations for the interfacial stress and particle velocity produced by the impact described in Sect. 10.3 and illustrated in Figs. 10.3 and 10.4.

10.4.2. Consider the experiment depicted below. The sapphire crystal responds elastically and is characterized by the properties $\rho_R = 3988\,\mathrm{kg/m^3}$ and $C_0 = 11{,}186\,\mathrm{m/s}$. A laser velocity interferometer can be configured to record the particle velocity history at the interface between the aluminum alloy sample and the sapphire. The recorded interface velocity is shown in the right panel of the figure. Using the theory of weak elastic–plastic waves, plot the X–t and $t_{11} - \dot{x}$ diagrams for the experiment and determine C_0 and C_B for the aluminum.

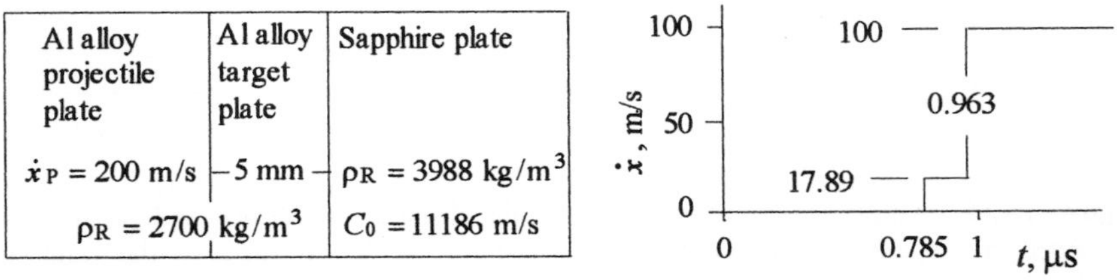

Figure for Exercise 10.4.2

10.4.3. Consider the experimental arrangement shown below in which a metal projectile impacts a stationary target of the same material (taken to be an elastic–plastic material that is adequately described by the theory of weak elastic–plastic deformation). The impact velocity is measured to be 136 m/s. A quartz plate configured as a stress gauge is bonded to the back of the target plate. The stress history at the interface of the target plate and the sample is measured by the gauge and the record obtained is shown. Describe the procedure for interpreting the experimental result

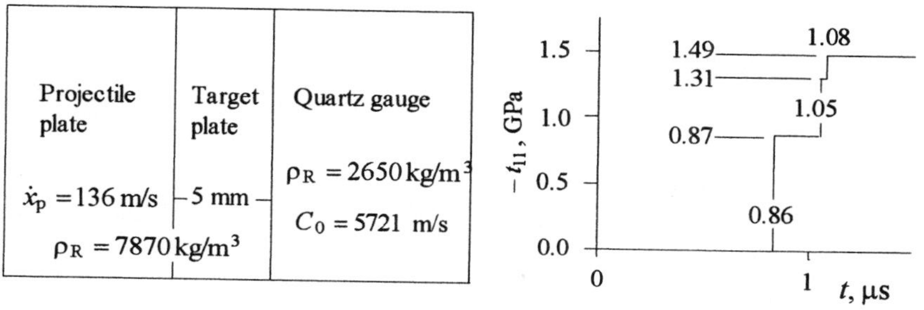

Figure for Exercise 10.4.3.

10.4.4. Calculate the shape of the stress pulse produced when a compressive stress $-t_{11} = 10$ GPa is suddenly applied to an aluminum alloy 6061-T6 half-space and removed after 2 μs. The aluminum is modeled as a non-hardening elastic–plastic material characterized by the parameters $\rho_R = 2703\,\mathrm{kg/m^3}$, $C_B = 5190$ m/s, $S = 1.338$, $\mu_R = 27.6$ GPa, $\mu_1 = 0.065\,\mathrm{GPa^{-1}}$, $\mu_2 = 0$, $Y_0 = 0.29$ GPa, and $v_{11}^{HEL} = 3.680 \times 10^4$.

10.4.5. Calculate the compression waveform produced when a stress $-t_{11} = 5$ GPa is suddenly applied to a hardening elastic–plastic material characterized by the parameters $h = 10$ and $n = 2$ in addition to the parameters given for the previous exercise.

CHAPTER 11

Porous Solids

Porous materials consist of a solid constituent intermingled with voids. Examples include natural materials such as soils, and a variety of manufactured materials such as lightly compacted powders and powder mixtures. They are of interest because the shock transition provides experimenters access to states of high pressure and very high temperature, because the shock compression process offers a means of producing new and unusual materials, and because they are unusually effective as shock attenuators.

In this chapter the porous material is treated as a continuum, so it is necessary that the characteristic scale of its microstructure (void diameter, etc.) be small relative to the size of objects to be investigated. It is often necessary to distinguish between the solid constituent of the porous material and the porous material as a whole. To facilitate this distinction we call the solid constituent the *parent matter*. General aspects of wave propagation in porous materials (e.g. waveforms, and shock attenuation) can be analyzed without detailed consideration of deformation occurring on the scale of the pore size, but details of the microstructure and the response of the material at this scale are important when bonding of powder particles, chemical reaction, etc. are of interest.

It is clear that the pore-collapse process is quite complicated, being associated with irregular motions, very large strains, and high temperatures in the neighborhood of each void. These local inhomogeneities lead to chemical, physical, and metallurgical effects that are not observed when the parent solid is placed in the same average thermodynamic state. The experimental literature addressing chemical, physical, and metallurgical effects peculiar to shock-compressed porous materials is extensive [50,83]. Chemical reactions and phase transformations occurring as a result of shock compression of porous materials, usually lightly-compacted powders, have led to the production of a variety of unusual materials. Initiation of detonation in porous explosives is often a result of the localized high temperatures produced during shock compression [8,27, 89].

The important effect of mesoscale aspects of porous material compaction upon chemical and metallurgical responses of these materials has motivated development of a number of theories that bridge the gap between those in which the material is modeled as a homogeneous continuum and fully resolved mesoscopic models. In general these theories, or computer simulations, are intended to capture void-collapse processes, the interactions among grains of a compacted powder, etc., along with the localized heating, material mixing, chemical reaction, and other phenomena that are important to understanding and using the unique responses of porous materials to shock compression.

Use of shock compression of porous materials for determination of the equation of state of materials at high temperature and pressure has been discussed in [67,68,104]. A variety of issues regarding the physical and chemical response of porous materials to compression by weaker shocks have been discussed in [10,33] and a comprehensive account of the mesomechanical aspects of shock compaction of porous materials has been presented by Nesterenko [83].

11.1 Materials of Very Low Density: Snowplow Model

Let us consider a highly porous material of low strength. Light, dry snow (ice crystals, the parent matter, separated by void space) is an example of such a material. If we were to compact this material with a piston in a confining cylinder, we would observe a stress–volume response like the compression curve of Fig. 11.1. The low-stress, gently sloping part of the curve is produced during compaction of the pores. The parent matter is not significantly compressed during this stage of the process. As the compaction proceeds, the pores are eliminated and further decrease in volume must arise through compression of the parent matter. This requires application of much greater stress, corresponding to the steep portion of the curve. Indeed, we can suppose that this latter stage of the process is essentially the same as if we began the process with a non-porous sample of the parent matter.

Let us idealize the curve by a horizontal and a vertical segment, as indicated by the broken lines on the graph. This idealization corresponds to a material that can be compacted to a non-porous state without application of significant stress,

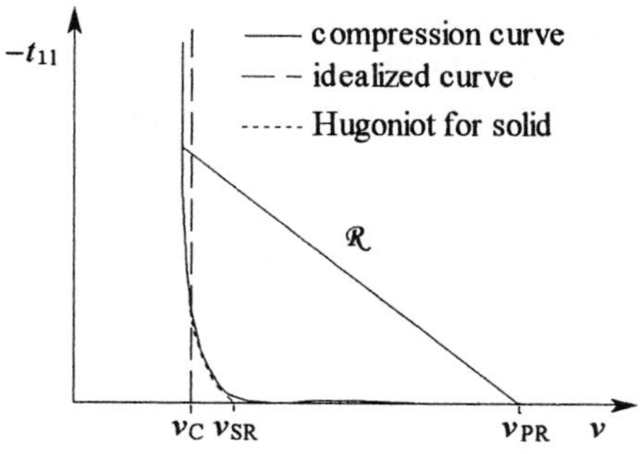

Figure 11.1. Compaction curves for a low-density porous material.

and that can be considered incompressible when fully compacted. Let us now consider propagation of a shock into this idealized material when it is in the reference state $\mathcal{S}^- = \{t_{11}^- = 0, \dot{x}^-, v^- = v_R, \varepsilon^- = 0\}$. Suppose that the material initially occupies the halfspace $x = X \geq 0$. Let a compressive stress step be applied to the boundary at $t = 0$, and let the magnitude of this stress be such that the particles acquire the constant velocity $\dot{x}^+ > 0$. At the later time t, the boundary $X = 0$ will have moved to the place $x = \dot{x}^+ t$ and the shock will be at the place $x = u_S t$. Since the material behind the shock is fully compacted it has the specific volume v_{SR} (the specific volume of the parent matter at the ambient condition). The material layer that was initially of thickness $\Delta X = (u_S - \dot{x}^-)t$ now has the reduced thickness $\Delta x = (u_S - \dot{x}^+)t$. Since the mass of the layer is unchanged by compaction, we have

$$u_S = \frac{v_R}{v_R - v_{SR}} \dot{x}^+ - \frac{v_{SR}}{v_R - v_{SR}} \dot{x}^-, \tag{11.1}$$

exactly the result that would have been obtained by application of the jump condition of Eq. 2.110_1. The driving stress, which can be calculated directly by a process similar to that used above, or from Eq. 2.110_2, is

$$-t_{11}^+ = \frac{1}{v_R - v_{SR}} (\dot{x}^+ - \dot{x}^-)^2, \tag{11.2}$$

and the specific internal energy of the material behind the shock is given by

$$\varepsilon^+ = \tfrac{1}{2}(\dot{x}^+ - \dot{x}^-)^2 = \tfrac{1}{2}(v_R - v_{SR})(-t_{11}^+). \tag{11.3}$$

By combining Eqs. 11.1 and 11.2 we find that

$$u_S = v_R \left(\frac{-t_{11}^+}{v_R - v_{SR}} \right)^{1/2} + \dot{x}^-, \tag{11.4}$$

which shows how the Eulerian shock velocity depends upon the slope of the Rayleigh line for the shock. It is easy to see from Fig. 11.1 that weak shocks, analyzed according to this model, propagate very slowly relative to the velocity of a shock of the same pressure jump in the parent solid.

When this model is used, it is usually assumed that the compacted material does not expand when the stress is relieved. This means that an unloading process would proceed along the vertical line in the figure. The Lagrangian speed of this wave would normally be given by Eq. 9.10_1, $C_L = [v_R\, \partial(-t_{11})/\partial v]^{1/2}$, which gives an infinite value for the model under discussion. The corresponding Eulerian wavespeed is also infinite.

Loading by a Flat-topped Pressure Pulse. Let us consider what happens when the boundary of a porous body occupying the halfspace $x = X \geq 0$ is subjected to the stress history $-t_{11}(t) \equiv p(t)$ given by

$$p = \begin{cases} 0, & t < 0 \\ p^+ > 0, & 0 \le t \le t^* \\ 0, & t > t^*. \end{cases} \tag{11.5}$$

In this model the compacted material forms a rigid layer that is subject to the pressure p^+ and is moving at the velocity

$$\dot{x}^+ = [(v_R - v_{SR})\, p^+]^{1/2}.$$

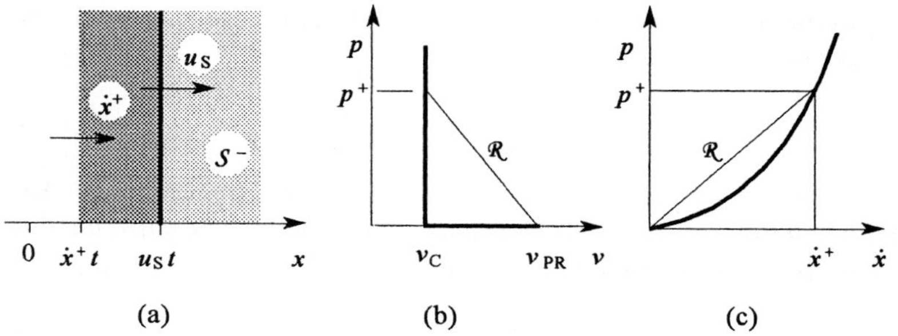

Figure 11.2. Shock propagating into a porous body. The configuration at $t > 0$ is shown in (a). The p–v and p–$\dot{x}$ Hugoniots are shown in (b) and (c), respectively.

When the stress is suddenly removed from the boundary at $t = t^*$ this fact is instantaneously communicated throughout the compacted layer and to the interface with the undisturbed material by a right-propagating shock of infinite velocity. This shock decelerates the compacted material to zero velocity. Because of the assumed locking response of the material, the compacted layer does not expand upon pressure release. The Eulerian x–t diagram for this loading–release process is given in Fig. 11.3.

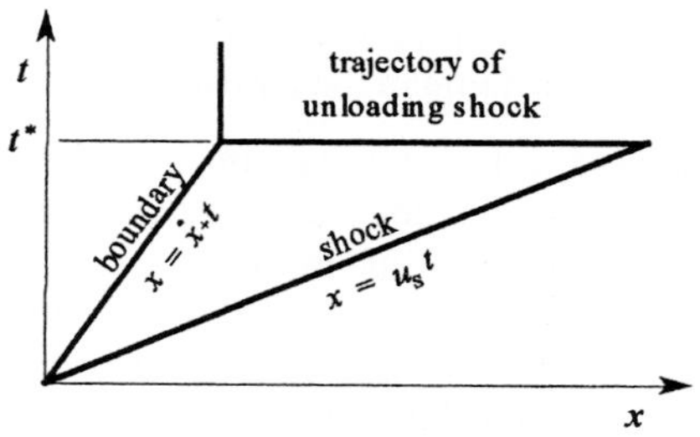

Figure 11.3. Eulerian x–t diagram for response to a flat-topped pulse introduced into a locking porous solid.

The specific internal energy of the compacted layer is

$$\varepsilon^{+} = \tfrac{1}{2}(\dot{x}^{+})^{2} = \tfrac{1}{2}(v_{R} - v_{C})\,p^{+} \tag{11.6}$$

when it is under pressure. At this time it also has the specific kinetic energy $\varepsilon^{+} = (\dot{x}^{+})^{2}/2$. When the pressure is relieved the motion ceases and this kinetic energy is converted to internal energy so the specific internal energy ε^{++} in the compacted, but depressurized, material is given by

$$\varepsilon^{++} = (\dot{x}^{+})^{2} = 2\varepsilon^{+}\,. \tag{11.7}$$

It is easy to see that this is exactly the work done upon the boundary divided by the mass of the compacted material (both for a unit-area cross section). The residual temperature of the compacted material is increased over the initial temperature of the material by the amount

$$\Delta\theta = \theta^{++} - \theta^{-} = \frac{1}{C^{p}}(\dot{x}^{+})^{2} = \frac{1}{C^{p}}(v_{R} - v_{SR})\,p^{+}\,, \tag{11.8}$$

where C^{p} is the specific heat at zero pressure and the temperature θ^{++}. All of the work done by the pressure pulse applied to the boundary is converted to heat.

Porous materials are often used as structural overlays to attenuate pressure pulses, and this analysis shows how the attenuation is accomplished. An indication of the temperatures achieved by a shock–release process in porous metals is given in Fig. 11.4. It is often inferred from Eq. 11.6_2 or 11.8_2 that an enormous specific internal energy can be imparted to highly distended porous materials if they are subjected to shocks of high pressure. This is true, of course, but it is important to realize that these highly distended materials are very soft and must be impacted at very high velocity if a high pressure is to be produced.

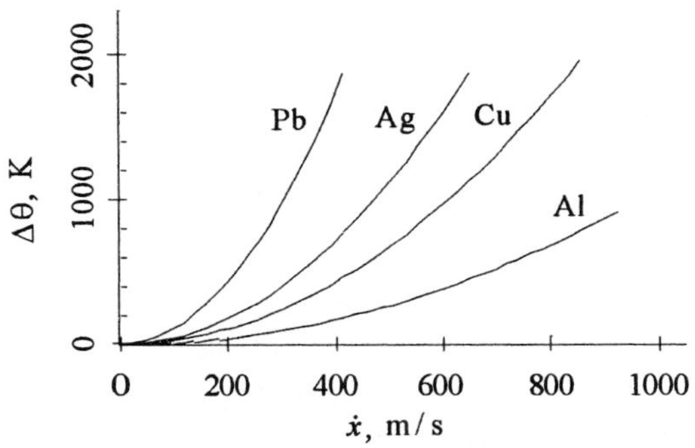

Figure 11.4. Temperature change in several porous metals accelerated by shocks of various strengths, as measured by particle velocity jump. When the shock strength is measured in this way, the temperature is independent of the initial porosity. Representative constant specific heats are used.

11.2 Strong Shocks

In the foregoing section, we considered the response of highly porous material to rather weak shocks—those strong enough to compact a very soft material but not strong enough to compress the parent matter significantly. In this section, we consider shocks that are far stronger than necessary to compact the porous material. We shall characterize the materials considered by simple thermodynamic properties so that explicit results can be obtained and the basic responses illustrated most directly. This restricts us to states of moderate pressure and temperature. Work conducted at the highest attainable pressures has been reviewed by Trunin et al. [104] and many important references are cited in this work. The objective of shock-compression experiments conducted in the high-pressure regime is to determine thermodynamic properties of the material in states of higher temperature and mass density than can be produced by other means.

Our first task is to determine the Hugoniot of a porous material from known properties of the parent matter. The Hugoniots of the porous material and the parent matter are as indicated qualitatively in Fig. 11.5.

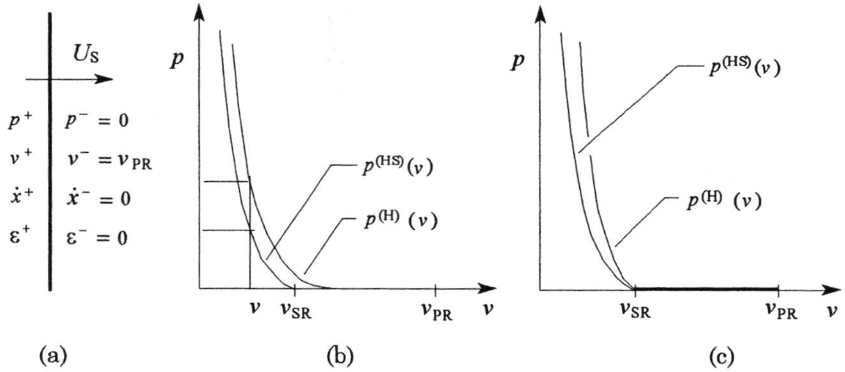

Figure 11.5. (a) Shock propagating into a porous material at rest in its reference configuration. (b) Pressure–volume Hugoniot for the porous material, $p^{(H)}(v)$, and for its parent matter, $p^{(HS)}(v)$. (c) Idealized pressure–volume Hugoniot for the porous material.

As suggested by Fig. 11.5b, the specific volume of the porous material at $p = 0$ is v_R and the specific volume of the parent solid at $p = 0$ and the same reference temperature is v_{SR}. Since we are considering pressures much greater than that required to compact the porous material to v_{SR}, we shall assume that the compaction pressure is zero so that the higher-volume portion of the Hugoniot of the porous material lies along the axis in the interval $v_{SR} \le v \le v_R$, as shown in Fig. 11.5c.

The Hugoniot of the porous material that has been fully compacted and for which $p > 0$ can be determined by calculating its pressure offset from the Hugoniot for the parent solid. The specific internal energy, $\varepsilon^{(H)}(v)$, of the porous material and that of the parent solid, $\varepsilon^{(HS)}(v)$, are related to the Hugoniot by the appropriate Rankine–Hugoniot equation:

$$\begin{aligned} \varepsilon^{(H)}(v) &= \varepsilon_R + \tfrac{1}{2} p^{(H)}(v)\,(v_R - v) \\ \varepsilon^{(HS)}(v) &= \varepsilon_{SR} + \tfrac{1}{2} p^{(HS)}(v)\,(v_{SR} - v). \end{aligned} \tag{11.9}$$

These two Hugoniots can be related by the Mie–Grüneisen equation

$$p^{(H)}(v) = p^{(HS)}(v) + \frac{\gamma(v)}{v}\left[\varepsilon^{(H)}(v) - \varepsilon^{(HS)}(v)\right]. \tag{11.10}$$

Since the analysis is restricted to the range in which the material is fully compacted, the Grüneisen coefficient used is that for the parent matter. Substitution of Eqs. 11.9 into Eq. 11.10 gives

$$p^{(H)}(v) = \frac{1 - \dfrac{\gamma(v)}{2v}(v_{SR} - v)}{1 - \dfrac{\gamma(v)}{2v}(v_R - v)}\, p^{(HS)}(v), \tag{11.11}$$

where we have neglected the surface energy distinguishing the porous and solid materials at $p = 0$ and $\theta = \theta_R$ so that $\varepsilon_R = \varepsilon_{SR}$. When the Hugoniot of the parent matter is given by Eq. 3.12, Eq. 11.11 takes the more specific form

$$p^{(H)}(v) = \frac{(\rho_{SR}\, C_B)^2 (v_{SR} - v)}{[1 - \rho_{SR}\, S\,(v_{SR} - v)]^2}\left(\frac{2v - \gamma(v)\,(v_{SR} - v)}{2v - \gamma(v)\,(v_R - v)}\right). \tag{11.12}$$

In applying this equation, $\gamma(v)$ is often taken to have the form

$$\gamma(v) = \gamma_{SR}\,\frac{v}{v_{SR}}, \tag{11.13}$$

where γ_{SR}, Grüneisen's coefficient for the parent matter in its reference state, is a material constant. Some representative Hugoniots obtained from Eqs. 11.12 and 11.13 are shown in Fig. 11.6.

Note that the value of $p^{(H)}(v)$ given by Eq. 11.12 is indeterminate at $v = v_{SR}$ when the reference specific volume has the value

$$v_R = v_{SR}\left(1 + \frac{2}{\gamma_{SR}}\right). \tag{11.14}$$

When v_R exceeds this value the slope of the Hugoniot changes sign: The specific volume of the shock-compressed porous material *exceeds* the reference specific volume of the parent solid. This behavior is observed for highly porous materials, but the temperature of these states is very high and neither Eq. 11.13 nor any other temperature-independent equation for Grüneisen's coefficient provides an adequate representation of γ (see [67,68,104] and [11, Appendix]). The simple theory of this chapter is restricted to states for which Eq. 11.13 is adequate.

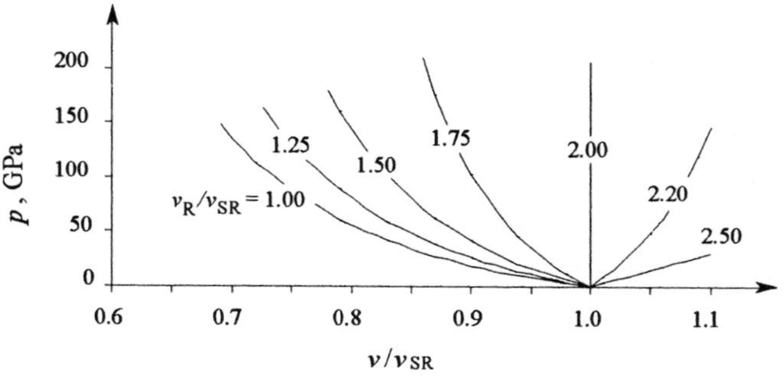

Figure 11.6. Hugoniot curves for distended copper calculated using Eqs. 11.12 and 11.13. Not shown is a horizontal segment of each curve that extends along the line $p = 0$ from the value of v_R/v_{SR} for the curve in question to the value of $v/v_{SR} = 1$.

When a solid material and a highly distended porous sample of the same material are subjected to the same shock pressure, the internal energy imparted to the porous material is very much greater than that imparted to the solid, as suggested by the drawing of Fig. 11.7 (according to Eq. 11.11, the shaded areas represent the increase in internal energy). Corresponding to the high internal energy in the porous material, we have a high thermal contribution to the pressure. This means that the cold contribution to the pressure in the porous material is less than in the solid material and the specific volume is correspondingly greater, as indicted on the figure. For highly distended materials subjected to high pressure, the thermal component can be larger than the total pressure. This means that the cold component must be tensile, and the specific volume greater than that of the parent solid at $p = 0$ as is predicted by Eq. 11.12. It is important to realize that, although the qualitative effect described is observed, the confidence one can place in the numerical values produced by this analysis is low by virtue of the extrapolations involved (the solid Hugoniot is extrapolated into the tensile region and the porous Hugoniot is far removed from this solid Hugoniot).

Measurement of a single Hugoniot provides very little thermodynamic information. When a parent material can be prepared as samples of various degrees of porosity, shock-compression experiments providing Hugoniots covering an area of the $p-v$ plane can be performed. In principle, fitting a surface to these data yields an equation of state $\varepsilon = \varepsilon(p, v)$ for the range of state variables covered. In practice, this has not proven practical, but such data are useful for guiding the development of theoretical equations of state. For example, the ratio $(p^{(H_1)} - p^{(H_2)})/(\varepsilon^{(H_1)} - \varepsilon^{(H_2)})$ calculated from two closely spaced Hugoniots provides an estimate of Grüneisen's coefficient for various values of specific volume. Since compression of highly porous materials by strong shocks produces very high temperatures, experiments of this sort are useful for investigating the properties of matter at temperatures and densities that cannot be produced by other means.

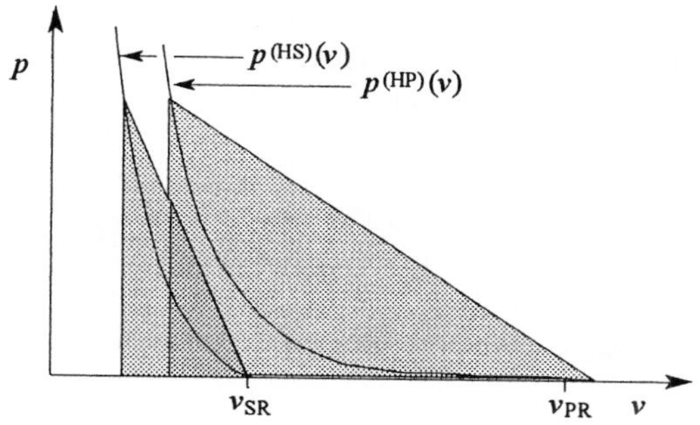

Figure 11.7. Hugoniots for solid and porous samples of the same parent matter, showing the relative thermal energy imparted by the shock.

The temperature of material in states on the Hugoniot is calculated for porous materials in the same way as for the parent matter by solving Eq. 5.142,

$$\frac{d\theta^{(H)}(v)}{dv} + \varphi(v)\theta^{(H)}(v) = \kappa(v),$$

with

$$\varphi(v) = \frac{\gamma(v)}{v} = \frac{\gamma_{SR}}{v_{SR}} \quad \text{and} \quad \kappa(v) = \frac{1}{2C_R^v}\left[p^{(H)}(v) + (v_R - v)\frac{dp^{(H)}(v)}{dv}\right]. \tag{11.15}$$

The Hugoniot of the porous material, $p^{(H)}(v)$, is related to that of the parent matter by Eq. 11.11 or the more specific form, Eq. 11.12.

The solution of this equation is

$$\theta^{(H)}(v) = \chi(v)\left[c + \int_{v_R}^{v} \frac{\kappa(v')}{\chi(v')} dv'\right], \tag{11.16}$$

where c is a constant of integration and

$$\chi(v) = \exp\left[\frac{\gamma_{SR}}{v_{SR}}(v_R - v)\right].$$

Since the Hugoniot lies along the line $p = 0$ on the interval $v_{SR} \le v \le v_R$, $\kappa(v) = 0$ in this range and Eq. 11.16 can be written

$$\theta^{(H)}(v) = \chi(v)\left[c + \int_{v_{SR}}^{v} \frac{\kappa(v')}{\chi(v')}dv'\right].$$

Because the compaction pressure is taken to be zero, no work is done during the compaction process and the temperature remains at its initial value, $\theta = \theta_R$, throughout the range $v_{SR} \le v \le v_R$. Therefore, the constant of integration is determined to be $c = \theta_R / \chi(v_{SR})$ and

$$\theta^{(H)}(v) = \chi(v)\left[\frac{\theta_R}{\chi(v_{SR})} + \int_{v_{SR}}^{v} \frac{\kappa(v')}{\chi(v')}dv'\right]. \tag{11.17}$$

The specific entropy is obtained from Eq. 5.136, which has the solution

$$\eta^{(H)}(v) = \eta_R + \int_{v_{SR}}^{v} \frac{C_R^v \psi(v')}{\theta^{(H)}(v')}dv'. \tag{11.18}$$

Finally, we note that the specific internal energy is given by the usual jump equation

$$\varepsilon^{(H)}(v) = \varepsilon_R + \tfrac{1}{2} p^{(H)}(v)(v_R - v). \tag{11.19}$$

The quantities η_R and ε_R appearing in these equations are values of the entropy and specific internal energy in the reference state. A particular example of the result of this analysis is given in Fig. 11.8.

Isentropes originating at points on the Hugoniot of porous materials are needed for analysis of decompression wave propagation. These isentropes are calculated in the same way as was done in Chapt. 5 for a compressible fluid. The isentrope is related to the Hugoniot for the porous material by the Rankine–Hugoniot equation

$$p^{(\eta)}(v) = p^{(H)}(v) + \frac{\gamma_{SR}}{v_{SR}}\left[\varepsilon^{(\eta)}(v) - \varepsilon^{(H)}(v)\right], \tag{11.20}$$

where we have adopted the equation $\gamma(v) = \gamma_{SR}\, v / v_{SR}$ for Grüneisen's coefficient and where the energy functions are given by

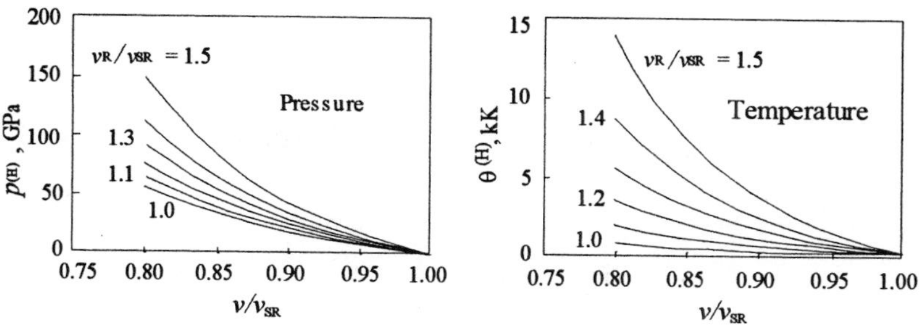

Figure 11.8. Hugoniots for solid and porous copper.

$$\varepsilon^{(\eta)}(v) = \varepsilon^{+} - \int_{v^{+}}^{v} p^{(\eta)}(v', \eta^{+})\, dv'$$
$$\varepsilon^{(H)}(v) = \varepsilon_{R} + \tfrac{1}{2}(v_{R} - v)\, p^{(H)}(v), \tag{11.21}$$

with the + superscripts designating values at the point of intersection of the isentrope and the Hugoniot. Substitution of Eqs. 11.21 into Eq. 11.20 and differentiation of the result leads to the differential equation

$$\frac{dp^{(\eta)}(v)}{dv} + \frac{\gamma_{SR}}{v_{SR}} p^{(\eta)}(v) = \kappa(v), \tag{11.22}$$

where

$$\kappa(v) = \frac{\gamma_{SR}}{2 v_{SR}} p^{(H)}(v) + \left[1 - \frac{\gamma_{SR}}{2}\left(\frac{v_R}{v_{SR}} - \frac{v}{v_{SR}}\right)\right] \frac{dp^{(H)}(v)}{dv}. \tag{11.23}$$

The solution of this equation that satisfies the initial condition $p^{(\eta)}(v^{+}) = p^{+}$ is

$$p^{(\eta)}(v, \eta^{+}) = \chi(v)\left[p^{+} + \int_{v^{+}}^{v} \frac{\kappa(v')}{\chi(v')} dv'\right]. \tag{11.24}$$

The temperature at points on this isentrope is given by

$$\theta^{(\eta)}(v, \eta^{+}) = \theta^{+} \exp\left[\gamma_{SR}\left(\frac{v^{+}}{v_{SR}} - \frac{v}{v_{SR}}\right)\right].$$

Some results of this analysis are given in Fig. 11.9.

11.3 Shocks of Moderate Strength: The $p-\alpha$ Theory

The first section of this chapter addressed the response of porous materials of very low strength and high porosity to loading by weak shocks. In this case, the shock produces compaction of the porous material but is not strong enough to

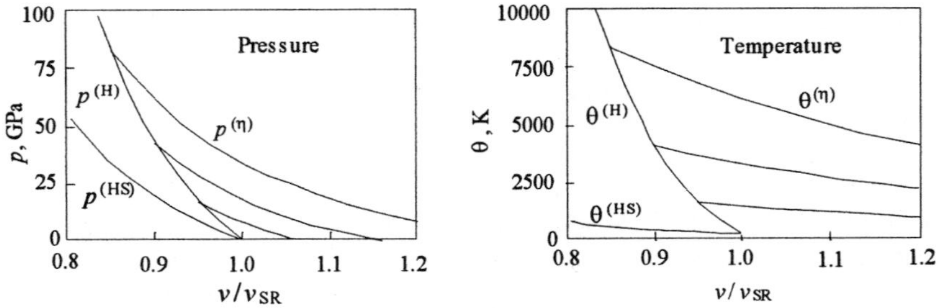

Figure 11.9. Hugoniot and decompression isentropes for copper distended to the state $v_R / v_{SR} = 1.5$.

cause significant compression of the parent matter. The second section addressed the effect of shocks of such great strength that details of the compaction process could be ignored and attention focused upon compression and decompression of the fully compacted material.

Now we consider the more difficult intermediate case in which both compaction and compression processes are important and interest is focused upon partially compacted states. The responses that we shall study have been investigated in considerable detail by theoretical, experimental, and computational methods, and the results obtained are the subject of a comprehensive, critical review by Nesterenko [83]. The materials that we shall consider are usually formed from weakly compacted granules of a ductile metal. The material responds elastically at very low stresses but yields in response to both pressure and shear stress when these quantities exceed critical values. In the analysis of this section we shall assume that the granules of the parent matter slide over one another freely enough that the macroscopic shear stress can be neglected and we are left with a theory involving only pressure. As the applied pressure increases the pressure–volume curve exhibits a discontinuity in slope that we identify as an elastic limit or pressure yield point. The observed yielding is attributed to the onset of inelastic compaction produced by rearrangement and plastic deformation of the granules of parent matter in response to the applied pressure. Beyond the yield point, the material also responds thermoelastically to the application of pressure, with this response being partly a decrease in void volume and partly compression of the parent matter.

The p–α theory that forms the basis of the discussion of this section was originally developed by Herrmann [54]. Because of the low tensile strength of these materials, consideration is restricted to states of compression. In this model the pores are taken into account by expressing the volume of a unit mass of the porous material as the sum of the volume of the parent solid and the volume of the voids. At this level of description no account is taken of the size or shape of individual voids, the distribution of void sizes, or any of the many mesoscale features that are apparent upon examination of the structure of typical materials to which the theory is applied. The objective in developing a model upon this basis is to take advantage of our knowledge of the thermodynamic properties of the parent solid when developing a description of the behavior of the porous material.

Thermoelastic Response. Applications of the p–α model lie primarily in the low-pressure region because, in most cases, compaction is essentially complete at a pressure of a few GPa. The state of the partially compacted material is described in terms of three independent variables: the specific volume, v, the specific entropy, η, and an additional variable α, called the *porosity*, defined by the equation

$$\alpha = \frac{v}{v_S}. \tag{11.25}$$

In this equation v_S is the specific volume of the parent solid at the prevailing thermodynamic state of the porous body and α can be interpreted as an internal state variable. It is a trivial but important observation that, at $p = 0$, the void volume can have any positive value and the specific volume and porosity of the material stand in the relation $\alpha = v / v_S$ that defines the porosity.

In addressing the thermoelastic compaction, we describe the parent solid by the specific internal energy function

$$\varepsilon = \varepsilon_S (v_S, \eta), \tag{11.26}$$

which serves as a thermodynamic potential for the pressure and temperature equations of state

$$p = -\frac{\partial \varepsilon_S (v_S, \eta)}{\partial v_S} \quad \text{and} \quad \theta = \frac{\partial \varepsilon_S (v_S, \eta)}{\partial \eta}. \tag{11.27}$$

The basic premise of the p–α theory is that conversion of the parent solid into a porous material at the same values of v_S and η is accomplished without altering its internal energy (thereby neglecting the surface energy, for example). This means that Eq. 11.26 describes *both* the parent matter *and* the porous material.

To express the equation of state of the porous material in terms of its specific volume, v, we substitute $v_S = v / \alpha$ into Eq. 11.26, obtaining

$$\varepsilon = \varepsilon(v, \eta, \alpha) = \varepsilon_S(v / \alpha, \eta) \tag{11.28}$$

as its equation of state. Menikoff [78] has recently modified Eq. 11.28 to include a term that depends on α, but we do not pursue that theory here. The pressure equation of state derived from Eq. 11.28_2 is

$$p(v,\eta,\alpha)=-\frac{\partial\varepsilon(v,\eta,\alpha)}{\partial v}=-\left.\frac{\partial\varepsilon_S}{\partial v_S}\right|_\eta\frac{\partial v_S}{\partial v}=-\left.\frac{1}{\alpha}\frac{\partial\varepsilon_S}{\partial v_S}\right|_\eta=\frac{1}{\alpha}p_S(v_S,\eta),\quad(11.29)$$

and the temperature equation of state is

$$\theta(v,\eta,\alpha)=\frac{\partial\varepsilon(v,\eta,\alpha)}{\partial\eta}=\frac{\partial\varepsilon_S(v/\alpha,\eta)}{\partial\eta}=\theta(v/\alpha,\eta)\,.\qquad(11.30)$$

In Herrmann's original development of the $p-\alpha$ model he assumed that the pressure in the porous material and in its solid portion were the same. Later, Carroll and Holt [19] suggested that Herrmann's postulate be replaced with the modified equation $p=p_S/\alpha$ because this equation leads to a better representation of experimental data. The development leading to Eq. 11.29 provides a thermodynamic justification for the postulate of Carroll and Holt.

Let us pursue a $p-\alpha$ theory based upon the complete Mie–Grüneisen equation of state for the parent solid. The reference state for the parent solid is taken to be that for which $p=0$, $v_S=v_{SR}$, and $\eta=\eta_R$. To simplify the analysis we assume that $\gamma_S(v_S)/v_S=\gamma_{SR}/v_{SR}$ and $C_S^v(\eta)=C_{SR}^v$. These simplifications are acceptable in view of the modest thermal effects in the range of states over which we intend to apply the theory. With these restrictions, the specific internal energy function describing the parent matter is given by Eq. 5.94,

$$\varepsilon_S(v_S,\eta)=\varepsilon_R+\varepsilon_S^{(\eta)}(v_S;\eta_R)+C_{SR}^v\,\theta_R\,\chi_c(v_S)[\omega_c(\eta)-1],\qquad(11.31)$$

where

$$\chi_c(v_S)=\exp\left[\frac{\gamma_{SR}}{v_{SR}}(v_{SR}-v_S)\right]\quad\text{and}\quad\omega_c(\eta)=\exp\left[\frac{\eta-\eta_R}{C_{SR}^v}\right].\qquad(11.32)$$

Substitution of Eq. 11.31 into Eq. 11.28 yields the specific internal energy function

$$\varepsilon(v,\eta,\alpha)=\varepsilon_R+\varepsilon_S^{(\eta)}(v/\alpha;\eta_R)+C_{SR}^v\,\theta_R\,\chi_c(v/\alpha)[\omega_c(\eta)-1]\qquad(11.33)$$

for the porous material. The pressure equation of state associated with Eq. 11.33 is

$$p(v,\eta,\alpha)=\frac{1}{\alpha}\left[p_S^{(\eta)}(v/\alpha;\eta_R)+\sigma_R\,\chi_c(v/\alpha)[\omega_c(\eta)-1]\right],\qquad(11.34)$$

where $\sigma_R=\rho_R\,\gamma_R\,C_{SR}^v\,\theta_R$. The temperature equation of state is

$$\theta(v,\eta,\alpha)=\theta_R\,\chi_c(v/\alpha)\,\omega_c(\eta),\qquad(11.35)$$

which can be solved to yield the equation

$$\eta(v/\alpha,\theta) = \eta_R + C_{SR}^v \ln\left(\frac{\theta}{\theta_R\,\chi_c(v/\alpha)}\right) \tag{11.36}$$

for the specific entropy.

Grüneisen's coefficient for the porous material is most conveniently determined using the thermodynamic derivative

$$\gamma = -\frac{v}{\theta_R}\frac{\partial\theta(v,\eta,\alpha)}{\partial v}\bigg|_\eta . \tag{11.37}$$

From the temperature equation of state, Eq. 11.35, we find that Grüneisen's coefficient for the porous material is given by

$$\gamma(v,\alpha) = \frac{\gamma_{SR}}{v_{SR}}\frac{v}{\alpha}. \tag{11.38}$$

Pressure-induced Yielding. In addition to the thermoelastic response just discussed, a porous material can respond to applied pressure by undergoing an inelastic compaction. As with the shear-induced yielding considered in metal plasticity, this compaction is caused by the applied pressure, but does not contribute to the pressure. When the pressure is removed, only the thermoelastic contribution to the decrease of specific volume is recovered.

One widely used approach to development of a theory such as we are considering is to suppose that the response of the porous material is the same as that of a hollow sphere of the parent matter that has the porosity of the porous material, and to calculate the response of this hollow sphere to imposition of a uniform external pressure using the ordinary methods of stress analysis [19,83]. This yields explicit equations relating the specific volume and the porosity of the sphere to the pressure. The hollow-sphere model predicts yielding, compaction of the pore and compression of the parent matter when pressure is applied. It also predicts the response to removal of the pressure. This model is informative of the issues we face in predicting the response of porous materials, but the assumed deformation is not at all similar to that experienced by actual porous materials, as shown by both experimental observations and numerical simulations in which details of the microstructure of the porous body are resolved.

Inelastic compaction occurs when the applied pressure exceeds a critical value. Boade [13] proposed that the critical pressure can be expressed as a function of α by the empirical equation

$$p_Y(\alpha) = p_{Y0} - \hat{p}\ln\left(\frac{\alpha-1}{\alpha_{Y0}-1}\right), \tag{11.39}$$

in which α_{Y0} and p_{Y0} are the values of porosity and pressure at which yielding begins and $\hat{p}$ is an additional material constant. Polynomial and other forms of the function $p_Y(\alpha)$ have also been used. Because of our intention to apply the theory to shock-compression phenomena, we shall identify p_{Y0} and α_{Y0} with the Hugoniot elastic-limit parameters. When a pressure in excess of $p_Y(\alpha)$ is applied, the porosity decreases to a value at which the material can support the applied pressure. Since the equilibrium pressure is entirely of thermoelastic origin, it is given by Eq. 11.34 or the equivalent equation

$$p(v,\theta,\alpha)=\frac{1}{\alpha}\left[\, p_S^{(\eta)}(v/\alpha;\eta_R)+\sigma_R\,[(\theta/\theta_R)-\chi_c(v/\alpha)]\right] \tag{11.40}$$

obtained when Eq. 11.36 is used to replace the specific entropy by the temperature. When we equate the pressures given by this equation and by Eq. 11.39 we obtain the equation

$$\frac{1}{\alpha}\left[\, p_S^{(\eta)}(v/\alpha;\eta_R)+\sigma_R\,[(\theta/\theta_R)-\chi_c(v/\alpha)]\right]= p_{Y0}-\hat{p}\ln\left(\frac{\alpha-1}{\alpha_{Y0}-1}\right), \tag{11.41}$$

relating v, α, and θ in material that is undergoing inelastic compaction. When θ is specified this becomes an equation relating v and α, and substitution of this relation into Eq. 11.40 yields an isotherm for compaction of the porous material.

As an example of the foregoing procedure, we consider compaction of the porous copper material described in Table 14.1. The isotherms $\alpha^{(\theta)}(v,\theta_R)$ and $p^{(\theta)}(v,\theta_R)$ that follow from the analysis are shown in Fig. 11.10.

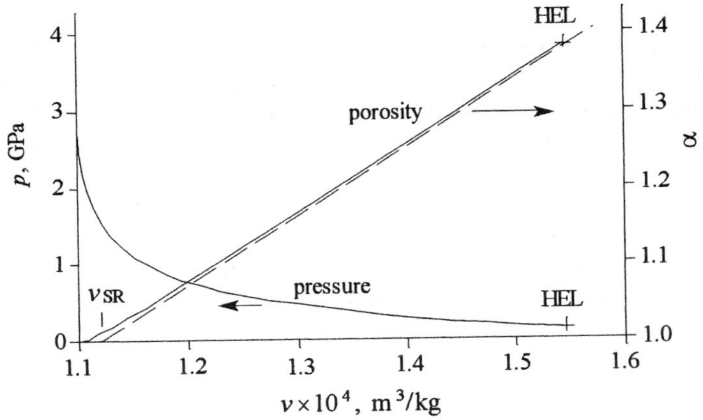

Figure 11.10. Pressure and porosity isotherms for the porous copper material described in Table 14.1. The broken line represents the function $\alpha(v)=v/v_{SR}$, which gives the values of α for which the pressure vanishes.

Table 11.1. Characteristics of a porous copper product studied by Boade [13]

Parent matter [77]	Porous material
$\rho_{SR} = 8930\,\text{kg/m}^3$	$\rho_R = 6430\,\text{kg/m}^3$
$C_B = 3940\,\text{m/s}$	$\alpha_R = 1.389$
$S = 1.489$	$p^{HEL} = 0.135\,\text{GPa}$
$\gamma_{SR} = 1.99$	$v^{HEL} = 1.546 \times 10^{-4}\,\text{m}^3/\text{kg}$
$C_{SR}^{v} = 392.7\,\text{J/(kg K)}$	$\alpha^{HEL} = 1.382$
$\theta_R = 293\,\text{K}$	$\hat{p} = 0.394\,\text{GPa}$
$\varepsilon_R = 77{,}732\,\text{J/kg}$	$\varepsilon^{HEL} = 77{,}792\,\text{J/kg}$

The Hugoniot. When Eqs. 11.33 and 11.34 are combined to eliminate $\omega_c(\eta)$ we obtain the pressure equation of state in the form

$$p(v,\eta,\alpha) = \frac{1}{\alpha}\left[p_S^{(\eta)}(v/\alpha;\eta_R) + \frac{\gamma_{SR}}{v_{SR}}[\varepsilon(v,\eta,\alpha) - \varepsilon_S^{(\eta)}(v/\alpha;\eta_R)]\right]. \tag{11.42}$$

Evaluation of this equation at the specific volume v on the Hugoniot yields the result

$$p^{(H)}(v) = \frac{1}{\alpha}p_S^{(\eta)}(v/\alpha;\eta_R) + \frac{\gamma_{SR}}{\alpha v_{SR}}\left[\varepsilon^{(H)}(v) - \varepsilon_S^{(\eta)}(v/\alpha;\eta_R)\right]. \tag{11.43}$$

The functions $p^{(H)}(v)$ and $\varepsilon^{(H)}(v)$ are related by the Rankine–Hugoniot equation

$$\varepsilon^{(H)}(v,\alpha) = \varepsilon^{HEL} + \tfrac{1}{2}[p^{(H)}(v,\alpha) + p^{HEL}](v^{HEL} - v), \tag{11.44}$$

where the HEL quantities are those for the porous material. Substitution of Eq. 11.44 into Eq. 11.43 yields the Hugoniot

$$p^{(H)}(v) = \left\{\frac{1}{\alpha}p_S^{(\eta)}(v/\alpha;\eta_R) - \frac{\gamma_{SR}}{\alpha v_{SR}}\varepsilon_S^{(\eta)}(v/\alpha;\eta_R) + \frac{\gamma_{SR}}{\alpha v_{SR}}\left[\varepsilon^{HEL} + \frac{1}{2}p^{HEL}(v^{HEL} - v)\right]\right\}\left[1 - \frac{\gamma_{SR}}{2\alpha v_{SR}}(v^{HEL} - v)\right]^{-1} \tag{11.45}$$

for the porous material.

It is convenient to choose the reference state so that this Hugoniot is centered on the HEL. Evaluation of Eq. 11.45 at the HEL gives

$$p^{HEL} = \frac{1}{\alpha^{HEL}}\left[p_S^{(\eta)}\left(v_S^{HEL};\eta_S^{HEL}\right) - \frac{\gamma_{SR}}{v_{SR}}\varepsilon_S^{(\eta)}\left(v_S^{HEL};\eta_S^{HEL}\right) + \frac{\gamma_{SR}}{v_{SR}}\varepsilon^{HEL}\right], \tag{11.46}$$

where $v_{\mathrm{S}}^{\mathrm{HEL}} = \alpha^{\mathrm{HEL}} p^{\mathrm{HEL}}$, and we are led to choose the isentrope that passes through this HEL. Therefore, we must have

$$\begin{aligned} \frac{1}{\alpha^{\mathrm{HEL}}} p_{\mathrm{S}}^{(\eta)}\left(v_{\mathrm{S}}^{\mathrm{HEL}}; \eta_{\mathrm{S}}^{\mathrm{HEL}}\right) &= p^{\mathrm{HEL}} \\ \varepsilon_{\mathrm{S}}^{(\eta)}\left(v_{\mathrm{S}}^{\mathrm{HEL}}; \eta_{\mathrm{S}}^{\mathrm{HEL}}\right) &= \varepsilon^{\mathrm{HEL}}. \end{aligned} \tag{11.47}$$

Our next task is establishing the specific form of the isentrope. In the general spirit of this volume, we shall use an isentrope based upon the principal Hugoniot for the parent matter, Eq. 5.176:

$$p_{\mathrm{S}}^{(\eta)}(v_{\mathrm{S}}; \eta_{\mathrm{S}}^{\mathrm{HEL}}) = \chi_{\mathrm{c}}(v_{\mathrm{S}}) \left[p_{\mathrm{S}}^{\mathrm{HEL}} + \int_{v_{\mathrm{S}}^{\mathrm{HEL}}}^{v_{\mathrm{S}}} \frac{\kappa_{\mathrm{c}}(v')}{\chi_{\mathrm{c}}(v')} dv' \right], \tag{11.48}$$

where

$$\begin{aligned} \chi_{\mathrm{c}}(v_{\mathrm{S}}) &= \exp\left[\frac{\gamma_{\mathrm{SR}}}{v_{\mathrm{SR}}} (v_{\mathrm{S}}^{\mathrm{HEL}} - v_{\mathrm{S}}) \right] \\ \kappa_{\mathrm{c}}(v_{\mathrm{S}}) &= \frac{\gamma_{\mathrm{SR}}}{2 v_{\mathrm{SR}}} p^{(\mathrm{HS})}(v_{\mathrm{S}}) + \left[1 - \frac{\gamma_{\mathrm{SR}}}{2 v_{\mathrm{SR}}} (v_{\mathrm{S}}^{\mathrm{HEL}} - v_{\mathrm{S}}) \right] \frac{dp^{(\mathrm{HS})}(v_{\mathrm{S}})}{dv_{\mathrm{S}}}. \end{aligned} \tag{11.49}$$

We shall adopt Eq. 3.12 as the representation of $p^{(\mathrm{HS})}(v_{\mathrm{S}})$, so

$$p^{(\mathrm{HS})}(v_{\mathrm{S}}) = \frac{(\rho_{\mathrm{SR}}\, C_{\mathrm{B}})^2\, (v_{\mathrm{SR}} - v_{\mathrm{S}})}{[1 - \rho_{\mathrm{SR}}\, S\, (v_{\mathrm{SR}} - v_{\mathrm{S}})]^2}. \tag{11.50}$$

The specific internal energy on the isentrope, as given by Eq. 5.132, is

$$\varepsilon_{\mathrm{S}}^{(\eta)}(v_{\mathrm{S}}, \eta_{\mathrm{S}}^{\mathrm{HEL}}) = \varepsilon_{\mathrm{S}}^{\mathrm{HEL}} - \int_{v_{\mathrm{S}}^{\mathrm{HEL}}}^{v_{\mathrm{S}}} p_{\mathrm{S}}^{(\eta)}(v', \eta_{\mathrm{S}}^{\mathrm{HEL}}) dv'. \tag{11.51}$$

It is apparent from the equations that these isentropes pass through the HEL. Equation 11.48 can now be rewritten in the form

$$p_{\mathrm{S}}^{(\eta)}(v_{\mathrm{S}}, \eta_{\mathrm{S}}^{\mathrm{HEL}}) = \chi_{\mathrm{c}}(v_{\mathrm{S}}) \left[p_{\mathrm{S}}^{\mathrm{HEL}} + \int_{v_{\mathrm{S}}^{\mathrm{HEL}}}^{v_{\mathrm{S}}} \frac{\kappa_{\mathrm{c}}(v')}{\chi_{\mathrm{c}}(v')} dv' \right], \tag{11.52}$$

and Eq. 11.51 becomes

$$\varepsilon_{\mathrm{S}}^{(\eta)}(v_{\mathrm{S}}; \eta_{\mathrm{S}}^{\mathrm{HEL}}) = \varepsilon^{\mathrm{HEL}} - \int_{v_{\mathrm{S}}^{\mathrm{HEL}}}^{v_{\mathrm{S}}} p_{\mathrm{S}}^{(\eta)}(v'; \eta_{\mathrm{S}}^{\mathrm{HEL}}) dv'. \tag{11.53}$$

In applying the foregoing equations one may assume that compression from the initial state to the Hugoniot elastic limit is isentropic, so that $\eta_S^{HEL} = \eta_R$.

Our next task is determination of α as a function of v on the Hugoniot. Since states on the Hugoniot are in equilibrium, the pressure is the same as the critical pressure for inelastic compaction as given, in this case, by Eq. 11.39. Equating these two pressures gives the equation

$$p^{HEL} - \hat{p}\ln\left(\frac{\alpha-1}{\alpha^{HEL}-1}\right) = \left\{\frac{1}{\alpha}p_S^{(\eta)}(v/\alpha;\eta_S^{HEL}) - \frac{\gamma_{SR}}{\alpha v_{SR}}\varepsilon_S^{(\eta)}(v/\alpha;\eta_S^{HEL}) + \frac{\gamma_{SR}}{\alpha v_{SR}}\left[\varepsilon^{HEL} + \frac{1}{2}p^{HEL}(v^{HEL}-v)\right]\right\}\left[1 - \frac{\gamma_{SR}}{2\alpha v_{SR}}(v^{HEL}-v)\right]^{-1}, \tag{11.54}$$

relating α and v on the Hugoniot. It must be solved numerically, but the result is the Hugoniot $\alpha = \alpha^{(H)}(v)$ and, with this, Eq. 11.45 can be evaluated to give $p^{(H)}(v)$. Pressure and porosity Hugoniots for the material described in Table 14.1 are illustrated in Fig. 11.11.

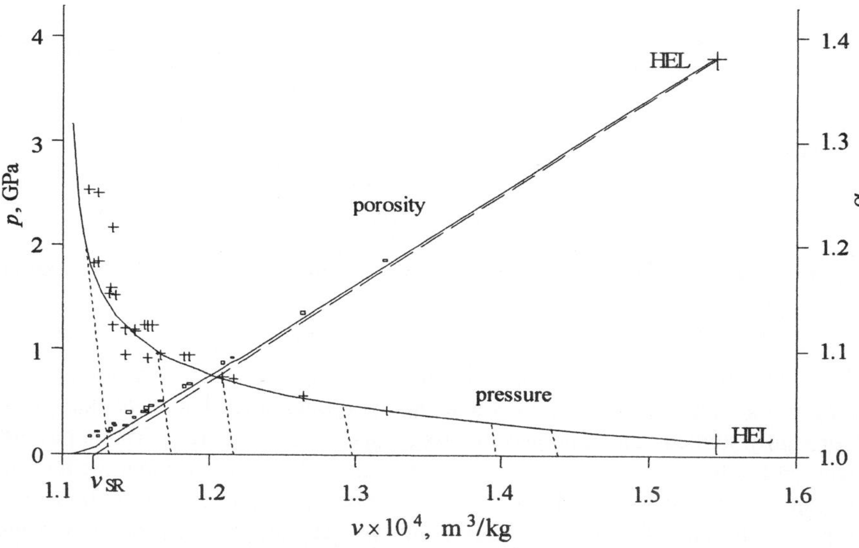

Figure 11.11. Pressure and porosity Hugoniots for the material described in Table 14.1. The curves are derived from the theory presented in this section and the plotted points are from Boade's measurements, but with the latter having been re-evaluated (producing only slight changes) using the theory of this section. The broken line represents the function $\alpha(v) = v/v_{SR}$, which gives the values of $\alpha(v)$ for which the pressure vanishes. The steep dotted lines are decompression isentropes.

Finally, from Eq. 11.35 we obtain the temperature Hugoniot

$$\theta^{(H)}(v) = \theta_R\, \chi_c(v/\alpha) + \frac{v_{SR}}{C_{SR}^{v}\, \gamma_{SR}} \left[\alpha\, p^{(H)}(v) - p_S^{(\eta)}(v/\alpha;\, \eta_S^{HEL}) \right], \tag{11.55}$$

and, from Eq. 11.36 we obtain the specific entropy Hugoniot

$$\eta^{(H)}(v) = \eta^{HEL} + C_{SR}^{v} \ln\left(\frac{\theta^{(H)}(v)}{\theta_R\, \chi_c(v/\alpha)} \right). \tag{11.56}$$

Example temperature and specific entropy Hugoniots for the material described in Table 14.1 are illustrated in Fig. 11.12.

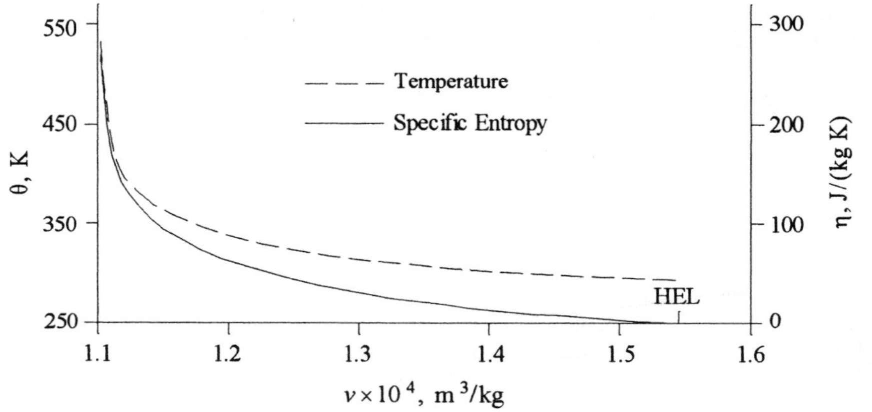

Figure 11.12. Temperature and specific entropy Hugoniots for the material described in Table 14.1.

The Decompression Isentrope. Decompression processes undergone by porous materials are very important in many applications, but have been less thoroughly studied than compression processes. When a ductile or friable porous material has been fully compacted it seems reasonable to suppose that the loss of porosity is irrecoverable and that a decompression process will follow the isentrope for the parent solid. For many materials, decompression processes from partially compacted states couldinvolve both expansion of the parent matter and at least some recovery of the porosity. It is to be expected that the expansion will proceed along the decompression isentrope from the Hugoniot state $\mathcal{S}^{+}$ to a state $\mathcal{S}^{++}$ at which the pressure is zero. The specific entropy of material in the state $\mathcal{S}^{+}$ is determined by evaluating Eq. 11.55 for the temperature θ^{+} on the Hugoniot and then Eq. 11.56 for the specific entropy, η^{+}, at this point on the Hugoniot and also on a decompression isentrope. The pressure isentrope can be obtained from Eq. 11.34 if an equation analogous to Eq. 11.39 is available to permit separation of $v_S = v/\alpha$ into its factors v and α. In the absence of such an equation one can still determine the variables characterizing the state $\mathcal{S}^{++}$. This is often sufficient because the isentrope is quite steep and can be satisfactorily approximated by a straight line connecting the states $\mathcal{S}^{+}$ and $\mathcal{S}^{++}$.

Let us consider determination of the state $\mathcal{S}^{++}$ in which $p = p^{++} = 0$ and $\eta = \eta^{+} = \eta^{++}$. Evaluation of Eq. 11.34 at this state gives

$$p_S^{(\eta)}(v_S^{++};\eta_R) + \sigma_R\,\chi_c(v_S^{++})[\omega_c(\eta^{++}) - 1] = 0, \tag{11.57}$$

which can be solved for v_S^{++}. With this information in hand, Eq. 11.33 can be evaluated to yield the value ε^{++}. We shall now consider a shock transition from $\mathcal{S}^{++}$ to $\mathcal{S}^{+}$ in the approximation that the shock is weak enough that the Hugoniot centered on $\mathcal{S}^{++}$ can be identified with the decompression isentrope that we seek. The jump conditions of interest are

$$\begin{aligned} (\rho_{SR}\,U_S)^2\,(v^{++} - v^{+}) &= p^{+} \\ \tfrac{1}{2}p^{+}\,(v^{++} - v^{+}) &= \varepsilon^{+} - \varepsilon^{++}. \end{aligned} \tag{11.58}$$

These equations can be rewritten in the forms

$$v^{++} = v^{+} + \frac{2(\varepsilon^{+} - \varepsilon^{++})}{p^{+}} \tag{11.59}$$

and

$$U_S^2 = \frac{(p^{+})^2}{2\rho_{SR}^2\,(\varepsilon^{+} - \varepsilon^{++})}, \tag{11.60}$$

from which we can calculate v^{++} and the velocity, U_S, of the decompression wave. Several decompression isentropes are shown on the graph of Fig. 11.11. When, as in the present example, the decompression path is much steeper than the Hugoniot a decompression wave will overtake the shock very rapidly, which causes much more rapid attenuation than is the case with shocks in non-porous materials. One application of porous materials involves placing them in a layer over a structure in order to attenuate stress pulses that might otherwise cause spallation of the structural material.

As noted previously, decompression of porous solids has been less thoroughly investigated experimentally than have compression processes. Because the present example is calculated upon the basis of Boade's experimental investigation, it is important to mention that his decompression measurements indicate a process that follows more closely along the Hugoniot than do the curves calculated here. It may be that the best way to determine decompression paths of porous materials is through pulse-attenuation measurements.

Steady Compaction Waveforms. Shocks introduced into porous materials usually evolve toward structured waves, which may become steady. The thickness of these waves depends upon the time required for the pores to close that, in turn, depends upon both the pressure and the dimension of the pores. The shock thickness is often significant in comparison to propagation distances of interest and cannot be ignored. To capture this phenomenon it is necessary to develop an equation that describes the evolution of the porosity toward equilibrium. Analysis of evolving waveforms requires use of numerical methods for solving the partial differential equations of conservation of mass, momentum, and energy in conjunction with constitutive equations such as we have considered in previous parts of this section and an evolutionary equation for the porosity. Nevertheless, it is informative to study the structure of steady waveforms as we did in the case of elastic–viscoplastic waves. In a steady wave propagating in the $+X$ direction, the dependent variables are all functions of the single independent variable

$$Z = X - U_S\, t\,. \tag{11.61}$$

The conservation equations 2.124 expressing the pressure, particle velocity, and specific internal energy as functions of $v(Z)$ can be written

$$\begin{aligned} p(Z) &= p^{\mathrm{HEL}} + (\rho_R U_S)^2 [v^{\mathrm{HEL}} - v(Z)] \\ \dot{x}(Z) &= \dot{x}^{\mathrm{HEL}} + \rho_R U_S\, [v^{\mathrm{HEL}} - v(Z)] \\ \varepsilon(Z) &= \varepsilon^{\mathrm{HEL}} + p^{\mathrm{HEL}}\, [v^{\mathrm{HEL}} - v(Z)] + \tfrac{1}{2}(\rho_R U_S)^2 [v^{\mathrm{HEL}} - v(Z)]^2 . \end{aligned} \tag{11.62}$$

The specific internal energy, pressure and temperature equations of state have been given in the previous paragraphs. To complete the constitutive description of the material we must adopt an evolutionary equation for the porosity. We shall consider the simplest equation that might prove effective,

$$\dot{\alpha} = -\frac{\alpha - \alpha_{\mathrm{eq}}(v, \eta)}{\tau}, \tag{11.63}$$

where the material constant, τ, is a characteristic compaction time. According to this equation, α approaches its equilibrium value, $\alpha_{\mathrm{eq}}(v, \eta)$, at a rate that is proportional to the difference between its current and equilibrium values. As we know, and can see from Eq. 11.62_1, the state point for a specific material particle moves along the Rayleigh line as the steady wave passes over the particle. Therefore, we need to determine values of α and $\alpha_{\mathrm{eq}}(v, \eta)$ as functions of the independent variable Z that characterizes points on the Rayleigh line. For a steady structured shock, the equilibrium compaction is $\alpha^{(\mathrm{H})}(v)$ and the current value of α is the value of this variable on the Rayleigh line at the same specific volume, $\alpha^{(\mathcal{R})}(v)$.

Let us consider determination of the function $\alpha^{(\mathcal{R})}(v)$. We have defined a Hugoniot curve as the locus of states that can be reached by a shock transition from a given initial state. As we consider steady waves we are led to alter this definition to say that a Hugoniot curve is the locus of states reached in a steady wave transition from the given initial state. Since the transition from the initial state to any state in a steady waveform satisfies the same equations as a shock transition to the state in question, we see that a steady compression process can be interpreted as a sequence of transitions from the initial state to states on members of a family of Hugoniots with each member of the family being characterized by a fixed value of α. These are called *constant-porosity Hugoniots.* Each of these Hugoniots is given by Eq. 11.45 when the latter is evaluated for the constant value of α characterizing the Hugoniot. A graph of several of these Hugoniots is given in Fig. 11.13.

Rayleigh lines are represented by Eq. 11.62_1. The values of α and v corresponding to points of intersection of the constant-porosity Hugoniots with the Rayleigh line form the function $\alpha^{(\mathcal{R})}(v)$. This function has been calculated numerically for the Rayleigh lines shown in Fig. 11.13. In these cases $\alpha^{(\mathcal{R})}(v)$ is very nearly a linear function as v ranges from v^{HEL} to its value at the intersection of the Rayleigh line and the equilibrium Hugoniot. For the calculations to be presented, the linear approximation to $\alpha^{(\mathcal{R})}(v)$ has been used.

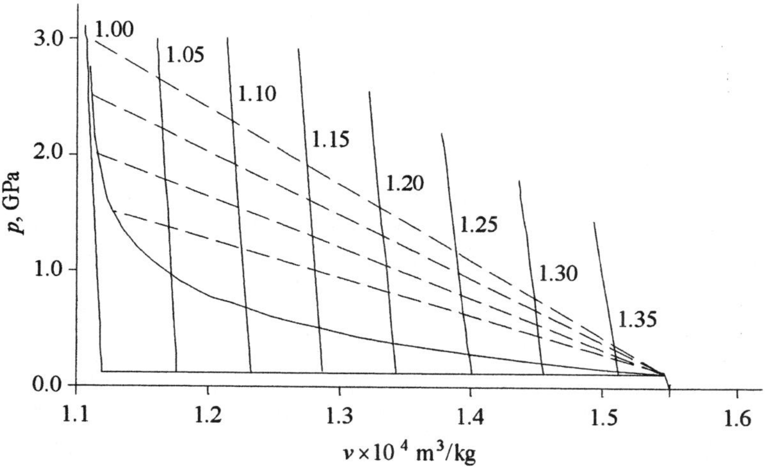

Figure 11.13. Partial-compaction and equilibrium Hugoniot curves for the material described in Table 14.1. Each of the steep curves is a constant-porosity Hugoniot for the value of α shown near the curve. Also shown is the equilibrium Hugoniot and Rayleigh lines for several steady waves.

In the steady wave, i.e., along $\mathcal{R}$, we have

$$\frac{d\alpha}{dZ} = \frac{d\alpha^{(\mathcal{R})}(v)}{dv}\frac{dv}{dZ} \qquad (11.64)$$

and, since $\dot{\alpha} = -U_S\, d\alpha/dZ$, Eq. 11.63 becomes

$$\frac{dv}{dZ} = \frac{\alpha^{(\mathcal{R})}(v) - \alpha^{(H)}(v)}{U_S\, \tau\, [d\alpha^{(\mathcal{R})}(v)/dv]} . \tag{11.65}$$

This equation is immediately integrable to yield the solution

$$Z = \int \frac{\alpha^{(\mathcal{R})}(v) - \alpha^{(H)}(v)}{U_S\, \tau\, [d\alpha^{(\mathcal{R})}(v)/dv]}\, dv + \text{const.}, \tag{11.66}$$

where the integration is performed numerically and the constant of integration is chosen to place the half-amplitude point near $Z = 0$. When the function $v(Z)$ given by Eq. 11.66 is substituted into Eqs. 11.62 we obtain pressure, particle velocity, and specific internal energy waveforms. Three pressure waveforms calculated in this way are illustrated in Fig. 11.14. A noteworthy feature of these waveforms is the slow initial rise followed by a much more rapid rise of the higher amplitude portion of the waveform. This is a consequence of the fact that the Hugoniot rises slowly until compaction is nearly complete and then rises much more rapidly, leading to a large difference between $\alpha^{(\mathcal{R})}(v)$ and $\alpha^{(H)}(v)$ for values of v near full compaction.

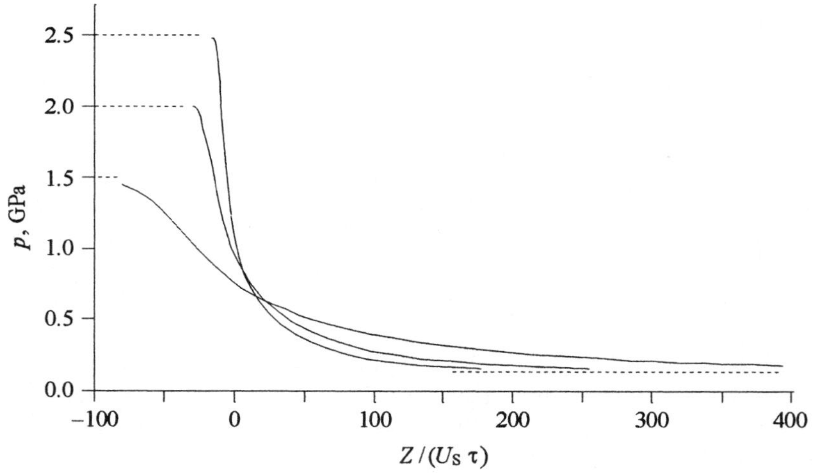

Figure 11.14. Steady waveforms calculated for the material described in Table 14.1.

CHAPTER 12

Spall Fracture

Solids fracture when subjected to tensile stress of sufficient magnitude. When this stress is produced as a result of the interaction of shocks it is usually of large magnitude and short duration, leading to formation of a diffuse distribution of microcracks or voids in the interior of a material body. This process is called *spall fracture* or *spallation*. Complete fracture results from coalescence of individual microfractures to form a plane of separation spanning the body.

From the time of the earliest observations of spall phenomena, investigators have been interested in its manifestations at the microscale. Although all levels of damage are of interest, the most revealing scientific information is obtained from samples exhibiting very low levels of damage. Material samples that have been recovered after being subjected to a spall test are sectioned to reveal the damage and permit description of its morphology. In some materials, described as brittle, the damage takes the form of a distribution of planar cracks. Damage in materials described as ductile takes the form of a distribution of rounded voids. In most materials the morphology is distinctly in the form of cracks or voids when the damage level is quite low. Some materials exhibit damage morphology that is intermediate between that of sharply defined cracks and nearly-spherical voids, and the distinction is further moderated at larger damage levels when crack opening becomes pronounced and voids coalesce into irregular arrays that resemble cracks. Microscopic observations occasionally disclose the site at which the damage was initiated and the mechanism by which it evolved. An understanding of the microscopic mechanisms underlying the spall process is essential to selection of materials that have desirable spall behavior and helpful to development of mathematical models of damage accumulation leading to spall. Many photomicrographs of cross sections of spall-damaged materials are included in [4,31].

The simplest case in which a wave interaction leads to fracture occurs when two sufficiently strong plane decompression waves collide to produce a region of tensile stress in the interior of a material body. Levels of spall damage caused by this tensile stress field range from formation of a few isolated microfractures to complete separation of a layer of material from the remainder of the body. Experimental investigations show that a damage distribution develops gradually through processes of nucleation of individual defects (or activation of existing defects) followed by growth of these microfractures. In the later stages of the process, individual microfractures coalesce to form larger fractures. Experiments can be configured so that the damage accumulation process is arrested at any stage of its evolution.

Because of the high intensity and short duration of tensile stress application associated with spall phenomena the damage produced begins with nucleation of cracks or voids in concentrations of the order of $10^4/\text{mm}^3$ and average volume of the order of 10^{-6} mm^3, leading to void fractions of the order of 0.01. The damage then increases through growth of the nucleate microfractures to form damage distributions having void fractions exceeding 0.1. The void fraction is often adopted as a quantitative measure of damage, but other quantities such as the decrease in material strength have also been used. There is an extensive metallurgical literature describing the morphology and statistical properties of the damage distribution [4,30,81].

Early investigations of spall fracture addressed development of a *spall criterion*, i.e., a mathematical rule for deciding whether or not a material body will fracture under given conditions of stress, strain, temperature, and other continuum field variables. It is implicit in the concept of a spall criterion that spallation is a discrete event that either does, or does not, occur. Some specific level of damage is identified as that constituting a spall and the criterion provides the means to determine whether or not the material spalled. Modern theories of failure are based on the observation that damage develops gradually by a process of nucleation, growth, and coalescence of microfractures distributed within the volume of the material. Continuum theories have been developed to explain the process of damage evolution observed in shock-wave experiments. These theories take into account the effect of the evolving damage on the mechanical properties of the material and, thus, on the stress field that drives the spall process. The conventional concept of spall as a discrete phenomenon described by a criterion for its occurrence has been supplanted by the newer concept of continuous damage accumulation described by evolutionary equations, but the ideas and language of the discrete model are still in widespread use and spall criteria are adequate for solution of many practical problems.

There is an extensive literature on dynamic fracture and fragmentation. Books on the subject include [4,31,49,63,81,87].

12.1 Experimental Means of Producing Spall Fracture

In this section we briefly describe some situations in which spall fracture can occur. Practical solution of spall problems is almost always achieved through numerical simulation, but consideration of idealized situations that admit simple analytical solution provides the essential background for understanding the wave-propagation phenomena that result in spall fracture. The discussion of this section is carried out in the context of complete spall described by a critical normal stress criterion.

12.1.1 Plate-impact Experiment

In an experimental arrangement commonly used to study spallation, a projectile plate of thickness L_P is impacted on a target plate of the same material that has thickness L_T, greater than L_P, as shown in Fig. 12.1.

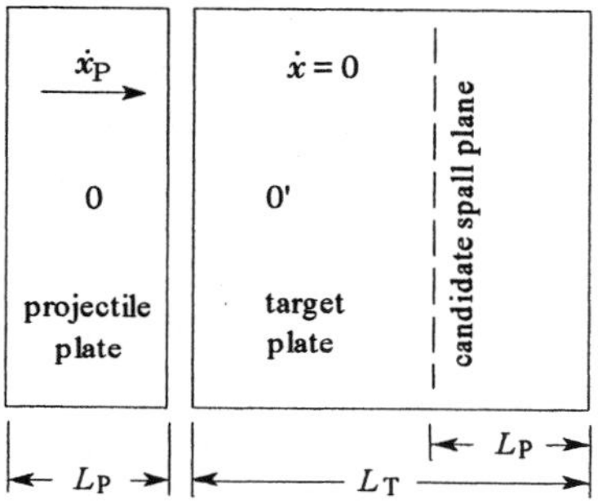

Figure 12.1. Plate-impact configuration for conducting a spall experiment. Note that this schematic greatly exaggerates the thickness of the plates relative to their lateral extent.

Spallation Due to Colliding Decompression Shocks. For the present discussion, let us suppose that the material is characterized by a simple normal Hugoniot and that the decompression wave can be approximated by a shock, as was done in Chap. 3. The interactions of interest have been analyzed in Sect. 3.7.3 and are illustrated in Figs. 12.2 and 12.3. The impact, at the velocity $\dot{x}_P$, produces a compressive stress accelerating the material to the right of the impact interface to the velocity $\dot{x}_P/2$ and decelerating the material to the left of the interface to this same velocity (see Fig. 12.2a). When the left-propagating compression wave reaches the unrestrained surface of the impactor plate it is reflected, leaving behind it a region (region 2) that is at rest in a state in which $t_{11} = 0$. This decompression wave passes through the impact interface without interaction since the impactor and target plates are of the same material and no tension is present to cause separation at the interface. Meanwhile, the right-propagating compression wave reflects from the back surface of the target plate, accelerating the material near this surface to the velocity $\dot{x}_P$ (see Fig. 12.2b).

The critical event occurs when the converging decompression waves meet. Since the material in region 3 is moving to the right, whereas the material in region 2 is at rest, the material will tend to separate. This tendency is resisted by tensile stress that develops at the plane of the interaction (see Fig. 12.2c). It is possible that this tension is sufficient to cause spallation, in which case the fract-

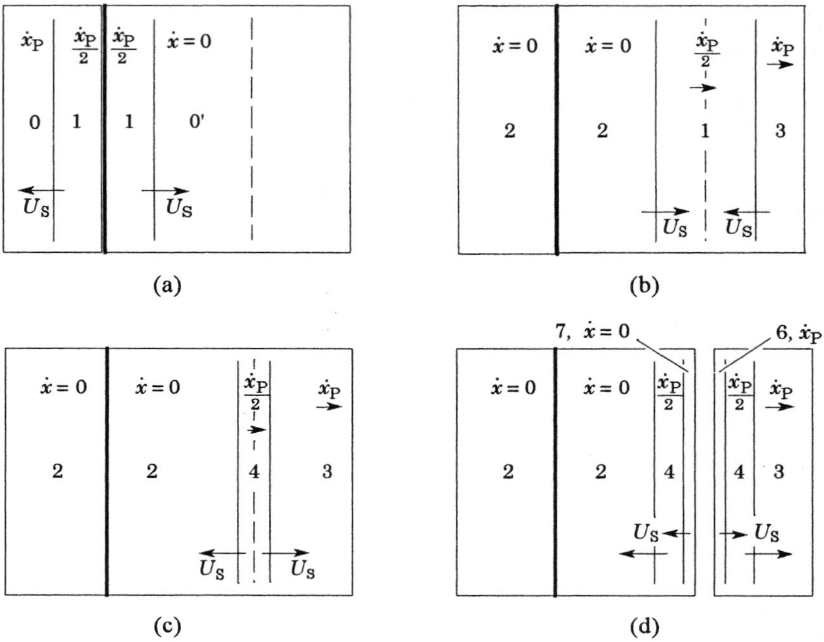

Figure 12.2. Wave interactions and spall formation resulting from collision of an impactor plate with a thicker target plate. (a) At the time of impact shocks propagate from the impact interface into each of the plates. (b) These shocks reflect from the unrestrained surfaces and the reflected shocks collide in the interior of the target plate. (c) In the region between the two waves arising from this latter interaction the material is in a state of tension. If this tension is great enough, fracture occurs. In drawing the figure it was supposed that the fracture occurred following a brief incubation period. Part (d) of the figure shows the fractured target plate.

ure will occur at the plane on which the tensile stress first appears. This plane, called the *candidate spall plane*, lies at a distance to the left of the back surface of the target that is equal to the thickness of the impactor (Fig. 12.2d shows the state of the assembly after spallation). Figure 12.3 is drawn on the premise that the tension produced in the target plate is insufficient to cause spallation, and shows several wave interactions following the original tensile excursion. In this case, tension is present on the candidate spall plane for the time required for one shock reverberation in the impactor plate, with its application being terminated by arrival of the stress-relief wave producing region 5 of the X–t plane. If spallation is to occur, it must do so within the brief time interval during which the material is in tension (state 4).

Experimental observation shows that spallation does not occur instantaneously upon application of tension in excess of the spall strength. A case in which spallation occurs after a brief incubation period is illustrated in the diagrams of

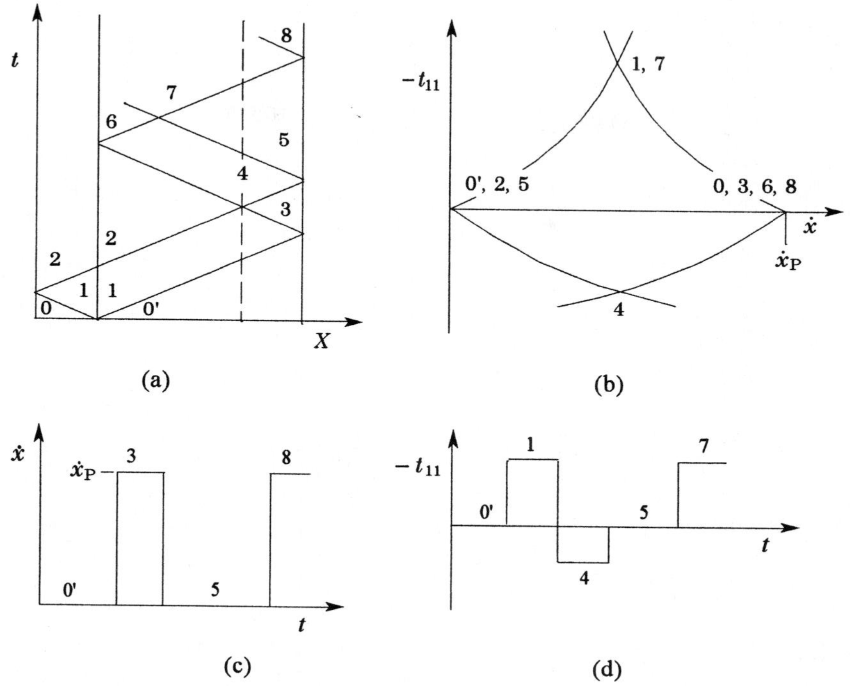

Figure 12.3. Diagrams illustrating (a) the $X-t$ trajectories of shocks and (b) the $t_{11}-\dot{x}$ states achieved in a spall experiment conducted below the fracture threshold. The history of the free-surface velocity of the target is shown in part (c) of the figure and the stress history at the candidate spall plane is shown in part (d) of the figure.

Fig. 12.4. Region 4 develops during the delay between decompression wave collision and spallation. When the spall occurs the material separates and recompression waves originate at the spall plane, leading to formation of regions 5 and 6 in the target. Several additional wave interactions occur, as illustrated in the figure. Figure 12.4d shows the stress history at the spall plane, including both the compression phase and the tension prevailing before spallation. The drop in free-surface velocity that occurs when the left boundary of the tensile region reaches the back surface of the target is called a *pull-back signal*. The maximum duration of the part of the free-surface velocity history that is associated with region 5 is $(2L_T - L_P)/C_S$, where C_S is the shock speed. When the free-surface velocity is zero for this amount of time it indicates that spallation did not occur. Examination of the free-surface velocity history illustrated in Fig. 12.4c shows that the time interval between the pull-back signal and arrival of the recompression wave is much less than was the case when spallation did not occur. This is because the recompression wave originates at the spall plane rather than the impact interface, and thus travels a smaller distance.

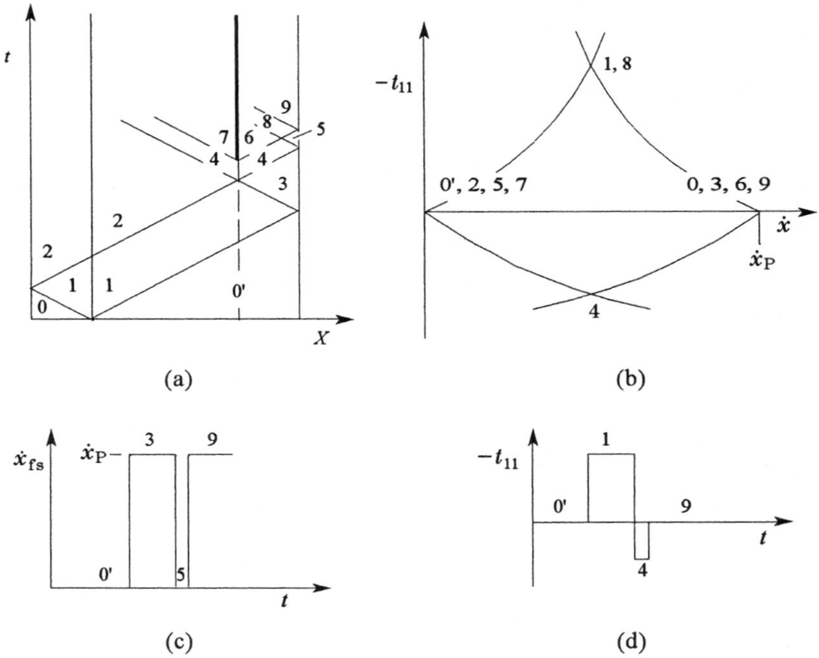

Figure 12.4. The process illustrated in this figure is the same as in Fig. 12.3 except that spallation occurs at the candidate spall plane after the material has been under tension for a brief interval. (a) X–t diagram. (b) $t_{11}-\dot{x}$ Hugoniots. (c) particle velocity history at the back surface of the target plate. (d) Stress history at the spall plane. The decrease in the duration of region 5 on the free-surface velocity history graph of this figure from that on the graph of Fig. 12.3 is an indication that spall occurred in the present case.

Spallation Due to Colliding Simple Decompression Waves. In Sect. 9.6 we considered a plate-impact problem rather like the one just discussed, except that the interacting decompression waves were analyzed as centered simple waves rather than being approximated by shocks. The X–t diagram for this problem, drawn on the premise that fracture did not occur, was presented in Fig. 9.19. Stress histories showing the gradual development of tension on planes X_A, X_C, and X_D are shown in Fig. 9.20. The maximum tension arising in this example is 9.1 GPa, a value well in excess of the reported 2.4 GPa spall strength of uranium. In order to extend the analysis of Sect. 9.6 to capture the effect of spallation we begin by assuming that a complete spall forms instantaneously when the tension at some plane reaches the critical value. Examination of the solution given in Sect. 9.6 shows that any given level of tension between zero and the maximum value achieved is first attained at a point on the characteristic CD. From the solution given previously, we find that the critical value $p = -\sigma_S = -2.4$ GPa is attained at $X = X_S = 5.91$ mm and $t = t_S = 4.95\ \mu s$.

At this time the material spalls and recompression shocks form and propagate in both directions from the spall plane. The material behind each of these shocks is recompressed to the state of zero pressure on the Hugoniot centered on the isentrope at the pressure $p = -\sigma_S$. The next step is to determine the trajectory and strength of the right-propagating recompression shock. We shall restrict attention to this shock because our objective is to calculate the effect of the spall formation on the velocity history of the back surface of the target plate. For this analysis we shall adopt the weak-shock approximation in which the Hugoniot and isentrope coincide. The X–t diagram for this case is given in Fig. 12.5.

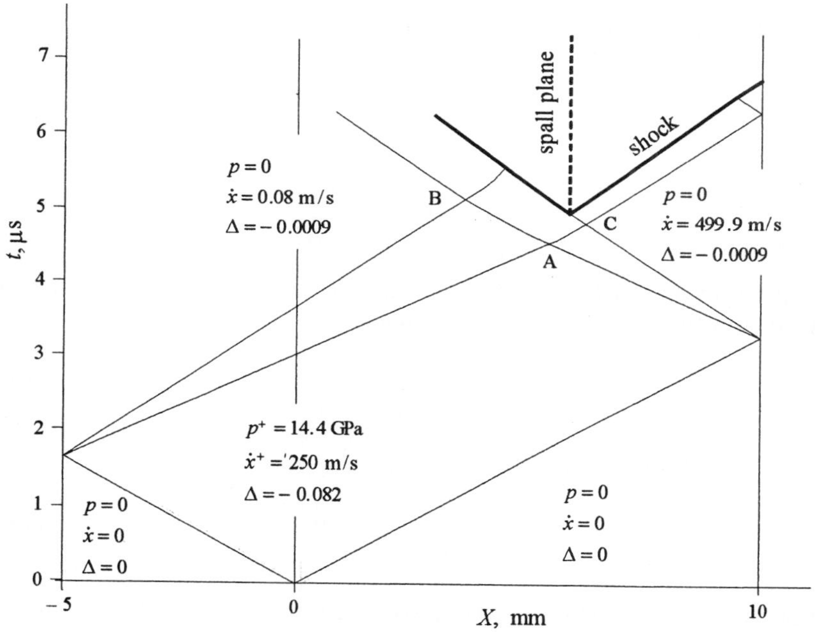

Figure 12.5. The X–t diagram of Fig. 9.19, modified for the effect of spallation at the time, t, and plane, X, at which the tension first exceeds the spall strength $\sigma_S = 2.4$ GPa.

Let us describe the recompression shock trajectory parametrically by the equations

$$X = X_S(\Delta) \quad \text{and} \quad t = t_S(\Delta)\,. \tag{12.1}$$

The ray characteristics in the right-propagating simple wave emanating from the interaction zone are given by

$$X - X_{CD}(\Delta) = C_L(\Delta)[t - t_{CD}(\Delta)]\,, \tag{12.2}$$

where we are interested only in the range $\Delta_S \le \Delta \le \Delta^{(2)}$. The functions $X_{CD}(\Delta)$

and $t_{\mathrm{CD}}(\Delta)$, which form a parametric representation of the curve CD are known from the numerical analysis of the interaction region.

On the shock trajectory we have

$$dX_{\mathrm{S}} = U_{\mathrm{S}}(\Delta)\,dt_{\mathrm{S}}, \tag{12.3}$$

or

$$\frac{dX_{\mathrm{S}}(\Delta)}{d\Delta} = U_{\mathrm{S}}(\Delta)\frac{dt_{\mathrm{S}}(\Delta)}{d\Delta}. \tag{12.4}$$

From the shock-jump conditions we have

$$\begin{aligned} U_{\mathrm{S}}[\![\Delta]\!] &= [\![\dot{x}]\!] \\ \rho_{\mathrm{R}}\, U_{\mathrm{S}}[\![\dot{x}]\!] &= [\![p]\!], \end{aligned} \tag{12.5}$$

so

$$U_{\mathrm{S}}^2 = \frac{1}{\rho_{\mathrm{R}}}\frac{[\![p]\!]}{[\![\Delta]\!]}. \tag{12.6}$$

The pressure behind the shock is zero and, since we are approximating the Hugoniot by the isentrope, the compression is $\Delta^{(2)}$. Therefore, we have

$$U_{\mathrm{S}}^2 = \frac{-p^{(\eta)}(\Delta)}{\rho_{\mathrm{R}}(\Delta^{(2)} - \Delta)}. \tag{12.7}$$

We also have the usual equation

$$C_{\mathrm{L}}(\Delta) = \left(\frac{1}{\rho_{\mathrm{R}}}\frac{dp^{(\eta)}(\Delta)}{d\Delta}\right)^{1/2} \tag{12.8}$$

for the soundspeed, where the isentrope is the one passing through the point p^+, Δ^+ on the Hugoniot. When $\gamma(v)/v = \gamma_{\mathrm{R}}/v_{\mathrm{R}}$ this isentrope is given by

$$p^{(\eta)}(\Delta) = \chi_{\mathrm{c}}(\Delta)\left\{ p^+ + \int_{\Delta^+}^{\Delta} \frac{\kappa_{\mathrm{c}}(\Delta')}{\chi_{\mathrm{c}}(\Delta')}\,d\Delta' \right\}, \tag{12.9}$$

with

$$\begin{aligned} \chi_{\mathrm{c}}(\Delta) &= \exp\left[\gamma_{\mathrm{R}}(\Delta - \Delta^+)\right] \\ \kappa_{\mathrm{c}}(\Delta) &= \left(1 - \frac{\gamma_{\mathrm{R}}}{2}\Delta\right)\frac{d\,p^{(\mathrm{H})}(\Delta)}{d\Delta} - \frac{\gamma_{\mathrm{R}}}{2}p^{(\mathrm{H})}(\Delta). \end{aligned} \tag{12.10}$$

When the Hugoniot is given by

$$p^{(\mathrm{H})}(\Delta) = \frac{\rho_{\mathrm{R}} C_{\mathrm{B}}^2 \Delta}{(1 - S\Delta)^2} \tag{12.11}$$

we have

$$\kappa_{\mathrm{c}}(\Delta) = \frac{\rho_{\mathrm{R}} C_{\mathrm{B}}^2}{(1 - S\Delta)^3} [1 + (S - \gamma_{\mathrm{R}})\Delta], \tag{12.12}$$

and

$$C_{\mathrm{L}}(\Delta) = \left\{ \frac{1}{\rho_{\mathrm{R}}} \left[p^{(\eta)}(\Delta) + \kappa_{\mathrm{c}}(\Delta) \right] \right\}^{1/2}$$
$$\frac{dC_{\mathrm{L}}(\Delta)}{d\Delta} = \frac{1}{2\rho_{\mathrm{R}} C_{\mathrm{L}}(\Delta)} \left\{ \gamma_R^2\, p^{(\eta)}(\Delta) + \gamma_{\mathrm{R}}\, \kappa_{\mathrm{c}}(\Delta) + \frac{d\kappa_{\mathrm{c}}(\Delta)}{d\Delta} \right\}, \tag{12.13}$$

with

$$\frac{d\kappa_{\mathrm{c}}(\Delta)}{d\Delta} = \frac{\rho_{\mathrm{R}} C_{\mathrm{B}}^2}{(1 - S\Delta)^4} [3S + (S - \gamma_{\mathrm{R}})(1 + 2S\Delta)]. \tag{12.14}$$

At points on the shock trajectory the pressure, compression, and particle velocity have the values associated with the ray characteristic intersecting it at that point. Evaluation of Eq. 12.2 at points on the shock trajectory yields the result

$$X_{\mathrm{S}}(\Delta) - X_{\mathrm{CD}}(\Delta) = C_{\mathrm{L}}(\Delta)[t_{\mathrm{S}}(\Delta) - t_{\mathrm{CD}}(\Delta)], \tag{12.15}$$

and differentiation of this equation gives

$$\frac{dX_{\mathrm{S}}(\Delta)}{d\Delta} - \frac{dX_{\mathrm{CD}}(\Delta)}{d\Delta} = [t_{\mathrm{S}}(\Delta) - t_{\mathrm{CD}}(\Delta)] \frac{dC_{\mathrm{L}}(\Delta)}{d\Delta}$$
$$+ C_{\mathrm{L}}(\Delta) \left[\frac{dt_{\mathrm{S}}(\Delta)}{d\Delta} - \frac{dt_{\mathrm{CD}}(\Delta)}{d\Delta} \right]. \tag{12.16}$$

Substitution of Eq. 12.3 into this result gives

$$\frac{dt_{\mathrm{S}}(\Delta)}{d\Delta} + \varphi(\Delta) t_{\mathrm{S}}(\Delta) = \psi(\Delta), \tag{12.17}$$

where

$$\varphi(\Delta) = \frac{1}{C_{\mathrm{L}}(\Delta) - U_{\mathrm{S}}(\Delta)} \frac{dC_{\mathrm{L}}(\Delta)}{d\Delta}, \tag{12.18}$$

and

$$\psi(\Delta) = \frac{\dfrac{dX_{CD}(\Delta)}{d\Delta} - C_L(\Delta)\dfrac{dt_{CD}(\Delta)}{d\Delta} - t_{CD}(\Delta)\dfrac{dC_L(\Delta)}{d\Delta}}{U_S(\Delta) - C_L(\Delta)}, \tag{12.19}$$

where we have used Eq. 12.4 to eliminate $dX_S/d\Delta$. Equation 12.17 is the familiar linear first-order ordinary differential equation and its solution for $t_S(\Delta)$ is easily reduced to quadrature. When this has been done, the function $X_S(\Delta)$ can be obtained from Eq. 12.15.

A practical problem is calculation of the derivatives $dX_{CD}(\Delta)/d\Delta$ and $dt_{CD}(\Delta)/d\Delta$. This is best accomplished by analytical differentiation of a function fit to the values of $X_{CD}(\Delta)$ and $t_{CD}(\Delta)$ that were calculated numerically. These functions can be fit quite accurately by the polynomials

$$\begin{aligned} X_{CD}(\Delta) &= X_E + C(\Delta - \Delta_F)(\Delta - \Delta_D) + \frac{X_C - X_D}{\Delta_C - \Delta_D}(\Delta - \Delta_D) \\ t_{CD}(\Delta) &= t_D + D(\Delta - \Delta_C)(\Delta - \Delta_D) + \frac{t_C - t_D}{\Delta_C - \Delta_D}(\Delta - \Delta_D), \end{aligned} \tag{12.20}$$

where C and D are adjustable coefficients. For the example at hand an adequate fit is obtained using $C = 0.024$ and $D = 1.86$. The derivatives of these functions are

$$\begin{aligned} \frac{dX_{CD}(\Delta)}{d\Delta} &= \frac{X_C - X_D}{\Delta_C - \Delta_D} - C(\Delta_F - \Delta_D) + 2C\Delta \\ \frac{dt_{CD}(\Delta)}{d\Delta} &= \frac{t_C - t_D}{\Delta_C - \Delta_D} - D(\Delta_F - \Delta_D) + 2D\Delta. \end{aligned} \tag{12.21}$$

With all of the foregoing results, the shock trajectory and strength at points within the transmitted simple wave are readily calculated by integrating Eq. 12.17. It should be noted that it is difficult to achieve high accuracy in the foregoing calculation because the shock trajectory and the characteristic curves have almost the same slope, so calculation of the intersection is subject to significant uncertainty.

Calculation of the extension of the shock trajectory through the reflection region is more complicated, but this can also be done numerically. Because the portion of the shock trajectory that lies within this region is so short, great accuracy is not required, and we shall omit details of the calculation. For our purpose, the important result is the time at which the trajectory intersects the boundary of the target plate. For the example under discussion this is $t = 6.625$ μs.

An important result of the foregoing calculation is the particle-velocity history of the back surface of the target plate. This function can be measured with great accuracy in spall experiments, and permits one to determine whether or not a spall formed in a given experiment and, if a spall formed, to estimate the spall stress. These data frequently provide the most important means of evaluating the accuracy of numerical simulations of the process.

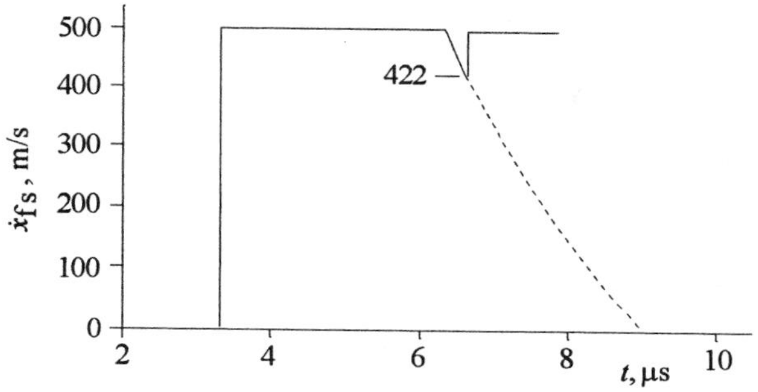

Figure 12.6. Velocity history of the unrestrained back surface of the target plate for the example problem. The solid line is the result of the analysis for the case in which spall occurs as illustrated in Fig. 12.5 and the dotted extension of the curve is the path that would have been followed had spall not occurred. The dip in the solid curve, called a *pullback signal*, is the indication that spall occurred.

Spallation Due to Colliding Elastic–Plastic Decompression Waves. In Sect. 10.1.7. we considered the problem of impact of steel plates using a small-deformation theory of elastoplastic material response. As in the case of elastic plates, wave reflections lead to colliding decompression waves and development of a region of tensile stress in the interior of the thicker plate. In the example considered the maximum tensile stress reached almost 4 GPa, a value sufficient to cause spallation in many steels. As shown in Fig. 10.22, a time interval of approximately 0.5 μs is required for development of this tension. When it is assumed that spallation occurs instantaneously, but after a brief incubation period, when the tensile stress reaches 3 GPa and the analysis is carried beyond the time at which the spall forms the results presented in Figs. 12.7 and 12.8 are obtained.

12.1.2. Explosive Loading Experiment

Plane spalls can be produced by means other than plate impact. One important case is spallation produced when an explosive charge is detonated in contact with a material sample. The difference between the plate-impact experiment discussed in the previous section and an explosive loading experiment is that the former introduces a flat-topped compression pulse into the target whereas the latter produces a pulse in which the compression rises suddenly to a peak and immediately begins a gradual fall. The peak compressive stress produced by explosive loading is almost always far in excess of the spall strength. Detailed sol-

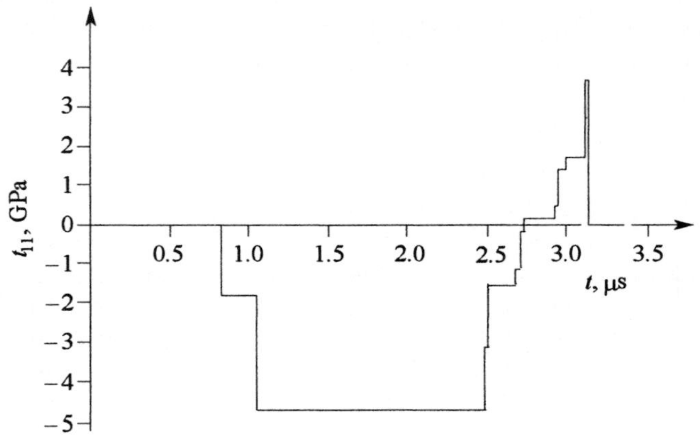

Figure 12.7. Stress history at $X = 9.96$ mm (the plane at which tension exceeding 3 GPa first appears) showing the gradual increase in tension that occurs before a spall forms. This analysis is a continuation to times after spallation of the calculation presented in Sect. 10.1.7.

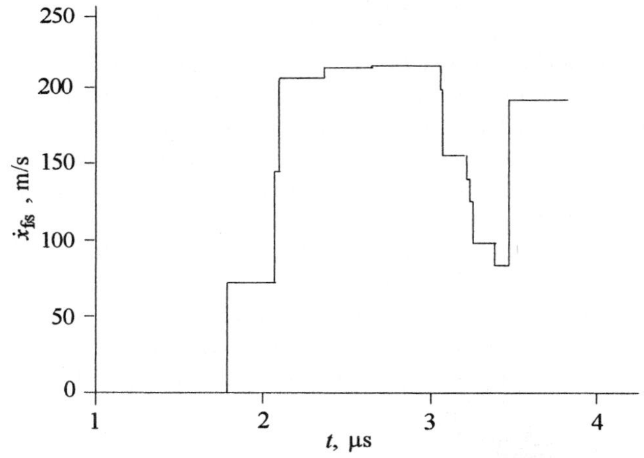

Figure 12.8. Free-surface velocity history resulting from a collision of elastic–plastic steel plates. The impactor plate is 5 mm thick and moving at a velocity of 230 m/s. The stationary target plate is 10 mm thick. This analysis is a continuation of the wave-propagation calculation presented in Sect. 7.7 and is conducted on the premise that a complete spall forms instantaneously, but after a short incubation period, when the tensile stress exceeds 3 GPa..

ution of the explosive loading problem can only be accomplished by numerical means, but the basic issues are easily illustrated by a simple graphical analysis of the interaction of a triangular pulse with an unrestrained surface of a body exhibiting a linear elastic response. Triangular compression pulses can also arise in plate-impact experiments in which the propagation distance is sufficient for the occurrence of attenuation, as discussed in Sect. 9.5.2.

The graphical solution for the linear interaction of a triangular wave with an unrestrained surface is illustrated in Fig. 12.9. This figure shows the wave within the plate moving toward the right and the virtual wave moving toward the left, initially outside the plate, but eventually entering the material. As discussed previously, the compression pulse incident on the unrestrained surface of the plate is reflected as a tensile pulse. The beginning of the reflection process is shown in Fig. 12.9c and continues in Fig. 12.9d. The illustration is made for the case that the spall strength is σ_S, as shown in the figure. When the reflection process has reached the stage that the peak tension attains this value, the plate spalls at the plane where the criterion is first satisfied, and a layer of material flies away to the right. When this occurs, a new unrestrained surface is formed and what is left of the right-propagating triangular pulse reflects from this surface in just the same way that the original reflection occurred. The first stage of this process is shown in Fig. 12.9e. Since the peak compressive stress in the portion of the incident wave remaining after formation of the first spall exceeds the spall strength, a second spall will form when the peak tension in the reflected pulse again reaches the spall strength, as shown in Fig. 12.9f. For the case at

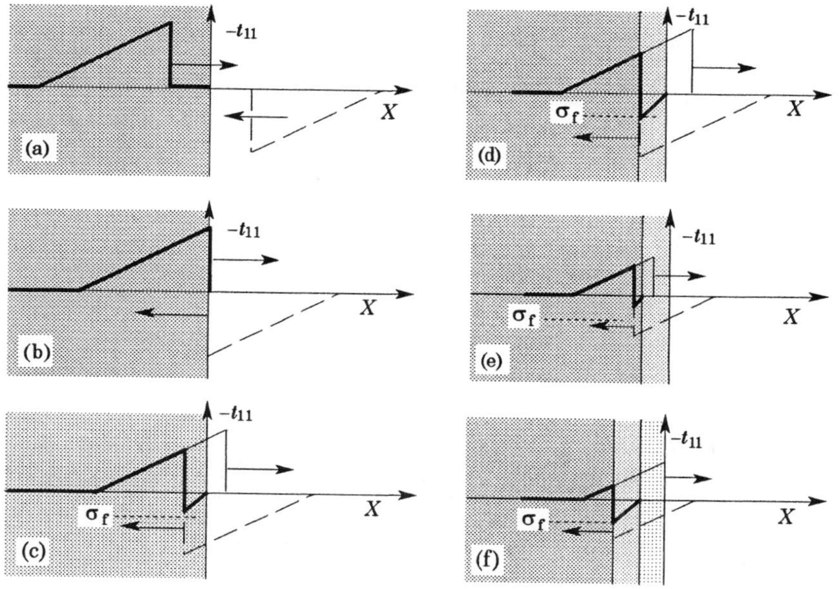

Figure 12.9. Graphical solution of reflection of a triangular pulse from an unrestrained boundary, illustrating the production of multiple spalls.

hand, the remaining compressive pulse is too weak to cause further spallation, but a stronger pulse could produce more spalls. Formation of multiple spalls is characteristic of explosive loading, but occurs in other cases as well.

Stress histories produced by reflection of a triangular compression wave are illustrated in Fig. 12.10. As can be seen from the figure, the peak tension at any given plane, X, is produced in an instantaneous transition from a compressed or unstrained state. At planes that are closer to the unrestrained surface than one-half of the length of the initial triangular pulse the peak tensile stress attained is less than the full pulse amplitude, but this stress is sustained for a finite time interval that decreases with increasing distance from the surface. The maximum tension is largest on planes for which the duration of its application is least.

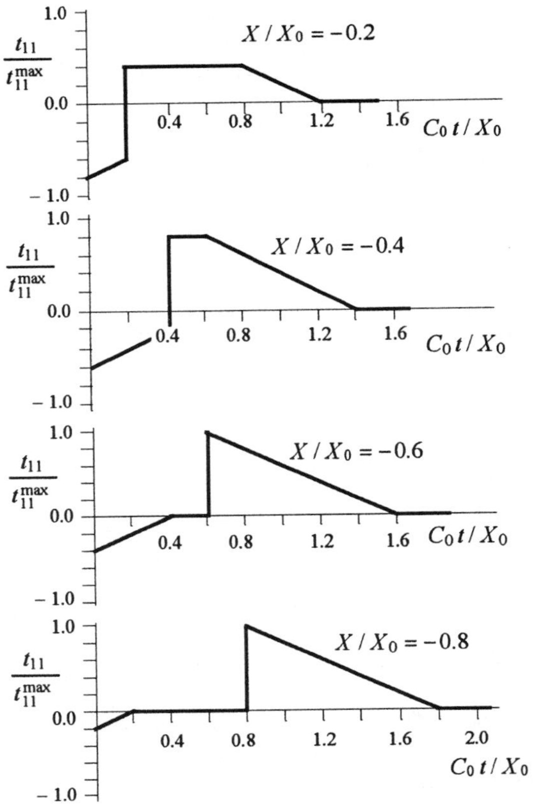

Figure 12.10. Stress histories at various planes in a plate in which a falling triangular compression pulse of length X_0 that is propagating in the $+X$ direction at the velocity C_0 is being reflected from an unrestrained surface beginning at $t = 0$.

Flat-topped compression pulses evolve to a triangular form at sufficiently large propagation distances (see Fig. 9.18). In this regime they are reflected from an unrestrained surface in the same way as triangular pulses introduced by explosives, and can cause spallation as discussed above.

12.1.3 Pulsed-radiation Absorption Experiment

An interesting case arises when we consider evolution of the thermal stress field produced by absorption of a burst of radiant energy, as when a plate of stained glass is exposed to a laser pulse. In this section we restrict attention to the simple case in which a modest amount of energy is instantaneously deposited in accordance with Lambert's law of absorption. In this case, the initial energy density distribution is an exponential function that decays with distance into the plate:

$$\varepsilon = \varepsilon_0 \exp(-kX),$$

as illustrated in Fig. 12.11. In the foregoing equation k is the optical absorption coefficient of the material.

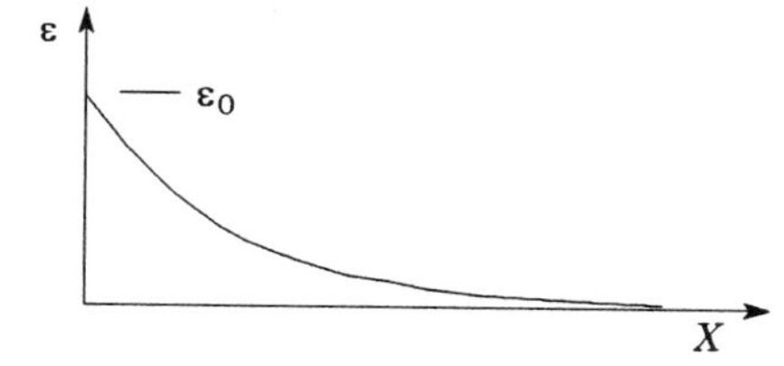

Figure 12.11. Energy deposited into a body when its surface is uniformly illuminated by a brief laser pulse. The compressive thermal stress distribution that results is, in linear approximation, of the same form.

When this problem is analyzed using the linear theory of thermoelasticity one finds that the compressive thermal stress at the instant following the deposition is

$$-t_{11}(X,0) = \frac{\gamma_R}{v_R}\varepsilon(X,0).$$

The disturbance produced by an initial stress state takes the form of left- and right-propagating pulses each having the form of the initial stress distribution but of one-half the amplitude of this distribution. Since the region of the initial stress distribution is adjacent to an unrestrained boundary, the interaction with this boundary must also be taken into account. This is done by adding to the two pulses already mentioned a right-propagating virtual pulse that ensures satisfaction of the boundary condition. These three pulses are shown (after propagating a short distance) in Fig. 12.12a. The solution to the problem at a given time is the sum of these three pulses as they exist at that time, and is shown in Fig. 12.12b. This figure shows that a region of tension has developed near the front face of the plate. As the solution evolves the peak tension rises until it reaches one-half the value of the peak compression in the initial thermal stress distribution. If this value exceeds the spall strength, a *front-surface spall* will form. As with the explosive loading example, multiple spalls occur when the initial compressive stress is sufficiently large. The stress histories at various planes within the material are shown in Fig. 12.13.

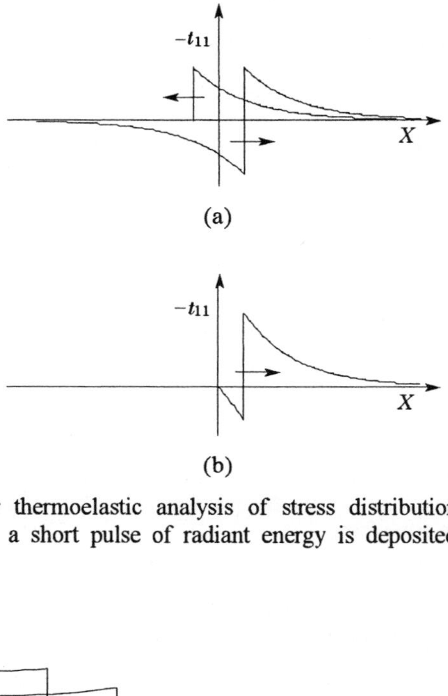

Figure 12.12 Linear thermoelastic analysis of stress distributions that arise in an experiment in which a short pulse of radiant energy is deposited into an absorbing material.

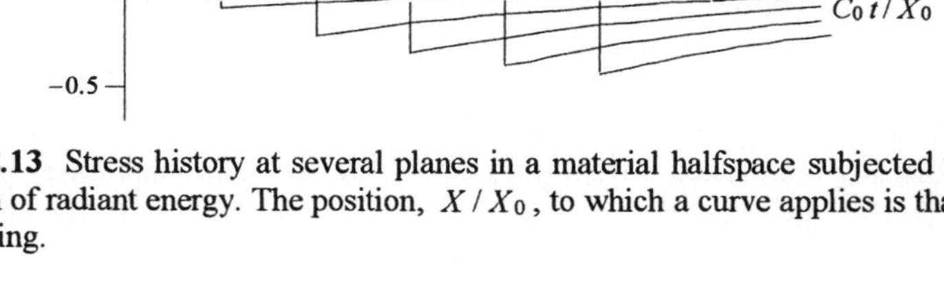

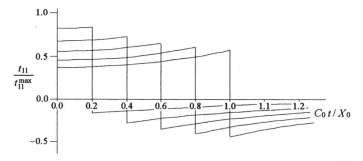

Figure 12.13 Stress history at several planes in a material halfspace subjected to rapid deposition of radiant energy. The position, X / X_0, to which a curve applies is that of the zero crossing.

12.2 Criteria for Spall-damage Accumulation

Detailed experimental observations have shown that cracks or voids are nucleated (or existing defects activated) at a critical level of tension that may depend on temperature and is characteristic of the material being studied. Often the critical stress for nucleation of microcracks or voids is equal to the fracture stress measured in quasistatic tests. Once nucleated, a crack or void grows at a finite rate that depends on the thermomechanical state as well as material parameters. In the following sections we shall discuss equations that model this process.

12.2.1 Simple Damage-accumulation Criteria

When spallation (however defined) is assumed to occur at a critical value of the tensile stress, this critical value is called the *spall strength*. The criterion just cited, called the *critical normal stress criterion*, can be expressed by the equation

$$t_{11} = \sigma_S, \tag{12.22}$$

where $\sigma_S > 0$, the *spall strength*, is a constant (or a function of temperature) that is regarded as a material property. It is implicit in the statement of this criterion that spallation is assumed to occur instantaneously. The critical stress criterion, although useful as a guide, is not consistent with detailed experimental observations. It has long been known, for example, that the amount of tension that a material can withstand increases as the duration of its application decreases. When sections of material samples that have been subjected to a given tensile stress for varying amounts of time are examined it is clear that increased levels of damage are produced when the stress is imposed for longer times. If we assume that damage, $\mathscr{D}$, accumulates at a constant rate that depends on the applied stress, then

$$\dot{\mathscr{D}}(X,t) = \frac{\mathscr{D}_0}{\tau(t_{11})}, \tag{12.23}$$

where $\mathscr{D}_0$ is a material constant and $\tau(t_{11})$ is the characteristic time for damage accumulation in the material when the stress is t_{11}. If the stress history at the point X is known, Eq. 12.23 can be integrated to yield the damage history

$$\mathscr{D}(X,t) = \mathscr{D}_0 \int_{-\infty}^{t} \frac{dt'}{\tau(t_{11}(X,t'))} \tag{12.24}$$

at this point. A special case of this equation is the criterion of Tuler and Butcher, in which $\tau(t_{11})$ varies inversely with the λ power of the excess of the tensile stress over a threshold σ_0: $\tau(t_{11}) = \tau_0 [(t_{11} - \sigma_0) + |t_{11} - \sigma_0|)/(2\sigma_0)]^{-\lambda}$ so that Eq. 12.24 takes the form

$$\mathscr{D}(X,t) = \frac{\mathscr{D}_0}{\tau_0}\int_{-\infty}^{t}\left[\frac{t_{11}(X,t') - \sigma_0 + \left|t_{11}(X,t') - \sigma_0\right|}{2\sigma_0}\right]^{\lambda} dt' . \tag{12.25}$$

The function $\tau(t_{11})$ used in this equation has been constructed so as to remove from consideration the part of the stress history that lies below the tension threshold for damage accumulation. As this criterion has normally been applied, the integral evaluated for each value of X using calculated stress histories and a specific level of damage (i.e. value of $\mathscr{D}$) is selected as characteristic of a spall.

The criterion of Eq. 12.25 is an example of what we may call simple damage-accumulation criteria. Within this class of criteria damage accumulates linearly in time at a given level of tension but at a rate that increases with an increase of tension beyond the threshold level. The stress history is calculated without regard to any damage that may have accumulated and the process of damage accumulation is assumed to proceed independently of any previously accumulated damage. The development of modern theories of damage accumulation is motivated by the need to remove these latter two restrictions.

12.2.2 Compound Damage-accumulation Criteria

When a material sample has been weakened by some accumulation of damage, it has been found that additional damage accumulates at a higher rate even if the tension is maintained constant. This means that $\dot{\mathscr{D}}$ depends on both the stress $\mathbf{t}$ and $\mathscr{D}$ itself:

$$\dot{\mathscr{D}} = \varphi(\mathbf{t}, \mathscr{D}) . \tag{12.26}$$

It is interesting to consider this equation in the special case that the dependence on $\mathscr{D}$ is linear. This can be interpreted as an exact relation or as a low-damage approximation to a more general equation. In either case, we have

$$\dot{\mathscr{D}} = \varphi_0(\mathbf{t}) + \varphi_1(\mathbf{t})\,\mathscr{D} . \tag{12.27}$$

When the stress is held constant, $\mathbf{t} = \mathbf{t}^*$, the solution of Eq. 12.27 is

$$\mathscr{D}(t) = \frac{\varphi_0(\mathbf{t}^*)}{\varphi_1(\mathbf{t}^*)}\left[\exp(\varphi_1(\mathbf{t}^*)t) - 1\right],$$

a damage-accumulation history quite different from the constant rate of accumulation of the simple damage-accumulation equations.

Example: The SRI International Nucleation and Growth Model. Equation 12.27 is of the same form as one inferred from experimental measurements of rates of nucleation and growth of voids in ductile metals [4, Chap. 7]. It was found that the voids in material specimens that had experienced a low level of spall damage were distributed in size (measured by the radius, R) according to the equation

$$\mathcal{N} = \mathcal{N}_0 \exp(-R/R_1), \tag{12.28}$$

where $\mathcal{N}_0$ is the total number of voids in a unit reference-state volume of material, $\mathcal{N}$ is the number of these voids having a radius greater than R, and R_1 is a characteristic size. Integration of this distribution gives the total void fraction (volume of void space in a unit reference-state volume of material) of these $\mathcal{N}_0$ voids as

$$\mathcal{V} = 8\pi \mathcal{N}_0 R_1^3 . \tag{12.29}$$

The experiments also showed that the radius, R, of each of the $\mathcal{N}_0$ voids grew at a rate proportional to its current value, with the proportionality factor being a function of the pressure (since the voids are observed to be spherical, it is reasonable to assume that their growth is controlled by the spherical component of the stress):

$$\dot{R} = f_G(p) R , \tag{12.30}$$

where

$$f_G(p) = \frac{|p - p_G| - (p - p_G)}{8\eta} , \tag{12.31}$$

with $p_G < 0$. This means that the characteristic radius R_1 also grows at this rate:

$$\dot{R}_1 = f_G(p) R_1 . \tag{12.32}$$

The void fraction associated with the $\mathcal{N}_0$ voids, given by Eq. 12.29, grows at the rate

$$\dot{\mathcal{V}}_G = 8\pi \mathcal{N}_0 3R_1^2 \dot{R}_1 = 3 f_G(p) \mathcal{V} . \tag{12.33}$$

There is an additional contribution to the increase of void fraction that arises from nucleation of new voids. The sizes of these voids were assumed to be distributed in the same way as the existing $\mathcal{N}_0$ voids, except that the characteristic radius had the constant value R_N. The $\Delta\mathcal{N}$ voids nucleated in the time increment Δt contribute the increment $\Delta\mathcal{V}_N$ to the void fraction, where

$$\Delta\mathcal{V}_N = 8\pi \Delta\mathcal{N} R_N^3 . \tag{12.34}$$

Since this increment of void fraction accumulated in the time interval Δt, we have

$$\dot{\mathcal{V}}_N = \lim_{\Delta t \to 0} \frac{\Delta\mathcal{V}_N}{\Delta t} = 8\pi R_N^3 \lim_{\Delta t \to 0} \frac{\Delta\mathcal{N}}{\Delta t} = 8\pi R_N^3 \dot{\mathcal{N}} . \tag{12.35}$$

The experimental results showed that voids were nucleated in a unit reference-state volume of material at the rate

$$\dot{\mathcal{N}} = C_{\mathrm{N}}\left[\exp\left(\frac{|p - p_{\mathrm{N}}| - (p - p_{\mathrm{N}})}{2p_1}\right) - 1\right]. \tag{12.36}$$

When the rate equation for the growth of existing voids, Eq. 12.33, is combined with that for growth of void fraction due to nucleation of new voids, Eq. 12.36, we obtain the evolutionary equation

$$\dot{\mathcal{V}} = \dot{\mathcal{V}}_{\mathrm{N}} + \dot{\mathcal{V}}_{\mathrm{G}} = 8\pi R_{\mathrm{N}}^3\, \dot{\mathcal{N}}(p) + 3 f_{\mathrm{G}}(p)\mathcal{V} \tag{12.37}$$

for the void fraction. In this equation, $\dot{\mathcal{N}}(p)$ is given by Eq. 12.36 and $f_{\mathrm{G}}(p)$ is given by Eq. 12.31. If we identify $\mathcal{V}$ with the damage $\mathcal{D}$ of Eq. 12.27 we see that Eq. 12.37 is a special case of Eq. 12.27.

Damage-accumulation criteria similar to the one just described have also been developed for cracking of brittle materials [4, Sect. 7.2].

12.3 Continuum Theory of Deformation and Damage Accumulation

In the range of stress, strain rate, temperature, etc. encountered in the study of spall phenomena, undamaged ductile materials are usually described by theories of viscoplasticity that are valid for finite deformations. As noted previously, damage accumulated in a material can be expected to affect its continuum-mechanical properties and, therefore, the process by which further damage accumulates. A comprehensive theory of spallation must account for this interaction of the wave-propagation and damage-accumulation processes. Several theories that do this have been developed and shown to yield results that are in good agreement with experimental observations [4,36,37,41,69]. Compound damage-accumulation equations of the class discussed in the preceding section comprise an important part of the theory that we seek. The applications that motivate the development of this section involve avoidance of significant spall damage to structures, so we shall limit our attention to low levels of damage.

Usually, a process leading to spall fracture begins with a compression wave propagating in undamaged material. This part of the problem is one of elastic–viscoplastic wave propagation that can be analyzed using the theory presented in Sect. 7.3, so the remainder of this section addresses the response of a ductile material as a low level of spall damage accumulates.

Kinematics. To extend the theory of Sect. 7.3 to include effects of spall dilatation we incorporate a contribution $\mathbf{F}^{\mathrm{d}}$ into the decomposition 7.54, which then becomes

$$\mathbf{F} = \mathbf{F}^{\mathrm{e}}\mathbf{F}^{\mathrm{d}}\mathbf{F}^{\mathrm{p}}. \tag{12.38}$$

As before, the plastic flow is taken to be isochoric, so that

$$\det \mathbf{F}^{\mathrm{p}} = 1. \tag{12.39}$$

Because the voids produced during spallation of ductile materials are essentially spherical, we take the deformation that accompanies void formation to be an isotropic dilatation, in which case $\mathbf{F}^{\mathrm{d}}$ has the form

$$\mathbf{F}^{\mathrm{d}} = F^{\mathrm{d}}\mathbf{1}. \tag{12.40}$$

The velocity gradient associated with Eq. 12.38 is

$$\mathbf{l} = \mathbf{l}^{\mathrm{e}} + \mathbf{l}^{\mathrm{d}} + \mathbf{l}^{\mathrm{p}}, \tag{12.41}$$

where

$$\mathbf{l} = \dot{\mathbf{F}}\overset{-1}{\mathbf{F}}, \tag{12.42}$$

and

$$\begin{aligned}
\mathbf{l}^{\mathrm{e}} &= \dot{\mathbf{F}}^{\mathrm{e}}\overset{-1}{\mathbf{F}}{}^{\mathrm{e}} \\
\mathbf{l}^{\mathrm{d}} &= \mathbf{F}^{\mathrm{e}}\dot{\mathbf{F}}^{\mathrm{d}}\overset{-1}{\mathbf{F}}{}^{\mathrm{d}}\overset{-1}{\mathbf{F}}{}^{\mathrm{e}} = (\dot{F}^{\mathrm{d}}/F^{\mathrm{d}})\mathbf{1} \\
\mathbf{l}^{\mathrm{p}} &= \mathbf{F}^{\mathrm{e}}\mathbf{F}^{\mathrm{d}}\dot{\mathbf{F}}^{\mathrm{p}}\overset{-1}{\mathbf{F}}{}^{\mathrm{p}}\overset{-1}{\mathbf{F}}{}^{\mathrm{d}}\overset{-1}{\mathbf{F}}{}^{\mathrm{e}} = \mathbf{F}^{\mathrm{e}}\mathbf{\Lambda}^{\mathrm{p}}\overset{-1}{\mathbf{F}}{}^{\mathrm{e}},
\end{aligned} \tag{12.43}$$

with

$$\mathbf{\Lambda}^{\mathrm{p}} = \dot{\mathbf{F}}^{\mathrm{p}}\overset{-1}{\mathbf{F}}{}^{\mathrm{p}}. \tag{12.44}$$

Damage Accumulation. We shall define the spall damage as the void volume fraction, $\mathcal{V}$, in the plastically deformed and damaged, but unstressed, material. Since the plastic deformation is isochoric, the volume of void space, v_{D}, formed in a unit reference-state volume of the material is related to $\mathcal{V}$ by the equation

$$v_{\mathrm{D}} = v_{\mathrm{R}}(1+\mathcal{V}). \tag{12.45}$$

The contribution these voids make to the deformation gradient is

$$\mathbf{F}^{\mathrm{d}} = (v_{\mathrm{D}}/v_{\mathrm{R}})^{1/3}\mathbf{1} = (1+\mathcal{V})^{1/3}\mathbf{1}, \tag{12.46}$$

so we can write

$$\dot{\mathbf{F}}^{\mathrm{d}} = \tfrac{1}{3}(1+\mathcal{V})^{-2/3}\,\dot{\mathcal{V}}\,\mathbf{1}\,, \tag{12.47}$$

where $\dot{\mathcal{V}}$ is given by Eq. 12.37.

Stress Relation. The stress relation for spall damaged-material must reflect the reduction in elastic stiffness that results from the presence of voids. In addition to the limitation on deviatoric stress that a ductile material can sustain, the pressure that can be supported by material in which voids are present is also limited by void growth or collapse to rather small values in both tension and compression. For this reason, the stresses in a damaged body are adequately described by Eq. 6.19,

$$t_{ij} = (B^{\eta} - \tfrac{2}{3}\mu^{\eta})\widetilde{E}^{\mathrm{e}}_{kk}\,\delta_{ij} + 2\mu^{\eta}\,\widetilde{E}^{\mathrm{e}}_{ij} - \rho_{\mathrm{R}}\,\theta_{\mathrm{R}}\,\gamma(\eta - \eta_{\mathrm{R}})\,\delta_{ij}\,, \tag{12.48}$$

describing linear thermoelastic response. The strain tensor appearing in this equation is calculated from $\mathbf{F}^{\mathrm{e}}$ by the equation

$$\widetilde{E}^{\mathrm{e}}_{ij} = \tfrac{1}{2}(F^{\mathrm{e}}_{k\alpha}F^{\mathrm{e}}_{k\beta} - \delta_{\alpha\beta})\,\delta_{\alpha i}\,\delta_{\beta j}\,. \tag{12.49}$$

Because Eq. 12.48 is being applied to a material containing voids, it is necessary that the coefficients B^{η}, μ^{η}, and γ capture the effect of the voids. Several investigators have proposed equations expressing the elastic moduli as functions of the void fraction. As one would expect, the bulk and shear moduli decrease from their values for the undamaged material as the void fraction increases. The decrease is linear in $\mathcal{V}$ for small values of $\mathcal{V}$, but the coefficient of the linear term differs among the various available formulae. For the present purpose it is adequate to adopt the forms

$$B^{\eta}(\mathcal{V}) = B^{\eta}_{\mathrm{R}}(1 - c_{\mathrm{B}}\mathcal{V}) \quad \text{and} \quad \mu^{\eta}(\mathcal{V}) = \mu^{\eta}_{\mathrm{R}}(1 - c_{\mu}\mathcal{V})\,, \tag{12.50}$$

where B^{η}_{R} and μ^{η}_{R} are isentropic moduli for the undamaged material and c_{B} and c_{μ} are positive material constants. For Grüneisen's coefficient we adopt Eq. 11.38, which can be written

$$\gamma(v, \mathcal{V}) = \gamma_{\mathrm{R}}\,[(v/v_{\mathrm{R}}) - \mathcal{V}\,]\,. \tag{12.51}$$

The void fraction, $\mathcal{V}$, can be expected to change as a result of both accumulating spall damage and the elastic deformation of the damaged material. However, we shall neglect the latter effect because the linear elastic response of hollow spheres to external pressure (or tension) is such that the void fraction changes very little even though the outer diameter of the sphere is changed significantly.

Viscoplastic Flow. We now turn to calculation of plastic deformation of the damaged material. We shall adopt the equations in Sect. 7.3, but with modifications to account for the effect of a dilute distribution of voids. The use of a plasticity theory based upon dislocation-mechanical considerations is particularly appropriate in the present case in which we must also deal with void growth because experimental observations show that this growth can be explained by the same dislocation-mechanical processes that produce the viscoplastic flow. When an edge dislocation encounters a void, the vacant half-plane of atoms associated with it may enter the void and contribute to its growth [92,93].

The presence of a dilute concentration of voids does not alter the mechanism of viscoplastic deformation. As before, we identify slip planes by their normal and Burgers vectors $\mathbf{N}^{(k)}$ and $\mathbf{B}^{(k)}$, respectively, in the plastically deformed configuration. The development of a distribution of voids in the material does not alter the overall crystallographic orientation or the Burgers vector of the material, which means that $\mathbf{N}^{(k)}$ and $\mathbf{B}^{(k)}$ transform from the plastically deformed configuration to the current configuration as though the material were undamaged. This means that the components of these vectors in the current configuration are

$$n_i^{(k)} = J\, N_\Gamma^{(k)}\, \delta_{\Gamma\alpha}\, \overset{-1}{F}{}^{\mathrm{e}}_{\alpha i} \quad \text{and} \quad b_i^{(k)} = F^{\mathrm{e}}_{i\alpha}\, \delta_{\alpha\Gamma}\, B_\Gamma^{(k)}, \tag{12.52}$$

where $J = \det \mathbf{F}^{\mathrm{e}} = 1 + \operatorname{tr} \mathbf{E}^{\mathrm{e}} + \cdots$ and the vector magnitudes are

$$n^{(k)} = N^{(k)} \left[1 + \widetilde{E}^{\mathrm{e}}_{\alpha\alpha} - (N^{(k)})^{-2}\, \widetilde{E}^{\mathrm{e}}_{\alpha\beta}\, N_\Gamma^{(k)}\, \delta_{\Gamma\alpha} N_\Delta^{(k)}\, \delta_{\Delta\beta} \right] \tag{12.53}$$

and

$$b^{(k)} = B^{(k)} \left[1 + (B^{(k)})^{-2}\, \widetilde{E}^{\mathrm{e}}_{\alpha\beta}\, B_\Gamma^{(k)}\, \delta_{\Gamma\alpha} B_\Delta^{(k)}\, \delta_{\Delta\beta} \right] \tag{12.54}$$

to first order in $\mathbf{E}^{\mathrm{e}}$.

The dislocation velocity in the current configuration is given by Eq. 7.112, just as it was for the undamaged material but, for a given strain, the magnitude of the shear traction, and thus the dislocation velocity, will be less than for the undamaged material because of the lower elastic moduli. The dislocation velocity in the plastically deformed configuration is related to that in the current configuration by the equation

$$V_{X^*}^{(k)} = (b^{(k)} / B^{(k)})\, V_x^{(k)}. \tag{12.55}$$

The ratio $b^{(k)} / B^{(k)}$ inferred from Eq. 12.54 is to be substituted into Eq. 12.55 to determine the relationship between the dislocation velocities in the plastically deformed and current configurations.

As shown in Chap. 7, we have the equation

$$\dot{\mathbf{F}}^{\mathrm{p}} \overset{-1}{\mathbf{F}}{}^{\mathrm{p}} = \sum_{k=1}^{n} \mathcal{N}_{k\mathrm{M}} V_{X^*}^{(k)}(\tau^{(k)}) \, \mathbf{B}^{(k)} \otimes \overline{\mathbf{N}}^{(k)} \tag{12.56}$$

for the components of $\mathbf{F}^{\mathrm{p}}$.

We must now determine the shear stress that produces the dislocation motion. The shear traction $\boldsymbol{\tau}^{(k)}$ in the direction $\mathbf{b}^{(k)}$ on the plane characterized by $\overline{\mathbf{n}}^{(k)}$ is given by Eq. 7.111 and, as before, the magnitude of this vector is $\tau^{(k)}$. The dislocation velocity on this slip system in the current configuration is given by Eq. 7.112.

The evolutionary equations 7.113 for the back stress, and 7.115 for the dislocation density and the mobile fraction are the same as for the undamaged material.

Uniaxial Deformation. In the special case of uniaxial deformation we have

$$\begin{aligned} \mathbf{F} &= \mathrm{diag}\left\| F_{\mathrm{L}},\ 1,\ 1 \right\| \\ \mathbf{F}^{\mathrm{p}} &= \mathrm{diag}\left\| F_{\mathrm{L}}^{\mathrm{p}},\ F_{\mathrm{T}}^{\mathrm{p}},\ F_{\mathrm{T}}^{\mathrm{p}} \right\| \\ \mathbf{F}^{\mathrm{d}} &= (1+\mathcal{V})^{1/3}\,\mathrm{diag}\left\| 1,\ 1,\ 1 \right\| \\ \mathbf{F}^{\mathrm{e}} &= \mathrm{diag}\left\| F_{\mathrm{L}}^{\mathrm{e}},\ F_{\mathrm{T}}^{\mathrm{e}},\ F_{\mathrm{T}}^{\mathrm{e}} \right\|. \end{aligned} \tag{12.57}$$

Since $\det \mathbf{F}^{\mathrm{p}} = 1$, the longitudinal and transverse components of $\mathbf{F}^{\mathrm{p}}$ are related by the equation

$$F_{\mathrm{L}}^{\mathrm{p}}\,(F_{\mathrm{T}}^{\mathrm{p}})^2 = 1 \quad \text{or} \quad F_{\mathrm{T}}^{\mathrm{p}} = (F_{\mathrm{L}}^{\mathrm{p}})^{-1/2}. \tag{12.58}$$

By Eq. 12.38 we have

$$F_{\mathrm{L}} = F_{\mathrm{L}}^{\mathrm{e}}\, F^{\mathrm{d}}\, F_{\mathrm{L}}^{\mathrm{p}}, \tag{12.59}$$

and, from Eq. 12.57_1,

$$F_{\mathrm{L}} = (1+\mathcal{V})^{1/3} F_{\mathrm{L}}^{\mathrm{e}}\, F_{\mathrm{L}}^{\mathrm{p}}. \tag{12.60}$$

Finally, from Eqs. 12.38 and 12.58 we obtain the result

$$F_{\mathrm{T}} = 1 = (1+\mathcal{V})^{1/3} F_{\mathrm{T}}^{\mathrm{e}}\, F_{\mathrm{T}}^{\mathrm{p}} = (1+\mathcal{V})^{1/3} F_{\mathrm{T}}^{\mathrm{e}}\,(F_{\mathrm{L}}^{\mathrm{p}})^{-1/2}, \tag{12.61}$$

or

$$F_{\mathrm{T}}^{\mathrm{e}} = (1+\mathcal{V})^{-1/3}\,(F_{\mathrm{L}}^{\mathrm{p}})^{1/2}. \tag{12.62}$$

With these results, Eqs. $12.57_{1,2,4}$ become

$$\mathbf{F} = \mathrm{diag}\left\| (1+\mathcal{V})^{1/3} F_{\mathrm{L}}^{\mathrm{e}} F_{\mathrm{L}}^{\mathrm{p}},\ 1,\ 1 \right\|$$

$$\mathbf{F}^{\mathrm{p}} = \mathrm{diag}\left\| F_{\mathrm{L}}^{\mathrm{p}},\ (F_{\mathrm{L}}^{\mathrm{p}})^{-1/2},\ (F_{\mathrm{L}}^{\mathrm{p}})^{-1/2} \right\| \qquad (12.63)$$

$$\mathbf{F}^{\mathrm{e}} = \mathrm{diag}\left\| F_{\mathrm{L}}^{\mathrm{e}},\ (1+\mathcal{V})^{-1/3}(F_{\mathrm{L}}^{\mathrm{p}})^{1/2},\ (1+\mathcal{V})^{-1/3}(F_{\mathrm{L}}^{\mathrm{p}})^{1/2} \right\|,$$

with

$$F_{\mathrm{L}}^{\mathrm{e}} = \frac{v/v_{\mathrm{R}}}{(1+\mathcal{V})^{1/3} F_{\mathrm{L}}^{\mathrm{p}}}, \qquad (12.64)$$

where we have used the equation $\det \mathbf{F} = F_{\mathrm{L}} = v/v_{\mathrm{R}}$.

The elastic strain is related to the deformation gradient components by the equation

$$\widetilde{\mathbf{E}}^{e} = \mathrm{diag}\left\| \widetilde{E}_{\mathrm{L}}^{\mathrm{e}},\ \widetilde{E}_{\mathrm{T}}^{\mathrm{e}},\ \widetilde{E}_{\mathrm{T}}^{\mathrm{e}} \right\| = \tfrac{1}{2}\mathrm{diag}\left\| (F_{\mathrm{L}}^{\mathrm{e}})^2 - 1,\ (F_{\mathrm{T}}^{\mathrm{e}})^2 - 1,\ (F_{\mathrm{T}}^{\mathrm{e}})^2 - 1 \right\|. \qquad (12.65)$$

Substitution of Eqs. 12.64 into Eq. 12.65 yields the elastic strain components

$$\widetilde{E}_{\mathrm{L}}^{\mathrm{e}} = \frac{1}{2}\left[\left(\frac{v/v_{\mathrm{R}}}{(1+\mathcal{V})^{1/3} F_{\mathrm{L}}^{\mathrm{p}}}\right)^{2} - 1\right] \quad \text{and} \quad \widetilde{E}_{\mathrm{T}}^{\mathrm{e}} = \frac{1}{2}\left[\frac{F_{\mathrm{L}}^{\mathrm{p}}}{(1+\mathcal{V})^{2/3}} - 1\right]. \qquad (12.66)$$

The nonzero stress components, given by Eq. 12.48, are

$$t_{11} = (B^{\eta} + \tfrac{4}{3}\mu^{\eta})\widetilde{E}_{\mathrm{L}}^{\mathrm{e}} + 2(B^{\eta} - \tfrac{2}{3}\mu^{\eta})\widetilde{E}_{\mathrm{T}}^{\mathrm{e}} - \rho_{\mathrm{R}}\,\theta_{\mathrm{R}}\,\gamma_{\mathrm{R}}\,[(v/v_{\mathrm{R}}) - \mathcal{V}](\eta - \eta_{\mathrm{R}})$$

$$t_{22} = (B^{\eta} - \tfrac{2}{3}\mu^{\eta})\widetilde{E}_{\mathrm{L}}^{\mathrm{e}} + 2(B^{\eta} + \tfrac{1}{3}\mu^{\eta})\widetilde{E}_{\mathrm{T}}^{\mathrm{e}} - \rho_{\mathrm{R}}\,\theta_{\mathrm{R}}\,\gamma_{\mathrm{R}}\,[(v/v_{\mathrm{R}}) - \mathcal{V}](\eta - \eta_{\mathrm{R}}), \qquad (12.67)$$

with $t_{22} = t_{33}$ because of the uniaxial symmetry. The pressure is given by

$$p(\widetilde{\mathbf{E}}^{\mathrm{e}}, \eta) = -B^{\eta}(\mathcal{V})\,(\widetilde{E}_{\mathrm{L}}^{\mathrm{e}} + 2\widetilde{E}_{\mathrm{T}}^{\mathrm{e}}) + \rho_{\mathrm{R}}\,\theta_{\mathrm{R}}\,\gamma_{\mathrm{R}}\,[(v/v_{\mathrm{R}}) - \mathcal{V}](\eta - \eta_{\mathrm{R}}). \qquad (12.68)$$

As for any isotropic material, the shear stress achieves its maximum absolute value, $|\tau_{45°}|$, on planes lying at 45° to the x axis, with

$$\tau_{45°} = \tfrac{1}{2}(t_{11} - t_{22}) = \mu(\widetilde{E}_{\mathrm{L}}^{\mathrm{e}} - \widetilde{E}_{\mathrm{T}}^{\mathrm{e}}). \qquad (12.69)$$

We shall assume that all of the slip occurs on these planes and that its Burgers vector is in the direction of the maximum shear traction vector which we designate $\boldsymbol{\tau}_{45°}$. The unit normal vector characterizing these slip planes in the current configuration is

$$\bar{\mathbf{n}}(\varphi_1) = \tfrac{1}{\sqrt{2}}(1, \cos\varphi_1, \sin\varphi_1), \quad 0 \le \varphi_1 < 2\pi. \qquad (12.70)$$

The Burgers vector is taken to have the direction of the maximum shear traction vector on these planes, so

$$\mathbf{b}(\varphi_1) = \tfrac{1}{\sqrt{2}} b(1, -\cos\varphi_1, -\sin\varphi_1) \qquad 0 \le \varphi_1 < 2\pi , \tag{12.71}$$

and the shear traction vector is given by

$$\boldsymbol{\tau}_{45^\circ} = \tau_{45^\circ} \bar{\mathbf{b}}(\varphi_1) = \tau_{45^\circ} \tfrac{1}{\sqrt{2}} (1, -\cos\varphi_1, -\sin\varphi_1), \quad 0 \le \varphi_1 < 2\pi . \tag{12.72}$$

The images of $\bar{\mathbf{n}}$ and $\mathbf{b}$ in the plastically deformed configuration are given (to first order in $\widetilde{\mathbf{E}}^{\mathrm{e}}$) by

$$\bar{\mathbf{N}} = \tfrac{1}{\sqrt{2}}\left(1+\tfrac{1}{2}(\widetilde{E}_{\mathrm{L}}^{\mathrm{e}} - \widetilde{E}_{\mathrm{T}}^{\mathrm{e}}), [(1-\tfrac{1}{2}(\widetilde{E}_{\mathrm{T}}^{\mathrm{e}} - \widetilde{E}_{\mathrm{T}}^{\mathrm{e}})]\cos\varphi_1, [(1-\tfrac{1}{2}(\widetilde{E}_{\mathrm{T}}^{\mathrm{e}} - \widetilde{E}_{\mathrm{T}}^{\mathrm{e}})]\sin\varphi_1\right), \tag{12.73}$$

and

$$\mathbf{B} = \tfrac{1}{\sqrt{2}} b\left(1-\widetilde{E}_{\mathrm{L}}^{\mathrm{e}}, -(1-\widetilde{E}_{\mathrm{T}}^{\mathrm{e}})\cos\varphi_1, -(1-\widetilde{E}_{\mathrm{T}}^{\mathrm{e}})\sin\varphi_1\right), \tag{12.74}$$

where

$$b = B\,[1+\widetilde{E}_{\mathrm{L}}^{\mathrm{e}} + \widetilde{E}_{\mathrm{T}}^{\mathrm{e}}] . \tag{12.75}$$

Proceeding in the same manner as for the undamaged material we have, in analogy to Eq. 7.129,

$$\boldsymbol{\Lambda}^{\mathrm{P}} = \tfrac{1}{2}\pi B\, \mathcal{N}_{\mathrm{M}}\, V_{X^*}(\tau_{45^\circ})\, \mathrm{diag}\,\| \, 2, \; -1, \; -1 \, \| \tag{12.76}$$

to within quadratic terms in $\widetilde{\mathbf{E}}^{\mathrm{e}}$. Therefore, we have

$$\dot{F}_{\mathrm{L}}^{\mathrm{p}} = \pi B\, \mathcal{N}_{\mathrm{M}}\, V_{X^*}(\tau_{45^\circ})\, F_{\mathrm{L}}^{\mathrm{p}} . \tag{12.77}$$

The dislocation velocity in the current and plastically deformed configurations are related by the equation

$$V_x(\tau_{45^\circ}) = (b/B)\, V_{X^*}(\tau_{45^\circ}) = [1+\widetilde{E}_{\mathrm{L}}^{\mathrm{e}} + \widetilde{E}_{\mathrm{T}}^{\mathrm{e}}]\, V_{X^*}(\tau_{45^\circ}), \tag{12.78}$$

so

$$\dot{F}_{\mathrm{L}}^{\mathrm{p}} = \pi B\, \mathcal{N}_{\mathrm{M}}\, [1+\widetilde{E}_{\mathrm{L}}^{\mathrm{e}} + \widetilde{E}_{\mathrm{T}}^{\mathrm{e}}]\, V_x(\tau_{45^\circ})\, F_{\mathrm{L}}^{\mathrm{p}} . \tag{12.79}$$

Analysis of spall problems set in terms of the foregoing theory can only be achieved by full numerical simulation. Such calculations are beyond the scope of this book, but results obtained using several related theories have been reported in [34,36,41,65].

CHAPTER 13

Steady Detonation Waves

Chemical reactions may be initiated when a sufficiently strong shock is introduced into a material of metastable composition. If these reactions produce gaseous products and are (in aggregate) strongly exothermic, the material is called an *explosive*. The chemical energy released upon passage of the shock provides the energy required for its steady propagation, replacing (or adding to) the energy supplied by the work done on the boundary that supports steady shock propagation in a nonreactive material. A chemically-supported shock is called a *detonation shock*. A *detonation wave* is a shock followed by a *chemical reaction zone* and a region of unsteady flow, often in the form of a centered simple decompression wave.

In this chapter we shall discuss two simple models of the steady detonation process. The first and simplest, the Chapman–Jouguet (CJ) theory, is based upon the premise that the chemical reaction is so rapid that it can be assumed to occur instantaneously. The waveform associated with this simplest case is illustrated in Fig. 13.1. The second model, represented by the Zel'dovich–von Neumann–Döring (ZND) theory, is an extension of the CJ theory that takes finite reaction rates into account. These theories provide a sound basis for understanding detonation wave propagation and an appropriate point of departure for conducting further research.

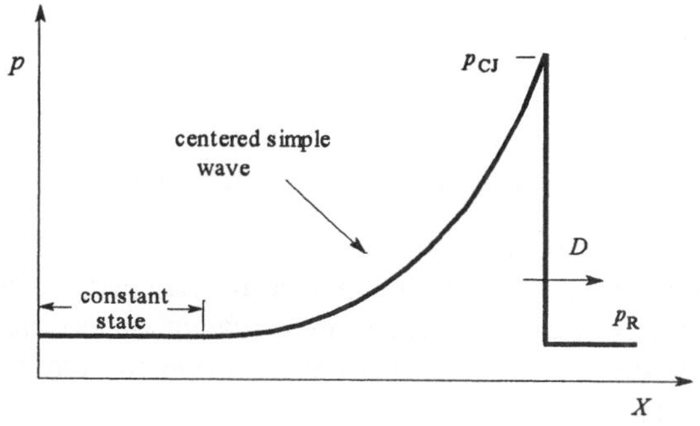

Figure 13.1. Illustration of the pressure waveform predicted by the Chapman–Jouguet theory.

A *steady detonation* is steady in the sense that neither the jumps nor the shock velocity vary as the wave propagates and the reaction zone is translated forward at the shock velocity but unchanged in form. Even in the case of a steady detonation, the decompression wave following the reaction zone is unsteady.

Explosives can be gases, liquids, or solids but the detonation products are entirely, or almost entirely, gaseous. Gaseous explosives and the detonation process they undergo are of both scientific interest and practical importance so they have been widely studied. Among the reasons gas detonations are of scientific interest is that they can be studied in chemical mixtures such as hydrogen and oxygen for which the thermodynamic and chemical properties are well understood. It is also important that the initial thermodynamic state of a gaseous explosive can be controlled by changing the initial pressure and temperature. The chemical energy liberated by reacting a unit mass of material can be decreased by adding a nonreactive diluent such as nitrogen or argon. Finally, the pressures encountered in studying gas detonations are low enough to permit repeated use of experimental apparatus, making investigations easier and less expensive for gases than for condensed explosives.

Molecules of solid and liquid explosives are usually quite complex and the chemical processes encountered in a detonation are correspondingly complex. Details of these chemical processes are often poorly understood and uncertainty regarding the composition of the reaction products carries over to uncertainty regarding the chemical energy liberated by the reaction and the equation of state of the reaction products. Technologically important solid explosives propagate detonation waves at velocities in the range of 5000–9000 m/s and produce pressures of 20–40 GPa, conditions under which measurements are often difficult and uncertain.

Even a small portion of detonation physics is much too broad a subject to cover in this chapter. Our discussion is limited to plane, steady detonation waves. Detonation physics and technology are discussed more comprehensively in [3,20,27,42,43,71].

13.1 The Chapman–Jouguet (CJ) Detonation

In the most idealized view of a detonation, each particle of the explosive undergoes an instantaneous transition from its initial (unreacted) form to reaction products as the shock passes. The equation of state of the detonation products is entirely different from that of the unreacted explosive and the internal energy of the products includes the chemical energy liberated by the reactions. The shock at the front of a detonation wave obeys the same jump conditions that describe nonreactive shocks and, as with the nonreactive case, these conditions provide three constraints relating the five variables p^+, $\dot{x}^+$, v^+ (or $\rho^+ = 1/v^+$), ε^+, and D, where we write $U_S \equiv D$ in conformity with the practice of detonation physics. The detonation products are described by a Hugoniot curve relating any pair of the above variables in the states that can be reached by a shock transition from the unreacted state. The Hugoniot for the detonation products is centered on the state of the unreacted material into which the shock is propagating but is offset from the center point by an amount that depends upon the chemical energy released by the reaction. The three jump conditions and the Hugoniot curve suffice to determine the state behind the shock to within one variable, which we have regarded as a measure of the shock strength. In the case of nonreactive materials, steady shock propagation is sustained by energy supplied by forces applied to material at the boundary behind the shock and the shock strength is determined by the imposed boundary condition. The situation differs in the present case because a detonation shock is sustained by chemical energy rather than by forces imposed on the boundary. The new concepts introduced in this section concern establishment of relations between chemical energy release and shock strength.

Let us restrict attention to the case in which the unreacted explosive is at rest in its reference state: $\dot{x}^- = 0$, $p^- = p_R$, $v^- = v_R$, and $\varepsilon^- = \varepsilon_R$. The jump conditions of Eq. 2.113 can be written

$$D\left(1-\frac{v^+}{v_R}\right) = \dot{x}^+$$

$$\rho_R D\,\dot{x}^+ = p^+ - p_R \tag{13.1}$$

$$\rho_R D\,(\varepsilon^+ - \varepsilon_R) + \tfrac{1}{2}\rho_R D\,(\dot{x}^+)^2 = p^+\,\dot{x}^+.$$

The usual manipulation of Eqs. 13.1 gives the equation

$$D = \dot{x}^+\left(1-\frac{v^+}{v_R}\right)^{-1} = \frac{p^+ - p_R}{\rho_R\,\dot{x}^+} \tag{13.2}$$

for the detonation velocity, the equation

$$p^+ - p_R = \rho_R\,D^2\left(1-\frac{v^+}{v_R}\right) \tag{13.3}$$

describing the Rayleigh line, the Rankine–Hugoniot equation

$$\varepsilon^+ = \varepsilon_R + \frac{1}{2}(p^+ + p_R)v_R\left(1-\frac{v^+}{v_R}\right) \tag{13.4}$$

relating the thermodynamic variables, and the equation

$$(\dot{x}^+)^2 = (p^+ - p_R)(v_R - v^+) \tag{13.5}$$

giving the particle velocity in terms of thermodynamic variables.

Since the detonation products form a gas, the equation of state of these products is one appropriate to a gas. We shall use the ideal gas theory discussed in Sect. 5.2 since it provides a convenient and widely used model. In this case, the equation of state of the products takes the form

$$\varepsilon(p, v) = \frac{pv}{\Gamma - 1} - q\,, \tag{13.6}$$

where $q > 0$ is the internal energy that is liberated by the chemical reaction of a unit mass of explosive (The heat of reaction, q, is not to be confused with the heat-flux vector component designated by the same symbol that was introduced in Chap. 2).

Some comment is necessary regarding the parameter q in Eq. 13.6. When atoms bind together to form compounds, a certain amount of energy (called the *standard heat of formation* and listed in chemical tables) is required to form the molecules. When a compound is unaltered during a thermodynamic process it is not necessary to take account of this heat of formation because the reference energy state is arbitrary. When the chemical binding changes, however, energy equal to the heat of formation of the reactants minus the heat of formation of the products is liberated and must be taken into account in analyzing the process. In the case of explosives, decomposition of the explosive molecules and recombination of the atoms to form detonation products results in a net decrease in the energy binding the atoms into molecules, so the heat of formation of the detonation products is less than that of the explosive. It is customary to set the reference energy of the unreacted explosive so that its internal energy is given by the equation of state, for example, $\varepsilon = C_{\mathrm{R}}^{v}\theta$ in the case of an ideal gas. Using this same reference state for the detonation products means that the internal energy of this material is that given by the equation of state for the products plus the heat of formation of the reactant explosive minus the heat of formation of the detonation products. When the detonation products are at the same pressure and specific volume as that of the unreacted explosive (i.e., are at the same temperature in the case of an ideal gas), the internal energy density of the detonation products is less than that of the explosive by a positive amount q, leading to the equation of state 13.6 for the detonation products if these products form an ideal gas.

When Eq. 13.6 is substituted into the Rankine–Hugoniot equation the result can be written

$$\left(\frac{p^+}{p_{\mathrm{R}}} + \mu^2\right)\left(\frac{v^+}{v_{\mathrm{R}}} - \mu^2\right) = 1 - \mu^4 + \frac{2\mu^2 q}{p_{\mathrm{R}}\, v_{\mathrm{R}}}\,, \tag{13.7}$$

where $\mu^2 \equiv (\Gamma - 1)/(\Gamma + 1)$. This reaction product Hugoniot is centered on the state of the unreacted explosive but does not pass through this state. At the specific volume v_R the pressure on the Hugoniot exceeds the center-point value, p_R, by the amount $\rho_R(\Gamma-1)q$.

As noted previously, the foregoing equations are not sufficient to determine all five of the variables that describe the detonation shock. Some means must be found to complete the solution. To begin, let us consider the Hugoniot and Rayleigh lines shown in the p–v plot of Fig. 13.2. We know that the state immediately behind the detonation shock must lie on both the Hugoniot and the Rayleigh line corresponding to the detonation velocity. The lowest-velocity Rayleigh line shown does not intersect the Hugoniot, so it cannot correspond to a solution. The intermediate-velocity Rayleigh line intersects the Hugoniot at exactly one point, designated CJ and called the *Chapman–Jouguet*, or *CJ, point*; it corresponds to a unique solution to the problem. The highest-velocity Rayleigh line intersects the Hugoniot at two points, designated S and W, respectively, for two possible solutions called the *strong detonation* and the *weak detonation*.

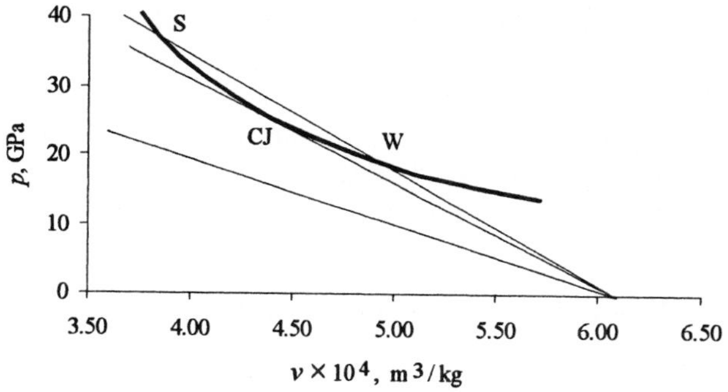

Figure 13.2. Detonation product Hugoniot and three Rayleigh lines. The lower Rayleigh line does not intersect the Hugoniot, the middle line is tangent to the Hugoniot at the point designated CJ, and the upper line intersects it twice at points designated S and W. The Hugoniot is drawn for the explosive TNT prepared at the density $\rho_R = 1600$ kg/m³. The equation of state is characterized by $\Gamma = 2.6$ and $q = 4.9 \times 10^6$ J/kg.

Since the CJ state lies on both the Rayleigh line and the Hugoniot for the detonation products, we can eliminate the pressure from this pair of equations (Eqs. 13.3 and 13.7), producing the result

$$1 = \frac{p_R v_R}{2\mu^2 q}\left(1 - \frac{v_{CJ}}{v_R}\right)\left[\frac{D_{CJ}^2}{p_R v_R}\left(\frac{v_{CJ}}{v_R} - \mu^2\right) - (1+\mu^2)\right]. \quad (13.8)$$

Before solving this equation it is convenient to introduce the dimensionless variables

$$\overline{D}_{\mathrm{CJ}}^2 = \frac{D_{\mathrm{CJ}}^2}{q}, \quad \overline{v}_{\mathrm{CJ}} = \frac{v_{\mathrm{CJ}}}{v_{\mathrm{R}}}, \quad e = \frac{p_{\mathrm{R}} v_{\mathrm{R}}}{q}, \tag{13.9}$$

so that it can be written

$$\overline{D}_{\mathrm{CJ}}^2 \overline{v}_{\mathrm{CJ}}^2 - (1+\mu^2)(\overline{D}_{\mathrm{CJ}}^2 + e)\overline{v}_{\mathrm{CJ}} + [2\mu^2 + \mu^2 \overline{D}_{\mathrm{CJ}}^2 + (1+\mu^2)e] = 0. \tag{13.10}$$

Solving for $\overline{v}_{\mathrm{CJ}}$ yields the result

$$\begin{aligned} \overline{v}_{\mathrm{CJ}} = \frac{1}{2\overline{D}_{\mathrm{CJ}}^2} \Big\{ & (1+\mu^2)(\overline{D}_{\mathrm{CJ}}^2 + e) \\ & \pm \left[(1+\mu^2)^2 (\overline{D}_{\mathrm{CJ}}^2 + e)^2 - 4\overline{D}_{\mathrm{CJ}}^2 [2\mu^2 + \mu^2 \overline{D}_{\mathrm{CJ}}^2 + (1+\mu^2)e] \right]^{1/2} \Big\}. \end{aligned} \tag{13.11}$$

As discussed previously, the CJ state corresponds to the detonation velocity for which a unique solution is obtained. Accordingly, we select D_{CJ}^2 so that the square root in Eq. 13.11 vanishes:

$$(1-\mu^2)^2 \overline{D}_{\mathrm{CJ}}^4 - 2[4\mu^2 + (1-\mu^4)e]\overline{D}_{\mathrm{CJ}}^2 + (1+\mu^2)^2 e^2 = 0. \tag{13.12}$$

The solution of this equation is

$$\begin{aligned} \overline{D}_{\mathrm{CJ}}^2 &= \frac{4\mu^2}{(1-\mu^2)^2} \left\{ \left[1 + \frac{1}{4\mu^2}(1-\mu^4)e \right] + \left[1 + \frac{1}{2\mu^2}(1-\mu^4)e \right]^{1/2} \right\} \\ &= (\Gamma^2 - 1)\left[1 + \left(1 + \frac{2\Gamma}{\Gamma^2 - 1} e \right)^{1/2} \right] + \Gamma e, \end{aligned} \tag{13.13}$$

where the positive square root has been taken to obtain a positive value for the squared detonation velocity.

The CJ value of $\overline{v}$ can now be obtained from Eq. 13.11 which, by virtue of the way in which D was chosen, takes the simple form

$$\overline{v}_{\mathrm{CJ}} = \frac{1}{2}(1+\mu^2)\left(1 + \frac{e}{\overline{D}_{\mathrm{CJ}}^2} \right) = \frac{\Gamma}{\Gamma + 1}\left(1 + \frac{e}{\overline{D}_{\mathrm{CJ}}^2} \right). \tag{13.14}$$

The CJ pressure lies at the point on the detonation-product Hugoniot at which $\overline{v} = \overline{v}_{\mathrm{CJ}}$:

$$\frac{p_{\mathrm{CJ}}}{p_{\mathrm{R}}} \equiv \overline{p}_{\mathrm{CJ}} = \frac{1 - \mu^2 \overline{v}_{\mathrm{CJ}} + (2\mu^2/e)}{\overline{v}_{\mathrm{CJ}} - \mu^2}. \tag{13.15}$$

Finally, substitution of this result into Eq. 13.5 gives

$$\dot{x}_{CJ}^2 = \frac{(1-\bar{v}_{CJ})q}{\bar{v}_{CJ}-\mu^2}\left[2\mu^2+(1+\mu^2)(1-\bar{v}_{CJ})e\right]. \tag{13.16}$$

It is useful to consider the magnitude of the various quantities appearing in the foregoing equations. The detonation products are very hot gases and a typical value of μ^2 is about 0.1. For reference conditions near $p_R = 1\,\text{atm.}$, $p_R v_R$ is only a few per cent of q for a typical gaseous explosive, i.e., $0 < p_R v_R / q \ll 1$. When terms proportional to p_R are ignored, D_{CJ}^2 and q stand in the relation $D_{CJ}^2 = 2(\Gamma^2 - 1)q$ so these two quantities are of comparable magnitude, i.e., $D_{CJ}^2 / q \approx 1$. We also know that v / v_R varies in the approximate range $0.1 < v/v_R < 1$. In view of these facts it is usually reasonable to neglect terms of higher order than the first in e. In this case, Eq. 13.13 takes the much simpler form

$$D_{CJ}^2 = 2(\Gamma^2 - 1)q + 2\Gamma p_R v_R + \cdots \tag{13.17}$$

when the original variables are restored. Similarly, substitution of Eq. 13.13 into Eq. 13.14 and introduction of the same approximation gives

$$\bar{v}_{CJ} = \frac{\Gamma}{\Gamma+1} + \frac{\Gamma}{2(\Gamma+1)(\Gamma^2-1)}\frac{p_R v_R}{q} + \ldots. \tag{13.18}$$

Substituting $\bar{v}_{CJ}$ as given by Eq. 13.14 into Eq. 13.15 and discarding higher-order terms yields the approximate result

$$p_{CJ} = 2\rho_R(\Gamma - 1)\,q + \frac{2\Gamma+1}{\Gamma+1}p_R + \ldots. \tag{13.19}$$

Finally, similar manipulation of Eq. 13. 16 produces the result

$$\dot{x}_{CJ} = \frac{D_{CJ}}{\Gamma+1} + \ldots. \tag{13.20}$$

In the limit $p_R \to 0$, Eqs. 13.17, 13.18, and 13.19 take the even simpler forms

$$D_{CJ}^2 = 2(\Gamma^2 - 1)q\,, \tag{13.21}$$

$$\bar{v}_{CJ} = \frac{\Gamma}{\Gamma+1}, \tag{13.22}$$

and

$$p_{CJ} = 2\rho_R(\Gamma - 1)q\,. \tag{13.23}$$

By substituting these results into the jump conditions we can obtain

$$\dot{x}_{CJ}^2 = 2\mu^2 q$$

$$D_{CJ}^2 = (\Gamma + 1)\, v_R\, p_{CJ} \tag{13.24}$$

$$\dot{x}_{CJ} = \frac{1}{\Gamma + 1} D_{CJ}.$$

A simple calculation based on the foregoing equations shows that

$$D_{CJ} = C_{CJ} = c_{CJ} + \dot{x}_{CJ}, \tag{13.25}$$

where

$$C_{CJ}^2 = (v_R / v)^2 \Gamma\, p_{CJ}\, v_{CJ} \text{ and } c_{CJ}^2 = \Gamma\, p_{CJ}\, v_{CJ} \tag{13.26}$$

are, respectively, the squared Lagrangian and Eulerian soundspeeds at the CJ state. Equation 13.25, which holds exactly for a general equation of state, can be regarded as defining the CJ point. On this basis, the CJ point is also identified as the *sonic point* in the flow.

Since gaseous explosives cannot exist at zero pressure, the effect of initial pressure cannot be ignored when developing a theory for detonation of these materials. Nevertheless, the limiting cases for small or vanishing initial pressure agree with the exact results to within a few per cent for gas detonations when the initial pressure is 1 atm. The results for the limiting case $p_R \to 0$ are those usually seen in the solid-explosives literature because the equation of state for the unreacted material is meaningful for $p = 0$ and initial pressures of ~1 atm are entirely negligible in comparison with the CJ pressure.

Some properties of four common explosives are given in Table 13.1.

Table 13.1. Properties of some common high explosives[a]

Common name	Chemical formula	ρ_R kg/m^3	q MJ/kg	D_{CJ} m/s	p_{CJ} GPa
TNT	$C_7H_5N_3O_6$	1640	5.40	6930	21
Nitroglycerine	$C_3H_5N_3O_9$	1600	6.19	7700	25
PETN	$C_3H_5N_4O_{12}$	1670	6.28	8260	31
HMX	$C_4H_8N_8O_8$	1890	6.19[c]	9110	42

[a] Data are from various sources and are intended for illustration only.

It is important to note that one need not restrict attention to detonations propagating at the CJ speed. We have seen that no solution exists for $D < D_{CJ}$. When $D > D_{CJ}$, there are solutions corresponding to strong and weak detonations, respectively. The pressure behind the strong and weak detonation shocks is given by

$$p_S = p_{CJ}\left(\frac{D}{D_{CJ}}\right)^2\left\{1+\left[1-\left(\frac{D_{CJ}}{D}\right)^2\right]^{1/2}\right\}$$
$$p_W = p_{CJ}\left(\frac{D}{D_{CJ}}\right)^2\left\{1-\left[1-\left(\frac{D_{CJ}}{D}\right)^2\right]^{1/2}\right\}. \tag{13.27}$$

Values of v_S, v_W, and other parameters characterizing strong and weak detonation points can be obtained by application of the jump conditions.

The strong detonation is discussed in the following section, but analysis of the weak detonation is much more difficult and must await our discussion of the ZND theory in Sect. 13.2.

13.1.1 Strong Detonation

In this section we consider the case in which a sustained pressure in excess of the CJ pressure is suddenly applied to the boundary of an explosive. When the applied pressure results from impact it is calculated as the point of intersection of the $p-\dot{x}$ Hugoniots for the impactor and the explosive products.

The $p-\dot{x}$ Hugoniot for the detonation product is obtained by using Eq. 13.5 to eliminate v/v_R from Eq. 13.7, yielding the Hugoniot

$$p^+ = (\Gamma+1)\frac{(\dot{x}^+)^2}{2v_R} + (\Gamma-1)\frac{q}{v_R}, \tag{13.28}$$

when terms proportional to p_R/p^+ are neglected. When the impact pressure exceeds the CJ pressure a strong detonation is produced and the state behind the detonation shock is uniform and characterized by the strong detonation parameters

$$\frac{v^+}{v_R} = 1 - \frac{(\dot{x}^+)^2}{p^+ v_R}$$
$$D = \frac{p^+ v_R}{\dot{x}^+}. \tag{13.29}$$

This detonation is stable in the same sense as a nonreactive shock, i.e., the Lagrangian soundspeed in the region behind the shock exceeds the shockspeed. This also means that a strong detonation cannot persist in the absence of sustained application of pressure at the boundary because any decompression wave will overtake and attenuate the detonation shock.

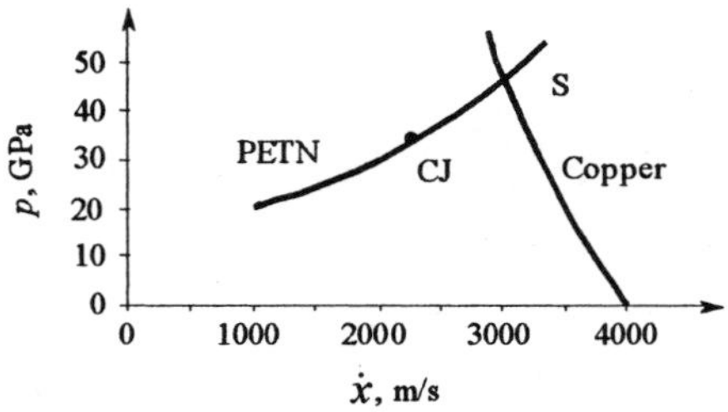

Figure 13.3. Pressure–particle-velocity diagram for an overdriven (strong) detonation produced in the explosive PETN by impacting it with a copper projectile plate moving at the velocity $\dot{x}_P = 4000\,\text{m/s}$.

13.1.2 Taylor Decompression Wave

Let us now consider the flow behind the detonation shock for the case in which the supporting pressure, p_B , is less than p_{CJ} . In this case the decompression process will proceed along the isentrope through the CJ state and decreasing in pressure to p_B . This decompression wave is called a *Taylor wave.* The equation for the decompression isentrope is $pv^\Gamma = p_{CJ}(v_{CJ})^\Gamma$. The Taylor-wave analysis is one that is usefully carried out in both the Lagrangian and Eulerian frames. The former is often the most useful for solution of problems arising in applications and is appropriate for interpreting experimental observations made using pressure or particle-velocity gauges embedded in the explosive. Many experiments conducted to study detonation physics produce data in the form of flash-radiographic images. Since these images give spatial positions of wavefronts, embedded tracer particles, etc. they are most easily interpreted in the context of an Eulerian analysis.

Lagrangian Analysis. The X–t diagram for this problem is shown in Fig. 13.4. Since the state behind the detonation shock is constant, the following wave is a centered simple wave and, as seen from Eq. 13.25$_1$, the leading characteristic of this wave lies along the shock trajectory. The trailing characteristic advances at a slower rate given by this same equation evaluated at the boundary pressure p_B and the associated specific volume v_B . In Chap. 9 we obtained the solution for the simple wave fields.

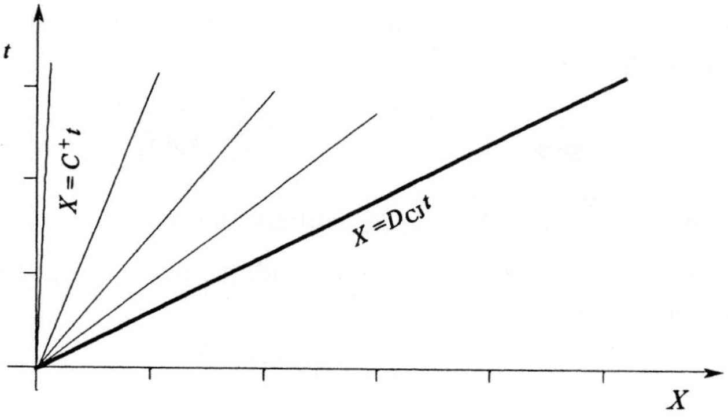

Figure. 13.4. Lagrangian space–time diagram for a Chapman–Jouguet detonation wave.

The field equations to be solved are 9.11. We seek a solution that is a function of the single variable

$$Z = \frac{X}{t}, \tag{13.30}$$

in which case the equations take the form

$$\begin{aligned} Z\,\dot{x}_Z + C_L^2\left(\frac{v}{v_R}\right)_Z &= 0 \\ Z\left(\frac{v}{v_R}\right)_Z + \dot{x}_Z &= 0, \end{aligned} \tag{13.31}$$

where $C_L = C_L(v/v_R)$. When the second of these equations is substituted into the first, we obtain

$$[Z^2 - C_L^2]\,\dot{x}_Z = 0, \tag{13.32}$$

which has a nontrivial solution only if

$$Z = \pm C_L. \tag{13.33}$$

To proceed, we need to determine the form of $C_L(v/v_R)$, the Lagrangian isentropic soundspeed in the detonation products. From Eq. 9.10 we have

$$C_L^2 = -(v_R)^2 \frac{dp^{(\eta)}(v)}{dv}, \tag{13.34}$$

where

$$p^{(\eta)}(v) = p_{CJ}\left(\frac{v_{CJ}}{v}\right)^{\Gamma} \tag{13.35}$$

is the isentrope through the CJ point. When this isentrope is substituted into Eq. 13.34 we obtain the relations

$$C_{\mathrm{L}} = D_{\mathrm{CJ}}\left(\frac{v_{\mathrm{CJ}}}{v}\right)^{(\Gamma+1)/2} = D_{\mathrm{CJ}}\left(\frac{p}{p_{\mathrm{CJ}}}\right)^{(\Gamma+1)/2\Gamma}, \tag{13.36}$$

for the Lagrangian soundspeed in the decompression wave.

Substituting these results into Eq. 13.33 (with the positive root taken for the right-propagating wave under consideration) gives

$$\begin{aligned} v &= v_{\mathrm{CJ}}\left(D_{\mathrm{CJ}}\frac{t}{X}\right)^{2/(\Gamma+1)} \\ p &= p_{\mathrm{CJ}}\left(\frac{1}{D_{\mathrm{CJ}}}\frac{X}{t}\right)^{2\Gamma/(\Gamma+1)} . \end{aligned} \tag{13.37}$$

Equations 13.31_1 and 13.33 can be combined to give

$$\frac{d\dot{x}}{dZ} + \left[C_{\mathrm{L}}\left(\frac{v}{v_{\mathrm{R}}}\right)\right]\frac{d}{dZ}\left(\frac{v}{v_{\mathrm{R}}}\right) = 0, \tag{13.38}$$

which can be integrated to give

$$\dot{x} = \dot{x}_{\mathrm{CJ}}\frac{\Gamma+1}{\Gamma-1}\left[\frac{2\Gamma}{\Gamma+1}\left(\frac{v_{\mathrm{CJ}}}{v}\right)^{(\Gamma-1)/2} - 1\right]. \tag{13.39}$$

Substitution of Eq. 13.37_1 into Eq. 13.39 gives the solution

$$\dot{x} = \dot{x}_{\mathrm{CJ}}\frac{\Gamma+1}{\Gamma-1}\left[\frac{2\Gamma}{\Gamma+1}\left(\frac{1}{D_{\mathrm{CJ}}}\frac{X}{t}\right)^{(\Gamma-1)/(\Gamma+1)} - 1\right] \tag{13.40}$$

in Lagrangian coordinates. The simple wave region in which this solution applies extends from the detonation shock to the point where the pressure has decreased to the value p_{B} that is imposed on the boundary. The trailing characteristic of the decompression fan is given by

$$X = C^{+}t, \tag{13.41}$$

where C^{+} is obtained by evaluating Eq. 13.36_2 at $p = p_{\mathrm{B}}$:

$$C^{+} = D_{\mathrm{CJ}}\left(\frac{p_{\mathrm{B}}}{p_{\mathrm{CJ}}}\right)^{(\Gamma+1)/2\Gamma} . \tag{13.42}$$

The characteristic coordinate lies in the range

$$C^{+} \le \frac{X}{t} \le D_{\mathrm{CJ}} . \tag{13.43}$$

In practical applications it is usually permissible to neglect the pressure, p_B, in comparison with p_{CJ}. The equations given remain valid in this case and $C^+ = 0$, so the trailing characteristic is given by $X = 0$.

Results obtained from Eqs. 13.37 and 13.40 are plotted in Fig. 13.5.

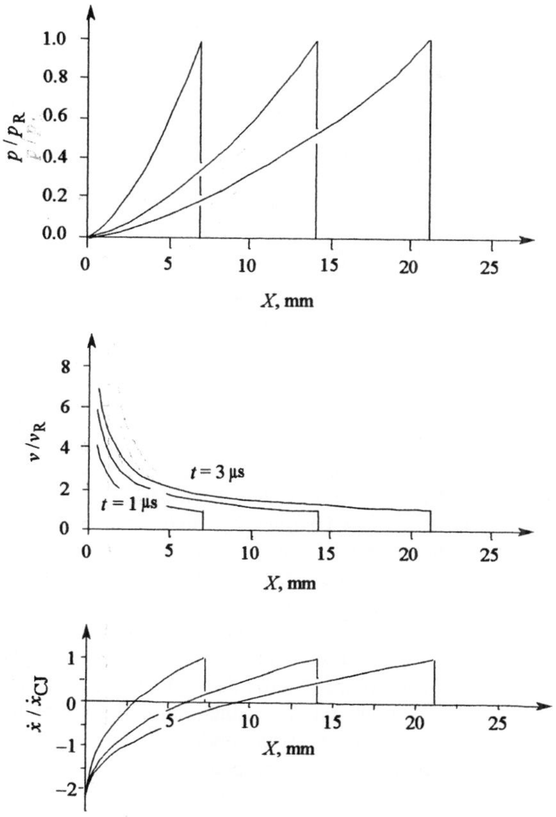

Figure 13.5. Lagrangian waveforms at 1, 2, and 3 μs after introduction of a CJ detonation wave into an explosive characterized by the parameters $D_{CJ} = 7000$ m/s and the ratio of specific heats of the detonation products, $\Gamma = 2.8$.

Eulerian Analysis. In this section we shall analyze the Taylor wave in an Eulerian framework. Included is one complication not discussed in the Lagrangian analysis just presented: When the detonation products expand into the atmosphere to the left of the explosive, a shock is introduced into this gas. The x–t diagram for the wave is shown in Fig. 13.6.

The analysis of this problem proceeds exactly the same as the shock-tube analysis given in Sect. 9.2. To solve this problem, one simply replaces the parameters p^- and ρ^- with the CJ values of the corresponding fields and inserts the value Γ_A for the ratio of specific heats of the background atmosphere and the value Γ_{DP} for the ratio of specific heats of the detonation product gas.

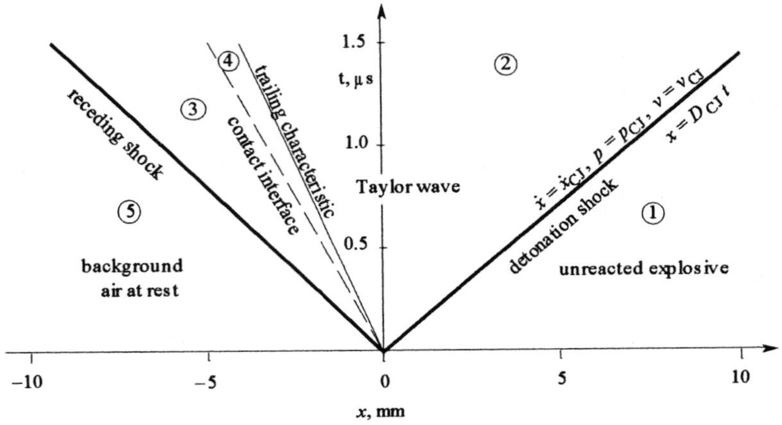

Figure 13.6. Eulerian space–time diagram for a Chapman–Jouguet detonation wave. The figure is drawn for TNT having an initial density of 1640 kg/m³, a CJ detonation velocity of 6930 m/s, a CJ pressure of 21 GPa, a CJ particle velocity of 1848 m/s, and a CJ density of 2236 kg/m³. The unreacted explosive in region 1 on the diagram is at rest at $p = 0$. The material to the left of the explosive is air ($\Gamma_A = 1.4$) at standard conditions, $p = 101325\,\text{Pa}$ and $\theta = 298.15\,\text{K}$. The Taylor decompression wave occupies region 2, bounded by the detonation shock and its trailing characteristic. To the left of the decompression wave we have regions 3 and 4 in which the pressure and particle velocity fields are uniform, but the density and the ratio of specific heats are discontinuous at the contact interface separating the air from the detonation products. Finally, this region of uniform pressure and particle velocity is separated from the undisturbed air by a receding shock. The field quantities are designated p, ρ, and $\dot{x}$ in region 2, p^+, ρ_A^+, and $\dot{x}^+$ in region 3, p^+, ρ_{DP}^+, and $\dot{x}^+$ in region 4, and p^-, ρ^-, and $\dot{x} = 0$ in region 5.

The isentrope defining the Taylor wave is (see Eq. 9.126)

$$\rho = \rho_{CJ}\left(\frac{p}{p_{CJ}}\right)^{1/\Gamma_{DP}}, \tag{13.44}$$

and the particle velocity at the point in this wave at which the pressure is p is given by (see Eq. 9.126)

$$x = \dot{x}_{CJ} - \frac{2c_{CJ}}{\Gamma_{DP} - 1}\left[1 - \left(\frac{p}{p_{CJ}}\right)^{(\Gamma_{DP}-1)/(2\Gamma_{DP})}\right]. \tag{13.45}$$

Since the characteristics are given by $z = x/t = c_L + \dot{x}$ and

$$c_L = c_{CJ}\left(\frac{p}{p_{CJ}}\right)^{(\Gamma_{DP}-1)/(2\Gamma_{DP})}, \tag{13.46}$$

one obtains the solution as functions of x and t in the form (see Eq. 9.134)

$$p = p_{CJ}\left(\frac{\Gamma_{DP}-1}{\Gamma_{DP}+1}\frac{z-\dot{x}_{CJ}}{c_{CJ}}+\frac{2}{\Gamma_{DP}+1}\right)^{2\Gamma_{DP}/(\Gamma_{DP}-1)} \tag{13.47}$$

and (see Eq. 9.133)

$$\dot{x} = \frac{1}{\Gamma_{DP}+1}\left[(\Gamma_{DP}-1)\dot{x}_{CJ}+2(z-c_{CJ})\right]. \tag{13.48}$$

In these relations $z = x/t$ occupies the range

$$\dot{x}^+ + c_L^+ \leq z \leq D_{CJ}, \tag{13.49}$$

where $\dot{x}^+$ and c_L^+ are to be obtained by matching this solution to the fields behind the receding air shock.

The pressure and density behind the air shock are related by the polytropic gas Hugoniot, which can be written

$$p^+ = p^-\frac{(\Gamma_A+1)\rho_A^+ - (\Gamma_A-1)\rho^-}{(\Gamma_A+1)\rho_A^- - (\Gamma_A-1)\rho_A^-}, \tag{13.50}$$

or

$$\rho_A^+ = \rho_A^-\frac{(\Gamma_A+1)p^+ + (\Gamma_A-1)p^-}{(\Gamma_A+1)p^- + (\Gamma_A-1)p^+}. \tag{13.51}$$

From the jump conditions, Eqs. 2.110$_{1,2}$, the Eulerian shock velocity and the particle velocity behind the receding shock are given by

$$u_S = \frac{\rho_A^+\,\dot{x}^+}{\rho_A^+ - \rho_A^-} \tag{13.52}$$

and

$$\dot{x}^+ = -\left[\frac{p^-}{\rho_A^-}\left(\frac{p^+}{p^-}-1\right)\left(1-\frac{\rho_A^-}{\rho_A^+}\right)\right]^{1/2}, \tag{13.53}$$

or, using Eq. 13.51,

$$\dot{x}^+ = -\left[\frac{p^-}{\rho_A^-}\left(\frac{p^+}{p^-}-1\right)\frac{2(p^+-p^-)}{(\Gamma_A+1)p^+ + (\Gamma_A-1)p^-}\right]^{1/2}. \tag{13.54}$$

Since the pressure and particle velocity must be continuous at the contact interface, these quantities are the same in regions 3 and 4. The Taylor wave extends behind the detonation shock to the point where the pressure has decreased from its CJ value to p^+. Evaluation of Eqs. 13.44 and 13.45 at this point gives

$$\rho_{DP}^{+} = \rho_{CJ}\left(\frac{p^+}{p_{CJ}}\right)^{1/\Gamma_{DP}} \tag{13.55}$$

and

$$\dot{x}^+ = \dot{x}_{CJ} - \frac{2c_{CJ}}{\Gamma_{DP}-1}\left[1-\left(\frac{p^+}{p_{CJ}}\right)^{(\Gamma_{DP}-1)/(2\Gamma_{DP})}\right]. \tag{13.56}$$

When these equations are adjoined to Eqs. 13.51, 13.52, and 13.54 we obtain five equations from which the field values p^+, $\dot{x}^+$, ρ_{DP}^+, ρ_A^+, and u_S can be determined. This solution is best obtained numerically. An example solution is illustrated in Fig. 13.7

When the pressure of the background gas is neglected there is no receding shock and the pressure vanishes on the trailing characteristic of the Taylor wave. In this case the solution is just as given above, with $p^+ = 0$.

13.2 Zel'dovich–von Neumann–Döring (ZND) Detonation

In developing the Chapman–Jouguet theory it was assumed that the chemical reaction occurred instantaneously in a shock transition from the initial state of the unreacted explosive to a point on the Hugoniot of the detonation products. In this section we introduce the Zel'dovich–von Neumann–Döring (ZND) theory of detonation, in which the reaction proceeds to completion at a finite rate after being initiated by passage of the shock. The waveform associated with this theory is depicted in Fig. 13.8.

The ZND theory immediately presents problems that do not arise in the Chapman–Jouguet analysis. In particular, we shall need the Hugoniot for the unreacted explosive, the equation of state of the partially reacted material, and a kinetic equation describing the rate at which the reaction proceeds. In each case, determining these equations is difficult, involving both chemical and physical considerations. Practical analyses of detonations are almost always carried out in the Chapman–Jouguet context, but understanding of detonation physics requires consideration of reaction processes.

We begin by seeking a solution to this problem in the form of a steady structured wave. As we have seen, the transition from an initial state to any state occurring in a steady wave satisfies the same conditions as a shock transition be-

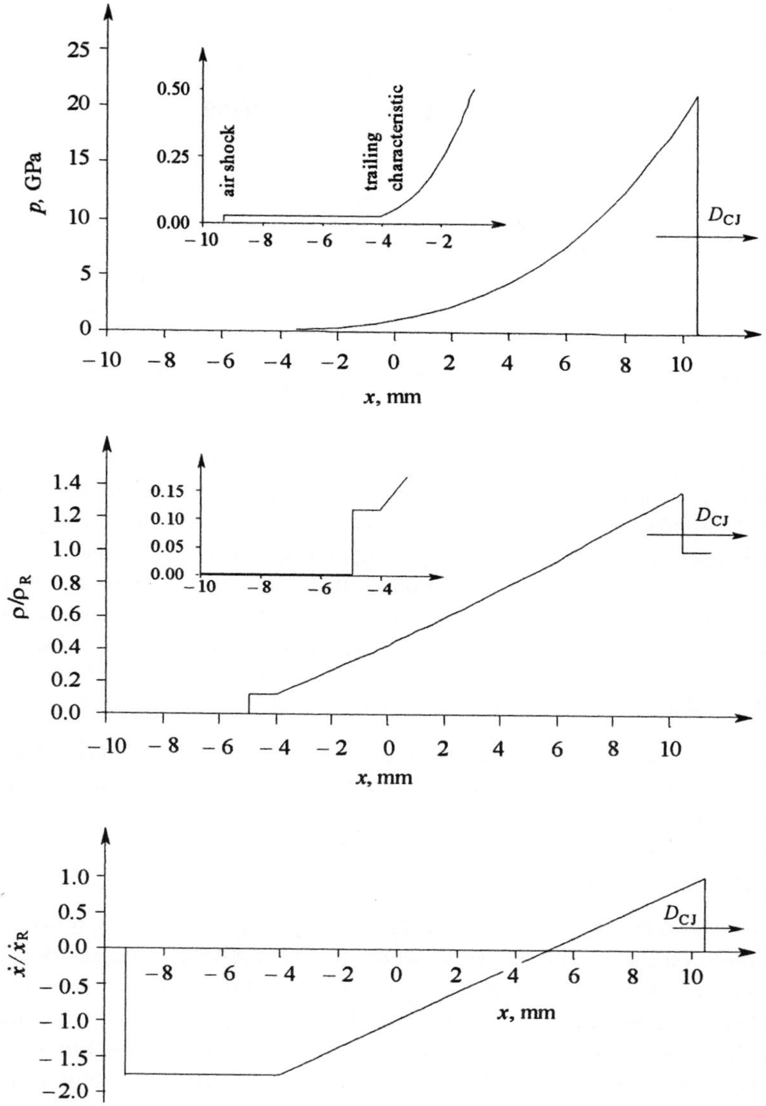

Figure 13.7. Eulerian waveforms at 1.5 μs after introduction of a CJ detonation wave into TNT characterized by the parameters $\rho_R = 1640\,\text{kg/m}^3$ $D_{CJ} = 6930\,\text{m/s}$ and $\Gamma_{DP} = 2.7$. The inset graphs are drawn to a scale that shows the effect of the detonation upon the background air. Clearly this is a strong air blast, but the pressure is negligible in comparison to the CJ pressure.

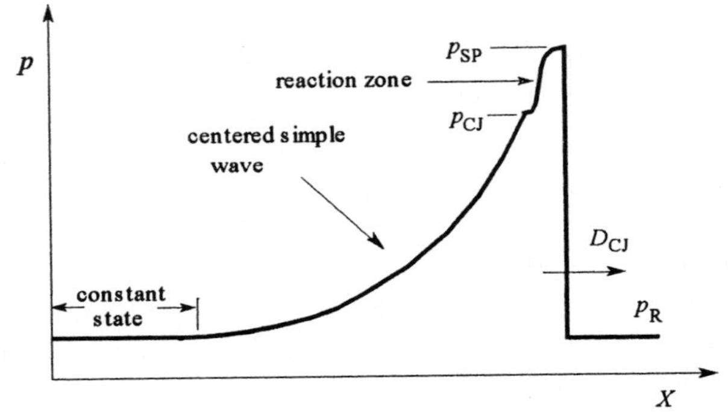

Figure 13.8. Detonation waveform including the resolved reaction zone that distinguishes the ZND theory from the CJ theory.

tween the two states. The steady-wave transition from the unreacted material to the fully reacted detonation products can be analyzed in exactly the same way as in the Chapman–Jouguet theory, and leads to the same results. Strong, weak, and CJ states can be identified. The propagation speed and the pressure, specific volume, and particle velocity at the point in the steady waveform at which the reaction is complete attain the same values as in the Chapman–Jouguet theory. The point in pursuing the ZND theory is not simply to calculate this final state. Rather, interest centers upon determination of the structure of the reaction zone, and this is done by means of a steady wave analysis.

Since the state immediately behind the ZND detonation shock is one in which the explosive is compressed but has not yet reacted, it lies on the Hugoniot of the unreacted explosive. This state, in which the unreacted explosive is compressed by a shock propagating at the velocity D, is called the *spike point* (see Fig. 13.9, on which the spike point is designated SP).

We shall consider a single irreversible reaction converting the explosive into detonation products. The degree to which this reaction has progressed is characterized by a variable λ, ranging from 0 to 1, called the *extent of reaction*. This variable is introduced into the equation of state for the detonation products to characterize both the varying characteristics of the reactant–product mixture and the degree to which the chemical energy has been liberated. Its value increases continuously over its range as the reaction zone propagates past a given material particle. The chemical energy liberated by reaction varies from none at $\lambda = 0$ to the heat of complete reaction, q, at $\lambda = 1$.

We shall conduct our discussion of the ZND theory in the context of gaseous explosives such as a methane–air mixture because this case lends itself to the simplest presentation. The internal energy function for the mixture comprising the partially reacted explosive gas and the detonation products is taken in the form of Eq. 13.6, except that the heat of complete reaction is liberated in proportion to the value of λ at the material particle under consideration:

$$\bar{\varepsilon}(v, p, \lambda) = \frac{pv}{\Gamma - 1} - q\lambda . \tag{13.57}$$

In order to use this equation of state for all values of λ the reaction must preserve the number of moles of gas and the heat capacity must be constant and have the same value for the reactants and products. For example, the methane–oxygen reaction, $CH_4 + 2O_2 \rightarrow CO_2 + 2H_2O$, begins and ends with three moles of gas and the average molecular weight is unchanged. In contrast, the hydrogen–oxygen reaction, $2H_2 + O_2 \rightarrow 2H_2O$, begins with three moles but ends with two, necessitating use of a mixture theory to obtain the function $\bar{\varepsilon}(v, p, \lambda)$. Although Eq. 13.57 is of limited applicability, it lends itself to a simple exposition of the theory.

As with all steady waves, the process follows the Rayleigh line of Eq. 13.3. In the present case the detonation shock produces a transition from the initial state to the spike point and the reaction occurs in a smooth steady decompression wave connecting the spike point to the CJ point. The internal energy density change satisfies the Rankine–Hugoniot equation, Eq. 13.4, throughout this process.

Substituting Eq. 13.57 into Eq. 13.4 yields a result like Eq. 13.7, but with q replaced by λq. We write this equation in the form

$$p^{(\mathrm{H})}(\bar{v}, \lambda) = p_{\mathrm{R}} \frac{1 - \mu^2 \bar{v} + (2\mu^2 q\lambda / p_{\mathrm{R}} v_{\mathrm{R}})}{\bar{v} - \mu^2}, \tag{13.58}$$

where $\bar{v} = v / v_{\mathrm{R}}$.

The result of evaluating Eq. 13.58 for a specific value of λ is a Hugoniot curve for material in the partially reacted state. This invokes a slightly generalized definition of a Hugoniot. We have previously defined a Hugoniot as the locus of endstates achievable by a shock transition from a given initial state. We now generalize this definition to say that a Hugoniot is the locus of states realized by a steady wave transition from the given initial state. This means that states in the reaction zone arise through transition from the given initial state to points on a continuum of Hugoniots, called *partial-reaction Hugoniots*, each of which is characterized by a value of λ. These Hugoniots are centered on the state $(p_{\mathrm{R}}, v_{\mathrm{R}})$, but are offset from this point by an amount that depends upon λq. A plot of several partial-reaction Hugoniots is given in Fig. 13.9.

The detonation shock is a transition from the reference state to the spike point, the intersection of the Rayleigh line and the Hugoniot for the unreacted explosive. This intersection lies at the point

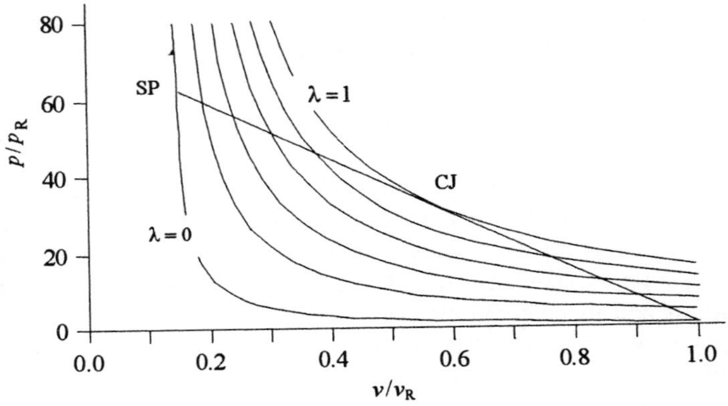

Figure 13.9. Partial-reaction Hugoniots for a stoichiometric methane–oxygen detonation. The parameters used are: $p_R = 101{,}325\,\text{Pa}$, $v_R = 0.9012\,\text{kg/m}^3$, $\Gamma = 1.3$, and $q = 4.566\,\text{MJ/kg}$. The spike point is designated SP.

$$\frac{v_{SP}}{v_R} = \frac{\Gamma-1}{\Gamma+1} + \frac{2\Gamma}{\Gamma+1}\frac{p_R v_R}{D^2}$$
$$\frac{p_{SP}}{p_R} = \frac{2D^2}{(\Gamma+1)p_R v_R} - \frac{\Gamma-1}{\Gamma+1}. \tag{13.59}$$

The particular state attained on each of these Hugoniots falls at the intersection of the Hugoniot and the Rayleigh line for the steady wave. When the equation of state $\bar{\varepsilon}$ is given, Eqs. 13.58 and 13.3 can be solved for the pressure and specific volume at any point in the wave.

Corresponding values of p and v in the waveform lie at the intersection of the partial-reaction Hugoniot and the Rayleigh line. Equating the pressure given by Eq. 13.3 with that given by Eq. 13.58 yields the expression

$$\lambda = \frac{1}{2\mu^2 q}(1-\bar{v})\left[D_{CJ}^2(\bar{v}-\mu^2) - p_R v_R(1+\mu^2)\right] \tag{13.60}$$

for the extent of reaction as a function of $\bar{v}$.

Conventional chemical-kinetic equations express $\dot{\lambda}$ in terms of the thermodynamic state variables, so we will need the material derivative of Eq. 13.60:

$$\dot{\lambda} = \frac{1}{(\Gamma-1)q}\left[\Gamma(D_{CJ}^2 + p_R v_R) - (\Gamma+1)D_{CJ}^2\,\bar{v}\right]\dot{\bar{v}}. \tag{13.61}$$

In the steady wave the fields are functions of $Z = X - D_{CJ} t$ so we have $\dot{\bar{v}} = -D_{CJ}\, d\bar{v}/dZ$ and the foregoing equation can be written

$$dZ = \frac{D_{CJ}}{(\Gamma - 1) q \dot{\lambda}} \left[(\Gamma + 1) D_{CJ}^2 \bar{v} - \Gamma (D_{CJ}^2 + p_R v_R) \right] d\bar{v}. \tag{13.62}$$

This equation can be integrated to give the steady waveform $\bar{v} = \bar{v}(Z)$ if we can express $\dot{\lambda}$ as a function of $\bar{v}$.

A typical kinetic relation for a simple chemical reaction is the first-order *Arrhenius equation*

$$\dot{\lambda} = k(1 - \lambda) \exp\left(-\frac{\varepsilon^{\dagger}}{\mathcal{R} \theta} \right), \tag{13.63}$$

where k and $\varepsilon^{\dagger}$ are constants called the *frequency factor* and *activation energy*, respectively. To use this relation, we need to determine the temperature θ in terms of the value of $\bar{v}$ at points in the wave. Since the state point in a steady wave moves along the Rayleigh line, $p(v) = p_R + D_{CJ}^2 (1 - v)/v_R$, substitution of this pressure into the equation of state $\theta = pv/\mathcal{R}$ gives the temperature as a function of $\bar{v}$ in the wave:

$$\theta(\bar{v}) = \frac{\bar{v}}{\mathcal{R}} \left[p_R v_R + D_{CJ}^2 (1 - \bar{v}) \right]. \tag{13.64}$$

An equation for the $\bar{v}(Z)$ in the steady reaction zone can now be obtained by substituting Eq. 13.64 into Eq. 13.63 and that result into Eq. 13.62. Integration of this equation by numerical means then gives the waveform. Some results of this analysis are shown in Fig. 13.10

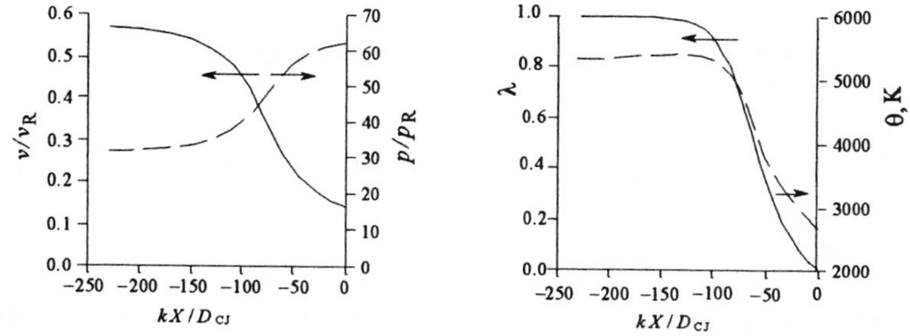

Figure 13.10 Waveforms showing reaction zone structure are plotted as functions of dimensionless distance behind the detonation shock. The calculations are made using the parameters $p_R = 101{,}325\,\text{Pa}$, $v_R = 0.9012$, $\theta_R = 293\,\text{K}$, $\Gamma = 1.3$, $\varepsilon^{\dagger}/\mathcal{R} = 15{,}000\,\text{K}$, and $q = 4.566 \times 10^6$ J/kg for a methane–oxygen detonation. Waveforms in the reaction zone.

13.3 Weak Detonation

Examination of Fig. 13.9 shows that the ZND analysis can be applied to calculation of the reaction-zone structure for a strong detonation in just the same way as for a CJ detonation because the Rayleigh line intersects a partial-reaction Hugoniot at each point of the steady wave between the spike point and the strong detonation point on the $\lambda = 1$ Hugoniot. This is not true for points on the Rayleigh line that lie between its two intersections with this Hugoniot so waveform analysis cannot be carried out in this region. This is one of several arguments for the nonexistence of weak detonations, but it only means that these detonations do not exist in the case of the single irreversible chemical reaction discussed in this chapter.

Analyses of multiple reactions (for example, a rapid exothermic reaction followed by a slower endothermic reaction), show that partial-reaction Hugoniots may exist that bridge the region between the strong and weak detonation states, thus opening the possibility of realizing a weak detonation. Effects of viscosity, or slightly divergent flow can also yield weak detonations and the same is likely true if transverse waves or other phenomena remove energy from the nominal uniaxial flow. There is reason to believe that weak detonation occurs in most explosives of technological importance (see [44, Chap. 2]). Fickett and Davis [43, Chap. 5] have analyzed a number of cases in which weak detonations occur. The variety and complexity of these cases indicates that analysis of the chemical process must go far beyond one or a few reactions. The chemical process is also complicated by the inhomogeneity of the temperature and deformation fields at the mesoscale. All of this means that even a comprehensive analysis conducted in the context of transient reactive flows may fail to capture important features of the actual detonation process.

13.4 Closing Remarks on Detonation Phenomena

More than some of the preceding chapters, this chapter on detonation waves presents a very idealized view of the subject. The art and science of explosive materials and detonation processes have been the subjects of investigation for some hundreds of years. The discussion of this chapter has been limited to a brief account of the interface between this subject and the theory of propagation of shocks in nonreactive materials. Among the important issues that were not addressed are

i. *Initiation Processes.* An important aspect of explosive behavior is the process by which a detonation is initiated. We have seen that a detonation propagates at a constant velocity that is characteristic of the explosive and produces a transition of the material to its product state. A detonation does not ordinarily originate in this fully developed condition, and an important aspect of detonation physics concerns the process by which a detonation is initiated. It is well known that stimuli much weaker than the detonation itself are adequate to initiate a reaction that can grow in intensity until a detonation wave is formed.

Solid explosives may be in the form of castings solidified from the molten material. In this case they are aggregates of molecular crystals (TNT is of this form). In other cases (e.g., Composition B) they are made by stirring granules of a high melting point explosive (e.g. RDX) into molten explosive having a lower melting temperature (e.g., TNT). A third kind of explosive is prepared by coating granules of the material with a polymeric binder and pressing the resulting mixture into a dense block of material (PBX–9404 is of this form). Experimental observation shows that chemical reaction is initiated in these solid explosives by shocks of strength far below the CJ pressure. Calculation of the temperature of the unreacted explosive at points on its Hugoniot at which shock-induced initiation is observed yields values at which the explosive is found to be stable for long periods.

It is conjectured that the initiation process is strongly affected by mechanical inhomogeneity deriving from the complex microstructure of the material, including voids, grain boundaries, polymeric binders, etc. When a shock compresses material containing these inhomogeneities a microscopically non-uniform temperature field is produced. in particular, *hot spots* form at points of high deformation and the reaction begins at these spots and propagates outward from them. This reaction generates a wave of growing strength that eventually becomes a detonation. This initiation problem has been studied extensively.

ii. *Reaction Mechanisms and Rates.* In even the simplest case of detonation of gaseous explosives the temperatures and pressures are higher than most widely studied processes and the reaction proceeds more rapidly than in flames or other common chemical processes involving the materials. It is usually inferred that the reaction proceeds as a composite of several (often many) sub-reactions. In the case of solid explosives the reaction is further complicated by the microscopically non-uniform temperature field discussed in connection with the initiation process. Obviously, the overall reaction rate produced in this situation would depend on both the physical and chemical aspects of this model and a reaction rate law inferred from the chemistry alone is entirely inappropriate.

iii. *Realistic Equations of State of Detonation Products.* In the analyses presented in this chapter the detonation products were assumed to form an ideal gas. This is oversimplified and, indeed, the high values of Γ_{DP} usually used to fit data on the behavior of solid explosives exceed the limit of 5/3 derived by the methods of statistical mechanics. Many other equations of state have been proposed for detonation products, and several have been fit to observations of the behavior of a variety of important explosives. When the detonation products do not appear instantaneously and in the gaseous phase, one is faced with the matter of dealing with partially reacted material that, in the case of a solid explosive, is a mixture of gaseous products and granules of unreacted explosive. One is faced with modeling a reaction process that has both chemical and physical aspects.

iv. *Stability of Solutions.* The plane detonation waves that have been discussed in this chapter have been shown to be unstable in the sense that waves form and propagate in the plane of the detonation shock. These waves have been extensively studied for gaseous explosives and there is every reason to assume that they are present when solid explosives detonate.

As a result of these and many other aspects of detonation physics, the subject, although highly developed, is one in which applications are based as much on art as on science.

APPENDIX

Solutions to the Exercises

Chapter 2. Mechanical Principles

Exercise 2.7.1. Consider the rod, as shown. Defining λ by the equation $l = \lambda L$, as the Lagrangian description of the deformation we have

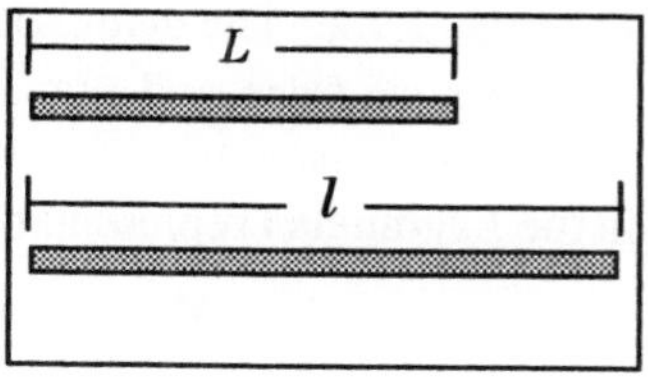

$$x_1 = X_1 + U(X_1) = X_1 + (\lambda - 1)X_1$$
$$x_2 = X_2, \quad x_3 = X_3,$$

where we have written the first line so that it is apparent that $U(X_1) = (\lambda - 1)X_1$. The Eulerian description takes the form

$$X_1 = x_1 - u(x_1) = x_1 - \left(1 - \frac{1}{\lambda}\right)x_1, \quad u(x_1) = \left(1 - \frac{1}{\lambda}\right)x_1 \;.$$
$$X_2 = x_2, \quad X_3 = x_3$$

From the Lagrangian representation:

$$\mathbf{F} = \text{diag}\| 1 + U_X, 1, 1 \| = \text{diag}\| \lambda, 1, 1 \|,$$

$$\mathbf{C} = \overset{\mathrm{T}}{\mathbf{F}}\mathbf{F} = \text{diag}\| \lambda^2, 1, 1 \|,$$

$$\mathbf{E} = \text{diag}\| \tfrac{1}{2}(\lambda^2 - 1), 0, 0 \|,$$

$$= \text{diag}\| (\lambda - 1) + O\big((\lambda - 1)^2\big), 0, 0 \|.$$

Note, in particular, that

$$E_{11} = (l - L)/L + \cdots = (\text{final length} - \text{initial length})/\text{initial length} + \cdots.$$

From the Eulerian representation:

$$\overset{-1}{\mathbf{F}} = \text{diag}\| 1/\lambda, 1, 1 \|$$

$$\mathbf{c} = \overset{-\mathrm{T}}{\mathbf{F}}\overset{-1}{\mathbf{F}} = \text{diag}\| 1/\lambda^2, 1, 1 \|$$

$$\mathbf{e} = \tfrac{1}{2}\text{diag}\| 1 - (1/\lambda^2), 0, 0 \|.$$

In this case,

$$e_{11} = (l - L)/l + \cdots = \text{(final length} - \text{initial length)/final length} + \cdots,$$

a result that differs from the Lagrangian strain measure. The two measures are the same to first order, but the difference is important at large strain. Either measure is a correct description of the kinematics, but the difference must be taken into account when interpreting the physics that is represented.

Exercise 2.7.2. The deformation in question can be written

$$x_1 = X_1 + (\tan\gamma) X_2, \quad x_2 = X_2, \quad x_3 = X_3$$

in the Lagrangian representation, or

$$X_1 = x_1 - (\tan\gamma) x_2, \quad X_2 = x_2, \quad X_3 = x_3$$

in the Eulerian representation. From this we get

$$\mathbf{F} = \begin{pmatrix} 1 & \tan\gamma & 0 \\ 0 & 1 & 0 \\ 0 & 0 & 1 \end{pmatrix}, \quad \mathbf{E} = \frac{1}{2}\begin{pmatrix} 0 & \tan\gamma & 0 \\ \tan\gamma & \tan^2\gamma & 0 \\ 0 & 0 & 0 \end{pmatrix}, \quad \mathbf{e} = \frac{1}{2}\begin{pmatrix} 0 & \tan\gamma & 0 \\ \tan\gamma & -\tan^2\gamma & 0 \\ 0 & 0 & 0 \end{pmatrix}.$$

When we view this deformation from the Lagrangian perspective, we are led to consider a plane $X_3 = \text{const.}$, as shown at the left in sketch (a) below. When we take the Eulerian view, we are led to consider the plane described by the equation $x_3 = \text{const.}$ as shown in the sketch at the right below. Prior to the deformation, this plane would be as shown in the sketch at the left of each diagram. After deformation, the line shown is rotated through the angle γ, as indicated at the right of each diagram. The deformation gradient and strain components are measures of this angle. The quantity F_{12} is the tangent of the angle, and the strain components E_{12} and e_{12} are each one-half of the tangent.

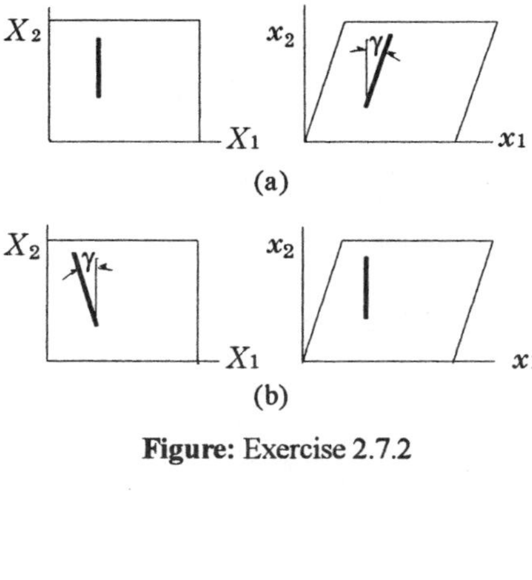

(a)

(b)

Figure: Exercise 2.7.2

We see that $E_{22} = -e_{22} = \frac{1}{2}\tan^2\gamma$. It is apparent from the drawing that the Lagrangian element is extended by the deformation whereas the Eulerian element is contracted. The extension in the former case is the same as the contraction in the latter. Consider the stretch of each of these elements, defined as $\lambda = l/L$, the deformed length divided by the reference-configuration length. In the Lagrangian case, $\lambda = (1+\tan^2\gamma)^{1/2}$ or $\tan^2\gamma = \lambda^2 - 1$, which we can write $E_{22} = \frac{1}{2}(\lambda^2-1) = (\lambda-1) + \frac{1}{2}(\lambda-1)^2$. From this we see that the small-deformation approximation is $E_{22} = (l-L)/L + \cdots$, the usual definition of a small Lagrangian strain. A similar analysis of the Eulerian case gives $e_{22} = \frac{1}{2}[1+(1/\lambda^2)] = (l-L)/l + \cdots$, the conventional Eulerian result.

Exercise 2.7.3. Let us express the motion in the forms

$$x_i = [X_I + U_I(\mathbf{X},t)]\,\delta_{Ii} \quad \text{and} \quad X_I = [x_i - u_i(\mathbf{x},t)]\,\delta_{iI}\,, \tag{A}$$

where $\mathbf{u}(\mathbf{x},0) = \mathbf{0}$ and $\mathbf{U}(\mathbf{X},0) = \mathbf{0}$, and where the Kronecker deltas are used simply to preserve the convention that upper-case subscripts are associated with the reference frame and the lower-case subscripts are associated with the spatial frame. Substituting either of these relations into the other shows that

$$u_i(x,t) = U_I(X,t)\delta_{Ii}\,, \tag{B}$$

i.e., the two displacement quantities are the same for corresponding values of $\mathbf{x}$ and $\mathbf{X}$. Differentiating Eq. A_1 yields

$$F_{iJ}(\mathbf{X},t) = \frac{\partial x_i}{\partial X_J} = \delta_{iJ} + \frac{\partial U_I}{\partial X_J}\delta_{Ii}\,, \tag{C}$$

we also have

$$\overset{-1}{F}_{Ji}(\mathbf{X},t) = \delta_{Ji} - \frac{\partial U_J}{\partial X_I}\delta_{Ii} + \ldots, \tag{D}$$

as one can see from the fact that $F_{iJ}\overset{-1}{F}_{Jj} = \delta_{ij} + \cdots$. The ellipsis denotes terms of order higher than one in $\partial U_I/\partial X_J$. We also have

$$\begin{aligned} C_{IJ} = F_{iI}F_{iJ} &= \left[\delta_{iI} + \frac{\partial U_K}{\partial X_I}\delta_{Ki}\right]\left[\delta_{iJ} + \frac{\partial U_L}{\partial X_J}\delta_{Li}\right] \\ &= \delta_{IJ} + \left[\frac{\partial U_I}{\partial X_J} + \frac{\partial U_J}{\partial X_I}\right] + \ldots \end{aligned} \tag{E}$$

and

$$E_{IJ} = \frac{1}{2}(C_{IJ} - \delta_{IJ}) = \frac{1}{2}\left[\frac{\partial U_I}{\partial X_J} + \frac{\partial U_J}{\partial X_I}\right] + \cdots = \widetilde{E}_{IJ} + \cdots, \tag{F}$$

where

$$\widetilde{E}_{IJ} = \frac{1}{2}\left[\frac{\partial U_I}{\partial X_J} + \frac{\partial U_J}{\partial X_I}\right] \tag{G}$$

is the linearized version of E_{IJ} and corresponds to the usual definition of strain in the context of linear elasticity. We also have

$$b_{ij} = F_{iJ}F_{jJ} = \delta_{ij} + \left[\frac{\partial U_I}{\partial X_J} + \frac{\partial U_J}{\partial X_I}\right]\delta_{Ii}\delta_{Jj} + \ldots, \tag{H}$$

so

$$e_{ij} = \frac{1}{2}(\delta_{ij} - \overset{-1}{b_{ij}}) = \frac{1}{2}\left[\frac{\partial U_I}{\partial X_J} + \frac{\partial U_J}{\partial X_I}\right]\delta_{Ii}\delta_{Jj} + \ldots, \tag{I}$$

where

$$\widetilde{e}_{ij} = \frac{1}{2}\left[\frac{\partial U_I}{\partial X_J} + \frac{\partial U_J}{\partial X_I}\right]\delta_{Ii}\delta_{Jj}. \tag{J}$$

Differentiating Eq. B gives

$$\frac{\partial U_I}{\partial X_J} = \frac{\partial u_i}{\partial x_j}\frac{\partial x_j}{\partial X_J}\delta_{iI} = \frac{\partial u_i}{\partial x_j}F_{jJ}\delta_{iI}, \tag{K}$$

and substitution for **F** from Eq. C gives

$$\frac{\partial U_I}{\partial X_J} = \frac{\partial u_i}{\partial x_j}\left[\delta_{jJ} + \frac{\partial U_K}{\partial X_J}\delta_{Kj}\right]\delta_{iI} = \frac{\partial u_i}{\partial x_j}\delta_{jJ}\,\delta_{iI} + \cdots, \tag{L}$$

showing that the two displacement gradients are the same to first order. Substitution of Eq. L into Eq. K gives

$$e_{ij} = \frac{1}{2}\left[\frac{\partial u_i}{\partial x_j} + \frac{\partial u_j}{\partial x_i}\right] + \cdots = \widetilde{e}_{ij} + \cdots,$$

where

$$\widetilde{e}_{ij} = \frac{1}{2}\left[\frac{\partial u_i}{\partial x_j} + \frac{\partial u_j}{\partial x_i}\right] \tag{M}$$

is the linearized version of e_{ij} and, as was the case with Eq. G, corresponds to the usual definition of strain in the context of linear elasticity. Comparing Eqs. G and K shows that

$$\widetilde{e}_{ij} = \widetilde{E}_{IJ}\,\delta_{Ii}\,\delta_{Jj}, \tag{N}$$

i.e., that $\widetilde{\mathbf{e}}$ and $\widetilde{\mathbf{E}}$ are the same, or, equivalently, that **e** and **E** are the same to first order.

For the particle velocity and acceleration we have

$$\begin{aligned} \dot{x}_i &= \frac{\partial}{\partial t}[X_I + U_I(\mathbf{X}, t)]\,\delta_{Ii} = \frac{\partial U_I}{\partial t}\delta_{Ii} \\ \ddot{x}_i &= \frac{\partial^2 U_I}{\partial t^2}\delta_{Ii}. \end{aligned} \tag{O}$$

We can also differentiate Eq. A_2, obtaining

We can also differentiate Eq. A_2, obtaining

$$\frac{\partial u_i}{\partial t} = \left[\delta_{ik} - \frac{\partial u_i}{\partial x_k}\right] \dot{x}_k .$$

When we multiply each side of this equation by $\delta_{il} - (\partial u_i / \partial x_l)$ and contract on i we obtain

$$\dot{x}_i = \frac{\partial u_i}{\partial t} + \dots . \qquad \text{(P)}$$

A second differentiation and elimination of higher-order terms gives

$$\ddot{x}_i = \frac{\partial^2 u_i}{\partial t^2} + \dots , \qquad \text{(Q)}$$

and comparison with Eq. O shows that particle velocity and acceleration can be computed to first order by time differentiation of either $\mathbf{U}(\mathbf{X}, t)$ or $\mathbf{u}(\mathbf{x}, t)$.

Finally, we have

$$\frac{\rho_R}{\rho} = \det \mathbf{F} = \det \left[\delta_{iJ} + \frac{\partial U_I}{\partial X_J} \delta_{Ii} \right] = 1 + \frac{\partial U_I}{\partial X_I} + \dots . \qquad \text{(R)}$$

Exercise 2.7.4. Lagrangian measures of deformation are calculated relative to the reference configuration, whereas Eulerian measures are calculated relative to the current configuration. Therefore, the Eulerian compression corresponding to the Lagrangian compression $\Delta = (v_R - v)/v_R$ would be $\delta = (v_R - v)/v$ and we have $\delta = \Delta/(1-\Delta)$ and $\Delta = \delta/(1+\delta)$.

Exercise 2.7.5. As suggested in the statement of the Exercise, a proof that the symmetry of the Cauchy stress tensor is necessary and sufficient for satisfaction of conservation of moment of momentum can be found in most elementary texts on elasticity or continuum mechanics. A simple indication of the plausibility of this result follows from the two-dimensional example illustrated below. We can see from this figure that the condition that the block of material not rotate under the influence of the shear stresses illustrated is that $\tau_{12} = \tau_{21}$.

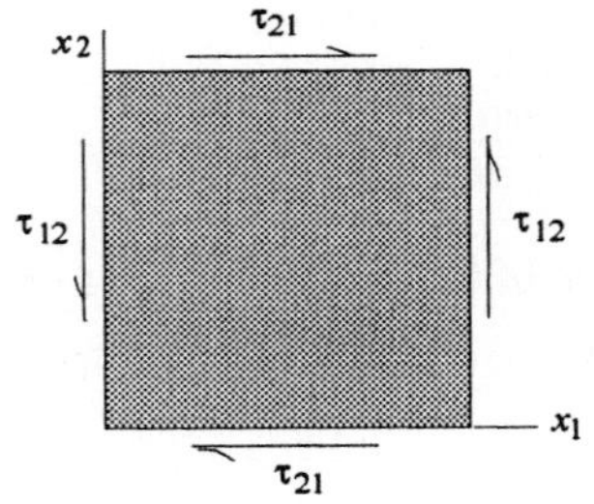

Figure: Exercise 2.7.5.

Exercise 2.7.6. The Eulerian form of the equation of conservation of mass is

$$\frac{\partial \rho(\mathbf{x},t)}{\partial t}+\frac{\partial[\rho(\mathbf{x},t)\dot{x}_i(\mathbf{x},t)]}{\partial x_i}=0\,. \tag{2.86$_1$}$$

From Eq. 2.15,

$$\frac{\partial \rho(\mathbf{X},t)}{\partial t}=\frac{\partial \rho(\mathbf{x},t)}{\partial t}+\frac{\partial \rho(\mathbf{x},t)}{\partial x_i}\dot{x}_i\,,$$

so

$$\frac{\partial \rho(\mathbf{x},t)}{\partial t}=\frac{\partial \rho(\mathbf{X},t)}{\partial t}-\frac{\partial \rho(\mathbf{x},t)}{\partial x_i}\dot{x}_i\,.$$

For the second term of Eq. 2.86_1 we have

$$\frac{\partial[\rho(\mathbf{x},t)\dot{x}_i(\mathbf{x},t)]}{\partial x_i}=\frac{\partial \rho(\mathbf{x},t)}{\partial x_i}\dot{x}_i(\mathbf{x},t)+\rho \overset{-1}{F}_{Ii}\frac{\partial \dot{x}_i(\mathbf{X},t)}{\partial X_I}\,,$$

so Eq. 2.86_1 becomes

$$\frac{\partial \rho(\mathrm{X},t)}{\partial t}+\rho \overset{-1}{F}_{Ii}\frac{\partial \dot{x}_i(\mathbf{X},t)}{\partial X_I}=0\,, \tag{2.91}$$

the desired result.

The Eulerian form of the equation of balance of momentum is

$$\frac{\partial t_{ij}(\mathbf{x},t)}{\partial x_j}-\rho\left(\frac{\partial \dot{x}_i(\mathbf{x},t)}{\partial t}+\dot{x}_j(\mathbf{x},t)\frac{\partial \dot{x}_i(\mathbf{x},t)}{\partial x_j}\right)=-\rho f_i\,. \tag{2.86$_2$}$$

The second term of this equation is just the material derivative of $\dot{x}$ the foregoing differential equation can be written

$$\partial t_{ij}(\mathbf{x},t)/\partial x_j-\rho\ddot{x}_i=-\rho f_i\,. \tag{A}$$

For the first term of this equation we have

$$\frac{\partial t_{ij}(\mathbf{x},t)}{\partial x_j}=\frac{\partial}{\partial x_j}\left(\frac{1}{J}F_{iM}F_{jN}\,T_{MN}\right)$$

$$=F_{iM}\frac{\partial}{\partial x_j}\left(\frac{1}{J}F_{jN}\right)T_{MN}+\frac{1}{J}F_{jN}\frac{\partial}{\partial x_j}\left(F_{iM}\,T_{MN}\right).$$

The first term of the right member of this equation vanishes according to the equation given in the hint, so

$$\frac{\partial t_{ij}(\mathbf{x},t)}{\partial x_j}=\frac{1}{J}F_{jN}\frac{\partial}{\partial x_j}\left(F_{iM}\,T_{MN}\right)=\frac{1}{J}F_{jN}\overset{-1}{F}_{Ij}\frac{\partial}{\partial X_I}\left(F_{iM}\,T_{MN}\right)$$

$$=\frac{1}{J}\frac{\partial}{\partial X_I}\left(F_{iM}\,T_{MI}\right).\tag{B}$$

Substitution of Eq B into Eq A leads to the required result.

The Eulerian form of the equation of balance of energy is

$$\rho\left(\frac{\partial\varepsilon}{\partial t}+\dot{x}_i\frac{\partial\varepsilon}{\partial x_i}\right)-t_{ij}\frac{\partial\dot{x}_i}{\partial x_j}=-\frac{\partial q_i}{\partial x_i}+\rho r\,.\tag{2.86$_3$}$$

The first term of this equation is just $\rho\dot{\varepsilon}$ so we have

$$\rho\dot{\varepsilon}-t_{ij}\overset{-1}{F}_{Ij}\frac{\partial\dot{x}_i(\mathbf{X},t)}{\partial X_I}=-\frac{\partial q_i(\mathbf{x},t)}{\partial x_i}+\rho r\,.\tag{C}$$

Defining the Lagrangian heat flux by the equation $q_i=(1/J)F_{iK}\,Q_K$ allows us to write

$$\frac{\partial q_i}{\partial x_i}=\frac{\partial}{\partial x_i}\left(\frac{1}{J}F_{iK}\,Q_K\right)=Q_K\frac{\partial}{\partial x_i}\left(\frac{1}{J}F_{iK}\right)+\frac{1}{J}\frac{\partial Q_I}{\partial X_I}=\frac{1}{J}\frac{\partial Q_I}{\partial X_I}\,.\tag{D}$$

For the second term of Eq. C we have

$$t_{ij}\overset{-1}{F}_{Ij}\frac{\partial\dot{x}_i(\mathbf{X},t)}{\partial X_I}=\left(\frac{\rho}{\rho_R}F_{iK}\,F_{jL}\,T_{KL}\right)\overset{-1}{F}_{Ij}\frac{\partial\dot{x}_i(\mathbf{X},t)}{\partial X_I}$$

$$=\frac{\rho}{\rho_R}F_{iK}\,T_{KI}\frac{\partial\dot{x}_i(\mathbf{X},t)}{\partial X_I},\tag{E}$$

and substitution of Eqs. D and E into Eq. C yields the result

$$\rho\dot{\varepsilon}-\overset{-1}{F}_{IK}\,T_{KI}\frac{\partial\dot{x}_i(\mathbf{X},t)}{\partial X_I}=-\frac{\partial Q_I(\mathbf{x},t)}{\partial X_I}+\rho r\,.\tag{F}$$

To obtain the required form of this equation, it remains to express $\partial x_i/\partial X_I$ in terms of $\partial E_{IJ}/\partial t$. We have

$$2E_{IJ} = \frac{\partial x_i}{\partial X_I}\frac{\partial x_i}{\partial X_J} - \delta_{ij},$$

so

$$2\frac{\partial E_{IJ}}{\partial t} = \frac{\partial^2 x_i}{\partial t \partial X_I}\frac{\partial x_i}{\partial X_J} + \frac{\partial x_i}{\partial X_I}\frac{\partial^2 x_i}{\partial t \partial X_J}$$

$$= \frac{\partial \dot{x}_i}{\partial X_I}F_{iJ} + \frac{\partial \dot{x}_i}{\partial X_J}F_{iI}$$

and

$$T_{IJ}\frac{\partial E_{IJ}}{\partial t} = F_{iJ}T_{IJ}\frac{\partial \dot{x}_i}{\partial X_I}. \tag{G}$$

Substitution of Eq. G into Eq. F gives the result sought.

Exercise 2.7.7. To make the required calculations, we begin by evaluating some kinematical quantities for the uniaxial motion $x = x(X, t)$, $x_2 = X_2$, $x_3 = X_3$. We find that $\mathbf{F} = \text{diag}\| F_{11},\ 1,\ 1\|$, $\mathbf{F}^{-1} = \text{diag}\ \|1/F_{11},\ 1,\ 1\|$, $\dot{x} = (\partial x/\partial t, 0, 0)$, $\det \mathbf{F} = F_{11} = \rho_R/\rho$.

Equation 2.90_1 is

$$\frac{\partial \rho}{\partial t} + \rho \overset{-1}{F}_{Ii}\frac{\partial \dot{x}_i}{\partial X_I} = 0. \tag{2.90$_1$}$$

Since the only nonvanishing component of $\partial \dot{x}_i/\partial X_I$ is $\partial \dot{x}/\partial X$ we have

$$\overset{-1}{F}_{Ii}\frac{\partial \dot{x}_i}{\partial X_I} = \frac{\rho}{\rho_R}\frac{\partial \dot{x}}{\partial X},$$

and Eq. 2.91_1 can be written in the required form.

Next, we consider Eq. 2.91_2,

$$\frac{\partial (F_{iK}T_{KJ})}{\partial X_J} - \rho_R\frac{\partial \dot{x}_i}{\partial t} = \rho f_i. \tag{2.90$_2$}$$

Use of Eq. 2.64 to replace **T** by **t** in the first term of this equation gives

$$\frac{\partial}{\partial X_J}\left(F_{iK}\,J\,\overset{-1}{F}_{Kj}\,\overset{-1}{F}_{Jk}\,t_{kj}\right) = \frac{\partial}{\partial X_J}\left(J\,\overset{-1}{F}_{Jk}\,t_{ki}\right) = t_{ki}\frac{\partial}{\partial X_J}\left(J\,\overset{-1}{F}_{Jk}\right) + J\,\overset{-1}{F}_{Jk}\frac{\partial t_{ki}}{\partial X_J}.$$

Following the hint, we see that the first term of the right member of this equation vanishes and Eq. 2.91_2 becomes

$$J\,\overset{-1}{F}_{Jk}\frac{\partial t_{ki}}{\partial X_J} - \rho_R\frac{\partial \dot{x}_i}{\partial t} = \rho f_i.$$

For the uniaxial motion the only nonvanishing component of the first term is

$$J\frac{1}{F_{11}}\frac{\partial t_{11}}{\partial X}=\frac{\partial t_{11}}{\partial X},$$

and substitution of this result into the foregoing equation yields the required result.

Equation 2.91_3 is

$$\rho_R\frac{\partial \varepsilon}{\partial t}-T_{IJ}\frac{\partial E_{IJ}}{\partial t}=-\frac{\partial Q_I}{\partial X_I}+\rho_R\, r.$$

Substitution of Eq. G from Exercise 2.7.6 into this equation yields the result

$$\rho_R\frac{\partial \varepsilon}{\partial t}-F_{iJ}\,T_{IJ}\frac{\partial \dot{x}_i}{\partial X_I}=-\frac{\partial Q_I}{\partial X_I}+\rho_R\, r,$$

and, by making the usual manipulations to replace **T** by **t**, we obtain the equation

$$\rho_R\frac{\partial \varepsilon}{\partial t}-\frac{\rho_R}{\rho}\overset{-1}{F}_{iK}\,t_{ki}\frac{\partial \dot{x}_i}{\partial X_I}=-\frac{\partial Q_I}{\partial X_I}+\rho_R\, r. \qquad \text{(A)}$$

For uniaxial motions the only nonvanishing component of $\partial \dot{x}_i/\partial X_J$ is the 11 component, $\partial \dot{x}/\partial X$, so Eq. A becomes

$$\rho_R\frac{\partial \varepsilon}{\partial t}-t_{11}\frac{\partial \dot{x}}{\partial X}=-\frac{\partial Q}{\partial X}+\rho_R\, r,$$

which is the required result.

Exercise 2.7.8. Equation 2.110_1 can be written in the form $\rho^+(u_S-\dot{x}^+)=\rho^-(u_S-\dot{x}^-)$. When this equation is substituted into Eq. 2.112 we obtain $U_S=\rho^+(u_S-\dot{x}^+)/\rho_R$, which is the required result.

Exercise 2.7.9. The Eulerian form of the jump equation for conservation of mass can be written

$$(\rho^+-\rho^-)u_S=\rho^+\dot{x}^+-\rho^-\dot{x}^-.$$

When we use Eq. 2.112 to replace u_S with U_S this becomes

$$(\rho^+-\rho^-)(\rho_R U_S+\rho^-\dot{x}^-)=\rho^-(\rho^+\dot{x}^+-\rho^-\dot{x}^-).$$

Performing the indicated multiplications, canceling like terms, and replacing ρ by v yields the required result. The same procedure applies for the jump equation for balance of momentum, but Eq. 2.110_1 must be used in the simplification process. The jump equation for balance of energy is treated similarly, but both Eq. 2.112 and the result of Exercise 2.7.8 are used.

Exercise 2.7.10. The total energy, $\mathcal{E}$, in a unit mass of material is the sum of the internal energy, ε, and the kinetic energy, $\frac{1}{2}\dot{x}^2$: $\mathcal{E} = \varepsilon + \frac{1}{2}\dot{x}^2$. Since the material ahead of the shock is at rest, its kinetic energy is zero and the change that takes place in the total energy upon passage of the shock is

$$\Delta\mathcal{E} = \varepsilon^+ - \varepsilon^- + \tfrac{1}{2}(\dot{x}^+)^2 .$$

Using Eq. 2.115,

$$[\![\varepsilon]\!] = \frac{1}{2}[\![\dot{x}]\!]^2 + \frac{-t_{11}^-}{\rho_R U_S}[\![\dot{x}]\!]$$

in the case that the material ahead of the shock is unstressed and at rest, we have

$$[\![\varepsilon]\!] = \tfrac{1}{2}(\dot{x}^+)^2 ,$$

showing that the increase in internal energy is just equal to the increase in kinetic energy—equipartition of energy—and, as we have seen, the total is the sum of the two terms is the change in total energy.

Exercise 2.7.11. The equations are derived by explicit consideration of the motion illustrated below. The material body is of unit cross-sectional area and thickness L. The principle of conservation of mass holds that the mass of the body, which is the sum of the masses of the part ahead of the shock and of the part behind the shock, is constant and equal to its value in the reference state. The mass of the material behind the shock is its mass density multiplied by the volume of material, $(u_S - \dot{x}^+)t$, plus the same quantity for the material ahead of the shock. This sum is to be equal to the mass at $t = 0$, $\rho^- L$. Accordingly,

$$\rho^+ (u_S - \dot{x}^+)t + \rho^- \big(L + (\dot{x}^- - u_S)t\big) = \rho^- L ,$$

or

$$(\rho^+ - \rho^-)u_S t - (\rho^+ \dot{x}^+ - \rho^- \dot{x}^-)t = 0 ,$$

which we write

$$[\![\rho]\!]u_S = [\![\rho\dot{x}]\!] . \qquad (2.110_1)$$

The principle of balance of momentum holds that the rate of change of momentum of a material body is equal to the applied force. The momentum of the material behind the shock is the momentum per unit volume, $\rho^+ \dot{x}^+$, and the volume of a unit cross section of the material is $(u_S - \dot{x}^+)t$ so the total momentum of this material is $\rho^+ \dot{x}^+ (u_S - \dot{x}^+)t$, and its rate of change is $\rho^+ \dot{x}^+ (u_S - \dot{x}^+)$. A similar calculation for the material ahead of the shock gives its rate of change of momentum as $\rho^- \dot{x}^- (\dot{x}^- - u_S)$, with the total rate of change of momentum being the sum of these quantities. The applied force on this column of material is $t_{11}^- - t_{11}^+$, so we have

$$\rho^+ \dot{x}^+ (u_S - \dot{x}^+) + \rho^- \dot{x}^- (\dot{x}^- - u_S) = t_{11}^- - t_{11}^+ ,$$

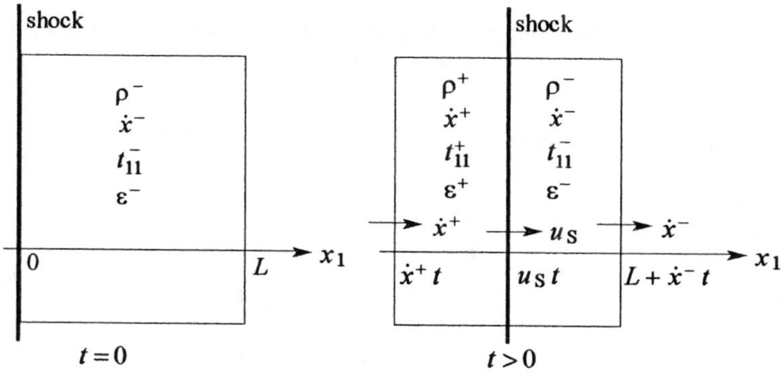

Figure: Exercise 2.7.11.

or

$$[\![\rho \dot{x}]\!] u_S = [\![\rho \dot{x}^2 - t_{11}]\!]. \qquad (2.110_2)$$

The energy of a unit volume of material is $\rho(\varepsilon + (\dot{x}^2/2))$. The principle of balance of energy holds that the rate at which this energy increases is equal to the power supplied, so we have

$$\rho^+[\varepsilon^+ + \tfrac{1}{2}(\dot{x}^+)^2](u_S - \dot{x}^+) + \rho^-[\varepsilon^- + \tfrac{1}{2}(\dot{x}^-)^2](\dot{x}^- - u_S) = -t_{11}^- \dot{x}^- + t_{11}^+ \dot{x}^+,$$

or

$$[\![\rho(\varepsilon + \tfrac{1}{2}\dot{x}^2)]\!] u_S = [\![\rho(\varepsilon + \tfrac{1}{2}\dot{x}^2)\dot{x} - t_{11}\dot{x}]\!]. \qquad (2.110_3)$$

Chapter 3. Plane Longitudinal Shocks

Exercise 3.8.1. The work, W, done on a unit area of the boundary is the product of the force applied and the distance moved: $W = -t_{11}^+ \dot{x}^+ t$. When we use Eq. 2.110_2 to express $-t_{11}^+$ in terms of other variables, this can be written $W = \rho^+ (\dot{x}^+)^2 (u_S - \dot{x}^+)t$. The additional energy imparted to a unit cross section of the material by the shock is the sum of the internal energy and the kinetic energy in the material behind the shock: $E = [\rho^+\varepsilon^+(u_S - \dot{x}^+) + \frac{1}{2}\rho^+(\dot{x}^+)^2]t$. When we relate ε^+ to the other variables using Eq. 2.110_3 and make the same substitution for $-t_{11}^+$ as before we obtain the result sought.

Exercise 3.8.2. From Eq. 3.12 we see that $[\![p]\!] \to \infty$ as $1 - \rho_R S[\![-v]\!] \to 0$. This latter limit is achieved when $\rho^+ = [S/(S-1)]\rho_R$. For $S = 1.5$, a typical value, $\rho_{max} = 3\rho_R$. The compression achieved by a strong shock is much less than would be achieved by isothermal or isentropic application of the same pressure. Be aware that the limiting compression is a mathematical consequence of the form of Eq. 3.12, but usually lies beyond the range of applicability of the empirical linear $U_S - \dot{x}$ Hugoniot.

Exercise 3.8.3. The Hugoniot diagram is shown below for the impact of copper on aluminum at the velocities of 0.5, 1.0, and 1.5 km/s.

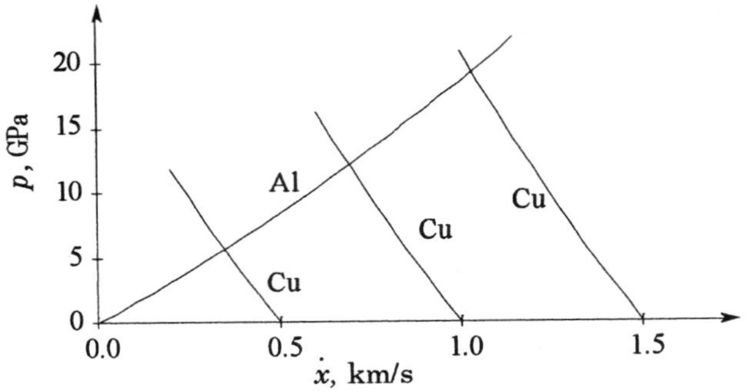

Figure: Exercise 3.8.3.

The Hugoniot curve for the target is

$$p^{+} = \rho_{RT}\,(C_{BT} + S_{T}\,\dot{x}^{+})\,\dot{x}^{+},$$

where the subscript T refers to the target-material value of the parameter. The Hugoniot curve for the projectile is

$$p^{+} = \rho_{RP}\,[C_{BP} + S_{P}\,(\dot{x}_{P} - \dot{x}^{+})]\,(\dot{x}_{P} - \dot{x}^{+}).$$

These two equations are easily solved for p^{+} and $\dot{x}^{+}$. The solution for $\dot{x}^{+}$ is

$$\dot{x}^{+} = \frac{1}{2\,(\rho_{RT}\,S_{T} - \rho_{RP}\,S_{P})}\Big\{(\rho_{RT}\,C_{T} + \rho_{RP}\,C_{P} + 2\,\rho_{RP}\,S_{P}\dot{x}_{P})\,\dot{x}_{P}$$

$$\pm\Big[(\rho_{RT}\,C_{T} + \rho_{RP}\,C_{P})^{2} + 4\,\rho_{RT}\,\rho_{RP}\,(C_{T}\,S_{T} + C_{P}\,S_{P} + 2\,S_{T}\,S_{P})\Big]^{1/2}\Big\},$$

where the ambiguous sign is resolved by the requirement that $\dot{x}^{+}$ be positive but less than $\dot{x}_{P}$. The value of p^{+} is then determined by substitution of this result into the Hugoniot of the target material. The results for the 1 km/s impact are $\dot{x}^{+} = 0.69$ km/s and $p^{+} = 12$ GPa .

Exercise 3.8.4. The Hugoniot curves for this problem are shown below and the pressure and particle velocity are calculated as for Exercise 3.8.2.

The pressures, particle velocities, etc. for this case are given in the following table.

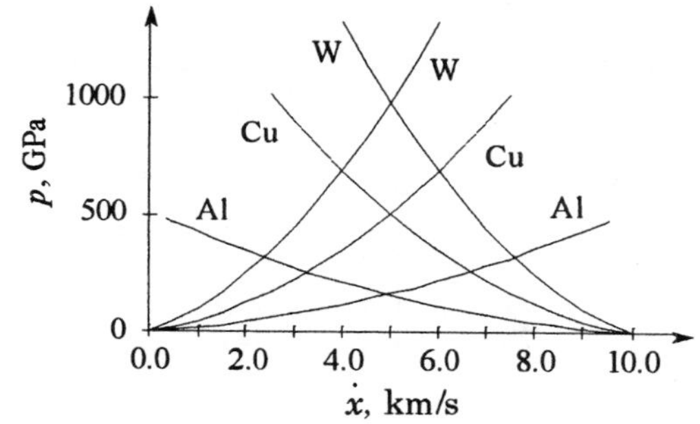

Figure: Exercise 3.8.4.

Tungsten projectile impacting a target at 10 km/s

Target material	$\dot{x}^+$ km/s	p^+ GPa	U_{ST} km/s	U_{SP} km/s	$\left.\frac{\rho-\rho_R}{\rho}\right\|_T$	$\left.\frac{\rho-\rho_R}{\rho}\right\|_P$
Al	7.58	327	15.5	−7.0	49%	35%
Cu	6.00	690	12.9	−9.0	47%	45%
W	5.00	982	10.2	−10.2	49%	49%

Exercise 3.8.5. The Hugoniot curves for this problem are shown below and the pressure and particle velocity are calculated as for Exercise 3.8.2.

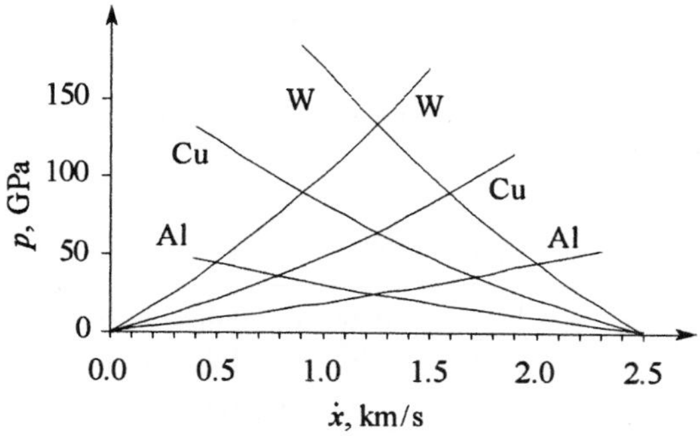

Figure: Exercise 3.8.5.

The pressures, particle velocities, etc. for this case are given in the following table.

Tungsten projectile impacting a target at 2.5 km/s

Target material	$\dot{x}^+$ km/s	p^+ GPa	U_{ST} km/s	U_{SP} km/s	$\left.\frac{\rho-\rho_R}{\rho}\right\vert_T$	$\left.\frac{\rho-\rho_R}{\rho}\right\vert_P$
Al	2.00	45	8.0	−4.7	25%	11%
Cu	1.59	90	6.3	−5.2	25%	18%
W	1.25	134	5.6	−5.6	22%	22%

Exercise 3.8.6. The Hugoniot curves for this problem are shown below and the pressure and particle velocity are calculated as for Exercise 3.8.2. The pressures, particle velocities, etc. for this case are given in the following table.

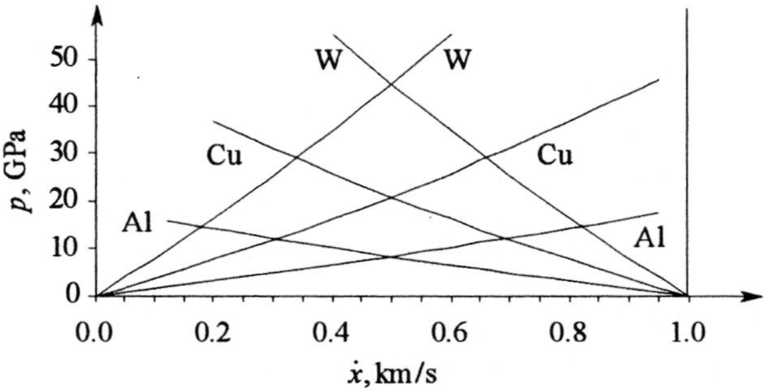

Figure: Exercise 3.8.6.

Tungsten projectile impacting a target at 1 km/s

Target material	$\dot{x}^+$ km/s	p^+ GPa	U_{ST} km/s	U_{SP} km/s	$\left.\frac{\rho-\rho_R}{\rho}\right\vert_T$	$\left.\frac{\rho-\rho_R}{\rho}\right\vert_P$
Al	0.82	14.7	6.43	−4.25	12.8%	4.2%
Cu	0.66	29.0	4.92	−4.45	13.4%	7.6%
W	0.50	44.7	4.65	−4.65	10.8%	10.8%

Exercise 3.8.7. The pressure–particle-velocity Hugoniot and X–t diagrams for the case of a high-impedance film on the back face of a low-impedance plate are shown below. When the incident shock encounters the film, transmitted and reflected shocks of higher pressure (state 2) are produced. Reverberation of the transmitted shock in the film leads to production of a sequence of states, as shown. The surface of the low-impedance plate is gradually decompressed and

accelerated to a state of zero stress and the velocity $\dot{x}_{fs}$ that would have been attained in the absence of the film. The plate/film interface is always in compression so there is no tendency for the film to separate from the substrate and a measurement of the velocity of the surface of the film can be considered to be the same as the free-surface velocity of the plate in the absence of the film. The time required to approach this equilibrium state decreases with decreasing thickness of the film.

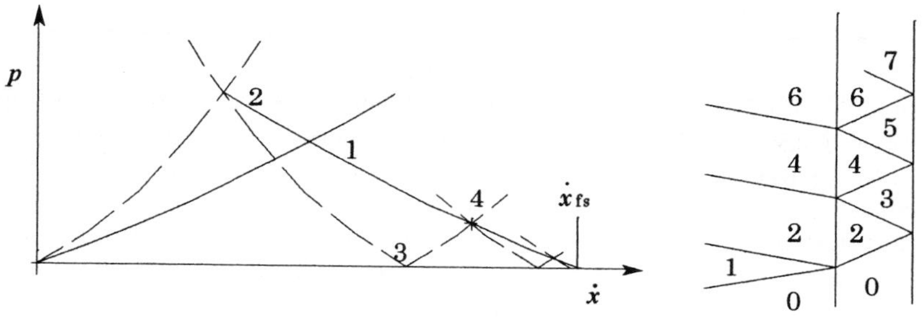

Figure. Exercise 3.8.7. Pressure–particle-velocity Hugoniot diagram and X–t diagram for the interactions produced when a shock propagating in a low-impedance plate encounters a thin high-impedance film on its surface. The Hugoniots of the plate material are shown as solid lines and those for the film are shown as broken lines.

Pressure–particle-velocity Hugoniot and $X-t$ diagrams for the case of a low-impedance film on the back face of a higher-impedance plate are shown below. When the incident shock encounters the low-impedance film, transmitted and reflected shocks of lower pressure (state 2) are produced. Reflection of the transmitted shock from the rear surface of the film accelerates this surface to a higher velocity (state 3) than would have been attained at the surface of the plate had the film not been there. If the film is free to separate from the plate, it will do so at the time of interaction producing state 4. If it is bonded to the plate sur-

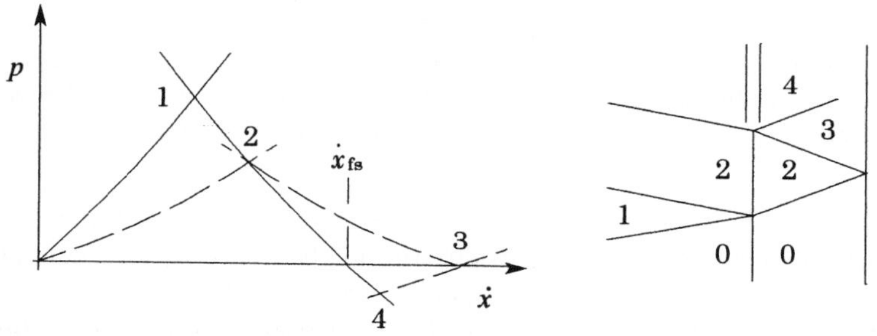

Figure. Exercise 3.8.7. Pressure–particle-velocity Hugoniot diagram and $X-t$ diagram for the interactions produced when a shock propagating in a high-impedance plate encounters a thin low impedance film on its surface. The Hugoniots of the plate material are shown as solid lines and those for the film are shown as broken lines.

face, a tensile stress (state 4) will be produced at the interface. If this stress exceeds the bond strength, the film will separate and be expelled from the plate.

It is clear from these analyses that a back-surface mirror or electrode placed on a sample should be of higher impedance than the sample.

Exercise 3.8.8. The pressure–particle-velocity and $X-t$ diagrams for this problem are given below. The incident shock takes the material from the state of zero stress and velocity to the state 1. The initial response of the gauge layer is a transition to state 2, but subsequent reverberations of the shock in the gauge layer bring the pressure to that of state 1, the pressure behind the incident shock. In principle, infinitely many reverberations are required for pressure equilibrium but the pressure rises essentially (say, within the gauge accuracy) to its equilibrium value within only a few reverberations. In materials that provide a good impedance match to the gauge insulator, fewer reverberations are required. A typical thickness of the gauge layer is 25 μm and a shock transit across this layer requires about 10 ns, so the gauge requires about 50 ns to respond to a shock.

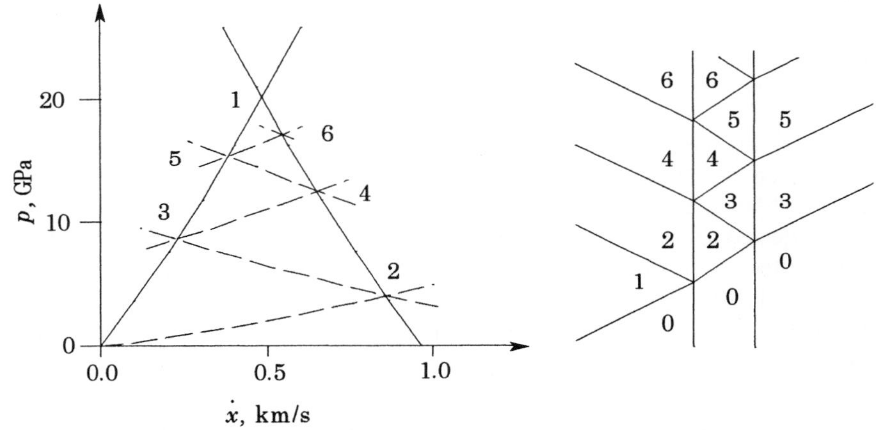

Figure. Exercise 3.8.8. Pressure–particle-velocity and $X-t$ diagrams for shock interaction with an embedded polymer-insulated manganin gauge. The solid lines are for a copper sample and the broken lines are for the polymeric gauge insulation.

Exercise 3.8.9. The required diagrams are shown in the figure below. This figure is drawn for the case in which the impactor and the backer plate are aluminum oxide monocrystals (sapphire) and the target plate is an X-cut α quartz crystal. The impact velocity is 85 m/s. As shown on the $p-\dot{x}$ diagram, the quartz plate experiences a sequence of shock compressions ultimately tending to produce the pressure that would have been realized by impact of the

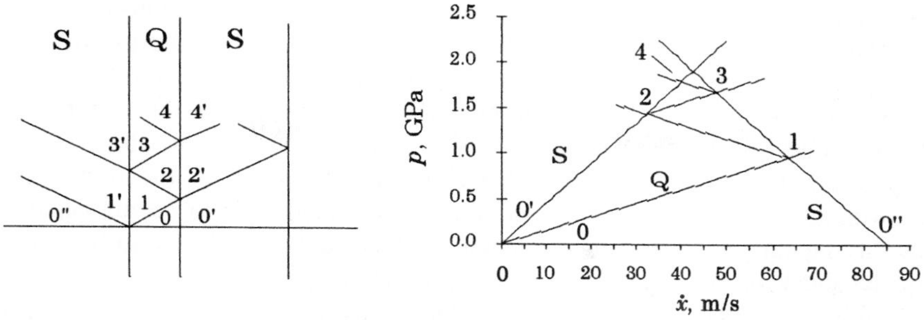

Figure: Exercise 3.8.9.

two sapphire plates. Although not important to this analysis, it is noteworthy that the quartz plate is piezoelectric. The current produced in a short circuit connecting electrodes on the two faces of the quartz provides a measure of the stress difference between these faces.

Exercise 3.8.10. The required diagrams are shown in the figure below.

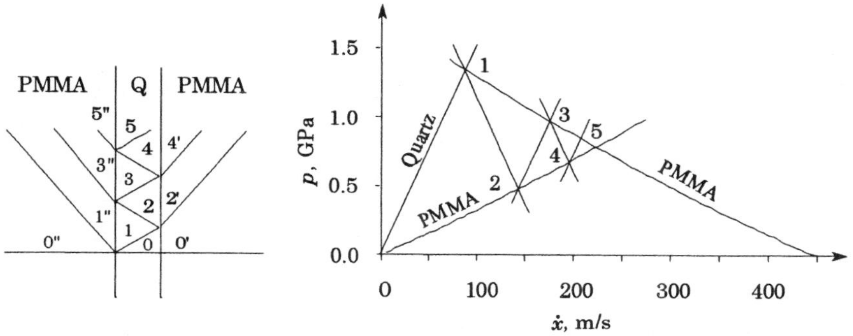

Figure: Exercise 3.8.10.

This figure is drawn for the case in which the impactor and the backer plate are made of the polymer PMMA (polymethyl methacrylate) and the target plate is an X-cut α-quartz crystal. The impact velocity is 450 m/s. As shown on the $p-\dot{x}$ diagram, each of the PMMA plates experiences a sequence of shock compressions ultimately tending to produce the pressure that would have been realized by impact of one on the other without the intervening quartz plate. When the quartz plate is configured as a gauge [1] the assembly can be used to measure the sequence of shocked states in the PMMA.

Exercise 3.8.11. A sketch of the physical layout, the $X-t$ diagram, and the $-t_{11}-\dot{x}$ Hugoniot diagram for the problem are shown in the figure below. The sketch of the physical configuration suggests an explosive lens arranged to drive a shock into a copper plate that is backed by a sample material, in this case, tungsten. The arrows labeled P_1, P_2, and P_3 denote electrical contact pins or other devices that detect the arrival of a shock wave. When the assembly is made, the depth of the well in which P_1 is located is measured carefully so that the shock velocity in the copper can be determined by dividing this distance by the time interval between the signal from P_1 and that from P_2. The Hugoniot for the copper standard is known, as shown in the lower panel of the figure, and knowledge of the shock velocity permits the Rayleigh line to be drawn, thus identifying state 1. Similar measurements permit the Rayleigh line to be drawn for the shock in the tungsten sample. Since the state in the tungsten lies along both this Rayleigh line and the Hugoniot for a left-propagating shock centered on state 1 in the copper, a point on the tungsten Hugoniot (not yet determined, but suggested by the broken line) has been determined. Use of explosive devices producing shocks of different strength, plates of other materials in place of the copper, etc., permits determination of other Hugoniot points for the sample material.

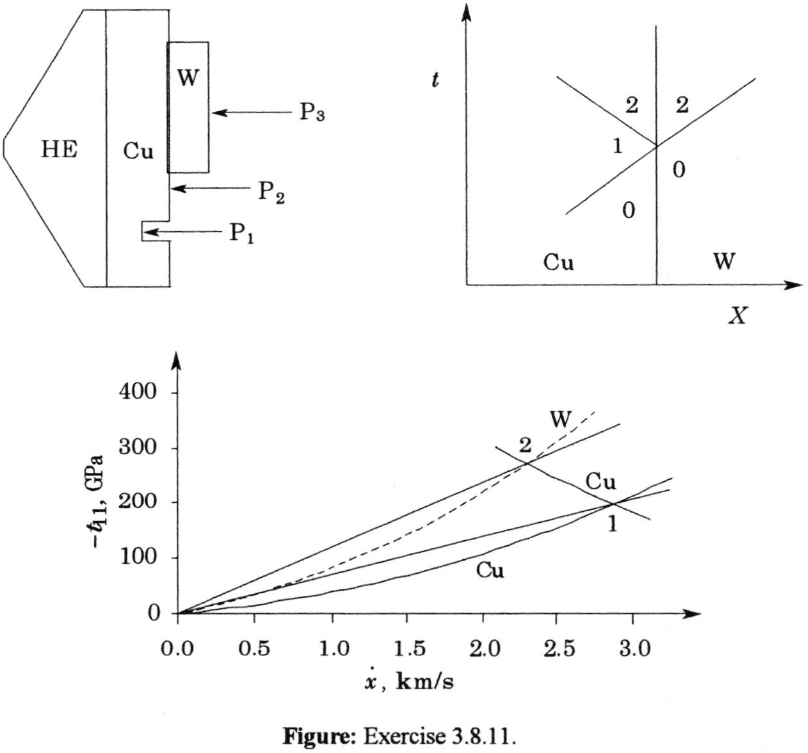

Figure: Exercise 3.8.11.

Exercise 3.8.12. This experiment may use an explosive driver like that suggested by the foregoing figure, except that the sample is placed in direct contact with the explosive. Pins are arranged so as to measure the shock velocity in the sample. In addition, another pin is placed at a small, carefully measured distance behind the sample. The velocity imparted to the unrestrained surface of the sample by the shock is determined by dividing the separation distance of the standoff pin by the time between the arrival of the shock at the surface of the sample and the arrival of the sample itself at the pin. The particle velocity of the material behind the shock is taken to be half of the measured free-surface velocity.

Exercise 3.8.13. The configuration at the time t is shown in the following figure. The work done on the boundary is the applied stress times the distance the boundary has moved, $-t_{11}\,\dot{x}^{+}t$. The kinetic energy of the material is one-half its mass times its velocity, $\frac{1}{2}(\rho_R\, U_S\, t)(\dot{x}^{+})^2$. The internal energy per unit mass (from the jump equation) is $\frac{1}{2}(\dot{x}^{+})^2$ so the total internal energy in the material at the time t is $\frac{1}{2}(\rho_R U_S\, t)(\dot{x}^{+})^2$. Energy balance requires that the work done on the boundary be equal to the sum of the internal and kinetic energy:

$$-t_{11}\,\dot{x}^{+}t = \frac{1}{2}(\rho_R\, U_S\, t)(\dot{x}^{+})^2 + \frac{1}{2}(\rho_R\, U_S\, t)(\dot{x}^{+})^2 = (\rho_R\, U_S\, t)(\dot{x}^{+})^2\,,$$

but the jump condition gives $-t_{11} = \rho_R\, U_S\, \dot{x}^{+}$ so the energy is seen to be in balance. One-half of the energy in the material is kinetic energy and one-half is internal energy.

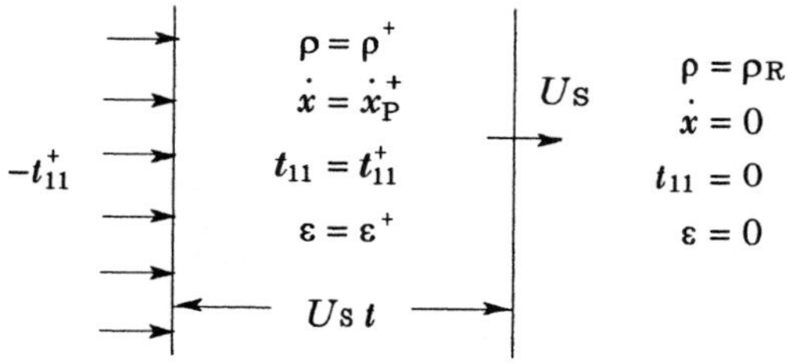

Figure. Exercise 3.8.13.

The energy increase in the material as the shock advances must be supplied by the stress applied to the boundary. A shock of constant amplitude cannot exist without this constant supply of energy. In Chap. 13 we shall see that a steady shock can be supported by release of chemical energy instead of a boundary stress. In this case the material is called an *explosive* and the chemically-supported shock is called a *detonation shock.*

Chapter 4. Material Response I: Principles

Exercise 4.4.1. The stress transforms as a tensor according to Eq. 4.5, $\mathbf{t}^* = \mathbf{Q}\mathbf{t}\mathbf{Q}$, the stress rate transforms by Eq. 4.6, $\dot{\mathbf{t}}^* = \dot{\mathbf{Q}}\mathbf{t}\mathbf{Q}^{-1} + \mathbf{Q}\dot{\mathbf{t}}\mathbf{Q}^{-1} - \mathbf{Q}\mathbf{t}\,\dot{\mathbf{Q}}$, and the velocity gradient transforms by Eq. 4.3, $\mathbf{l}^* = \dot{\mathbf{Q}}\,\mathbf{Q} + \mathbf{Q}\,\mathbf{l}\,\mathbf{Q}$. Since $\overset{\nabla}{\mathbf{t}} = \dot{\mathbf{t}} - \mathbf{t}\,\mathbf{l} - \mathbf{l}\,\mathbf{t}$, we have $\overset{\nabla}{\mathbf{t}}{}^* = \dot{\mathbf{t}}^* - \mathbf{t}^* \overset{T}{\mathbf{l}}{}^* - \mathbf{l}^* \mathbf{t}^*$, or

$$\overset{\nabla}{\mathbf{t}}{}^* = \dot{\mathbf{Q}}\mathbf{t}\mathbf{Q}^{-1} + \mathbf{Q}\dot{\mathbf{t}}\mathbf{Q}^{-1} - \mathbf{Q}\mathbf{t}\dot{\mathbf{Q}}^{-1} - \mathbf{Q}\mathbf{t}\mathbf{Q}^{-1}(\mathbf{Q}\dot{\mathbf{Q}} + \mathbf{Q}\overset{T}{\mathbf{l}}\mathbf{Q}^{-1}) - (\dot{\mathbf{Q}}\mathbf{Q}^{-1} + \mathbf{Q}\mathbf{l}\mathbf{Q}^{-1})\mathbf{Q}\mathbf{t}\mathbf{Q}^{-1}$$

$$= \mathbf{Q}\dot{\mathbf{t}}\mathbf{Q}^{-1} - \mathbf{Q}\mathbf{t}\overset{T}{\mathbf{l}}\mathbf{Q}^{-1} - \mathbf{Q}\mathbf{l}\mathbf{t}\mathbf{Q}^{-1} = \mathbf{Q}(\dot{\mathbf{t}} - \mathbf{t}\overset{T}{\mathbf{l}} - \mathbf{l}\mathbf{t})\mathbf{Q}^{-1} = \mathbf{Q}\overset{\nabla}{\mathbf{t}}\mathbf{Q}^{-1},$$

so we have the equation $\mathbf{t}^* = \mathbf{Q}\mathbf{t}\mathbf{Q}^{-1}$, which is the transformation equation for a tensor.

Exercise 4.4.2. The stress rate of Eq. 4.8 is the same as the Jaumann rate discussed in the previous exercise except for the additional term $(\mathrm{tr}\,\mathbf{d})t_{ij}$. Since $\mathbf{d}^* = \mathbf{Q}\mathbf{d}\mathbf{Q}^{-1}$ we have $\mathrm{tr}\,\mathbf{d}^* = \mathrm{tr}\,\mathbf{d}$ and, since $\mathbf{t}$ transforms as a tensor the same is true of $\overset{\square}{\mathbf{t}}$.

Exercise 4.4.3. The constitutive equation of interest is

$$T_{IJ} = C_{IJKL}\, E_{KL}\,. \tag{A}$$

If we subject the reference coordinate frame to the orthogonal transformation H_{IJ} the stress and strain tensors transform according to Eqs. 4.17:

$$T^*_{IJ} = \overset{-1}{H}_{IP}\, T_{PQ}\, H_{QJ} \quad \text{and} \quad E^*_{IJ} = \overset{-1}{H}_{IK}\, T_{KL}\, H_{LJ}\,. \tag{B}$$

The fact that the constitutive equation is invariant to this transformation means that

$$T^*_{IJ} = C_{IJKL}\, E^*_{KL}\,, \tag{C}$$

where the coefficient tensor **C** has the same components in both Eqs. A and C.

Substitution of Eqs. B into Eq. C yields the result

$$\overset{-1}{H}_{IP}\, T_{PQ}\, H_{QJ} = C_{IJKL}\, \overset{-1}{H}_{KC}\, E_{CD}\, H_{DL}\,. \tag{D}$$

Pre-multiplying each member of this equation by H_{AI}, post-multiplying by H^{-1}_{JB}, contracting, and use of Eq. A, gives

$$C_{ABCD}\, E_{CD} = H_{AI}\, C_{IJKL}\, \overset{-1}{H}_{KC}\, E_{CD}\, H_{DL}\, \overset{-1}{H}_{JB}\,,$$

or

$$(C_{ABCD} - H_{AI}\, H_{BJ}\, H_{CK}\, H_{DL}\, C_{IJKL})\, E_{CD} = 0\,,$$

since $\mathbf{H}^{-1} = \mathbf{H}^{\mathrm{T}}$. Because of the symmetry with respect to interchange of indices the coefficient of **E** in this equation vanishes and we see that **C** must satisfy the equation

$$C_{ABCD} = H_{AI}\, H_{BJ}\, H_{CK}\, H_{DL}\, C_{IJKL}\,.$$

The same procedure leads to analogous results for higher-order elastic moduli. Although working out the specific result for a given **H** is very tedious, tables of second-, third-, and fourth-order coefficients have been prepared for all crystal classes and for isotropic and transversely isotropic materials.

Exercise 4.4.4. For the motion we choose $x_i(\mathbf{X},t) = X_I^* \delta_{Ii} + A_{iI}(t)(X_I - X_I^*)$, where we require that $\mathbf{A}$ be invertible. With this we have $\mathbf{F} = \mathbf{A}(t)$ and $\dot{\mathbf{F}} = \dot{\mathbf{A}}(t)$. For the specific entropy we choose $\eta(\mathbf{X},t) = \eta^*(t) + B_I(t)(X_I - X_I^*)$ so we have $\dot{\eta}(\mathbf{X},t) = \dot{\eta}^*(t) + \dot{B}_I(t)(X_I - X_I^*)$. The temperature field requires a slightly different approach because we also need to provide for an arbitrary value of the spatial gradient $g_i = \theta_{,i}$ and its material derivative $\dot{g}_i$. Choose $\theta(\mathbf{X},t) = \theta^*(t) + c_i(t) A_{iI}(t)(X_I - X_I^*)$. From the motion we have $A_{Ii}^{-1}(x_i - x_i^*) = X_I - X_I^*$ so $\theta(x,t) = \theta^*(t) + c_i(t)(x_i - x_i^*)$, where $x_i^* = X_I^* \delta_{Ii}$. With this, $g_i(x,t) = c_i(t)$ and $\dot{g}_i(x,t) = \dot{c}_i(t)$.

Exercise 4.4.5. For an inviscid thermoelastic fluid we have

$$\varepsilon = \varepsilon(v,\eta,\mathbf{g}),\ \theta = \theta(v,\eta,\mathbf{g}),\ t_{ij} = -p(v,\eta,\mathbf{g})\delta_{ij},\ \mathbf{q} = \mathbf{q}(v,\eta,\mathbf{g}). \tag{A}$$

The Clausius–Duhem inequality to be satisfied is, in the form 4.6,

$$0 \ge \frac{1}{\theta}\dot{\varepsilon} - \dot{\eta} - \frac{1}{\rho\theta} t_{ij} l_{ij} + \frac{1}{\rho\theta^2} q_i g_i . \tag{B}$$

When we substitute Eqs. A into this inequality we obtain the new inequality

$$0 \ge \frac{1}{\theta}\left(\frac{\partial\varepsilon}{\partial v}\dot{v} + \frac{\partial\varepsilon}{\partial\eta}\dot{\eta} + \frac{\partial\varepsilon}{\partial g_i}\dot{g}_i\right) - \dot{\eta} + \frac{1}{\rho\theta} p l_{kk} + \frac{1}{\rho\theta^2} q_i g_i ,$$

but we have $l_{kk} = \dot{x}_{k,k} = \dot{J}/J = \dot{v}/v$ so the inequality becomes

$$0 \ge \frac{1}{\theta}\left(\frac{\partial\varepsilon}{\partial v} + p\right)\dot{v} + \left(\frac{1}{\theta}\frac{\partial\varepsilon}{\partial\eta} - 1\right)\dot{\eta} + \frac{1}{\theta}\frac{\partial\varepsilon}{\partial g_i}\dot{g}_i + \frac{1}{\rho\theta^2} q_i g_i .$$

If we choose $\mathbf{g} = \mathbf{0}$, $\dot{\mathbf{g}} = \mathbf{0}$, and $\dot{\eta} = 0$ we see that we must have

$$p = -\frac{\partial\varepsilon}{\partial v}$$

if the inequality is to be satisfied for an arbitrary value of $\dot{v}$. Similarly, we find that

$$\frac{\partial\varepsilon}{\partial\eta} = \theta,\quad \frac{\partial\varepsilon}{\partial g_i} = 0,\quad \text{and}\quad q_i g_i \le 0,$$

so we have $\varepsilon = \varepsilon(v,\eta)$ and $\partial\varepsilon(v,\eta)/\partial\eta = \theta$ and we are left with the requirement on $\mathbf{q}(v,\eta,\mathbf{g})$ that $q_i g_i \le 0$.

Exercise 4.4.6. For a nonconducting thermoelastic fluid the Clausius–Duhem inequality can be written

$$0 \geq \frac{1}{\theta}\left(\frac{\partial \varepsilon}{\partial v} + p\right)\dot{v} + \left(\frac{1}{\theta}\frac{\partial \varepsilon}{\partial \eta} - 1\right)\dot{\eta} + \frac{1}{\theta}\frac{\partial \varepsilon}{\partial \alpha}\dot{\alpha}_i \,.$$

Since $\varepsilon(v, \eta, \alpha) = \varepsilon_s(v/\alpha, \eta)$ the requirements imposed on ε are that

$$p = -\frac{1}{\alpha}\frac{\partial \varepsilon_s(v_s, \eta)}{\partial v_s}, \quad \theta = \frac{\partial \varepsilon_s(v_s, \eta)}{\partial \eta},$$

and we are left with the inequality

$$0 \geq -\frac{v}{\alpha^2}\frac{\partial \varepsilon_s(v/\alpha, \eta)}{\partial v_s}\dot{\alpha} = \frac{v\,p}{\alpha}\dot{\alpha} = \frac{v\,p}{\alpha}\frac{\alpha - \alpha_{eq}(v, \eta)}{\tau},$$

in which $v > 0$, $p > 0$, and $\alpha > 0$. During the collapse process, $\alpha - \alpha_{eq}(v, \eta) > 0$. Therefore, it is necessary and sufficient that $\tau > 0$ for satisfaction of the inequality.

Chapter 5. Material Response II: Inviscid Compressible Fluids

Exercise 5.7.1. To derive Eq. 5.30 we note that

$$\left.\frac{\partial \eta}{\partial v}\right|_\theta = \left.\frac{\partial \eta}{\partial p}\right|_\theta \left.\frac{\partial p}{\partial v}\right|_\theta .$$

The required result follows by substituting expressions for the derivatives from Table 5.1.

Exercise 5.7.2. The hint given in the exercise suggests that we consider derivatives of the function $p = p(v, \theta(v, \eta))$. Accordingly, we calculate

$$\left.\frac{\partial p}{\partial v}\right|_\eta = \left.\frac{\partial p}{\partial v}\right|_\theta + \left.\frac{\partial p}{\partial \theta}\right|_v \left.\frac{\partial \theta}{\partial v}\right|_\eta ,$$

and substitution from Table 5.1 yields the required result.

Exercise 5.7.3. The hint given in the exercise suggests that we consider derivatives of the function $\eta = \eta(\theta, v(p, \theta))$. Accordingly, we calculate

$$\left.\frac{\partial \eta}{\partial \theta}\right|_p = \left.\frac{\partial \eta}{\partial \theta}\right|_v + \left.\frac{\partial \eta}{\partial v}\right|_\theta \left.\frac{\partial v}{\partial \theta}\right|_p ,$$

and substitution from Table 5.1 yields the required result after making the substitution $\gamma C^v = v\beta B^\theta$ that follows from Eqs 5.26 and 5.29.

It is appropriate to establish the conditions under which a function of the form $\eta = \eta(\theta, v(p,\theta))$ can be obtained. The pressure response function $p = p(v,\eta)$ can be inverted to yield a relation of the form $v = v(p,\eta)$ since the derivative, $\partial p(v,\eta)/\partial\eta = -B^{\eta}/v$, is nonzero. The temperature response function $\theta = \hat{\theta}(v,\eta)$ can be inverted to give $\eta = \eta(v,\theta)$ since $\partial\hat{\theta}(v,\eta)/\partial\eta = \theta/C^{(v)} \neq 0$. Then, the pressure-response function can be written $p = p(v,\eta(v,\theta)) = p(v,\theta)$. Since $\partial p(v,\theta)/\partial v = -B^{\theta}/v \neq 0$ this can be inverted to give $v = v(p,\theta)$. Substitution of this expression into $\eta = \eta(v,\theta)$ gives $\eta = \eta(v(p,\theta),\theta)$, which is the function needed.

Exercise 5.7.4. This solution is begun by calculating B^{η}/B^{θ} and C^{p}/C^{v} from Eqs. 5.31 and 5.32.

Exercise 5.7.5. The principal Hugoniot for an ideal gas is given by Eq. 5.59:

$$p^{(\mathrm{H})}(v) = p_{\mathrm{R}} \frac{(\Gamma+1)\,v_{\mathrm{R}} - (\Gamma-1)\,v}{(\Gamma+1)\,v - (\Gamma-1)\,v_{\mathrm{R}}}. \tag{A}$$

This Hugoniot is transformed to a $P-\dot{x}$ Hugoniot by substituting this result into the jump equation

$$\dot{x}^2 = \left[(p-p_{\mathrm{R}})(v_{\mathrm{R}}-v)\right],$$

and solving the result for $p^{(\mathrm{H})}(\dot{x})$. The result is

$$p^{(\mathrm{H})}(\dot{x}) = p_{\mathrm{R}}\left\{1+\frac{\Gamma+1}{4p_{\mathrm{R}}v_{\mathrm{R}}}\dot{x}^2 + \dot{x}\left[\frac{\Gamma}{p_{\mathrm{R}}v_{\mathrm{R}}} + \left(\frac{\Gamma+1}{4p_{\mathrm{R}}v_{\mathrm{R}}}\right)^2 \dot{x}^2\right]^{1/2}\right\},$$

where the + sign was chosen to ensure that p became large for large $\dot{x}$.

Exercise 5.7.6. The second-shock Hugoniot is obtained by substituting Eq. A of the foregoing solution into Eq. 5.165 with $\gamma(v) = \Gamma - 1$. The result is

$$\frac{p^{(\mathrm{H}+)}(v)}{p_{\mathrm{R}}} = \frac{(\Gamma+1)-(\Gamma-1)\bar{v}}{(\Gamma+1)\bar{v}-(\Gamma-1)\bar{v}^{+}} + \frac{(\Gamma-1)(1-\bar{v})[(\Gamma+1)-(\Gamma-1)\bar{v}^{+}]}{[(\Gamma+1)\bar{v}-(\Gamma-1)\bar{v}^{+}][(\Gamma+1)\bar{v}^{+}-(\Gamma-1)\bar{v}^{+}]},$$

where $\bar{v} = v/v_R$ and $\bar{v}^{+} = v^{+}/v_R$.

Exercise 5.7.7. The principal Hugoniot for an ideal gas is given by Eq. 5.59:

$$p^{(\mathrm{H})}(v) = p_{\mathrm{R}} \frac{(\Gamma+1)\,v_{\mathrm{R}} - (\Gamma-1)\,v}{(\Gamma+1)\,v - (\Gamma-1)\,v_{\mathrm{R}}}. \tag{A}$$

Examination of this fraction shows that $p^{(\mathrm{H})}(v) \to \infty$ as $v \to v_{\mathrm{R}}(\Gamma-1)/(\Gamma+1)$, the limiting specific volume that can be achieved by shock compression of an ideal gas. The associated limiting density is

$$\rho_{\text{limit}} = \rho_{\text{R}} (\Gamma + 1)/(\Gamma - 1)\,. \tag{B}$$

Exercise 5.7.8. The solutions are: $B^{(\theta)} = p$, $B^{(\eta)} = \Gamma p$, $\beta = \mathcal{R}/pv = 1/\theta$, $\gamma = \mathcal{R}/C_{\text{R}}^{v} = \Gamma - 1$, $C^{(p)} = \Gamma C_{\text{R}}^{v}$. It is useful to note that $C^{v} = 3\mathcal{R}/2$ and $\gamma = 2/3$ for a monatomic gas. These values are approached rather closely when highly porous metals are compressed by strong shocks that cause their temperatures to rise to some tens of thousands of Kelvin [104].

Exercise 5.7.9. Demonstration of the truth of the assertions of the exercise is based upon analysis of Eq. 5.136,

$$\frac{d\eta^{(\text{H})}(v)}{dv} = \frac{1}{2\theta^{(\text{H})}(v)}\left\{p^{(\text{H})}(v) - p^{-} + (v^{-} - v)\frac{dp^{(\text{H})}(v)}{dv}\right\}. \tag{A}$$

We first consider the concave upward (normal) Hugoniot. Write Eq. A in the form

$$\frac{d\eta^{(\text{H})}(v)}{dv} = \frac{v^{-} - v}{2\theta^{(\text{H})}(v)}\left\{\frac{dp^{(\text{H})}(v)}{dv} - \frac{p^{-} - p^{(\text{H})}(v)}{v^{-} - v}\right\}. \tag{B}$$

To demonstrate the theorem for compression from p^{-} to some higher pressure we need to show that $d\eta^{(\text{H})}(v)/dv < 0$. The factor $(v^{-} - v)/2\theta^{(\text{H})}(v)$ in Eq. B is positive, so the sign of the derivative turns on the relationship between the two terms in braces. The term $(p^{(\text{H})}(v) - p^{-})/(v^{-} - v)$ is the slope of the straight line in the figure and the term $dp^{(\text{H})}(v)/dv$ is the slope of the tangent at the point *p*, *v*. In the case shown in which the Hugoniot is concave upward the tangent line is steeper, i.e., its slope is more negative than that of the secant line so the sum of these terms, and hence $d\eta^{(\text{H})}(v)/dv$ is negative. Since this is true for all values $v < v^{-}$, integration from any compressed state to the specific volume v^{-} gives a lower entropy density for the state p^{-}, v^{-} than for any compressed state. Accordingly, the entropy increases upon passage of a compression shock. Similar consideration of the other cases completes the proof of the theorem.

Exercise 5.7.10. Equation 5.142 is a first-order linear ordinary differential equation of a form that we shall meet quite often in this book. Substitution of Eq. 5.143 into Eq. 5.142 shows that the former is the solution of the latter that satisfies the initial condition $\theta^{(\text{H})}(v_{\text{R}}) = \theta^{-}$. Reduction of Eq. 5.143 to Eq. 5.146, when $\gamma(v) = \gamma_{\text{R}}\, v/v_{\text{R}}$ and $p^{(\text{H})}(v)$ is given by Eq. 3.12, is routine.

The integral is easily evaluated using the trapezoidal rule to calculate the area under the curve representing the function $f(v) = \kappa_{\text{c}}(v)/\chi_{\text{c}}(v)$. The idea is illustrated in the figure below.

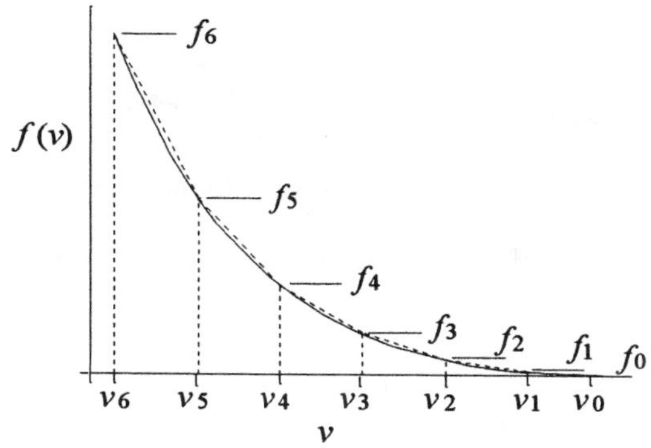

Figure: Exercise 5.7.10.

The value of the integral is approximately the sum of the areas of the trapezoids formed by the dotted lines:

$$\int_{v_0}^{v_n} f(v')\,dv' = \frac{1}{2}\sum_{k=0}^{n} (f_{i+1} + f_i)(v_{i+1} - v_i)\,.$$

The sum, and then the right member of Eq. 5.146, is easily calculated and a graph of the result prepared using a spreadsheet program. Note that the area is overestimated for concave upward curves and underestimated for concave downward curves. Books on numerical analysis present more refined ways of evaluating the integral.

Exercise 5.7.11. Transforming a known isentrope to one for a different value of the specific entropy is not as easy as transforming an isotherm to a different temperature, but it can be done following a procedure that is useful in transforming other thermodynamic curves. We begin by using the Mie–Grüneisen equation to relate the two isentropes of interest. We have

$$p^{(\eta)}(v;\eta^{**}) = p^{(\eta)}(v;\eta^{*}) + \frac{\gamma(v)}{v}\left[\varepsilon^{(\eta)}(v;\eta^{**}) - p^{(\eta)}(v;\eta^{*})\right]. \tag{A}$$

Differentiation of this equation with respect to v gives

$$\begin{aligned} &\frac{p^{(\eta)}(v;\eta^{**})}{dv} + \left[\frac{\gamma(v)}{v} - \frac{v}{\gamma(v)}\frac{d}{dv}\left(\frac{\gamma(v)}{v}\right)\right] p^{(\eta)}(v;\eta^{**}) \\ &\quad = \frac{p^{(\eta)}(v;\eta^{*})}{dv} + \left[\frac{\gamma(v)}{v} - \frac{v}{\gamma(v)}\frac{d}{dv}\left(\frac{\gamma(v)}{v}\right)\right] p^{(\eta)}(v;\eta^{*}), \end{aligned} \tag{B}$$

which is a linear, first-order ordinary differential equation that has the well-known solution

$$p^{(\eta)}(v;\eta^{**}) = \hat{\chi}(v)\int_{v_R^{**}}^{v} \frac{\kappa^{(\eta)}(v';\eta^{*})}{\hat{\chi}(v')}dv', \tag{C}$$

when the condition $p^{(\eta)}(v_R^{**};\eta^{**}) = 0$ is imposed. In this equation

$$\kappa^{(\eta)}(v;\eta^{*}) = \left[\frac{\gamma(v)}{v} - \frac{v}{\gamma(v)}\frac{d}{dv}\left(\frac{\gamma(v)}{v}\right)\right]p^{(\eta)}(v;\eta^{*}) + \frac{dp^{(\eta)}(v;\eta^{*})}{dv} \tag{D}$$

and

$$\hat{\chi}(v) = \exp\left\{-\int_{v_R^{**}}^{v}\left[\frac{\gamma(v')}{v'} - \frac{v'}{\gamma(v')}\frac{d}{dv'}\left(\frac{\gamma(v')}{v'}\right)\right]dv'\right\}. \tag{E}$$

The specific internal energy isentrope is obtained from the pressure isentrope by integration.

As presented, the foregoing equations solve the usual problem of moving an isentrope so that it crosses the $p=0$ axis at a specific volume, v_R^{**}, that is different from the value for the η^{*} isentrope. It does not give the specific entropy on the new isentrope except implicitly through v_R^{**}. To remedy this shortcoming, we must look at the pressure equation of state, Eq. 5.89, that follows from the complete Mie–Grüneisen equation of state. This equation is to be solved for the specific entropy value, η^{**}, that corresponds to zero pressure at $v = v_R^{**}$. We obtain the value from the equation

$$\int_{\eta^{*}}^{\eta^{**}} \omega(\eta')\,d\eta' = -\frac{p^{(\eta)}(v^{**};\eta^{*})}{\theta_R\,[\gamma(v^{**})/v^{**}]\,\chi(v^{**})}. \tag{F}$$

Since the right member of this equation is known, one simply evaluates the integral of the left member as a function of η^{**} until a value is reached that satisfies the equation.

As before, the problem is more easily solved if $\gamma(v)/v$ and C^{v} are constants. In this case, one simply evaluates Eqs. 5.94 for η^{*} and η^{**}, obtaining the equations

$$\begin{aligned}
\varepsilon^{(\eta)}(v,\eta^{**}) &= \varepsilon^{(\eta)}(v;\eta^{*}) + \theta_R\, C_R^{v}\,\chi_c(v)\left[\omega_c(\eta^{**})-1\right]\\
p^{(\eta)}(v,\eta^{**}) &= p^{(\eta)}(v;\eta^{*}) + \theta_R\, C_R^{v}\,\frac{\gamma_R}{v_R}\chi_c(v)\left[\omega_c(\eta^{**})-1\right]\\
\theta^{(\eta)}(v,\eta^{**}) &= \theta_R\,\chi_c(v)\,\omega_c(\eta^{**}),
\end{aligned} \tag{G}$$

relating the two isentropes.

Exercise 5.7.12. Recentered Hugoniots are needed for calculation of shocks propagating into material that has already been compressed by a shock. The second-shock Hugoniot is centered at the point on the principal Hugoniot corresponding to the state behind this first shock. A schematic illustration of the Hugoniot curves of interest is given in the following figure.

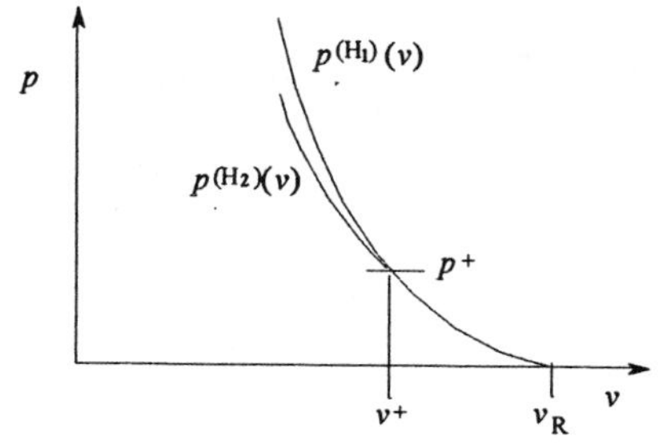

Figure. Exercise 5.7.12. Illustration of the relative positions of a principal Hugoniot and a second-shock Hugoniot centered on the state (p^+, v^+) produced by a shock transition from the reference state.

We suppose that the Hugoniot $p^{(H_1)}(v)$ is known. A shock transition from $p = 0$, $v = v_R$ to a state defined by $p = p^+$ on $p^{(H_1)}(v)$ is followed by a second transition from the state S^+ to a new, more highly compressed state. To determine this new state it is necessary to know the Hugoniot, $p^{(H_2)}(v)$, centered on the state S^+.

Application of the Rankine–Hugoniot equation to transitions from the reference state gives

$$\varepsilon^{(H)}(v) = \varepsilon_R + \tfrac{1}{2} p^{(H)}(v)\,(v_R - v)\,, \tag{A}$$

and, in particular, the transition from the reference state to the state S^+ gives

$$\varepsilon^+ = \varepsilon_R + \tfrac{1}{2} p^+ (v_R - v^+)\,. \tag{B}$$

Similar application to the transition from S^+ to some state p, v on the second-shock Hugoniot gives

$$\varepsilon^{(H_2)}(v) = \varepsilon^+ + \frac{1}{2}\left[p^{(H_2)}(v) + p^+\right](v^+ - v)\,. \tag{C}$$

Substitution of Eqs. A, B, and C into the Mie–Grüneisen equation, and solution for $p^{(H_2)}(v)$ gives the second-shock Hugoniot

$$p^{(H2)}(v) = \frac{p^{(H)}(v)\left[1 - \frac{\gamma(v)}{2v}(v_R - v)\right] + \frac{\gamma(v)}{2v} p^+ (v_R - v)}{1 - \frac{\gamma(v)}{2v}(v^+ - v)}. \quad \text{(D)}$$

An example of a second-shock Hugoniot is shown in the following figure.

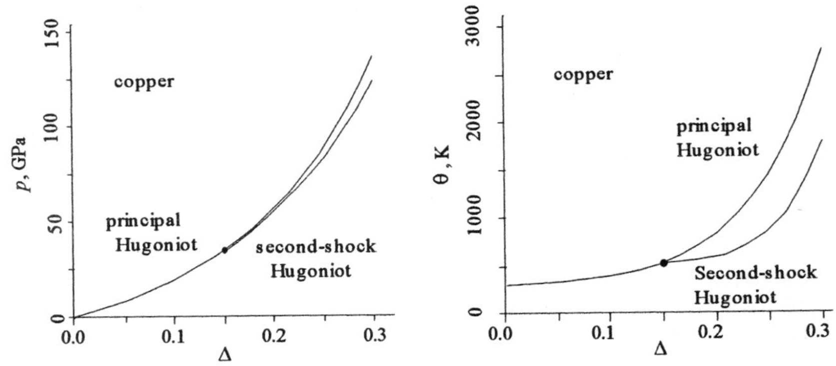

Figure: Exercise 5.7.12. Hugoniots for copper. The principal Hugoniots are centered on the reference state $\rho_R = 8930\,\text{kg/m}^3$, $\theta_R = 293\,\text{K}$ and the second-shock Hugoniots are centered on the principal Hugoniot at the point of 15% compression.

In Chap. 3 we calculated shock interactions on the premise that a second-shock $p - \dot{x}$ Hugoniot centered on a state of moderate compression differed little from the principal Hugoniot through this state. With the results of this exercise we can now address the accuracy of this approximation. To do this, we begin by calculating the $p - \dot{x}$ Hugoniots. The principal $p - \dot{x}$ Hugoniot is calculated in the usual way using Eq. 3.13_2. The second-shock $p - v$ Hugoniot is given by Eq. D. Substitution of the value v associated with one of these points into the jump equation $p = p^+ + (\dot{x} - \dot{x}^+)^2/(v^+ - v)$ and solution for $\dot{x}$ gives a point on the second-shock $p - \dot{x}$ Hugoniot. The principal and second-shock Hugoniots on a graph made in this way lie so close together that they are not easily separated on a plot of a size that can be included in this book. As an example, the second-shock Hugoniot for copper centered on a state of 10% compression falls below the principal $p - \dot{x}$ Hugoniot by about 3.5% when extended to 20% compression.

Chapter 6. Material Response III: Elastic Solids

Exercise 6.4.1. We have $(v/v_R)^{1/3} = 1 - \frac{1}{3}[1 - (v/v_R)] + \cdots$, as can be seen by calculating the cube of each side of the equation. An expansion accurate to within an error proportional to $[1-(v/v_R)]^3$ is obtained by adding the term $-\frac{1}{9}[1-(v/v_R)]^2$ to the previous result.

Exercise 6.4.2. The expansion is

$$\hat{\varepsilon}(\mathbf{E},\eta)=\hat{\varepsilon}(\mathbf{0},\eta_R)+\left.\frac{\partial\hat{\varepsilon}}{\partial E_{IJ}}\right|_R E_{IJ}+\left.\frac{\partial\hat{\varepsilon}}{\partial\eta}\right|_R(\eta-\eta_R)+\frac{1}{2}\left.\frac{\partial^2\hat{\varepsilon}}{\partial E_{IJ}\,\partial E_{KL}}\right|_R E_{IJ}E_{KL}$$

$$+\frac{1}{2}\left.\frac{\partial^2\hat{\varepsilon}}{\partial E_{IJ}\,\partial\eta}\right|_R E_{IJ}(\eta-\eta_R)+\frac{1}{2}\left.\frac{\partial^2\hat{\varepsilon}}{\partial\eta^2}\right|_R(\eta-\eta_R)^2$$

$$+\frac{1}{6}\left.\frac{\partial^3\hat{\varepsilon}}{\partial E_{IJ}\,\partial E_{KL}\,\partial E_{MN}}\right|_R E_{IJ}E_{KL}E_{MN}+\frac{1}{2}\left.\frac{\partial^3\hat{\varepsilon}}{\partial E_{IJ}\,\partial\eta^2}\right|_R E_{IJ}(\eta-\eta_R)^2$$

$$+\frac{1}{2}\left.\frac{\partial^3\hat{\varepsilon}}{\partial E_{IJ}\,\partial E_{KL}\,\partial\eta}\right|_R E_{IJ}E_{KL}(\eta-\eta_R)+\frac{1}{6}\left.\frac{\partial^3\hat{\varepsilon}}{\partial\eta^3}\right|_R(\eta-\eta_R)^3+\cdots.$$

The specific internal energy in the reference state, $\hat{\varepsilon}(\mathbf{0},\eta_R)$, is designated ε_R and the derivative $\partial\hat{\varepsilon}/\partial E_{IJ}$ is the stress T_{IJ}, which vanishes in the reference state. The derivative $\partial\hat{\varepsilon}/\partial\eta$ is the temperature, which is designated θ_R in the reference state. The second- and third-order elastic constants are defined on the third line of Eq. 6.6. Grüneisen's tensor is defined by the equation [98, Eq. 10.40]:

$$\gamma_{IJ}=-\frac{1}{\theta}\frac{\partial\theta(\mathbf{E},\eta)}{\partial E_{IJ}},$$

so we have

$$\left.\frac{\partial^2\hat{\varepsilon}}{\partial E_{IJ}\partial\eta}\right|_R=-\theta_R\,\gamma_{IJ}(\mathbf{0},0).$$

We also have [98, Eq. 10.5]

$$\frac{\partial^2\varepsilon(\mathbf{E},\eta)}{\partial\eta^2}=\frac{\partial\theta(\mathbf{E},\eta)}{\partial\eta}=\frac{\theta}{C^{\mathbf{E}}}.$$

For the derivative $\partial^3\hat{\varepsilon}(\mathbf{E},\eta)/\partial E_{IJ}\,\partial\eta^2$ we have

$$\frac{\partial^3\hat{\varepsilon}(\mathbf{E},\eta)}{\partial E_{IJ}\,\partial\eta^2}=\frac{1}{\rho_R}\frac{\partial^2 T_{IJ}(\mathbf{E},\eta)}{\partial\eta^2}=-\frac{\partial(\theta\,\gamma_{IJ})}{\partial\eta}=-\frac{\theta}{C^{\mathbf{E}}}\gamma_{IJ}-\theta\frac{\partial\gamma_{IJ}(\mathbf{E},\eta)}{\partial\eta}.$$

We shall assume that γ_{IJ} is a function of $\mathbf{E}$ alone, so that

$$\frac{\partial^3 \hat{\varepsilon}(\mathbf{E}, \eta)}{\partial E_{IJ}\, \partial \eta^2} = -\frac{\theta}{C^{\mathbf{E}}} \gamma_{IJ} .$$

Similar analyses lead to the forms given for the other two derivatives.

Exercise 6.4.3. The reason that one might prefer coefficients that involve temperature derivatives rather than entropy derivatives is that the former are more easily measured than the latter. The derivatives in question are

$$\frac{\partial C^{\eta}_{IJKL}(\mathbf{E}, \eta)}{\partial \eta} \quad \text{and} \quad \frac{dC^{\mathbf{E}}(\eta)}{d\eta} .$$

We have

$$\frac{\partial C^{\eta}_{IJKL}(\mathbf{E}, \eta)}{\partial \eta} = \frac{\partial C^{\eta}_{IJKL}(\mathbf{E}, \theta)}{\partial \theta} \frac{\partial \theta(\mathbf{E}, \eta)}{\partial \eta} = \frac{\theta}{C^{\mathbf{E}}} \frac{\partial C^{\eta}_{IJKL}(\mathbf{E}, \theta)}{\partial \theta}$$

and

$$\frac{dC^{\mathbf{E}}(\eta)}{d\eta} = \frac{\partial C^{\mathbf{E}}(\mathbf{E}, \theta)}{\partial \theta} \frac{\partial \theta(\mathbf{E}, \eta)}{\partial \eta} = \frac{\theta}{C^{\mathbf{E}}} \frac{\partial C^{\mathbf{E}}(\mathbf{E}, \theta)}{\partial \theta} .$$

Since the coefficients depend on the values of these derivatives at the reference state their use does not introduce θ as a variable in the equation of state.

Exercise 6.4.4. For uniaxial strain we have $\mathbf{F} = \mathrm{diag}\, \| F_{11}, 1, 1 \|$. This can be uniquely decomposed into the factors $\mathbf{F} = \mathbf{R}\,\mathbf{U}$, where $\mathbf{R}$ is orthogonal and $\mathbf{U}$ is symmetric and positive-definite. Since $\mathbf{F}$ is symmetric and positive-definite we see that $\mathbf{R}$ is the identity transformation.

Exercise 6.4.5. The required equation is obtained by making the substitutions indicated in the text, neglecting quadratic terms in $\tilde{\mathbf{E}}^{s}$, and use of the relation $\mathbf{R}^{-1} = \mathbf{R}^{T}$.

Exercise 6.4.6. Two stress–deformation relations that are important in analysis of longitudinal wave propagation are the Hugoniot and the isentrope. The specific issue addressed in this exercise is the degree to which these curves differ and, in particular, the jump in entropy that occurs across a shock.

The Rankine–Hugoniot jump condition 2.24 shows that the equation

$$H(-t_{11}, v) \equiv [\![\varepsilon]\!] + \tfrac{1}{2}(-t_{11}^{+} - t_{11}^{-})\, [\![v]\!] = 0 \qquad \text{(A)}$$

must be satisfied along a Hugoniot curve centered on $\mathcal{S}^{-} = \{\rho^{-}, t_{11}^{-}, \varepsilon^{-}, \dot{x}^{-}\}$. From the internal energy response function $\varepsilon = \hat{\varepsilon}(\mathbf{F}, \eta)$ we obtain

$$d\varepsilon = \frac{\partial \hat{\varepsilon}}{\partial F_{iJ}} dF_{iJ} + \frac{\partial \hat{\varepsilon}}{\partial \eta} d\eta = \frac{1}{\rho} \overset{-1}{F}_{Jj} t_{ji} \, dF_{iJ} + \theta \, d\eta \, .$$

For uniaxial motion, this equation takes the form

$$d\varepsilon = \theta \, d\eta + \frac{1}{\rho_R} t_{11} \, dF_{11} \; . \tag{B}$$

Because the Rankine–Hugoniot equation is written in terms of v it proves convenient to consider the internal energy response to be a function of this variable as well ($F_{11} = \rho_R v$, $E_{11} = (\rho_R^2 v^2 - 1)/2$), in which case Eq. B acquires the familiar form

$$d\varepsilon = \theta \, d\eta + t_{11} \, dv \, ,$$

or,

$$\frac{d\varepsilon}{dv} = t_{11} + \theta \frac{d\eta}{dv} . \tag{C}$$

Differentiating the Hugoniot of Eq. A and substituting from Eq. C we obtain

$$\begin{aligned} 0 = \frac{dH}{dv} &= \frac{d\varepsilon}{dv} + \frac{1}{2}(v - v^-) \frac{d(-t_{11})}{dv} + \frac{1}{2}(-t_{11} - t_{11}^-) \\ &= \theta \frac{d\eta}{dv} + \frac{1}{2}(t_{11} - t_{11}^-) + \frac{1}{2}(v - v^-) \frac{d(-t_{11})}{dv} . \end{aligned} \tag{D}$$

Evaluating this derivative at the center point S^-, we find that

$$\left. \frac{d\eta}{dv} \right|_{S^-} = 0 \, , \tag{E}$$

i.e., the rate of change of entropy with respect to specific volume along the Hugoniot at the center point is zero.

Differentiating Eq. D, we obtain

$$0 = \frac{d\theta}{d\eta} \frac{d\eta}{dv} + \theta \frac{d^2\eta}{dv^2} + \frac{1}{2}(v - v^-) \frac{d^2(-t_{11})}{dv^2} . \tag{F}$$

Evaluation of this result at the center point and using Eq. E yields

$$\left. \frac{d^2\eta}{dv^2} \right|_{S^-} = 0 \, . \tag{G}$$

Finally, let us take still another derivative. From Eq. F we obtain

$$0 = \frac{d^2\theta}{d\eta^2} \frac{d\eta}{dv} + 2 \frac{d\theta}{d\eta} \frac{d^2\eta}{dv^2} + \theta \frac{d^3\eta}{dv^3} + \frac{1}{2} \frac{d^2(-t_{11})}{dv^2} + \frac{1}{2}(v - v^-) \frac{d^3(-t_{11})}{dv^3} ,$$

which yields

$$\left.\frac{d^3\eta}{dv^3}\right|_{S^-} = -\frac{1}{2\theta^-}\left.\frac{d^2(-t_{11})}{dv^2}\right|_{S^-} . \tag{H}$$

The expanded form of the function $\eta(v)$ giving the value of the entropy along the Hugoniot is

$$\eta(v) = \eta(v^-) + \left.\frac{d\eta}{dv}\right|_{S^-} [\![v]\!] + \left.\frac{d^2\eta}{dv^2}\right|_{S^-} [\![v]\!]^2 + \left.\frac{d^3\eta}{dv^3}\right|_{S^-} [\![v]\!]^3 + \cdots,$$

but substitution of Eqs. E, G, and H into this expression shows that the jump in entropy across a shock is given by

$$[\![\eta]\!] = \frac{1}{12\theta^-}\left.\frac{\partial^2 t_{11}}{\partial v^2}\right|_{S^-} [\![v]\!]^3 + \cdots, \tag{I}$$

to within terms of fourth order in $[\![v]\!]$.

Let us consider expansion of the stress and temperature response functions in a neighborhood of a point (v^-, η^-):

$$\begin{aligned} t_{11} &= \hat{t}_{11}(v^-,\eta^-) + \left.\frac{\partial \hat{t}_{11}}{\partial v}\right|_{S^-} \Delta v + \left.\frac{\partial \hat{t}_{11}}{\partial \eta}\right|_{S^-} \Delta\eta + \frac{1}{2}\left.\frac{\partial^2 \hat{t}_{11}}{\partial v^2}\right|_{S^-} (\Delta v)^2 \\ &+ \left.\frac{\partial^2 \hat{t}_{11}}{\partial v\,\partial \eta}\right|_{S^-} \Delta v\Delta\eta + \frac{1}{2}\left.\frac{\partial^2 \hat{t}_{11}}{\partial \eta^2}\right|_{S^-} (\Delta\eta)^2 + \frac{1}{6}\left.\frac{\partial^3 \hat{t}_{11}}{\partial v^3}\right|_{S^-} (\Delta v)^3 \\ &+ \frac{1}{2}\left.\frac{\partial^3 \hat{t}_{11}}{\partial v^2\partial \eta}\right|_{S^-} (\Delta v)^2\Delta\eta + \frac{1}{2}\left.\frac{\partial^3 \hat{t}_{11}}{\partial v\partial \eta^2}\right|_{S^-} \Delta v(\Delta\eta)^2 + \frac{1}{6}\left.\frac{\partial^3 \hat{t}_{11}}{\partial \eta^3}\right|_{S^-} (\Delta\eta)^3 + \cdots \end{aligned}$$

$$\text{(J)}$$

$$\begin{aligned} \theta &= \hat{\theta}(v^-, \eta^-) + \left.\frac{\partial \hat{\theta}}{\partial v}\right|_{S^-} \Delta v + \left.\frac{\partial \hat{\theta}}{\partial \eta}\right|_{S^-} \Delta\eta + \frac{1}{2}\left.\frac{\partial^2 \hat{\theta}}{\partial v^2}\right|_{S^-} (\Delta v)^2 \\ &+ \left.\frac{\partial^2 \hat{\theta}}{\partial v\,\partial \eta}\right|_{S^-} \Delta v\,\Delta\eta + \frac{1}{2}\left.\frac{\partial^2 \hat{\theta}}{\partial \eta^2}\right|_{S^-} (\Delta\eta)^2 + \frac{1}{6}\left.\frac{\partial^3 \hat{\theta}}{\partial v^3}\right|_{S^-} (\Delta v)^3 \\ &+ \frac{1}{2}\left.\frac{\partial^3 \hat{\theta}}{\partial v^2\partial \eta}\right|_{S^-} (\Delta v)^2\Delta\eta + \frac{1}{2}\left.\frac{\partial^3 \hat{\theta}}{\partial v\,\partial \eta^2}\right|_{S^-} \Delta v(\Delta\eta)^2 + \frac{1}{6}\left.\frac{\partial^3 \hat{\theta}}{\partial \eta^3}\right|_{S^-} (\Delta\eta)^3 + \cdots, \end{aligned}$$

where $\Delta v \equiv v - v^-$, $\Delta\eta \equiv \eta - \eta^-$ and all of the derivatives are to be evaluated at the point (v^-, η^-).

Along the isentrope $\eta = \eta^-$, so the terms in $\Delta\eta \equiv \eta - \eta^-$ vanish and values of stress and temperature are given by the equations

$$t_{11} = t_{11}^{(\eta)}(v) = \hat{t}_{11}(v^-, \eta^-) + \left.\frac{\partial \hat{t}_{11}}{\partial v}\right|_{S^-} \Delta v + \frac{1}{2}\left.\frac{\partial^2 \hat{t}_{11}}{\partial v^2}\right|_{S^-} (\Delta v)^2 + \frac{1}{6}\left.\frac{\partial^3 \hat{t}_{11}}{\partial v^3}\right|_{S^-} (\Delta v)^3 + \cdots$$

(K)

$$\theta = \theta^{(\eta)}(v) = \hat{\theta}(v^-, \eta^-) + \left.\frac{\partial \hat{\theta}}{\partial v}\right|_{S^-} \Delta v + \frac{1}{2}\left.\frac{\partial^2 \hat{\theta}}{\partial v^2}\right|_{S^-} (\Delta v)^2 + \frac{1}{6}\left.\frac{\partial^3 \hat{\theta}}{\partial v^3}\right|_{S^-} (\Delta v)^3 + \cdots .$$

For the special process of a *shock transition* we know from Eq. I that $\eta - \eta^-$ is of the order $(v - v^-)^3$, so we can write t_{11} and θ in the forms

$$t_{11} = t_{11}^{(H)}(v, \eta) = \hat{t}_{11}(v^-, \eta^-) + \left.\frac{\partial \hat{t}_{11}}{\partial v}\right|_{S^-} [\![v]\!] + \frac{1}{2}\left.\frac{\partial^2 \hat{t}_{11}}{\partial v^2}\right|_{S^-} [\![v]\!]^2$$
$$+ \frac{1}{6}\left.\frac{\partial^3 \hat{t}_{11}}{\partial v^3}\right|_{S^-} [\![v]\!]^3 + \left.\frac{\partial \hat{t}_{11}}{\partial \eta}\right|_{S^-} [\![\eta]\!] + \cdots \qquad (L_a)$$

and

$$\theta = \theta^{(H)}(v, \eta) = \hat{\theta}(v^-, \eta^-) + \left.\frac{\partial \hat{\theta}}{\partial v}\right|_{S^-} [\![v]\!] + \frac{1}{2}\left.\frac{\partial^2 \hat{\theta}}{\partial v^2}\right|_{S^-} [\![v]\!]^2$$
$$+ \frac{1}{6}\left.\frac{\partial^3 \hat{\theta}}{\partial v^3}\right|_{S^-} [\![v]\!]^3 + \left.\frac{\partial \hat{\theta}}{\partial \eta}\right|_{S^-} [\![\eta]\!] + \cdots \qquad (L_b)$$

that are accurate to third order in $v - v^-$.

Comparison of Eqs. K and L_b yields the Hugoniot in the form

$$t_{11} = t_{11}^{(H)}(v) = t_{11}^{(\eta)}(v) + \left.\frac{\partial \hat{t}_{11}}{\partial \eta}\right|_{S^-} [\![\eta]\!] + \cdots = t_{11}^{(I)}(v) - \rho_R^2\, \gamma_{11}\, v^-\, \theta^-\, [\![\eta]\!] + \cdots$$

(M)

$$\theta = \theta^{(H)}(v) = \theta^{(\eta)}(v) + \left.\frac{\partial \hat{\theta}}{\partial \eta}\right|_{S^-} [\![\eta]\!] + \cdots = \theta^{(I)}(v) + \frac{\theta^-}{C_E} [\![\eta]\!] + \cdots,$$

where we have used the equations

$$\gamma_{11} \equiv -\frac{1}{\theta^-}\left.\frac{\partial^2 \hat{\varepsilon}}{\partial E_{11}\, \partial \eta}\right|_{S^-} = -\frac{1}{\rho_R^2 v^- \theta^-}\left.\frac{\partial \hat{t}_{11}}{\partial \eta}\right|_{S^-}$$

(N)

$$\frac{1}{C_E} \equiv \frac{1}{\theta^-}\left.\frac{\partial \hat{\theta}}{\partial \eta}\right|_{S^-}$$

to express the derivatives $\partial t_{11}/\partial \eta$ and $\partial \theta/\partial \eta$ in terms of Grüneisen's coefficient, γ_{11}, which we shall henceforth write simply γ, and C_E, the specific heat at constant strain.

Because the more readily available data are specific heats at constant pressure, we note the conversion

$$C_E = \frac{(C_T)^2}{C_T + (3\alpha)^2 B^\eta \rho^- \theta^-}, \tag{O}$$

where C_E and C_T are specific heats at constant strain and stress, respectively, $\alpha \equiv \alpha_{11}$ is the coefficient of linear thermal expansion, and B^η is the isentropic bulk modulus. When the stress and strain vanish, C_T and C_E correspond to specific heats at constant pressure and volume. Thermodynamic relations such as that of Eq. O are discussed by Thurston [98].

A variety of equations have been proposed for the dependence of γ on v. One can, of course, assume that γ is a constant. The assumption that $\rho\gamma$ is constant is also widely adopted, partly because it is a reasonable approximation and partly because it simplifies some calculations. For the present, we shall assume $\gamma = \gamma_R$, a constant, or $\rho\gamma = \rho_R \gamma_R$.

Substitution of Eq. I into Eqs. M yields equations for the $t_{11} - v$ and $\theta - v$ Hugoniots:

$$\begin{aligned} -t_{11}^{(H)}(v) &= -t_{11}^{(\eta)}(v) + \frac{1}{12}\rho_R^2 \gamma v^- \left.\frac{\partial^2 t_{11}}{\partial v^2}\right|_{S^-} [\![v]\!]^3 + \cdots \\ \theta^{(H)}(v) &= \theta^{(\eta)}(v) + \frac{1}{12 C_E} \left.\frac{\partial^2 t_{11}}{\partial v^2}\right|_{S^-} [\![v]\!]^3 + \cdots. \end{aligned} \tag{P}$$

For normal materials, those in which a compression shock is stable, $\partial^2 t_{11}/\partial v^2 < 0$, so we see that, in the $-t_{11} - v$ and $\theta - v$ planes, the Hugoniot lies above the isentrope in the direction of further compression and below the isentrope in the direction of decompression.

The results of this section show that the entropic effects of a weak shock are quite small. Indeed, the slope and curvature of the Hugoniot and the isentrope are the same at the center point; the curves differ only in the third derivative. This observation is often used to justify the use of a Hugoniot in an equation calling for an isentrope and vice versa. It may also justify approximating a recentered Hugoniot by the continuation of the principal Hugoniot through the new center point. Loosely stated, one may use a purely mechanical theory (i.e. one neglecting thermal effects) when the compression is not too great, as is usually the case for materials deformed within the elastic range

When the initial state S^- is the reference state, expressions in terms of derivatives such as $\partial t_{11}/\partial v$ that appear in the foregoing equations can be restated in terms of isentropic elastic constants such as those of Eqs. 6.5 by use of the relations

$$
\begin{aligned}
\hat{t}_{11}\big|_{S^-} &= 0 \\
\left.\frac{\partial \hat{t}_{11}}{\partial v}\right|_{S^-} &= \rho_R\, C_{11}^{\eta} \\
\left.\frac{\partial^2 \hat{t}_{11}}{\partial v^2}\right|_{S^-} &= \rho_R^2\, (3C_{11}^{\eta} + C_{111}^{\eta}) \\
\left.\frac{\partial^3 \hat{t}_{11}}{\partial v^3}\right|_{S^-} &= \rho_R^3\, (3C_{11}^{\eta} + 6C_{111}^{\eta} + 4C_{1111}^{\eta}),
\end{aligned}
\tag{Q}
$$

where the $C_{IJ\cdots MN}^{\eta}$ are elastic moduli of various orders (as defined by Brugger [15]), in Voigt notation.

Example: Results for the Linear $U_S - \dot{x}$ Hugoniot. In this section we consider the application of the foregoing results to a material for which the $U_S - \dot{x}$ Hugoniot has been determined to have the linear form $U_S = C_B + S\dot{x}$ discussed in Sect. 3.4. The $t_{11} - v$ Hugoniot corresponding to the foregoing relation is

$$
t_{11}^{+} = \frac{(\rho_R C_B)^2\, (v_R - v)}{[1 - \rho_R S(v_R - v)]^2}\,. \tag{R}
$$

Using Eqs. E, G, and H, we can show that

$$
\begin{aligned}
\left.\frac{\partial \hat{t}_{11}}{\partial v}\right|_{S^-} &= \left.\frac{d\hat{t}_{11}^{(H)}}{dv}\right|_{S^-} \\
\left.\frac{\partial^2 \hat{t}_{11}}{\partial v^2}\right|_{S^-} &= \left.\frac{d^2\hat{t}_{11}^{(H)}}{dv^2}\right|_{S^-} \\
\left.\frac{\partial^3 \hat{t}_{11}}{\partial v^3}\right|_{S^-} &= \left.\frac{d^3\hat{t}_{11}^{(H)}}{dv^3}\right|_{S^-} - \frac{\gamma \rho_R}{2} \left.\frac{d^2\hat{t}_{11}^{(H)}}{dv^2}\right|_{S^-},
\end{aligned}
\tag{S}
$$

so the derivatives in Eq. K_1 can be calculated by differentiating Eq. R. We obtain

$$
t_{11}^{(I)}(v) = (\rho_R C_B)^2 [\![v]\!] \left[1 - \rho_R S [\![v]\!] + \rho_R^2 S(9S - \gamma)[\![v]\!]^2 + \cdots\right]. \tag{T}
$$

This expression is not very accurate, even at rather low compressions, because the rational function of Eq. R is not well represented by the polynomial expression resulting from the expansion. Nevertheless, it is easy to see that the thermal correction relating the Hugoniot to the isentrope is small. A more accurate result might be obtained by simply applying the correction to Eq. R, in which case, we obtain

$$t_{11}^{(\mathrm{I})}(v)=\frac{(\rho_{\mathrm{R}}C_{\mathrm{B}})^2\,(v_{\mathrm{R}}-v)}{[1-\rho_{\mathrm{R}}\,S(v_{\mathrm{R}}-v)]^2}+\frac{1}{3}\gamma\,\rho_{\mathrm{R}}^4\,C_{\mathrm{B}}^2S(v_{\mathrm{R}}-v)^3+\cdots\,. \tag{U}$$

Chapter 7. Material Response IV: Elastic–Plastic and Elastic–Viscoplastic Solids

Exercise 7.4.1. For the problem at hand, $\mathbf{t}=\mathrm{diag}\,\|\,t_{11}, t_{22}, t_{22}\,\|$. Since the maximum shear stress is the same on all planes inclined at a given angle to the axis it is sufficient to consider only the plane having the normal vector $\mathbf{n}=(\cos\varphi, \sin\varphi, 0)$. The stress vector on this plane has components $t_j^{(\mathbf{n})}=t_{ij}n_i$ or, $\mathbf{t}^{(\mathbf{n})}=(t_{11}\cos\varphi, t_{22}\sin\varphi, 0)$ and the component of this stress that is normal to the plane is

$$\mathbf{t}^{(\mathbf{nn})}=(\mathbf{t}^{(\mathbf{n})}\cdot\mathbf{n})\mathbf{n}=(t_{11}\cos\varphi, t_{22}\sin\varphi, 0)=(t_{11}\cos\varphi+t_{22}\sin\varphi)(\cos\varphi, \sin\varphi, 0)\,.$$

The shear stress on this plane is $\mathbf{t}^{(\mathbf{nt})}=\mathbf{t}^{(\mathbf{n})}-\mathbf{t}^{(\mathbf{nn})}$, or

$$\mathbf{t}^{(\mathbf{nt})}=(t_{11}-t_{22})\sin\varphi\cos\varphi(\sin\varphi, -\cos\varphi, 0)\,,$$

and the magnitude of this stress is

$$|\mathbf{t}^{(\mathbf{nt})}|=(t_{11}-t_{22})\sin\varphi\cos\varphi\,. \tag{A}$$

The maximum of this quantity occurs when $d|\mathbf{t}^{(\mathbf{nt})}|/d\varphi=0$, which occurs when $\varphi=45°$. We find from Eq. A that $|\mathbf{t}^{(\mathbf{nt})}|_{\max}=\frac{1}{2}|(t_{11}-t_{22})|$.

Exercise 7.4.2. *Elastic Range.* The elastic strain is of the form $\widetilde{\mathbf{E}}=\mathrm{diag}\,(\widetilde{E}_{11}^{\mathrm{e}}, \widetilde{E}_{22}^{\mathrm{e}}, \widetilde{E}_{22}^{\mathrm{e}})$ and, using the linear stress relation, we have $t_{11}=(\lambda_{\mathrm{R}}+2\mu_{\mathrm{R}})\widetilde{E}_{11}^{\mathrm{e}}+2\lambda_{\mathrm{R}}\widetilde{E}_{22}^{\mathrm{e}}$ and $t_{22}=\lambda_{\mathrm{R}}\widetilde{E}_{11}^{\mathrm{e}}+2(\lambda_{\mathrm{R}}+\mu_{\mathrm{R}})\widetilde{E}_{22}^{\mathrm{e}}=0$. From the latter equation, we find that $\widetilde{E}_{22}^{\mathrm{e}}=-\{\lambda_{\mathrm{R}}/[2(\lambda_{\mathrm{R}}+\mu_{\mathrm{R}})]\}\widetilde{E}_{11}^{\mathrm{e}}$, and from this we find that $t_{11}=[(3\lambda_{\mathrm{R}}+2\mu_{\mathrm{R}})\mu_{\mathrm{R}}/(\lambda_{\mathrm{R}}+\mu_{\mathrm{R}})]\widetilde{E}_{11}^{\mathrm{e}}$ (Note: The coefficient $(3\lambda_{\mathrm{R}}+2\mu_{\mathrm{R}})\mu_{\mathrm{R}}/(\lambda_{\mathrm{R}}+\mu_{\mathrm{R}})$ is called *Young's modulus*). The limit of the elastic range occurs at $t_{11}=\chi Y$, or $\widetilde{E}_{11}^{\mathrm{e}}=[(\lambda_{\mathrm{R}}+\mu_{\mathrm{R}})/[\mu_{\mathrm{R}}(3\lambda_{\mathrm{R}}+2\mu_{\mathrm{R}})]\chi Y$ and $\widetilde{E}_{22}^{\mathrm{e}}=-\{\lambda_{\mathrm{R}}/[2\mu_{\mathrm{R}}(3\lambda_{\mathrm{R}}+2\mu_{\mathrm{R}})]\}\chi Y$. The pressure is $p=-\frac{1}{3}t_{kk}=-\frac{1}{3}t_{11}$ and the stress deviator is $\mathbf{t}'=\mathbf{t}+p\mathbf{1}=\mathrm{diag}\,\|\,t_{11}+p, p, p\,\|=\frac{1}{3}t_{11}\,\mathrm{diag}\,\|\,2,-1.-1\,\|$. The spherical part of the strain is $\vartheta=\frac{1}{3}\widetilde{E}_{kk}^{\mathrm{e}}=\frac{1}{3}(\widetilde{E}_{11}^{\mathrm{e}}+2\widetilde{E}_{22}^{\mathrm{e}})$ and the strain deviator is $\widetilde{\mathbf{E}}'^{\mathrm{e}}=\widetilde{\mathbf{E}}^{\mathrm{e}}-\vartheta\mathbf{1}=\frac{1}{3}(\widetilde{E}_{11}^{\mathrm{e}}-\widetilde{E}_{22}^{\mathrm{e}})\,\mathrm{diag}\,\|\,2,-1,-1\,\|$. Using the foregoing equations we can rewrite ϑ in the form $\vartheta=[\mu_{\mathrm{R}}/(3(\lambda_{\mathrm{R}}+\mu_{\mathrm{R}})]\widetilde{E}_{11}^{\mathrm{e}}$. From these equations we find that $p=-(3\lambda_{\mathrm{R}}+2\mu_{\mathrm{R}})\vartheta$, showing that the pressure, the spherical part of the stress, is proportional to the spherical part of the strain. The coefficient $\lambda_{\mathrm{R}}+\frac{2}{3}\mu_{\mathrm{R}}$ is called the *bulk modulus.* The stress deviator is $\mathbf{t}'=\mathbf{t}+p\mathbf{1}=-p\,\mathrm{diag}\,\|\,2,-1,-1\,\|$. We can write the stress component t_{11} as the sum of pressure and shear terms: $t_{11} = -p + t'_{11} = -(3\lambda_R + 2\mu_R)\vartheta + 2\mu_R \widetilde{E}'^{e}_{11}$. The maximum shear stress is $\tau_{45°} = \frac{1}{2}t_{11}$ and, as suggested by the notation, occurs on planes inclined at 45° to the axis of the rod.

Plastic Range. In the plastic range, $\widetilde{\mathbf{E}} = \widetilde{\mathbf{E}}^{e} + \widetilde{\mathbf{E}}^{p}$, with $\widetilde{E}^{p}_{11} + 2\widetilde{E}^{p}_{22} = 0$ because the plastic part of the strain is isochoric. Since the stress satisfies the yield condition throughout the range of plastic deformation, we have the elastic strains $\widetilde{E}^{e}_{11} = \{(\lambda_R + \mu_R)/[\mu_R(3\lambda_R + 2\mu_R)]\}\chi Y$ and $\widetilde{E}^{e}_{22} = -\{\lambda_R/[2\mu_R(3\lambda_R + 2\mu_R)]\}\chi Y$. The plastic strain components are $\widetilde{E}^{p}_{11} = \widetilde{E}_{11} - \widetilde{E}^{e}_{11}$ and $\widetilde{E}^{p}_{22} = -\frac{1}{2}\widetilde{E}^{p}_{11}$. For non-hardening materials Y is constant and $\widetilde{E}_{11}$ is specified as a boundary condition. for hardening materials Y increases as the rod is deformed and either the strain or the stress can be specified.

Exercise 7.4.3. The vector $\mathbf{n}^{(k)}$ has the components

$$n_i^{(k)} = (v/v_R)^{2/3} N_\Gamma^{(k)} \delta_{\Gamma\alpha} \overset{-1}{F}{}^{es}_{\alpha i}\,, \qquad \text{(A)}$$

as given by Eq. 7.106. The squared magnitude of this vector is

$$n_i^{(k)} n_i^{(k)} = (v/v_R)^{4/3} N_\Gamma^{(k)} N_\Delta^{(k)} \delta_{\Gamma\alpha}\delta_{\Delta\beta} \overset{-1}{F}{}^{es}_{\alpha i} \overset{-1}{F}{}^{es}_{\beta i}\,. \qquad \text{(B)}$$

We can write

$$F^{es}_{i\alpha} = R_{i\beta} \widetilde{U}^{es}_{\beta\alpha}\,, \qquad \text{(C)}$$

when $\mathbf{U}^{es}$ is small and, as with Eqs. 6.53 and 6.55,

$$\widetilde{U}^{es}_{\alpha\beta} = \delta_{\alpha\beta} + \widetilde{E}^{es}_{\alpha\beta} \quad \text{and} \quad F^{es}_{i\alpha} = R_{i\alpha} + R_{i\gamma}\widetilde{E}^{es}_{\gamma\alpha}\,. \qquad \text{(D)}$$

We also need the equation

$$\overset{-1}{F}{}^{es}_{\alpha i} = \overset{-1}{\widetilde{U}}{}^{es}_{\alpha\gamma} \overset{-1}{R}_{\gamma i}\,. \qquad \text{(E)}$$

From Eq. D we see that

$$\overset{-1}{\widetilde{U}}{}^{es}_{\alpha\beta} = \delta_{\alpha\beta} - \widetilde{E}^{es}_{\alpha\beta} \qquad \text{(F)}$$

to first order in $\widetilde{\mathbf{E}}^{es}$. With this,

$$\overset{-1}{F}{}^{es}_{\alpha i} = \overset{-1}{R}_{\alpha i} - \widetilde{E}^{es}_{\alpha\gamma} \overset{-1}{R}_{\gamma i}\,. \qquad \text{(G)}$$

Substitution of Eq. G into Eq. B yields the result

$$\begin{aligned} n_i^{(k)} n_i^{(k)} &= (v/v_R)^{4/3} N_\Gamma^{(k)} N_\Delta^{(k)} \delta_{\Gamma\alpha}\delta_{\Delta\beta} (\overset{-1}{R}_{\alpha i} - \widetilde{E}^{es}_{\alpha\gamma} \overset{-1}{R}_{\gamma i})(\overset{-1}{R}_{\beta i} - \widetilde{E}^{es}_{\beta\delta} \overset{-1}{R}_{\delta i}) \\ &= (v/v_R)^{4/3} N_\Gamma^{(k)} N_\Delta^{(k)} \delta_{\Gamma\alpha}\delta_{\Delta\beta} (\delta_{\alpha\beta} - 2\widetilde{E}^{es}_{\alpha\beta})\,, \end{aligned}$$

and the square root of this is the magnitude of $\mathbf{n}^{(k)}$ that, to first order, is

$$n^{(k)} = (v/v_{\mathrm{R}})^{2/3} N^{(k)} \left[1 - (N^{(k)})^{-2} N_{\Gamma}^{(k)} N_{\Delta}^{(k)} \delta_{\Gamma\alpha} \delta_{\Delta\beta} \widetilde{E}_{\alpha\beta}^{\mathrm{es}} \right]. \tag{H}$$

The magnitude of $\mathbf{b}^{(k)}$ is calculated in the same way.

Exercise 7.4.4. When restricted to uniaxial motions, Eq. 7.21 takes the form of the two equations

$$\frac{\partial \widetilde{E}'_{11}}{\partial t} = \frac{1}{2\mu_{\mathrm{R}}} \frac{\partial t'_{11}}{\partial t} + \frac{3\rho_{\mathrm{R}} \dot{W}^{\mathrm{p}}}{2Y^2} t'_{11} \quad \text{and} \quad \frac{\partial \widetilde{E}'_{22}}{\partial t} = \frac{1}{2\mu_{\mathrm{R}}} \frac{\partial t'_{22}}{\partial t} + \frac{3\rho_{\mathrm{R}} \dot{W}^{\mathrm{p}}}{2Y^2} t'_{22}. \tag{A}$$

For $\dot{W}^{\mathrm{p}}$ we have

$$\rho_{\mathrm{R}} \dot{W}^{\mathrm{p}} = \frac{\partial \widetilde{E}_{ij}^{\mathrm{p}}}{\partial t} t'_{ij} = \frac{\partial \widetilde{E}_{11}^{\mathrm{p}}}{\partial t} t'_{11} + 2 \frac{\partial \widetilde{E}_{22}^{\mathrm{p}}}{\partial t} t'_{22}. \tag{B}$$

Since $\widetilde{\mathbf{E}}^{\mathrm{p}}$ and $\mathbf{t}'$ are deviator tensors we have $\mathrm{tr}\,\mathbf{E}^{\mathrm{p}} = 0$ and $\mathrm{tr}\,\mathbf{t}' = 0$ so $\widetilde{E}_{22}^{\mathrm{p}} = -\frac{1}{2} \widetilde{E}_{11}^{\mathrm{p}}$ and $t'_{22} = -\frac{1}{2} t'_{11}$ Eq. B becomes

$$\rho_{\mathrm{R}} \dot{W}^{\mathrm{p}} = \frac{3}{2} \frac{\partial \widetilde{E}_{11}^{\mathrm{p}}}{\partial t} t'_{11} = 6 \frac{\partial \widetilde{E}_{22}^{\mathrm{p}}}{\partial t} t'_{22}. \tag{C}$$

Substituting the first of Eqs. C into Eq. A_1 gives

$$\frac{\partial \widetilde{E}'_{11}}{\partial t} = \frac{1}{2\mu_{\mathrm{R}}} \frac{\partial t'_{11}}{\partial t} + \frac{9}{4Y^2} \frac{\partial \widetilde{E}'^{\mathrm{p}}_{11}}{\partial t} (t'_{11})^2, \tag{D}$$

but $\widetilde{E}'_{11} = \widetilde{E}'^{\mathrm{e}}_{11} + \widetilde{E}'^{\mathrm{p}}_{11} = \widetilde{E}'^{\mathrm{e}}_{11} + \widetilde{E}^{\mathrm{p}}_{11}$ so Eq. D becomes

$$\frac{\partial \widetilde{E}'^{\mathrm{e}}_{11}}{\partial t} - \frac{1}{2\mu_{\mathrm{R}}} \frac{\partial t'_{11}}{\partial t} = -\frac{\partial \widetilde{E}^{\mathrm{p}}_{11}}{\partial t} + \frac{9}{4Y^2} \frac{\partial \widetilde{E}^{\mathrm{p}}_{11}}{\partial t} (t'_{11})^2.$$

The left member of this equation vanishes because the stress relation is $t'_{11} = 2\mu_R \widetilde{E}'^{\mathrm{e}}_{11}$. This leaves us with the equation

$$\left[\frac{9}{4Y^2} (t'_{11})^2 - 1 \right] \frac{\partial \widetilde{E}^{\mathrm{p}}_{11}}{\partial t} = 0. \tag{E}$$

By the yield condition $(t'_{11})^2 + 2(t'_{22})^2 = \frac{2}{3} Y^2$ or $\frac{9}{4}(t'_{11})^2 = 9(t'_{22})^2 = Y^2$ and we see that Eq. E, and, therefore Eq. A_1 is satisfied identically. A similar calculation leads to the same result for Eq. A_2.

Exercise 7.4.5. The deformation is illustrated in the following figure. The plastic part of the deformation gradient is

$$\mathbf{F}^{\mathrm{p}} = \mathrm{diag} \left\| F_{\mathrm{L}}^{\mathrm{p}}, F_{\mathrm{T}}^{\mathrm{p}}, F_{\mathrm{T}}^{\mathrm{p}} \right\| = \mathrm{diag} \left\| \lambda_{\mathrm{L}}, \lambda_{\mathrm{T}}, \lambda_{\mathrm{T}} \right\|,$$

where $\lambda_{\mathrm{L}} \lambda_{\mathrm{T}}^2 = 1$ since the plastic part of the deformation is isochoric. From the figure we see that

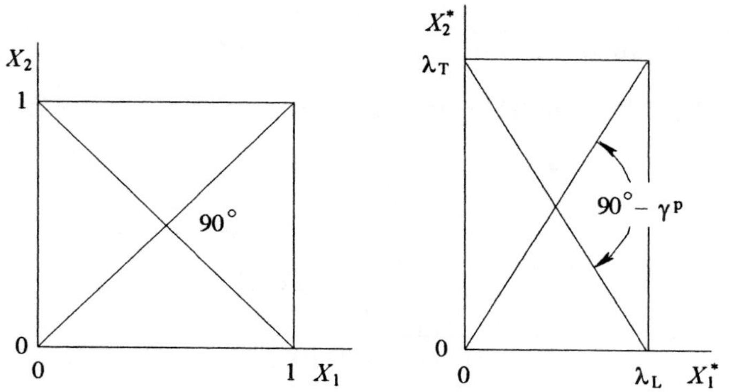

Figure: Exercise 7.4.5.

$$\tan\left[\frac{1}{2}\left(\frac{\pi}{2}-\gamma^{P}\right)\right]=\frac{\lambda_T}{\lambda_L}=(\lambda_L)^{-3/2}=(F_L^P)^{-3/2}. \tag{A}$$

We also have the trigonometric identity

$$\tan\left(\frac{\pi}{4}-\frac{\gamma^P}{2}\right)=\frac{1-\tan(\gamma^P/2)}{1+\tan(\gamma^P/2)}=(F_L^P)^{-3/2}, \tag{B}$$

where the second of these equations follows from Eq. A. Manipulation of this second equation yields the required result,

$$\gamma^P=2\arctan\left[-\frac{1-(F_L^P)^{3/2}}{1+(F_L^P)^{3/2}}\right]. \tag{C}$$

Exercise 7.4.6. *Plastic Range.* Consider the response just above the HEL. We have the hardening equation 7.81 which, for compressive deformations, can be written

$$Y=Y_0[1+h(-\gamma^P)^{1/n}], \tag{7.81}$$

where γ^P is given by Eq. 7.82_1. When Eq. 7.81 is solved for γ^P we obtain the result

$$-\gamma^P=\left[\frac{1}{h}\left(\frac{Y}{Y_0}-1\right)\right]^n. \tag{A}$$

We also have Eq. 7.79_2,

$$(F_{\rm L}^{\rm p})^{1/2} = (v/v_{\rm R})^{1/3}\left[1+\frac{Y}{6\mu_{\rm R}}\right]. \tag{7.79}$$

Differentiating Eq. 7.81 yields the result

$$\frac{dY}{dv} = -\frac{Y_0 h}{n}(-\gamma^{\rm p})^{(1/n)-1}\frac{d\gamma^{\rm p}}{dv}, \tag{B}$$

and from Eq. 7.82_1 we obtain

$$\frac{d\gamma^{p}}{dv} = \frac{3(F_{\rm L}^{\rm p})^{1/2}}{2[1+(F_{\rm L}^{\rm p})^3]}\frac{dF_{\rm L}^{\rm p}}{dv}, \tag{C}$$

so Eq. B becomes

$$\frac{dY}{dv} = -\frac{3Y_0 h}{2n}\left[\frac{1}{h}\left(\frac{Y}{Y_0}-1\right)\right]^{1-n}\frac{(F_{\rm L}^{\rm p})^{1/2}}{1+(F_{\rm L}^{\rm p})^3}\frac{dF_{\rm L}^{\rm p}}{dv}, \tag{D}$$

and, from Eq. 7.82_2 we obtain

$$\frac{dF_{\rm L}^{\rm p}}{dv} = \frac{2}{3}\frac{1}{v}F_{\rm L}^{\rm p} + \frac{1}{3\mu_{\rm R}}\left(\frac{v}{v_{\rm R}}\right)^{1/3}(F_{\rm L}^{\rm p})^{1/2}\frac{dY}{dv}. \tag{E}$$

Substitution of Eq. E into Eq. D and solution for dY/dv yields the result

$$\begin{aligned}\frac{dY}{dv} &= \frac{\dfrac{Y_0 h}{n}\dfrac{1}{v}\left[\dfrac{1}{h}\left(\dfrac{Y}{Y_0}-1\right)\right]^{1-n}\dfrac{(F_{\rm L}^{\rm p})^{3/2}}{1+(F_{\rm L}^{\rm p})^3}}{1+\dfrac{Y_0 h}{2n\mu_{\rm R}}\dfrac{F_{\rm L}^{\rm p}}{[1+(F_{\rm L}^{\rm p})^3]}\left[\dfrac{1}{h}\left(\dfrac{Y}{Y_0}-1\right)\right]^{1-n}\left(\dfrac{v}{v_{\rm R}}\right)^{1/3}} \\ &= \frac{\dfrac{Y_0 h}{n}\dfrac{1}{v}\dfrac{(F_{\rm L}^{\rm p})^{3/2}}{1+(F_{\rm L}^{\rm p})^3}}{\left[\dfrac{1}{h}\left(\dfrac{Y}{Y_0}-1\right)\right]^{n-1}+\dfrac{Y_0 h}{2n\mu_{\rm R}}\dfrac{F_{\rm L}^{\rm p}}{[1+(F_{\rm L}^{\rm p})^3]}\left(\dfrac{v}{v_{\rm R}}\right)^{1/3}}.\end{aligned} \tag{F}$$

We are interested in evaluating this equation at the HEL, where $Y = Y_0$, $F_{\rm L}^{\rm p} = 1$, and $v = v^{\rm HEL}$, in which case we have

$$\frac{dY}{dv} = -\frac{2\mu_{\rm R}}{v^{\rm HEL}}\left(\frac{v^{\rm HEL}}{v_{\rm R}}\right)^{-1/3}. \tag{G}$$

From Eq. 7.80_1 we have

$$\frac{d(-t_{11})}{dv} = p'(v) + \frac{2}{3}\frac{dY}{dv}, \tag{H}$$

and, at the HEL this becomes

$$\left.\frac{d(-t_{11})}{dv}\right|_{\text{HEL}+} = p'(v^{\text{HEL}}) - \frac{4\mu_{\text{R}}}{3\,v^{\text{HEL}}}\left(\frac{v^{\text{HEL}}}{v_{\text{R}}}\right)^{-1/3}. \tag{I}$$

Elastic Range. In the elastic range (i.e., below the HEL) we have the stress relation $-t_{11} = p(v) - \frac{4}{3}\tau(v)$, where $\tau(v) = -\mu_{\text{R}}\,[1-(v/v_{\text{R}})]$ and, with this, we have

$$\begin{aligned}\left.\frac{d(-t_{11})}{dv}\right|_{\text{HEL}-} &= p'(v^{\text{HEL}}) - \frac{4\,\mu_{\text{R}}}{3v_{\text{R}}} \\ &= p'(v^{\text{HEL}}) - \frac{4\,\mu_{\text{R}}}{3v^{\text{HEL}}}\left(\frac{v^{\text{HEL}}}{v_{\text{R}}}\right)^{-1/3}\left(\frac{v^{\text{HEL}}}{v_{\text{R}}}\right)^{4/3}.\end{aligned} \tag{J}$$

Comparison of Eqs. I and J shows that the stress–volume curve is slightly steeper immediately above the HEL than immediately below it. If we neglect the small factor that distinguishes Eq. J from Eq, I, we see that, for the hardening equation 7.81, the slope of the compression curve is continuous at the Hugoniot elastic limit. It is noteworthy that these results at the HEL are independent of the hardening parameters.

Chapter 8. Weak Elastic Waves

Exercise 8.5.1. In the absence of body force, the equation of motion 2.92_2 takes the form

$$\frac{\partial}{\partial t}(\rho_{\text{R}}\,\dot{x}) + \frac{\partial}{\partial X}(-t_{11}) = 0\,,$$

and the stress is given by Eq. 6.17_1, which can be written

$$t_{11} = (\lambda_{\text{E}} + 2\,\mu_{\text{E}})\,U_X$$

for isentropic motions. According to Eq. 2.29 we have $\dot{x} = U_t(X,t)$. Substituting into the equation of motion gives

$$U_{tt} = \frac{1}{\rho_{\text{R}}}(\lambda_{\text{E}} + 2\,\mu_{\text{E}})\,U_{XX}\,.$$

According to the definition of Eq. 8.5

$$C_0^2 = \frac{1}{\rho_R}(\lambda_E + 2\mu_E),$$

and we see that we have obtained the required result.

Exercise 8.5.2. Substitution of Eqs. 8.16 into Eq. 8.13 gives

$$U_{(1)}(X,t) = \frac{1}{2}\int_0^{X-C_0 t} [S(\xi) - V(\xi)]\,d\xi + \frac{1}{2}\int_0^{X+C_0 t} [S(\xi) - V(\xi)]\,d\xi,$$

where we have used Eq. 8.12 to show that the integration constants satisfy $U_{(1R)}(0) + U_{(1L)}(0) = 0$.

In the region $X - C_0 t \geq 0$ (with $X \geq 0$ and $t \geq 0$), we have $0 \leq X - C_0 t \leq X + C_0 t$ so the integral over the interval $[0, X + C_0 t]$ can be broken into integrals over the two intervals $[0, X - C_0 t]$ and $[X - C_0 t, X + C_0 t]$. The desired result then follows immediately.

Exercise 8.5.3. We are considering a slab of thickness X_1 that is undeformed and at rest. The solutions of Sect. 8.3.1 remain valid for this new domain for times $0 \leq t < X_1 / C_0$ prior to that at which the wave first encounters the boundary at $X = X_1$.

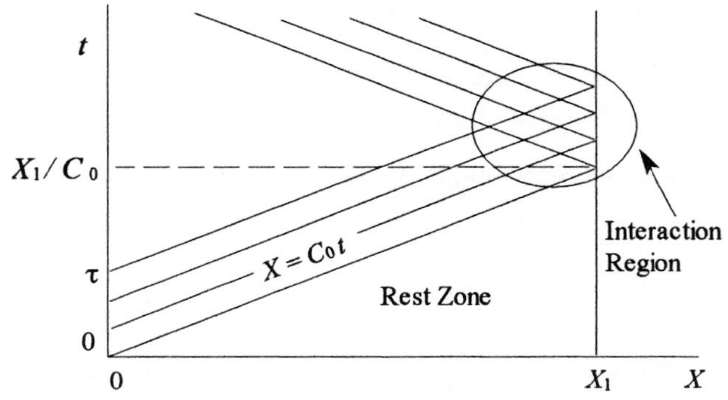

Figure: Exercise 8.5.3. Lagrangian space–time diagram of a pulse of finite duration interacting with a boundary.

The X–t diagram given above illustrates the wave field for the case in which a traction

$$t_{11} = \begin{cases} \rho_0 C_0^2 E(t), & 0 \leq t \leq \tau \\ 0, & t > \tau, \end{cases}$$

with $E(0)=E(\tau)=0$ is applied to the boundary at $X=0$. Focusing on the incident wave region, we have

$$U_1(X,t) = -C_0 \int_0^{t-(X/C_0)} E(\zeta)\, d\zeta, \quad C_0(t-\tau) \le X \le C_0 t .$$

The strain and particle velocity associated with this disturbance are given by

$$\frac{\partial U_1}{\partial X} = E\left(t - \frac{X}{C_0}\right)$$

$$\frac{\partial U_1}{\partial t} = -C_0 E\left(t - \frac{X}{C_0}\right).$$

Because the boundary at $X = X_1$ is immovable, a reflected wave must arise to cancel the motion transmitted to the boundary by the incident wave. When the incident wave encounters the boundary at $t = X_1/C_0$ the reflection process begins. Clearly, the reflected wave is a left-propagating wave having the property that it exactly offsets the motion that the incident wave would produce at X_1 if the material extended beyond the boundary at that point. If we imagine the material to extend beyond X_1 to $X = 2X_1$, then we see that a wave of the same temporal shape as the incident wave, but carrying a particle velocity of the opposite sign and propagating to the left from $X = 2X_1$ would effect the cancellation. The reflected wave U_L would therefore be such that

$$\frac{\partial U_L(X,t)}{\partial t} = +C_0 E\left(t - \frac{2X_1 - X}{C_0}\right), \quad \frac{2X_1 - X}{C_0} \le t \le \frac{2X_1 - X}{C_0} + \tau , \tag{A}$$

subject to the additional condition that t cannot exceed $2X_1/C_0$, the time at which the reflected disturbance first encounters the surface at $X = 0$.

The combined incident and reflected waves gives the velocity field

$$\frac{\partial U(X,t)}{\partial t} = -C_0 E\left(t - \frac{X}{C_0}\right) + C_0 E\left(t - \frac{2X_1 - X}{C_0}\right) \tag{B}$$

in the region where both are defined. Clearly, the particle velocity $\partial U/\partial t$ vanishes on $X = X_1$, thus satisfying the condition that the boundary does not move.

The displacement field associated with the reflected wave is obtained by integrating Eq. B. We obtain

$$U_L(X,t) = C_0 \int_0^{t-[(2X_1-X)/C_0]} E(\zeta) d\zeta, \quad 2X_1 - C_0 t \le X \le 2X_1 - C_0(t-\tau).$$

The strain and particle velocity associated with this reflected wave are given by

$$\frac{\partial U_{\mathrm{L}}}{\partial X} = E\left(t - \frac{2X_1 - X}{C_0}\right)$$
$$\frac{\partial U_{\mathrm{L}}}{\partial t} = C_0\, E\left(t - \frac{2X_1 - X}{C_0}\right).$$

The stress in the interaction region is

$$t_{11} = \frac{\partial U_1}{\partial X} + \frac{\partial U_{\mathrm{L}}}{\partial X} = \left[E\left(t - \frac{X}{C_0}\right) + E\left(t - \frac{2X_1 - X}{C_0}\right)\right], \tag{8.5}$$

and the value at the interface, $X = X_1$, is given by

$$t_{11}(X, t) = 2E\left(t - \frac{X_1}{C_0}\right), \qquad \frac{X_1}{C_0} \le t \le \frac{X_1}{C_0} + \tau.$$

Note that the wave reflection doubles the stress produced by the incident wave.

Exercise 8.5.4. The required procedure is the same as for the case of an unrestrained boundary, except that the virtual wave has the same sign as the incident wave.

Exercise 8.5.5. Let us consider the case in which a plate of thickness L that is moving to the right impacts a stack consisting of two plates, each of thickness L. All three plates are of the same material. The impact will result in application of a compressive stress of amplitude $t_{11} = -\rho_{\mathrm{R}}\, C_0\, \dot{x}_{\mathrm{p}}/2$ (where $\dot{x}_{\mathrm{p}}$ is the impact velocity) and duration $t = 2L/C_0$. The figure below shows the two target plates with the compression pulse entering at the impact interface (the projectile plate

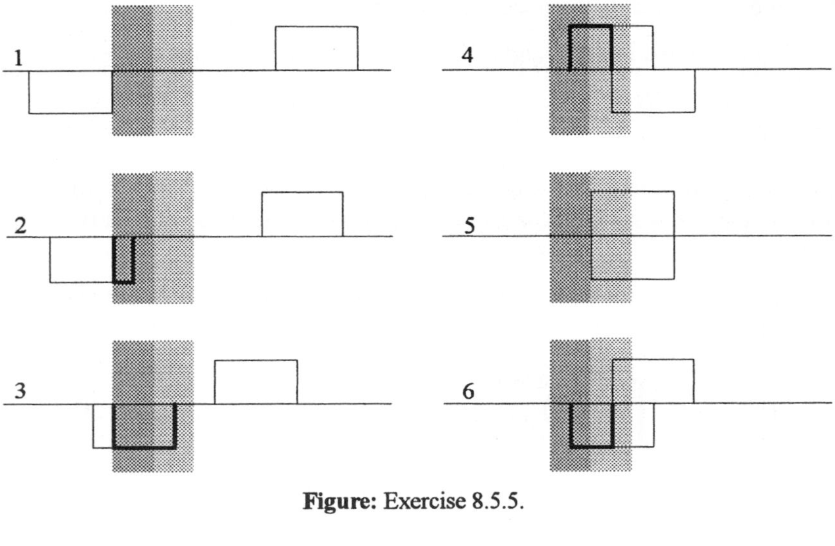

Figure: Exercise 8.5.5.

is not shown). The virtual pulse is shown to the right of the diagram; it is at the same distance from the unrestrained back face of the target as the incident pulse, is propagating to the left, and carries stress of equal magnitude but opposite sign to that of the applied stress, tensile in this case. The next two panels of the figure show the pulses as they are approaching the rear surface. The stress in the target is compressive, as shown by the heavy line. The compression pulse can propagate through the interface of the two target plates without interaction. In the fourth panel we see the two pulses in a condition of some overlap. Since the stress in the plates is the sum of the stress carried by the actual and the virtual pulses, it is zero at and near the rear boundary, as required by the boundary condition. The interface of the two target plates remains in compression, but the compressed region is narrowing. In the fifth panel the pulses are shown in a position of exact overlap; the plates are unstressed. Finally, the sixth panel shows the configuration after the pulses have passed partially through each other. The material is shown as having gone into tension. If the interface is bonded, this may be possible; otherwise the plates will separate. One can show that the back target plate will be ejected from the stack with the same velocity as the projectile plate had at impact. The motion of the other two plates will be arrested: The projectile plate and the first target layer remain motionless and in contact. When there are more layers in the target, the same thing happens: The back layer is ejected but the projectile and the other target layers remain motionless.

Exercise 8.5.6. The graphical solution for the particle velocity field is obtained in essentially the same way as for the stress field, except that the two pulses considered have the same sign (cf. Eq. 8.53) as compared to having the opposite sign (cf. Eq. 8.53) when the stresses were being determined. The following figure illustrates the calculation. Note that the particle velocity in the region near the unrestrained surface increases as a result of the reflection, and that it exceeds the velocity of particles further inside the boundary. When the particle velocity discontinuity reaches one of the boundaries separating the layers, the layer nearest the surface is ejected from the stack and the configuration becomes as illustrated in panel c of the figure. This is the same as the original configuration shown in panel a, except that the amplitude of the pulse is reduced. The same wave interactions will occur and another layer of material will be ejected. This process will continue until the pulse is completely attenuated.

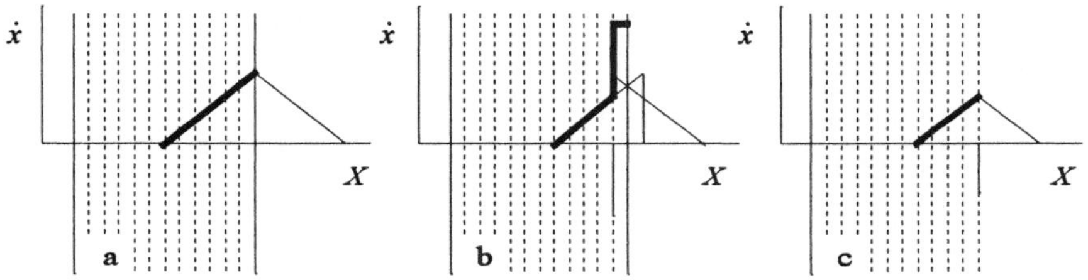

Figure: Exercise 8.5.6.

Exercise 8.5.7. The solution is obtained using three stress distributions. The initial distribution is formed from two distributions, each of which is one-half the amplitude of the original distribution. These pulses propagate in opposite directions, one into the material and one out of it. This latter pulse is compensated by a pulse of opposite sign that originates outside the material and propagates into it.

Chapter 9. Nonlinear Elastic Waves

Exercise 9.7.1. Examination of Eq. 9.62 shows that the third-order stress–strain equation becomes linear when C_{11} and C_{111} stand in the relationship indicated. As suggested in the statement of the problem, more can be said on the matter.

Exercise 9.7.2. The problem is illustrated in the following figure. When the incident shock propagating in the high-impedance material encounters the interface with the low-impedance material a shock is transmitted into this material and a centered decompression wave is reflected back into the high-impedance material. Since the interface is under compression, the materials will not separate and the pressure and particle velocity fields are continuous at this interface. The materials are different so we cannot expect the displacement gradient to be continuous. The shock transition in the low-impedance material must satisfy the jump conditions

$$\dot{x}^{++} + U_{\mathrm{SL}}\, G_{\mathrm{L}}^{++} = 0 \quad \text{and} \quad \rho_{\mathrm{RL}}\, U_{\mathrm{SL}}\, \dot{x}^{++} - p^{++} = 0\,. \tag{A}$$

The pressure and displacement gradient in the shock compressed material must also satisfy the Hugoniot relation

$$p^{++} = \frac{\rho_{\mathrm{RL}}\, C_{\mathrm{BL}}^2\, G_{\mathrm{L}}^{++}}{(1+S_{\mathrm{L}}\, G_{\mathrm{L}}^{++})^2}\,. \tag{B}$$

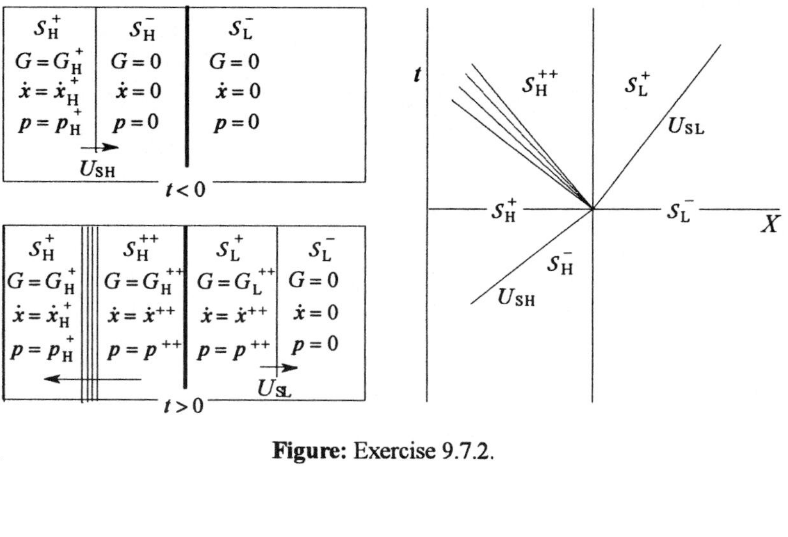

Figure: Exercise 9.7.2.

The centered simple decompression wave in the high-impedance material is analyzed as in Section 9.3.1, except that the pressure behind the wave is p^{++}, not zero. The key quantities that must be determined are those in the region between the shock and the simple wave. They are p^{++}, $\dot{x}^{++}$, G_{H}^{++}, G_{L}^{++}, and U_{SL}. The available equations are the jump conditions for the shock, Eqs. A, the Hugoniot relation, Eq. B, and the isentropic stress relation that, in the present case is the same as Eq. B, except that it relates p^{++} to G_{H}^{++} using the material properties ρ_{RH} and C_{BH},

Finally, we have Eq. 9.162 evaluated for the case at hand,

$$\dot{x}^{++} = \dot{x}_{\mathrm{H}}^{+} - \frac{2C_{\mathrm{BH}}}{S_{\mathrm{H}}}\left[\sqrt{\frac{1-S_{\mathrm{H}}G_{\mathrm{H}}^{++}}{1+S_{\mathrm{H}}G_{\mathrm{H}}^{++}}} - \tan^{-1}\sqrt{\frac{1-S_{\mathrm{H}}G_{\mathrm{H}}^{++}}{1+S_{\mathrm{H}}G_{\mathrm{H}}^{++}}} + \sqrt{\frac{1-S_{\mathrm{H}}G_{\mathrm{H}}^{+}}{1+S_{\mathrm{H}}G_{\mathrm{H}}^{+}}} - \tan^{-1}\sqrt{\frac{1-S_{\mathrm{H}}G_{\mathrm{H}}^{+}}{1+S_{\mathrm{H}}G_{\mathrm{H}}^{+}}}\right]. \tag{D}$$

The five equations, Eqs. A, B, C, and D can now be solved for the field quantities defining the state in the region between the waves and everything else follows.

Exercise 9.7.3. The particle velocity behind the receding centered simple decompression wave is given by Eq. 9.162 when $G=0$. Evaluation of this equation for several conditions produces the results given in the following table.

Escape velocities.

	Cu	Pb	Al alloy 2024	Be
$S =$	1.489	1.46	1.338	1.124
$C_{\mathrm{B}} =$	3940 m/s	2051 m/s	5328 m/s	7998 m/s
Δ^{+}	$\dot{x}_{\mathrm{esc}}/C_{\mathrm{B}}$	$\dot{x}_{\mathrm{esc}}/C_{\mathrm{B}}$	$\dot{x}_{\mathrm{esc}}/C_{\mathrm{B}}$	$\dot{x}_{\mathrm{esc}}/C_{\mathrm{B}}$
0.00	0.00	0.00	0.00	0.00
0.05	0.05	0.05	0.05	0.05
0.10	0.12	0.12	0.12	0.11
0.15	0.19	0.19	0.19	0.18
0.20	0.28	0.28	0.27	0.26
0.25	0.39	0.38	0.37	0.34
0.30	0.52	0.51	0.48	0.44

Exercise 9.7.4. For the simple wave we have

$$\dot{x}-\dot{x}^{-}=-\int_{G^{-}}^{G} C_{\mathrm{L}}(G')dG', \tag{A}$$

and, for the shock, we have

$$[\![\dot{x}]\!]^2=[\![-t_{11}]\!][\![-v]\!]=v_{\mathrm{R}}[\![t_{11}]\!][\![G]\!], \tag{B}$$

since $v=v_{\mathrm{R}}(1-G)$. Let us expand the integral in Eq. A in powers of $G-G^{-}$:

$$\int_{G^{-}}^{G} C_{\mathrm{L}}(G')\,dG'=C_{\mathrm{L}}(G^{-})\,(G-G^{-})+\tfrac{1}{2}C_{\mathrm{L}}'(G^{-})\,(G-G^{-})^2$$
$$+\tfrac{1}{6}C_{\mathrm{L}}''(G^{-})\,(G-G^{-})^3+\cdots,$$

so

$$\begin{aligned}\dot{x}-\dot{x}^{-}=&-C_{\mathrm{L}}(G^{-})\,(G-G^{-})-\tfrac{1}{2}C_{\mathrm{L}}'(G^{-})\,(G-G^{-})^2\\&-\tfrac{1}{6}C_{\mathrm{L}}''(G^{-})\,(G-G^{-})^3+\cdots,\end{aligned} \tag{C}$$

where the primes denote differentiation with respect to G.

Now, let us expand Eq. B in the same manner. Defining

$$\Phi(G)=v_{\mathrm{R}}\,[\![t_{11}]\!][\![G]\!],$$

we have

$$\frac{d}{dG}\Phi(G)=v_{\mathrm{R}}\left\{\frac{\partial\hat{t}_{11}}{\partial G}(G-G^{-})+(t_{11}-t_{11}^{-})\right\}$$

$$\frac{d^2}{dG^2}\Phi(G)=v_{\mathrm{R}}\left\{\frac{\partial^2\hat{t}_{11}}{\partial G^2}(G-G^{-})+2\frac{\partial\hat{t}_{11}}{\partial G}\right\}$$

$$\frac{d^3}{dG^3}\Phi(G)=v_{\mathrm{R}}\left\{\frac{\partial^3\hat{t}_{11}}{\partial G^3}(G-G^{-})+3\frac{\partial^2\hat{t}_{11}}{\partial G^2}\right\}$$

$$\frac{d^4}{dG^4}\Phi(G)=v_{\mathrm{R}}\left\{\frac{\partial^4\hat{t}_{11}}{\partial G^4}(G-G^{-})+4\frac{\partial^3\hat{t}_{11}}{\partial G^3}\right\},$$

so

$$\Phi(G^{-})=0$$
$$\Phi'(G^{-})=0$$
$$\Phi''(G^{-})=2v_{\mathrm{R}}\left.\frac{\partial\hat{t}_{11}}{\partial G}\right|_{S^{-}}$$

$$\Phi'''(G^-) = 3\nu_R \left.\frac{\partial^2 \hat{t}_{11}}{\partial G^2}\right|_{S^-}$$

$$\Phi^{iv}(G^-) = 4\nu_R \left.\frac{\partial^3 \hat{t}_{11}}{\partial G^3}\right|_{S^-},$$

and

$$[\![\dot{x}]\!]^2 = \nu_R \left.\frac{\partial \hat{t}_{11}}{\partial G}\right|_{S^-}(G^+ - G^-)^2 + \frac{\nu_R}{2}\left.\frac{\partial^2 \hat{t}_{11}}{\partial G^2}\right|_{S^-}(G^+ - G^-)^3 + \frac{\nu_R}{6}\left.\frac{\partial^3 \hat{t}_{11}}{\partial G^3}\right|_{S^-}(G^+ - G^-)^4 + \cdots. \tag{D}$$

To compare this result with Eq. C we need to express the derivatives of the stress-response function in terms of the wavespeed and its derivatives. We have

$$C_L^2(G) = \nu_R \frac{\partial \hat{t}_{11}(G, \eta^-)}{\partial G}$$

$$C_L(G)C_L'(G) = \frac{\nu_R}{2}\frac{\partial^2 \hat{t}_{11}(G, \eta^-)}{\partial G^2} \tag{E}$$

$$C_L(G)C_L''(G) + [C_L'(G)]^2 = \frac{\nu_R}{2}\frac{\partial^3 \hat{t}_{11}(G, \eta^-)}{\partial G^3},$$

which can be solved for the derivatives. When these are substituted into Eq. D we get

$$[\![\dot{x}]\!]^2 = C_L^2(G^-)(G^+ - G^-)^2 + C_L(G^-)C_L'(G^-)(G^+ - G^-)^3 + \tfrac{1}{3}\left[C_L(G^-)C_L''(G^-) + [C_L(G^-)]^2\right](G^+ - G^-)^4 + \cdots$$

Taking the square root of this, with sign appropriate to a decompression shock, we get

$$[\![\dot{x}]\!] = -C_L(G^-)(G^+ - G^-) - \tfrac{1}{2}C_L'(G^-)(G^+ - G^-)^2 - \left[\frac{1}{6}C_L''(G^-) + \frac{1}{24}\frac{[C_L'(G^-)]^2}{C_L(G^-)}\right](G^+ - G^-)^3 + \cdots \tag{F}$$

Comparing Eqs. B and F,we see that

$$\left.\dot{x}^+ - \dot{x}^-\right|_{\text{simple wave}} = [\![\dot{x}]\!]_{\text{shock}} + \frac{1}{24}\frac{[C_L'(G^-)]^2}{C_L(G^-)}(G^+ - G^-)^3 + \cdots.$$

Exercise 9.7.5. The required result is easily derived using the expansions given in the previous exercise.

Exercise 9.7.6. Two errors are introduced when a centered simple wave is approximated by a shock. First, the shock approximation does not capture the spreading of the simple wave that occurs with increasing propagation distance and, second, the decompression shock involves a non-physical decrease in entropy, whereas the simple wave is an isentropic process. As one can see from the figure, the spreading is more pronounced for the stronger waves and longer propagation distances because of the increased influence of the nonlinearity of the stress equation. An indication of the effect of the differing entropy on the particle velocity is given in Table 3.3. As discussed in Exercise 6.4.2, the entropic error is proportional to the cube of the shock strength, and thus is small for weak shocks.

Exercise 9.7.7. In a simple wave the forward characteristics are defined by t^* alone:

$$X = S(t^*)(t - t^*)\,, \tag{A}$$

where $S(t^*) = C_L(G(t^*))$. Any trajectory through a simple wave that depends on G and/or $\dot{x}$ is a function of t^* alone (and the point (X_0, t_0) at which the trajectory enters the wave).

Consider the $\xi = \text{const.}$ characteristic having its X-axis intercept at $X = X^*$. In the region ahead of the wave it is given by

$$X = -C_L(G^-)t + X^*\,. \tag{B}$$

Its intersection with the leading characteristic of the wave, $X = C_L(G^-)(t - t^-)$ is, at (X_0, t_0), given by

$$\begin{aligned} X_0 &= \tfrac{1}{2}\left[X^* - C_L(G^-)t^-\right] \\ t_0 &= \frac{X^* + C_L(G^-)t^-}{2C_L(G^-)}\,. \end{aligned} \tag{C}$$

Within the wave region this characteristic can be described parametrically by equations of the form

$$X = X(t^*), \quad t = t(t^*)\,. \tag{D}$$

The cross characteristic is defined by

$$\frac{dX}{dt} = -C_L(G(t^*)) = -S(t^*)\,. \tag{E}$$

Using Eq. D, we can write Eq. E in the form

$$\frac{dX}{dt^*} = -S(t^*)\frac{dt}{dt^*}. \tag{F}$$

At any point in the wave the variables X, t, and t^* stand in the relation given by Eq. A, so

$$\frac{dX}{dt^*} = S'(t^*)(t-t^*) + S(t^*)\left(\frac{dt}{dt^*} - 1\right), \tag{G}$$

where $S' = dS/dt^*$.

Using Eq. G allows us to write Eq. F in the form

$$\frac{dt}{dt^*} + \varphi(t^*)t = \psi(t^*), \tag{H}$$

where

$$\varphi(t^*)t = \frac{S'(t^*)}{2S(t^*)}, \quad \psi(t^*) = \frac{1}{2}\left[\frac{S'(t^*)t^*}{S(t^*)} + 1\right]. \tag{I}$$

this is a linear, first-order ordinary differential equation and has the solution

$$t(t^*) = \left(\frac{S(t^-)}{S(t^*)}\right)^{1/2}\left\{t_0 - t^- + t^*\left(\frac{S(t^*)}{S(t^-)}\right)^{1/2} - \frac{1}{2\sqrt{S(t^-)}}\int_{t^-}^{t^*}\sqrt{S(\vartheta)}\,d\vartheta\right\}, \tag{J}$$

where the constant of integration has been chosen so that $t = t_0$ when $t^* = t^-$.

The function $X = X(t^*)$ is obtained by substituting Eq. J into Eq. F and integrating. The result is

$$\begin{aligned} X(t^*) = X_0 + S(t^-)&\left[\left(\frac{S(t^*)}{S(t^-)}\right)^{1/2} - 1\right](t_0 - t^-) \\ &- \int_{t^-}^{t^*} S(\vartheta)\,d\vartheta + \frac{1}{2}\sqrt{S(t^*)}\int_{t^-}^{t^*}\sqrt{S(\vartheta)}\,d\vartheta. \end{aligned} \tag{K}$$

The parametric equations J and K for the characteristics can be simplified when a specific choice of $C_L(G)$ is made, but the characteristics are completely determined only when one specifies both the wavespeed and the boundary condition $G(t^*)$.

Exercise 9.7.8. From Eqs. 9.10 and 9.17 we have $C_L^2 = -(1/\rho_R^2)\,dp^{(\eta)}/dv$ and $c_L^2 = dp^{(\eta)}/d\rho = -(1/\rho^2)\,dp^{(\eta)}/dv$. Equating the two expressions for $dp^{(\eta)}/dv$ that follow from these equations and taking the square root yields the desired result.

Exercise 9.7.9. The issue turns on demonstrating that

$$\int_0^G C_L(G')\,dG' = -\int_{\rho_R}^{\rho} \frac{c_L(\rho')}{\rho'}\,d\rho' . \tag{A}$$

We have equations $G = (\rho_R/\rho) - 1$ so $G = 0 \Rightarrow \rho = \rho_R$, $dG = -(\rho_R/\rho^2)\,d\rho$, and $C_L = (\rho/\rho_R)c_L$, and we see that the right member of Eq. A is the same as the left member.

Exercise 9.7.10. From Eqs. 2.40 and 2.49 we have $\rho = \rho_R(1 - u_x)$ so $\rho_x = -\rho_R u_{xx}$ and $\rho_t = -\rho_R u_{xt}$. From Eq. 2.48 we have $\dot{x} = (\rho_R/\rho)u_t$ so $\dot{x}_t = -(\rho_R/\rho^2)\rho_t u_t + (\rho_R/\rho)u_{tt}$ and $\dot{x}_x = -(\rho_R/\rho^2)\rho_x u_t + (\rho_R/\rho)u_{tx}$. Substitution of these results into Eq. 9.14_1 shows that it is identically satisfied in linear approximation. Substitution into Eq. 9.14_2 and neglect of nonlinear terms yields the result $u_{tt} = c_L^2(\rho_R)u_{xx}$.

Chapter 10. Elastic–Plastic and Elastic–Viscoplastic Waves

Exercise 10.4.1. The solution is obtained by application of the jump condition of Eq. 2.113_2 to the advancing plastic wave in the target plate and the receding wave in the projectile plate. The jump condition for the plastic shock in the target is $-t_{11} = t_{11(\mathrm{T})}^{\mathrm{HEL}} - \rho_{\mathrm{R(T)}} C_{\mathrm{B(T)}}(\dot{x} - \dot{x}_{(\mathrm{T})}^{\mathrm{HEL}})$ and, in the projectile, we have $-t_{11} = t_{11(\mathrm{P})}^{\mathrm{HEL}} - \rho_{\mathrm{R(P)}} C_{\mathrm{B(P)}}(\dot{x} - \dot{x}_{(\mathrm{P})}^{\mathrm{HEL}})$. Since the longitudinal stress component and particle velocity in each material are the same at the impact interface, the stress given by these two equations can be equated and the resulting equation solved for $\dot{x}$, the result is

$$\dot{x}^+ = \frac{\rho_{\mathrm{R(P)}}(C_{\mathrm{O(P)}} + C_{\mathrm{B(P)}})\,\dot{x}_{(\mathrm{P})}^{\mathrm{HEL}} - \rho_{\mathrm{R(T)}}(C_{\mathrm{O(T)}} - C_{\mathrm{B(T)}})\,\dot{x}_{(\mathrm{T})}^{\mathrm{HEL}}}{\rho_{\mathrm{R(P)}} C_{\mathrm{B(P)}} + \rho_{\mathrm{R(T)}} C_{\mathrm{B(T)}}},$$

where we have used the equations $t_{11(\mathrm{T})}^{\mathrm{HEL}} = \rho_{\mathrm{R(T)}} C_{\mathrm{O(T)}}\,\dot{x}_{(\mathrm{T})}^{\mathrm{HEL}}$ and $t_{11(\mathrm{P})}^{\mathrm{HEL}} = \rho_{\mathrm{R(P)}} C_{\mathrm{O(P)}}\,\dot{x}_{(\mathrm{P})}^{\mathrm{HEL}}$. With this result in hand, we can return to either of the jump equations to calculate t_{11}^{+}.

Exercise 10.4.2. The X–t and $-t_{11} - \dot{x}$ diagrams for this problem are shown below. These diagrams, although shown in full in the figure, must be constructed in steps as the interpretation of the experiment evolves.

The first jump in the recorded $\dot{x}$ history occurs when the precursor shock arrives at the sample/window interface. The transit time of this shock through the 5-mm thick sample is 0.785 μs, so we find that $C_{\mathrm{O(Al)}} = 6369\ \mathrm{m/s}$. Since, for this shock, the jump condition is $-t_{11} = \rho_{\mathrm{R(Al)}} C_{\mathrm{O(Al)}}\,\dot{x}$ we know the slope of this part of the sample Hugoniot although we do not yet know the HEL. Because the shock impedance of the sapphire ($Z = 0.04\ \mathrm{GPa/(m/s)}$) is greater than the elastic

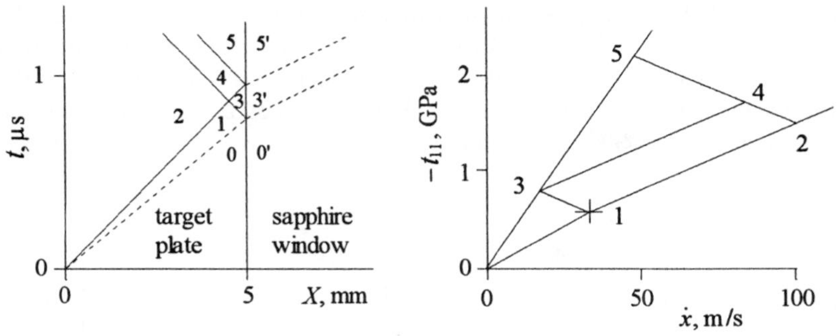

Figure: Exercise 10.4.2.

shock impedance of the aluminum alloy ($Z = 0.02\,\mathrm{GPa/(m/s)}$) the reflected wave creating region 3 of the X–t plane is one of compression and, therefore, a plastic wave propagating at a speed C_{B} that is yet to be determined. In the theory of weak elastic–plastic waves that we are using the Lagrangian plastic wavespeed is the same for all of the waves, so the transit time of a plastic wave across the target plate corresponds to the time of the second jump in the interface-velocity history and the plastic wavespeed is, therefore, $C_{\mathrm{B}} = 5192\,\mathrm{m/s}$. With this information we can determine the HEL for the sample material. The particle velocity in state 3, $\dot{x}^{(3)} = 17.89\,\mathrm{m/s}$, is known from the interferometer measurement and the corresponding stress is known from the sapphire Hugoniot. We have $-t_{11}^{(3)} = \rho_{\mathrm{R(S)}} C_{\mathrm{R(S)}} \dot{x}^{(3)} = 0.798\,\mathrm{GPa}$. We can now use the jump equations for the transition from state 1 to state 3 and the transition from state 0 to state 1 in the aluminum alloy to determine its HEL. When these equations are solved we obtain the values $t_{11}^{\mathrm{HEL}} = 0.578\,\mathrm{GPa}$ and $\dot{x}^{\mathrm{HEL}} = 33.63\,\mathrm{m/s}$. Since the HEL and C_{B} are now known, we can plot the plastic part of the aluminum alloy Hugoniot. State 2 falls on this Hugoniot at the particle velocity $\dot{x} = 100\,\mathrm{m/s}$ corresponding to one-half of the projectile velocity. The corresponding stress is found from the jump condition to be $-t_{11}^{(2)} = 1.51\,\mathrm{GPa}$. From the jump conditions we find that $-t_{11}^{(5)} = 2.21\,\mathrm{GPa}$ and $\dot{x}^{(5)\mathrm{L}} = 49.64\,\mathrm{m/s}$.

Because sapphire monocrystals are elastic to approximately 15 GPa, experiments of the sort just considered are often conducted at much higher impact velocities to study curvature of the plastic part of the Hugoniot. In this case, interpretation of the measurements is less direct. It is necessary to conduct experiments at various impact velocities to infer the dependence of C_{B} on the particle velocity by iterative interpretation of the results.

Exercise 10.4.3. The X–t and $-t_{11} - \dot{x}$ diagrams for this problem are shown below. These diagrams, although shown in full in the figures, must be constructed in steps as the interpretation of the experiment evolves.

The things that we know about the experiment are i) the Hugoniot for the quartz: $-t_{11} = \rho_R C_0 \dot{x}$, where $\rho_R = 2650\,\text{kg/m}^3$ and $C_0 = 5721\,\text{m/s}$, ii) the stress amplitude and arrival times of the three shocks recorded at the interface of the sample and the gauge, iii) the thickness and density of the sample material, $X_T = 5\,\text{mm}$ and $\rho_R = 7870\,\text{kg/m}^3$ and, iv) the projectile velocity, $\dot{x}_P = 136\,\text{m/s}$.

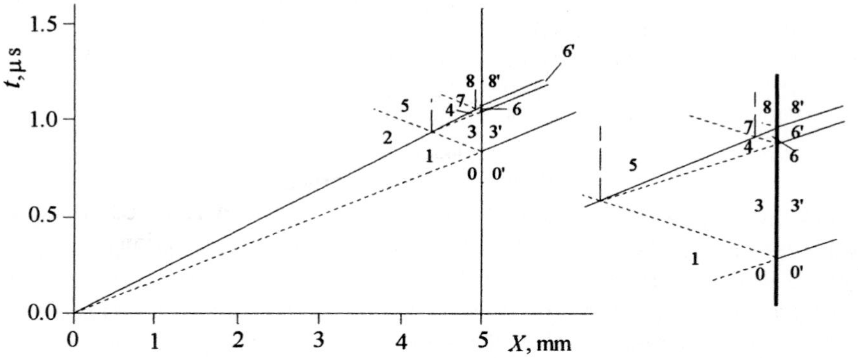

Figure: Exercise 10.4.3. The X–t diagram. The illustration at the right is an enlarged version of the important part of the full diagram.

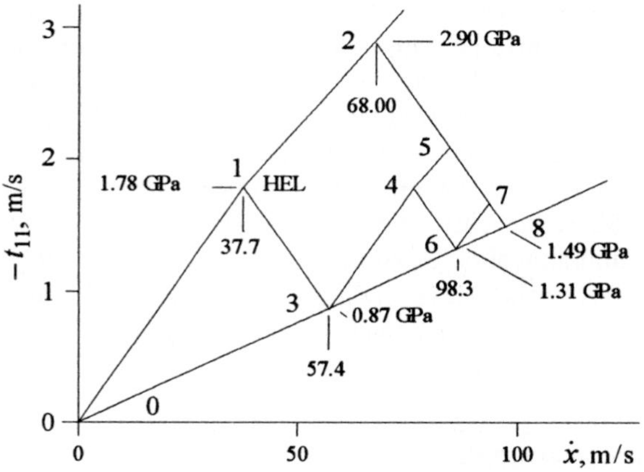

Figure: Exercise 10.4.3. $t_{11} - \dot{x}$ diagram

To interpret the experimental results we construct the X–t and $t_{11} - \dot{x}$ diagrams incrementally as we identify the various states produced. If the gauge record revealed only a single step, we would expect that the impact stress was below the HEL and that only an elastic wave was produced. Since the gauge record exhibits several steps, we may assume that an elastic–plastic waveform propagated through the sample and that transmitted and reflected waves resulted from its interaction with the gauge. From the gauge record we see that the transit time of the precursor shock through the 5-mm thick sample was 0.84 μs, so the elastic wavespeed in the sample material is $C_0 = 5952$ m/s . This permits us to plot the precursor trajectory on the X–t plot. The quartz Hugoniot and the slope of the elastic part of the sample Hugoniot can be plotted on the $t_{11} - \dot{x}$ diagram using the values of ρ_R and C_0 for the two materials. We also know the stress produced at the sample/quartz interface when the precursor arrives, which defines state 3 on the $t_{11} - \dot{x}$ plot. The Hugoniot elastic limit, state 1, can now be determined as the intersection of the reflected elastic Hugoniot through point 3 with the elastic Hugoniot for the right-propagating precursor shock. When the reflected precursor wave encounters the advancing plastic shock another interaction occurs, and elastic (and possibly also plastic) shocks propagate in both directions. The right-propagating elastic shock encounters the sample/quartz interface at the time of the second step in the recorded waveform, t=1.05 μs after the projectile impact. The elastic waves incident on, and reflected from, the plastic shock define region 3 on the X–t plot and their intersection lies on the plastic shock trajectory in this plane. Since the origin of the graph also lies on this trajectory, the trajectory can now be plotted. As we consider this X–t plot and the quartz gauge record, we see that the step in interfacial stress at 1.08 μs after the projectile impact must arise from a plastic shock formed when the elastic precursor encountered the plastic shock. The transition from state 3 to the (still undetermined) state 5 involves an elastic transition to the HEL stress (state 4), followed by a plastic transition to the final state 5. The wedge-shaped space between the reflected elastic and plastic shocks is designated region 4 on the $X - t$ plot. State 6 arises as the result of an elastic decompression shock from state 4. Analysis of the transition from region 2 to region 5 and from region 4 to region 5 identifies States 4 and 5 on the $t_{11} - \dot{x}$ plot and shows that region 5 comprises two sub-regions separated by a contact surface. Continuation of the analysis defines states 7 and 8, and interpretation of the experiment is complete.

Comparison of this experiment with the one discussed in the previous exercise shows that interpretation of the result is much easier when the material downstream from the sample is such that the reflected wave is one of compression rather than decompression.

Exercise 10.4.4. Since the elastic–plastic stress relation is independent of the rate at which the deformation occurs, waveforms can be calculated in the same way as was done for nonlinearly elastic materials, except that the several parts of the stress relation must be considered separately. Our first task is determination of the stress–strain path that is followed as the pulse passes a material point. This path is shown in Fig. 7.1 and in Fig. 10.1, which is a schematic illustration drawn to emphasize the nonlinearities of the path.

Stress–Strain Path

Elastic Compression. In the elastic range ($-t_{11} \le t_{11}^{\mathrm{HEL}}$) we shall use the theory of Sect. 6.3.2, which gives the stress relation in the form $t_{11} = t'_{11} - p$ where $t'_{11} = 2\mu_R \widetilde{E}^{s}_{11}$ with $\widetilde{E}^{s}_{11} = -\frac{2}{3}[1-(v/v_R)]$ and

$$p(v) = \frac{(\rho_R C_B)^2 (v_R - v)}{[1 - \rho_R S(v_R - v)]^2}. \tag{A}$$

Combining these equations, we have

$$-t_{11} = \frac{(\rho_R C_B)^2 (v_R - v)}{[1 - \rho_R S(v_R - v)]^2} + \frac{4\mu_R}{3}\left(1 - \frac{v}{v_R}\right). \tag{B}$$

When evaluated at the HEL this is a cubic equation that can be solved for v^{HEL} when t_{11}^{HEL} has been measured.

In solving this Exercise we shall use the stress relation of Eq. 6.61

$$-t_{11} = p(v) - \tfrac{4}{3}\mu(v)[1 - (v/v_R)].$$

Elastic–Plastic Compression. The stress on the elastic–plastic compression curve is given by the equation

$$-t_{11} = p^{(\text{H})}(v) + \tfrac{2}{3}Y_0, \tag{C}$$

where pressure is calculated using Eq. A and where Y_0 is the given constant value of the yield stress.

The plastic part of the deformation gradient in the material at states on this curve is given by Eq. 7.84:

$$(F_L^{\text{P}})^{1/2} = \left(\frac{v}{v_R}\right)^{1/3}\left[1 + \frac{-t_{11} - p^{(H)}(v)}{4\mu(v)}\right]. \tag{D}$$

In evaluating this equation it is important that the value of the shear modulus used be appropriate to the compressed state of the material when the compression is large. Equation 6.62 has been used to represent the function $\mu(v)$.

Elastic decompression. The elastic decompression process begins with the material at a given specific volume v^+, and the associated values of the stress t_{11}^+, and the plastic deformation gradient $F_L^{\text{p}+}$. The elastic decompression path is given by Eq. 7.78 with F_L^{p} held constant at the value $F_L^{\text{p}+}$. This process terminates and reverse yielding begins at the value v^{++} given as the solution of the equation $-t_{11}(v^{++}) = p(v^{++}) - \frac{2}{3}Y_0$.

Elastic–Plastic Decompression. For the non-hardening material under discussion the elastic–plastic decompression path begins at the stress $-t_{11}(v^{++})$ and specific volume v^{++}. It is given by the equation $-t_{11}(v) = p(v) - \frac{2}{3}Y_0$ and terminates at zero stress. The pressure is calculated using Eq. A and Y_0 is the given constant value of the yield stress. The specific volume at the zero-stress point on this path is obtained as the solution of the equation $p(v) = \frac{2}{3}Y_0$. We shall need the soundspeed at points on this curve, which is given by the equation

$$C^2 = -v_R^2 \frac{d(-t_{11}(v))}{dv} = -v_R^2 \frac{d(p(v))}{dv} = C_B^2 \frac{1+\rho_R S(v_R - v)}{[1-\rho_R S(v_R - v)]^3}. \tag{E}$$

Waveform Calculation

Precursor Shock. When the impact occurs a shock is introduced into the halfspace that is immediately split into an elastic precursor shock that is followed by a plastic shock transition to the state imposed on the boundary.

Applying the jump equations to the precursor shock, we obtain the equation

$$C_0 = v_R [t_{11}^{\mathrm{HEL}} / (v_R - v^{\mathrm{HEL}})]^{1/2} \tag{F}$$

for the precursor-shock velocity and the equation

$$\dot{x}^{\mathrm{HEL}} = \frac{1}{\rho_R C_0} t_{11}^{\mathrm{HEL}} \tag{G}$$

for the particle velocity amplitude of the precursor shock.

Elastic–Plastic Shock. For a normal non-hardening elastic–plastic solid the precursor shock will be followed by a plastic shock propagating at the Lagrangian velocity given by the equation

$$U_L = v_R \left[\frac{-t_{11}^{+} - t_{11}^{\mathrm{HEL}}}{v^{\mathrm{HEL}} - v^{+}} \right]^{1/2}. \tag{H}$$

The particle velocity of the material behind this shock is $\dot{x}^{+} = \dot{x}_P / 2$, where the projectile velocity $\dot{x}_P$ is given. The jump equation relating stress and particle velocity is

$$\rho_R U_L (\dot{x}^{+} - \dot{x}^{\mathrm{HEL}}) = -t_{11}^{+} - t_{11}^{\mathrm{HEL}}. \tag{I}$$

Equations C, H, and I (with $p^{(H)}(v^{+})$ given by Eq. A) comprise three equations that can be solved for the three quantities U_L, t_{11}^{+}, and v^{+}, completing the characterization of the plastic shock.

Elastic Decompression. An elastic decompression wave governed by the stress relation

$$-t_{11}(v) = p(v) - 4\mu(v)[1-(v/v_R)^{-1/3}(F_L^{p+})^{1/2}] \tag{J}$$

will take the form of a centered simple wave because the stress–volume path is concave upward. However, the curvature is slight and the stress range is relatively small so we shall approximate it as a decompression shock. The stress amplitude of this shock is $-t_{11}^{++}+t_{11}^{+}$ and its Lagrangian velocity is

$$C = v_R \left(\frac{-t_{11}^{++}+t_{11}^{+}}{v^{+}-v^{++}} \right)^{1/2} . \tag{K}$$

Elastic–Plastic Decompression Wave. The elastic–plastic decompression wave is a centered simple decompression wave calculated using the stress–volume path

$$-t_{11} = p(v) - \tfrac{2}{3} Y_0 , \tag{L}$$

where , as before, the pressure is given by Eq. A. The waveform is calculated as was done for an elastic material in Sect. 9.3.

Example Waveform. Results calculated for the problem given include the amplitude and propagation velocities of the several shocks and the shape of the centered simple elastic–plastic decompression wave. The $X-t$ plot and a temporal waveform are illustrated in the following figure.

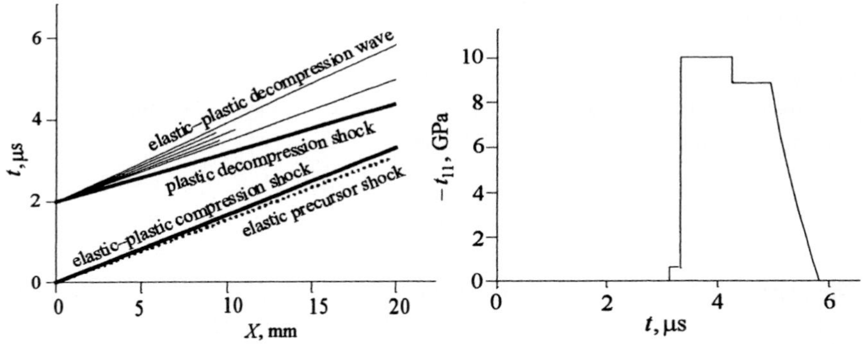

Figure: Exercise 10.4.4. The $X-t$ plot and temporal waveform for the position $X=20$ mm.

References

[1] Ahrens, T.J., "Equation of state," in [7], pp. 75–113.

[2] Al'tshuler, L.V., "Application of Shock Waves in Physics of High Pressures," *Sov. Phys.–Usp.* **8**, (1965), pp. 52–91.

[3] Anonymous, *Engineering Design Handbook. Principles of Explosive Behavior*, U.S. Army Material Command, Report AMCP 706-180, 1972.

[4] Antoun, Tarabay, Lynn Seaman, Donald R. Curran, Gennady I. Kanel, Sergey V. Razorenov, and Alexander V. Utkin, *Spall Fracture*, Springer, New York (2003).

[5] Asay, J.R., and L.M. Barker, "Interferometric Measurements of Shock-Induced Internal Particle Velocity and Spatial Variation of Particle Velocity." *J. Appl. Phys.* **45**, (1974), pp. 2540–2546.

[6] Asay, James R., and Lalit C. Chhabildas, "Paradigms and Challenges in Shock Wave Research," in [57], pp. 57–119.

[7] Asay, James R. and Mohsen Shahinpoor (eds), *High-Pressure Shock Compression of Solids*, Springer-Verlag, New York (1993).

[8] Baer, Mel R., "Mesoscale Modeling of Shocks in Heterogeneous Reactive Materials," in [56], pp. 321–356.

[9] Barker, L.M., M. Shahinpoor, and L.C. Chhabildas, "Experimental and Diagnostic Techniques," in [7], pp. 43–73.

[10] Batsanov, S.S., *Effects of Explosions on Materials—Modification and Synthesis Under High-Pressure Shock Compression*, Springer-Verlag, New York (1994).

[11] Bennett, L.S., K. Tanaka, and Y. Horie, "Constitutive Modeling of Shock-Induced Reactions," in [33], pp. 106–142.

[12] Benson, David J., "Numerical Methods for Shocks in Solids," in [56], pp. 275–319.

[13] Boade, R.R., "Principal Hugoniot, Second-Shock Hugoniot, and Release Behavior of Pressed Copper Powder," *J. Appl. Phys.* **41(11)**, (1970), pp. 4542–4551.

[14] Brendel, Rémi, "Independent Fourth-Order Elastic Coefficients for All Crystal Classes." *Acta Crystallogr.* **A35**, (1979), pp. 525–533.

[15] Brugger, K., Thermodynamic Definition of Higher-Order Elastic Coefficients," *Phys. Rev.* **133**, (1964), pp. A1611-A1612.

[16] Brugger, K., "Pure Modes for Elastic Waves in Crystals," *J. Appl. Phys.* **36(3)**, (1965), pp. 759–768.

[17] Cagnoux, J., and J.-Y. Tranchet, "Response of High-Strength Ceramics to Plane and Spherical Shock Waves," in [35], pp. 147–169.

[18] Callen, Herbert B., *Thermodynamics*, John Wiley, New York (1960).

[19] Carroll, M.M., and A.C. Holt, "Static and Dynamic Pore-Collapse Relations for Ductile Porous Materials," *J. Appl. Phys.* **43(4)**, (1972), pp. 1626–1636.

[20] Chéret, R., *Detonation of Condensed Explosives*, Springer-Verlag, New York (1993).

[21] Chhabildas, L.C., and R.A. Graham, "Developments in Measurement Techniques for Shock Loaded Solids," in *Techniques and Theory of Stress Measurements for Shock Wave Applications* (eds R.R. Stout, F.R. Norwood, and M.E. Fourney) Report AMD83, American Society of Mechanical Engineers, New York (1987), pp. 1–18.

[22] Clifton, R.J., "Plastic Waves: Theory and Experiment," in *Mechanics Today* **1**, (ed. S. Nemat-Nasser) Pergamon Press, New York, (1972), pp. 102–167.

[23] Clifton, R.J., and X. Markenscoff, *J. Mech. Phys. Solids* **29(3)**, (1981), pp. 227–251.

[24] Coleman, Bernard D., and Morton E. Gurtin, "Thermodynamics with Internal State Variables," *J. Chem. Phys.* **47**, (1967), pp. 597–613.

[25] Coleman, Bernard D., and Victor J. Mizel, "Existence of Caloric Equations of State," *J. Chem. Phys.* **40(4)**, (1964), pp. 1116–1125.

[26] Coleman, Bernard D., and Walter Noll, "The Thermodynamics of Elastic Materials with Heat Conduction and Viscosity," *Arch. Rational Mech. Anal.* **13**, (1963), pp. 167–178.

[27] Cooper, P.W., *Explosives Engineering*, VCH, New York (1996).

[28] Courant, R., and K.O. Friedrichs, *Supersonic Flow and Shock Waves*, Interscience Publishers, New York (1948).

[29] Dafalias, Y.F., "Issues on the Constitutive Formulation at Large Elastoplastic Deformations, Part 1: Kinematics and Part 2: Kinetics," *Acta Mechanica* **69**, (1987), pp. 119–138 and **73**, (1988), pp. 121–146.

[30] Davison, Lee, "Attenuation of Longitudinal Elastoplastic Pulses," in [35], pp. 277–327.

[31] Davison, Lee, D.E. Grady, and Mohsen Shahinpoor (eds), *High-Pressure Shock Compression of Solids II*, Springer-Verlag, New York (1996).

[32] Davison, L., and R.A. Graham, *Phys. Rep.* **55**, (1979), pp 255–379.

[33] Davison, Lee, Y. Horie, and Mohsen Shahinpoor, eds., *High-Pressure Shock Compression off Solids IV: Response of Highly Porous Solids To Shock Loading*, Springer-Verlag, New York (1997).

[34] Davison, Lee, and M.E. Kipp, "Calculation of Spall Damage Accumulation in Ductile Metals," in *High Velocity Deformation of Solids* (eds Kozo Kawata and Jumpei Shiori) Springer-Verlag, Berlin/Heidelberg, (1979), pp. 163–175.

[35] Davison, Lee, and Mohsen Shahinpoor (eds), *High-Pressure Shock Compression of Solids III*, Springer-Verlag, New York (1997).

[36] Davison, L., A.L. Stevens, and M.E. Kipp, "Theory of Spall Damage Accumulation in Ductile Metals," *J. Mech. Phys. Solids* **25**, (1977), pp. 11–28.

[37] Dienes, J.K., "A Unified Theory of Flow, Hot Spots, and Fragmentation with an application to Explosive Sensitivity," in [31], pp. 366–398.

[39] Drumheller, D.S., *Introduction to Wave Propagation in Nonlinear Fluids and Solids*, Cambridge University Press, Cambridge (1998).

[40] Duvall, G.E., and R.A. Graham, "Phase Transitions Under Shock-wave Loading," *Rev. Mod. Phys.* **49**, pp. 523–579.

[41] Eftis, John, "Constitutive Modeling of Spall Fracture," in [31], pp. 399–451.

[42] Engelke, Ray, and Stephen A. Sheffield, "Initiation and propagation of Detonation in Condensed-Phase High Explosives," in [35], pp. 171–239.

[43] Fickett, Wildon, and William C. Davis, *Detonation*, University of California Press, Berkeley (1979).

[44] Fortov, V.E., L.V. Al'tshuler, R.F. Trunin, and A.I. Funtikov, *High-Pressure Shock Compression of Solids VII—Shock Waves and Extreme States of Matter*, Springer, New York (2004).

[45] Funtikov, A.I., and M.N. Pavlovsky, "Shock Waves and Polymorphic Phase Transformations in Solids," in [44], pp. 197–224.

[46] Fritz, J.N., S.P. Marsh, W.J. Carter, and R.G. McQueen, "The Hugoniot Equation of State of Sodium Chloride in the Sodium Chloride Structure," in *Accurate Characterization of the High-Pressure Environment*, (ed. Edward C Lloyd) National Bureau of Standards Special Publication 326, U.S. Government Printing Office, Washington, (1971), pp. 201–217.

[47] Geringer, Hilda, "Ideal Plasticity," in *Handbuch der Physik* **VIa/3**, (ed C. Truesdell), Springer-Verlag, Berlin, (1973), pp. 403–533.

[48] Godwal, B.K., S.K. Sikka, and R. Chidambaram, *Phys. Rep.* **102**, (1983), pp. 122–197.

[49] Grady, D.E., *Fragmentation of Rings and Shells: The Legacy of N.F. Mott*, Springer, Heidelberg (2006).

[50] Graham, R.A., *Solids Under High-Pressure Shock Compression: Mechanics, Physics, and Chemistry*, Springer-Verlag, New York (1993).

[51] Graham, R.A., "Determination of Third- and fourth-Order Longitudinal Elastic Constants by Shock Compression Techniques—Application to Sapphire and Fused Quartz," *J. Acoust. Soc. Am.* **51**, (1972), pp. 1576–1581.

[52] Graham, R.A. and J.R. Asay, "Measurements of Wave Profiles in Shock-Loaded Solids," *High Temp. High Press.* **10**, (1978), pp. 355–390.

[53] Gray, G.T., III, "Influence of Shock-Wave Deformation on the Structure/Property Behavior of Materials," in [7], pp. 187–215.

[54] Herrmann, W., "Constitutive Equation for the Dynamic Compaction of Ductile Porous Materials," *J. Appl. Phys.* **10(6)**, (1969), pp. 2490–2499.

[55] Herrmann, W., D.L. Hicks, and E.G. Young, "Attenuation of Elastic–Plastic Stress Waves," in *Shock Waves and the Mechanical Properties of Solids* (eds. J.J. Burke and V. Weiss) Syracuse Univ. Press, Syracuse, NY, (1971), pp. 23–63.

[56] Horie, Yasuyuki (ed), *Shock Wave Science and Technology Reference Library* **2**, Springer, Berlin, Heidelberg (2007).

[57] Horie, Yasuyuki, Lee Davison, and Naresh N. Thadhani (eds.), *High Pressure Shock Compression of Solids VI; Old Paradigms and New Challenges*, Springer, New York (2003).

[58] Hugoniot, P.H., *Sur la Propagation du Mouvement Dans les Corps et Plus Spécialement Dans les Gaz Parfaits*, J. l'Ecole Polytechniqe, 57e cahier (1887), pp. 3–97 and 58e cahier (1899), pp. 1–125. Translated in *Classic Papers in Shock Compression Science* (eds J.N. Johnson and R Chéret), Springer, New York, (1998), pp 161–358.

[59] Johnson, J.N., “Micromechanical Considerations in Shock Compression of Solids,” in [35], pp. 217–264.

[60] Johnson, J.N., and L.M. Barker, *J. Appl. Phys.* **40**, (1969), p. 4321.

[61] Johnson, Wallace E., and Charles E. Anderson, Jr., “History and Application of Hydrocodes in Hypervelocity Impact,” *Int. J. Impact Eng.* **5**, pp. 423–439 (1987).

[62] Kalitkin, N.N., and L.V. Kuzmina, *Wide-Range Characteristic Thermodynamic Curves*, in [44], pp. 109–176.

[63] Kanel, G.I., S.V. Razorenov, and V.E. Fortov, *Shock-Wave Phenomena and the Properties of Condensed Matter*, Springer, New York (2004).

[64] Kelly, J.M., and P.P. Gillis, “Thermodynamics and Dislocation Mechanics,” *J. Franklin Inst.* **297(1)**, (1974), p. 59.

[65] Kipp, M.E., and Lee Davison, “Analyses of Ductile Flow and Fracture in Two Dimensions,” in *Shock Waves in Condensed Matter — 1981* (eds. W.J. Nellis, L. Seaman, and R.A. Graham), American Institute of Physics, New York, (1982), pp. 442–445.

[66] Kondo, K., “Magnetic Response of Powders to Shock Loading and Fabrication of Nanocrystalline Magnets,” in [33], pp. 309–330.

[67] Kormer, S.B., A.I. Funtikov, V.D. Urlin, and A.N. Kosesnikova, “Dynamic Compression of Porous Metals and the Equation of State with Variable Specific Heat at High Temperatures,” *Sov. Phys.–JETP* **15(3)**, (1962), pp. 477–488.

[68] Krupnikov, K.K., M.I. Brazhnik, and V.P. Krupnikova, “Shock Compression of Porous Tungsten,” *Sov. Phys.–JETP* **15(3)**, (1962), pp. 470–476.

[69] Lewis, M.W., and H.L. Schreyer, “A Thermodynamically Consistent Description of Dynamic Continuum Damage,” in [31], pp. 453–471.

[70] Lyon, S.P., and J.D. Johnson, *SESAME: The Los Alamos National Laboratory Equation of State Database*, Report LA-UR-92-3407, Los Alamos National Laboratory, Los Alamos, NM (1992).

[71] Mader, C.L., *Numerical Modeling of Detonations*, University of California Press, Berkeley (1979).

[72] Malvern, Lawrence E., *Introduction to the Mechanics of a Continuous Medium*, Prentice Hall, Englewood Cliffs, NJ (1969).

[73] Makarov, P.V., “Shear Strength and Viscosity of Metals in Shock Waves,” in [44], pp. 297–335.

[74] Marsh, S.P, *LASL Shock Hugoniot Data,* University of California Press, Berkeley (1980).

[75] Mashimo, Tsutomu, “Effects of Shock Compression on Ceramic Materials,” in [35], pp. 101–146.

[76] McGlaun, J.M., and P. Yarrington, “Large Deformation Wave Codes,” in [7], pp. 323–353.

[77] McQueen, R.G., S.P. Marsh, J.W. Taylor, J.N. Fritz, and W.J. Carter, “The Equation of State of Solids from Shock Wave Studies,” in *High Velocity Impact Phenomena* (ed R. Kinslow), Academic Press, New York (1970), pp. 293–417 with appendices on pp. 515–568.

[78] Menikoff, Ralph, “Empirical Equations of State for Solids,” in [56], pp. 143–188.

[79] Mescheryakov, Yuri I., "Meso–Macro Energy Exchange in Shock Deformed and Fractured Solids," in [57], pp. 169–213.

[80] Meyers, M.A., *Dynamic Behavior of Materials*, J. Wiley, New York (1994).

[81] Meyers, M.A., and C.T. Aimone, "Dynamic Fracture (Spalling) of Metals," *Prog. Mater. Sci.* **28**, (1983), pp. 1–96.

[82] Mindlin, R.D., and H.F. Tiersten, "Effects of Couple Stresses in Linear Elasticity," *Arch. Rational Mech. Anal.* **11**, (1962), pp. 415–448.

[83] Nesterenko, Vitali F., *Dynamics of Heterogeneous Materials*, Springer-Verlag, New York (2001).

[84] Nunziato, J.W., J.E. Kennedy, and D.R. Hardesty, "Modes of Shock Growth in the Initiation of Explosives," in: *Proceedings–Sixth Symposium (International) on Detonation*, August 24–27, 1976, Report ACR-221, Office of Naval Research–Department of the Navy, Arlington, Virginia, pp. 47–61.

[85] Nunziato, Jace W., Edward K. Walsh, Karl W. Schuler, and Lynn M. Barker, "Wave Propagation in Nonlinear Viscoelastic Solids," *Mechanics of Solids IV*, Handbuch der Physik, Vol. VIa/4 (ed C. Truesdell), Springer-Verlag, Berlin (1974), pp. 1–108.

[86] Rice, M.H., R.G. McQueen, and J.M. Walsh, "Compression of Solids by Strong Shock Waves," in: *Solid State Physics* **6**, (eds F. Seitz and D. Turnbull) Academic Press, New York (1958).

[87] Rinehart, J.S., and J. Pearson, *Behavior of Metals Under Impulsive Loads*, American Society for Metals, Cleveland (1954). Reprinted by Dover Publications, New York (1965).

[88] Sekine, T. "Shock Synthesis of New Materials," in [33], pp. 289–308.

[89] Sheffield, S.A., R.L. Gustavsen, and M.U. Anderson, "Shock Loading of Porous High Explosives," in [33], pp. 23–61.

[90] Sikka, S.K., B.K. Godwal, and R. Chidambaram, *Equation of State at High Pressure*, in [30], pp. 1–35.

[91] Steinberg, D.J., S.G. Cochran, and M.W. Guinan, "A Constitutive Model for Metals Applicable at High-Strain Rate," *J. Appl Phys.* **51(3)** (1980), pp. 1498–1533,.

[92] Stevens, A.L., Lee Davison, and W.E. Warren, "Spall Fracture in Aluminum Monocrystals: a Dislocation-Dynamics Approach," *J. Appl. Phys.* **43(12)** (1972), pp. 4922–4927.

[93] Stevens, A.L., Lee Davison, and W.E. Warren, "Void Growth During Spall Fracture of Aluminum Monocrystals," in *Proceedings of an International Conference on Dynamic Crack Propagation*, (ed George C. Sih), Noordhoff, Leyden (1973), pp. 37–48.

[94] Swegle, J.W., and D.E. Grady, "Shock Viscosity and the Prediction of Shock Wave Rise Times," *J. Appl. Phys.* **58 (2)** (1985), pp. 692–701.

[95] Thadhani, N.N., and T. Azawa, "Materials Issues in Shock-Compression-Induced Chemical Reactions in Porous Solids," in [33], pp. 257–288.

[96] Thompson, S.L., and H.S. Lawson "Improvements in the CHART D Radiation-Hydrodynamic Code III: Revised Analytic Equations of State," Report SC-RR-710714, Sandia National Laboratories, Albuquerque, NM (1972).

[97] Thurston, R.N., "Definition of a Linear Medium for One-Dimensional Longitudinal Motion," *J. Acoust. Soc. Am.* **45** (1969), pp. 1329–1341.

[98] Thurston, R.N., Waves in Solids, in *Handbuch der Physik*, Band VIa/4 (eds S. Flügge and C. Truesdell), Springer-Verlag, Berlin (1974).

[99] Thurston, R.N., H.J. McSkimin, and P. Andreach, Jr., "Third-Order Elastic Coefficients of Quartz," *J. Appl. Phys.* **37** (1966), pp. 267–275.

[100] Toupin, R.A., "Elastic Materials with Couple Stresses," *Arch. Rational Mech. Anal.* **11** (1962), pp. 385–414.

[101] Toupin, R.A., and B. Bernstein, "Sound Waves in Deformed Perfectly Elastic Materials. Acoustoelastic Effect," *J. Acoust. Soc. Am.* **33** (1961), pp. 216–225.

[102] Truesdell, C., and W. Noll, The Non-linear Field Theories of Mechanics, in *Handbuch der Physik*, Band III/3 (ed S. Flügge), Springer-Verlag, Berlin (1965).

[103] Truesdell, C., and R.A. Toupin, The Classical Field Theories, in *Handbuch der Physik*, Band III/1 (ed S. Flügge), Springer-Verlag, Berlin (1960).

[104] Trunin, R.F., K.K. Krupnikov, G.V. Simakov, and A.I. Funtikov, *Shock-Wave Compression of Porous Metals*, in [44], pp. 177–195.

[105] van Thiel, M., J. Shaner, and E. Salinas, *Compendium of Shock Wave Data,* Lawrence Radiation Laboratory Report UCRL-50108 (Vols. 1 & 2 Rev. 1 & Vol. 3), (1977).

[106] Wallace, Duane C., *Thermodynamics of Crystals*, John Wiley, New York (1972).

[107] Werne, Roger W., and James M. Kelly, "A Dislocation Theory of Isotropic Polycrystalline Plasticity," *Int. J. Eng. Sci.* **26** (1978), pp. 951–965.

[108] Whitham, G.B., *Linear and Nonlinear Waves*, John Wiley and Sons, New York (1974).

[109] Zel'dovich, Ya.B. and Yu.P. Raizer, *Physics of Shock Waves and High-Temperature Hydrodynamic Phenomena*, Academic Press, New York, Vol. 1 (1966), Vol. 2 (1967). Reprinted in a single volume by Dover Publications, Mineola, New York (2002).

[110] Zhernokletov, M.V., and B.L. Glushak, *Material Properties Under Intensive Dynamic Loading*, Springer, Berlin, Heidelberg (2006).

[111] Zurek, Anna K., and Marc André Meyers, "Microstuctural Aspects of Dynamic Failure," in [31], pp. 25–70.

Index

Printing: Krips bv, Meppel, The Netherlands
Binding: Stürtz, Würzburg, Germany